Nanobiosensors and Nanobioanalyses

Mun'delanji C. Vestergaard • Kagan Kerman
I-Ming Hsing • Eiichi Tamiya
Editors

Nanobiosensors and Nanobioanalyses

Springer

Editors
Mun'delanji C. Vestergaard
School of Materials Science
Japan Advanced Institute of Science and Technology
Ishikawa, Japan

I-Ming Hsing
Department of Chemical and Biomolecular Engineering
The Hong Kong University of Science and Technology
Kowloon, Hong Kong SAR

Kagan Kerman
Department of Physical and Environmental Sciences
University of Toronto Scarborough
Toronto, ON, Canada

Eiichi Tamiya
Graduate School of Engineering
Osaka University
Osaka, Japan

ISBN 978-4-431-56267-2 ISBN 978-4-431-55190-4 (eBook)
DOI 10.1007/978-4-431-55190-4
Springer Tokyo Heidelberg New York Dordrecht London

Printed on acid-free paper

Springer Japan KK is part of Springer Science+Business Media (www.springer.com)

Preface

Nanobiosensing and nanobioanalysis are the inevitable fruits of the tremendous advances made in the disciplines of materials development, technology, and engineering with application areas including medicine, the environment, food, pharmaceuticals, energy, and security. It would not be far-fetched to describe nanobiosensing and nanobioanalysis as a marriage of convenience across a number of disciplines, perfectly underscoring what a successful interdisciplinary exploitation and utilization of resources can achieve.

Nanomaterials have a number of outstanding physical and chemical properties such as enhanced chemical and biological activities, high molecular adsorption, catalytic effects, and mechanical strength. These properties of nanomaterials enable high sensitivity and selectivity in nanobiosensors and nanobioanalysis.

The book (1) introduces nanomaterials and their synthesis, (2) presents applications of nanomaterials and nanofabrication, (3) demonstrates that nanotechnologies can provide solutions to significant technical barriers and potentially revolutionize research in emerging areas, and (4) provides a good account of up-and-emerging nanobiotechnologies and outstanding challenges in taking these technologies to the point-of-care.

The book is divided into four parts. The first part consists of a single review in Chap. 1, which provides an overview of principles, analytical main players, and detection strategies in the emerging areas of nanobiosensing and nanobioanalysis.

The second part contains four chapters on up-and-emerging nanobiosensing architectures for increased selectivity, sensitivity, and multiplexing. In Chap. 2, Razeeb et al. describe the catalytic applications of vertically aligned nanowire-based sensors. In Chap. 3, the development of one- and two-dimensional nanostructures is discussed in detail by Liao et al. with biosensing applications. In Chap. 4, the emerging nanoimprinting technology that creates a wide range of nanostructures on surfaces is introduced for biosensors and biochips. Interferometic reflectance imaging is a novel approach with single particle detection capabilities as developed and discussed by Ünlü et al. in Chap. 5.

The third part of the book has seven chapters on nanomaterials commonly used in nanobioanalysis, with a focus on functionalization of the nanomaterials and their applications in nanobioanalysis using a range of detection platforms. The applications of gold nanoparticles as electrochemical and optical detection platforms are highlighted in Chap. 6. The nanocarbon film modified electrochemical sensors and biosensors are discussed in Chap. 7. The synthesis of hybrid metallic nanoparticles displayed significantly enhanced physical and chemical properties. Daniele et al. review the applications of hybrid metallic nanoparticles with carbon nanotubes, graphene, and organic conjugations in Chap. 8. The novel applications of nanobiosensing using a laser beam are described in Chap. 9. The principles and applications of semiconductor quantum dots are highlighted in Chaps. 10 and 11, respectively. The detection of protein phosphorylation using nanoparticles provided cost-effective and multiplexed platforms for drug discovery as reviewed by Martic and Kraatz in Chap. 12.

The last part, comprising five chapters, focuses on the applications of nanobiosensor technologies including current challenges in bringing them to the point-of-care and on the future outlook given the exponential growth in nanotechnology. Chapter 13 highlights the modification of screen-printed electrodes using carbon nanotubes. The emerging applications of nanoneedles for the detection of intracellular processes are discussed in Chap. 14. The chemical modification of plasmonic sensors enabled the detection of proteins and oncologic drugs in human serum as discussed in Chap. 15. Electrochemical scanning microscopy was utilized for imaging single cells and membrane proteins, as described in Chap. 16. Current advances and challenges in bringing field-effect transistors to the point-of-care are described in Chap. 17.

Nanobiosensing and nanobioanalysis already comprise an important area and are expected to grow in the next few years. While editing the book, we aimed to put together the fundamental and applied aspects of these highly interesting technologies for a broad audience including those involved in research, teaching, and learning. The book is also suited to those utilizing nanomaterials for applications targeting biomedical, military, industrial, environmental, and clinical work.

We are very fortunate to undertake this project. We wish to thank all the authors who have contributed high-quality chapters to the book, and we warmly acknowledge the gracious support of our families. Last, we extend our sincere thanks to Springer Japan for their excellent work in editing this book.

Ishikawa, Japan — Mun'delanji C. Vestergaard
Toronto, ON, Canada — Kagan Kerman
Kowloon, Hong Kong SAR — I-Ming Hsing
Osaka, Japan — Eiichi Tamiya

Contents

Contributors

W. Russ Algar Department of Chemistry, University of British Columbia, Vancouver, BC, Canada

Stephanie Burrs Agricultural and Biological Engineering Department, Institute of Food and Agricultural Sciences, University of Florida, Gainesville, FL, USA

Prachee Chaturvedi Agricultural and Biological Engineering Department, Institute of Food and Agricultural Sciences, University of Florida, Gainesville, FL, USA

Yit-Tsong Chen Department of Chemistry, National Taiwan University, Taipei, Taiwan

Institute of Atomic and Molecular Sciences, Academia Sinica, Taipei, Taiwan

Allison A. Cargill Department of Mechanical Engineering, Iowa State University, Ames, IA, USA

Jonathan C. Claussen Department of Mechanical Engineering, Iowa State University, Ames, IA, USA

Michael A. Daniele U.S. Naval Research Laboratory, Center for Bio-Molecular Science and Engineering, Washington, DC, USA

Shaowei Ding Department of Mechanical Engineering, Iowa State University, Ames, IA, USA

Pedro Estrela Department of Electronic and Electrical Engineering, University of Bath, Bath, UK

Maksudul Hasan Tyndall National Institute, University College Cork, Cork, Ireland

Vinci Wing Sze Hung Department of Physical and Environmental Sciences, University of Toronto Scarborough, Toronto, ON, Canada

Mamun Jamal Tyndall National Institute, University College Cork, Cork, Ireland

Dai Kato National Institute of Advanced Industrial Science and Technology, Ibaraki, Japan

Ryuzo Kawamura Biomedical Research Institute, National Institute of Advanced Industrial Science and Technology (AIST), Ibaraki, Japan

Kagan Kerman Department of Physical and Environmental Sciences, University of Toronto Scarborough, Toronto, ON, Canada

Heinz-Bernhard Kraatz Department of Chemistry, University of Toronto, Toronto, ON, Canada

Department of Physical and Environmental Sciences, University of Toronto Scarborough, Toronto, ON, Canada

Filiz Kuralay Department of Chemistry, Faculty of Science and Arts, Ordu University, Ordu, Turkey

Nan Li Department of Physical and Environmental Sciences, University of Toronto Scarborough, Toronto, ON, Canada

Chun-Da Liao Department of Chemistry, National Taiwan University, Taipei, Taiwan

Institute of Atomic and Molecular Sciences, Academia Sinica, Taipei, Taiwan

Yi-Ying Lu Department of Chemistry, National Taiwan University, Taipei, Taiwan

Institute of Atomic and Molecular Sciences, Academia Sinica, Taipei, Taiwan

Sanela Martic Department of Chemistry, Oakland University, Rochester, MI, USA

Jean-Francois Masson Department of Chemistry, University of Montreal, Montreal, QC, Canada

Alan Mathewson Tyndall National Institute, University College Cork, Cork, Ireland

Tomokazu Matsue Advanced Institute for Materials Research (WPI-AIMR), Tohoku University, Sendai, Japan

Graduate School of Environmental Studies, Tohoku University, Sendai, Japan

Eric McLamore Agricultural and Biological Engineering Department, Institute of Food and Agricultural Sciences, University of Florida, Gainesville, FL, USA

Chikashi Nakamura Biomedical Research Institute, National Institute of Advanced Industrial Science and Technology (AIST), Ibaraki, Japan

Osamu Niwa National Institute of Advanced Industrial Science and Technology, Ibaraki, Japan

Shrey Pathak Department of Electronic and Electrical Engineering, University of Bath, Bath, UK

María Pedrero Departamento de Química Analítica, Universidad Complutense de Madrid Avda, Madrid, Spain

Kafil M. Razeeb Tyndall National Institute, University College Cork, Cork, Ireland

Susana Campuzano Ruiz Departamento de Química Analítica, Universidad Complutense de Madrid, Madrid, Spain

Masato Saito Graduate School of Engineering, Osaka University, Osaka, Japan

Wan Wardatul Amani Wan Salim Department of Biotechnology Engineering (BTE), Kuliyyah of Engineering (KOE), IIUM, Gombak, Kuala Lumpur

Derin Sevenler Department of Biomedical Engineering, Boston University, Boston, MA, USA

Hitoshi Shiku Advanced Institute for Materials Research (WPI-AIMR), Tohoku University, Sendai, Japan

Graduate School of Environmental Studies, Tohoku University, Sendai, Japan

Yaron R. Silberberg Biomedical Research Institute, National Institute of Advanced Industrial Science and Technology (AIST), Ibaraki, Japan

Yasufumi Takahashi Advanced Institute for Materials Research (WPI-AIMR), Tohoku University, Sendai, Japan

Graduate School of Environmental Studies, Tohoku University, Sendai, Japan

Eiichi Tamiya Department of Applied Physics, Graduate School of Engineering, Osaka University, Osaka, Japan

Tien-Chun Tsai Department of Chemistry, National Taiwan University, Taipei, Taiwan

Institute of Atomic and Molecular Sciences, Academia Sinica, Taipei, Taiwan

Neşe Lortlar Ünlü Department of Biomedical Engineering, Boston University, Boston, MA, USA

School of Medicine, Bahçeşehir University, Istanbul, Turkey

M. Selim Ünlü Department of Biomedical Engineering, Boston University, Boston, MA, USA

Department of Electrical and Computer Engineering, Boston University, Boston, MA, USA

Mun'delanji C. Vestergaard School of Materials Science, Japan Advanced Institute of Science and Technology, Ishikawa, Japan

Miao Wu Department of Chemistry, University of British Columbia, Vancouver, BC, Canada

Keiichiro Yamanaka Photonics Advanced Research Center (PARC), Graduate School of Engineering, Osaka University, Osaka, Japan

Hiroyuki Yoshikawa Department of Applied Physics, Osaka University, Osaka, Japan

Sandy Shuo Zhao Department of Chemistry, University of Montreal, Montreal, QC, Canada

Part I
Introduction to Nanobiosensors and Nanobioanalyses

Chapter 1
Nanobiosensors and Nanobioanalyses: A Review

Mun'delanji C. Vestergaard and Eiichi Tamiya

Abstract This chapter provides an overview of nanobiosensors and nanobioanalyses. Both of these topics are the inevitable fruits of tremendous advances made within the disciplines of materials development, technology, engineering and application areas including medical, environmental, food, pharmaceutical, energy and security. In order to provide a comprehensive picture for readers new to the field, the chapter takes the reader back to when it all supposedly began, by providing a brief history of the advances and some challenges along the way. The more advanced reader can still benefit from not only recapping the old, but also being introduced to some of the new developments: in particular, the applications and challenges of developed nano-biosensors. The chapter is conveniently split into two main parts: we discuss nano-biosensors in the first half and the second presents nanobioanalyses. Throughout, we use examples of applications from our group and other research groups.

Keywords Nanobioanalyses • Nanobiosensors • Nanobiotechnology • Nanodevices • Nanomaterials

1.1 Introduction

Nanobiosensing and nanobionalyses are the inevitable fruits of the tremendous advances made within the disciplines of materials development, technology, engineering and application areas including medical, environmental, food, pharmaceutical, energy and security. In fact, it would not be far-fetched to describe nanobiosensing and nanobioanalyses as a polygamy-type marriage of a number of disciplines. It highlights, at least partly, what a successful inter-disciplinary exploitation and utilization of resources can achieve.

M.C. Vestergaard (✉)
School of Materials Science, Japan Advanced Institute of Science and Technology, 1-1 Asahidai, Nomi City, Ishikawa 923-1292, Japan
e-mail: munde@jaist.ac.jp

E. Tamiya
Department of Applied Physics, Graduate School of Engineering, Osaka University, 2-1, Yamadaoka, Suita, Osaka 565-0871, Japan

M.C. Vestergaard et al. (eds.), *Nanobiosensors and Nanobioanalyses*,
DOI 10.1007/978-4-431-55190-4_1

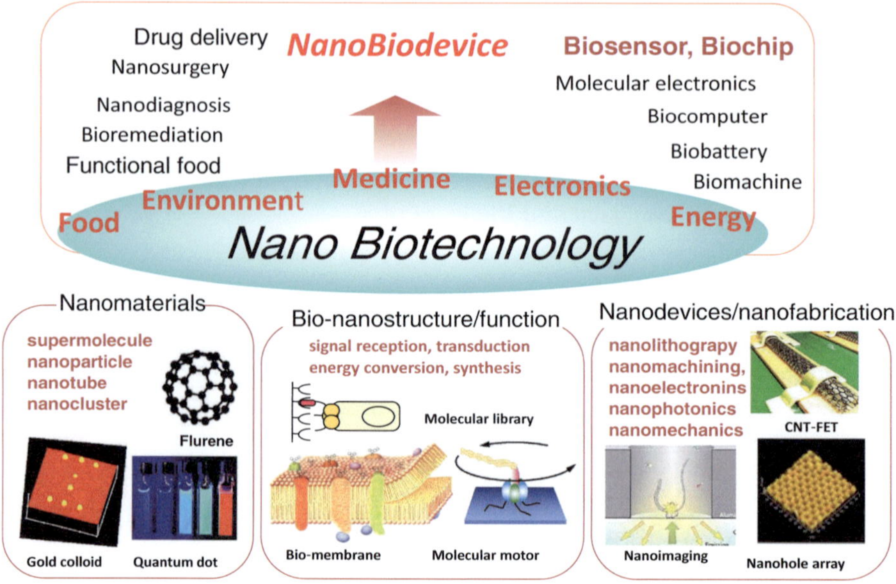

Fig. 1.1 Nanobiotechnology and nanobiodevices

Nanobiosensing and nanobioanalysis are about the detection and analysis of target molecules, using minuscule, often integrated and multiplexed, nano-sized sensing devices to achieve selective, yet extremely sensitive, detection limits. This has been made possible with the discovery, synthesis and utilization of materials for nanodevice fabrication whose morphological features on a nano-scale confer the desired sensitivity for detection. Figure 1.1 provides a good summation of what nanobiosensing and nanobioanalysis are, placed within the wider context of nanobiotechnology.

The incredible progress made in this area is demonstrated, at least to some extent, by the increase in the number of articles on nanobiosensors based on nanomaterials and nanodevices selected for presentation at the World Congress on Biosensors (Fig. 1.2).

1.1.1 The Birth of Nanotechnology

Nanotechnology has been described in many but fundamentally similar ways. The Oxford English Dictionary defines the term as "the branch of technology that deals with dimensions and tolerances of less than 100 nm, especially the manipulation of individual atoms and molecules". Whatmore (2006) described it as "… the design, characterization, production and application of structures, devices and systems by controlling shape and size at nanometre scale. Simply put, it is the study of manipulating materials/objects at an atomic or molecular level.

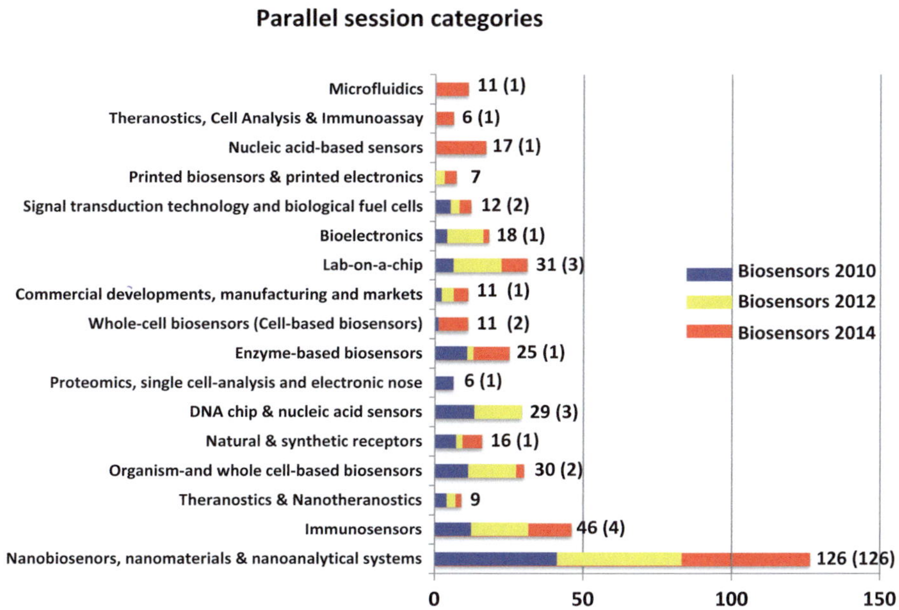

Fig. 1.2 Number and types of articles selected for oral presentation at the World Congress on Biosensors in 2010, 2012 and 2014. The numbers outside the brackets represent the total number of oral presentations. The bracketed values represent the total number of oral presentations with the term "nano" in the title

Many scientists trace the roots of nanotechnology to Richard Feynman, a physicist who in his lecture at an American Physical Society meeting in 1959 said "… I don't know how to do this on a small scale in a practical way, but I do know that computing machines are very large; they fill rooms. Why can't we make them very small, make them of little wires, little elements, and by little, I mean little? …". Feynman's lecture was entitled "There's plenty of room at the bottom", as he described the direct manipulation of individual atoms for building novel compounds and structures. This talk inspired the conceptual beginnings of the "top-down" and "bottom-up" terminologies in the field of nanotechology decades later. For their fundamental work in quantum electrodynamics with profound consequences for the physics of elementary particles, Feynman, jointly with Julian Schwinger and Sin-Itiro Tomonaga, received the Nobel Prize in Physics in 1965. Then again, in 1984, Feynman gave a talk which described the advances in technology since his famous 1959 lecture (in Whatmore 2006). However, the actual term "nanotechnology" was first coined by the late Norio Taniguchi in 1974 during his talk in which he discussed how the dimensional accuracy with which things are made had improved over time (Taniguchi 1996). However, it wasn't until 1981 that individual atoms could be visualized by Binnig and Rohrer, using a scanning electron microscope (Binnig et al. 1982). In 1996, the Nobel Prize in Chemistry was awarded jointly to Robert F. Curl Jr., Sir Harold W. Kroto and Richard E. Smalley for their discovery of fullerenes, a new form of carbon that was also called buckminsterfullerene or "buckyballs". Smalley was a leading advocate of nanotechnology throughout his life. On his death at the age of 62, the US Senate passed a resolution to honour Smalley, crediting him as the "Father of Nanotechnology". Since then, there have been huge leaps and bounds, with the early 2000s seeing the beginnings of commercial applications of nanotechnology. In 2005, a huge project on emerging technologies was established through a collaboration between the Woodrow Wilson International Center for Scholars and the Pew Charitable Trusts. This ambitious project promised to revolutionize all aspects of our lives, from the cars we drive and the medical treatment we receive to the clothes we wear (Fiorino 2010). Suffice to say that some of the promises have already been made manifest. However, what is most important is the on-going effort to explore, exploit and continue to make huge advances in this field and the many disciplines that have benefited from nanotechnology. In 2010, the Nobel Prize in Physics was awarded jointly to Andre Geim and Konstantin Novoselov for their groundbreaking experiments regarding the two-dimensional material graphene, the perfect atomic lattice. Since graphene is able to conduct electricity whilst being practically transparent, it has initiated a new wave of miniaturization in electronics that will possibly lead to the mass production of more affordable, thinner and faster mobile phones, solar cells, computers and biosensors in the future.

1.1.2 Nanomaterials in Nanobiosensors and Nanobioanalyses

Nanomaterials are chemical substances or materials that are synthesized at the billionth of a meter scale (10^{-9} m). They range from 1 to 100 nm in size. Nanomaterials for nanobiosensing and nanobioanalyses can perhaps be best classified into four main categories: (1) metal-based nanoparticles (NPs) including quantum dots (QDs), nanowires, nanorods, oxide nanoparticles and noble metals; (2) carbon-based NPs (including carbon nanotubes, graphene, fullerene and carbon dots); (3) dendrimers: nanosized polymers built from branched units; and (4) composites/hybrids: a combination of at least one nanomaterial. We will discuss the latter in the concluding section.

Briefly, noble metal NPs (Au and Ag) have unique thermal, electrical and optical properties that can and are being utilized in a range of nanobiosensors. In particular, their optical properties have received much attention in biosensor development (Hiep et al. 2009; Huang et al. 2011). Huang and colleagues (2011), in a recent review, discussed the photo-physical properties of matter, in particular gold NPs. They described what happens when light hits matter. The light can be (1) absorbed, (2) absorbed and re-emitted (fluorescence), (3) scattered at the same frequency as the incoming light (Mie or Rayleigh scattering), and (4) the local electromagnetic field of the incoming light can be enhanced. This leads to increase in any spectroscopic signals from the molecules at the material surface (surface-enhanced Raman scattering). These properties are even more enhanced in metal NPs because of unique interaction of light with free electrons in NPs (Huang et al. 2007). Metal NPs also have unique electronic properties. Generally, they have excellent conductivity and catalytic properties, making them suitable for fabrication of "electronic wires" to enhance the electron transfer between redox centers in proteins and electrode surfaces, and as catalysts to increase electrochemical reactions (Lan et al. 2011; Kerman and Kraatz 2009; Selvaraju et al. 2008). Gold and Ag NPs have been exploited in the fabrication of localised surface plasmon resonance (LSPR) chips for the detection of biomolecules (Endo et al. 2006; Vestergaard et al. 2008; Haes et al. 2004). These metal NPs show specific changes in their absorbance responses in the visible region of the spectrum on binding with various molecules such as nucleic acids or proteins. The electronic properties, in particular, of Au and Ag NPs have been employed as labels for detection of proteins and other target molecules (Kerman et al. 2007; Chen et al. 2007; Ambrosi et al. 2007). Semi-conducting materials such as QDs are semi-conductor nanocrystals with attractive optical properties for nanobiosensing. Upon absorbing light, semi-conductor QDs quickly re-emit the light but in a different color with longer wavelength, which is fluorescent. The wavelength of the light can be controlled by changing the shape and size of the nanocrystals (Gaponik et al. 2002; Banin and Millo 2003; Bailey and Nie 2003).

Dendrimers are man-made molecules about the size of an average protein, and have a branching shape. This shape gives dendrimers a vast surface area to which scientists can attach therapeutic agents or other biologically active molecules (Buhleier et al. 1978). Because of their highly branched structure, one dendrimer can carry a molecule that recognizes targeted cancer cells (for example), a therapeutic agent to kill those cells, and a molecule that recognizes the signals of cell death. Researchers hope to manipulate dendrimers to release their contents only in the presence of certain trigger molecules associated with cancer. Following drug release, the dendrimers may also report back whether they are successfully killing the targets. They have also been used for stabilization of other NPs (Hasanzadeh et al. 2014).

Carbon-based nanomaterials are synthesized from graphite, one of the oldest and perhaps most used natural materials. They possess very high mechanical strength, thermal and electrical conductivity as well as optical properties. Thus, they have been hugely exploited for various applications including nanobiosensing and nanobioanalysis (Cha et al. 2013). Carbon nanotubes (CNTs) are, so far, the most widely used of the carbon-based nanomaterials. They were first discovered by Iijima (1991) and have found a wide variety of applications. Graphene is one of the latest carbon nanomaterials, having been synthesized from graphite by Nobel Laureates Geim and Novoselov in 2004 (Novoselov et al. 2004). The oxidative product of graphene, graphene oxide (GO), has emerged as an even more attractive carbon nanomaterial as it is easily dispersed in the aqueous environment, and has hydrophilic functional groups which enable the oxide to be functionalized simply and covalently with biomolecules for the development of miniaturized nanobiosensors (Cha et al. 2013; Dreyer et al. 2010).

1.2 Nanobiosensors

1.2.1 Introduction to Biosensors: Basic Principles and Types of Biosensors

Biosensors were once defined as devices that respond to chemical species in biological samples or biological components. Now this rather misleading definition has been changed to "analytical devices composed of a biological recognition element directly interfaced to a signal transducer, which together relate the concentration of a targeted analyte or analytes to a measurable response". Some of the different types of nanobiosensors will be discussed in detail in the following chapters. To avoid repetition, here we simply present an overview (Fig. 1.3).

Biosensors are often classified according to either the type of recognition molecule and/or the transducer used. A biosensor would therefore be presented as "an optical biosensor for …"; "an immunosensor for …"; "an electrochemical DNA sensor for …", and so on. We will present a brief history of biosensor development, drawn from Palchetti and Mascini (2010). In 1916, a first account on the immobilization of protein adsorption of invertase on activated charcoal was reported. Leland Clark presented the glucose sensor in 1956, and provided the first description of a biosensor, an amperometric enzyme electrode for glucose in 1962. In 1969, the first potentiometric enzyme-based sensor for detection of urea was

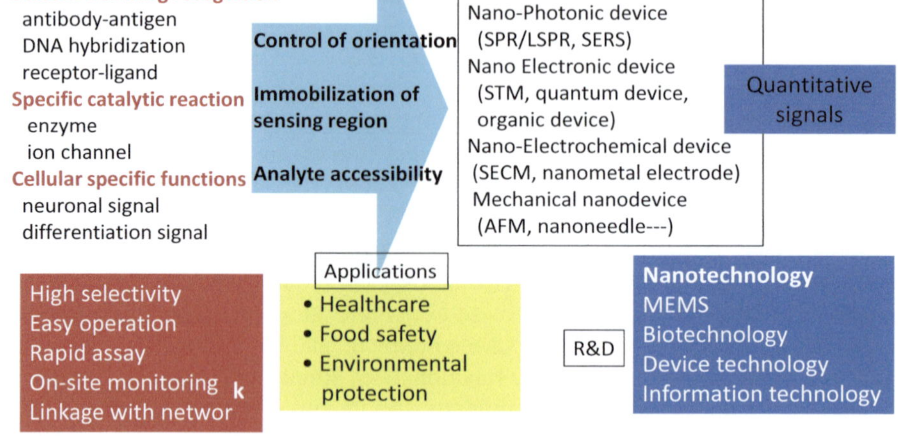

Fig. 1.3 The basic principle of nanospatial transducers

reported by Guilbauly and Montalvo. The 1970s and 1980s saw a huge range of biosensors: the ion-selective field effect transistor (ISFET) in 1970; the first commercial glucose biosensors by Yellow Springs Instruments (1972/1975); the first microbe-based biosensors and the first immunosensor (1975); the first bedside artificial pancrease by Miles Laboratories (1976); the first fiberoptic pH sensor for in vivo blood gases by Peterson (1980); the first fiberoptical biosensor for glucose (1982); the first surface plasmon resonance biosensor (1983); the launch of the MediSense ExacTech™ blood glucose biosensor (1987) and many more. In 1990, the Pharmacia BIACore SPR-based biosensor system was launched. The 1990s saw increased commercialization of hand-held biosensors. Since then, there has been an exponential exploitation of nanomaterials in biosensors: quantum dots, nanocantilevers, nanowires and carbon nanotubes, graphene, fullerene, carbon dots, metal nanoparticles, and many more.

It is important to note that the table is far from an exhaustive account of the progress made in biosensor development. We have chosen some of what we see as the representative landmarks.

Briefly, the majority of biosensors are either affinity-based or catalytic-based biosensors. Affinity-based biosensors are used to bind molecular species of interest, irreversibly and non-catalytically. Examples include antibodies, nucleic acids and hormone receptors. Catalytic-based sensors such as enzymes and microbiological cells recognize and bind a molecule of interest, followed by a catalyzed chemical conversion of that molecule to a product which is then detected. There is also a distinction between single-use disposable biosensors (for example, for pregnancy testing) and those that incorporate additional processing steps, needing bioanalytical support systems such as reagents, data processing and handling systems. Most of the work discussed in this book falls into the latter category. However, it is the less tedious and layman-level nanobiosensors which can be used at point of need that are the ultimate application device that we in this field aspire to.

1.2.2 Surface Modification

Surface modification is needed to increase sensitivity and selectivity of the fabricated sensor. It is also used to provide a means of miniaturization in an array format. The type of surface modification depends on the surface as well as the target biological recognition element. Most of these utilize a self-assembled monolayer (SAM), which is a single layer of molecules on a solid substrate, stabilized by hydrogen bonds, coulombic interactions (electrical forces), van der Waals forces (dispersion forces) or hydrophobic effects. SAMs can be formed from a wide variety of compounds such as fatty acids, organo-silicon or organo-sulphur compounds and on a wide range of surface materials such as aluminum oxide, silver, silicon oxide, glass and gold (Ulman 1996). Indeed, a SAM of Au NPs using alkyl-thiol chemistry has been widely exploited (Vestergaard et al. 2008; Du et al. 2009; Wu et al. 2011; Bertok et al. 2013). The structure of a thiol monolayer-protected Au NP at 1.1 Å resolution was reported by Jadzinsky et al. (2007). We refer the interested reader to excellent reviews on the formation and characterization of SAMs on Au surfaces by Love et al. (2005), Vericat et al. (2010), Gooding and Ciampi (2011), and Häkkinen (2012).

1.2.3 Versatility: Multiplexing and Integration

Another attractive feature of biosensors is versatility. The detector surface can be fabricated for single or multiple detection of target analytes (Ligler et al. 2003; Endo et al. 2006) (Fig. 1.4). The latter can also be done for single as well as simultaneous analyses. Being able to perform biomolecular detection requires the interrogation of several targets at a time; thus, high-throughput detection and multiplexing are much desired. Xiang and colleagues (2013) have developed a surface plasmon resonance (SPR) biosensor for multiple point detection associated with drug resistance in multidrug-resistant *Mycobacterium tuberculosis* (Xiang et al. 2013).

We refer the interested reader to a chapter by Ng and Chong (2011), which provides a very good account of the multiplexing capabilities of biosensors. Briefly, the authors discuss three types of multiplexing technologies: (1) solution-based, giving

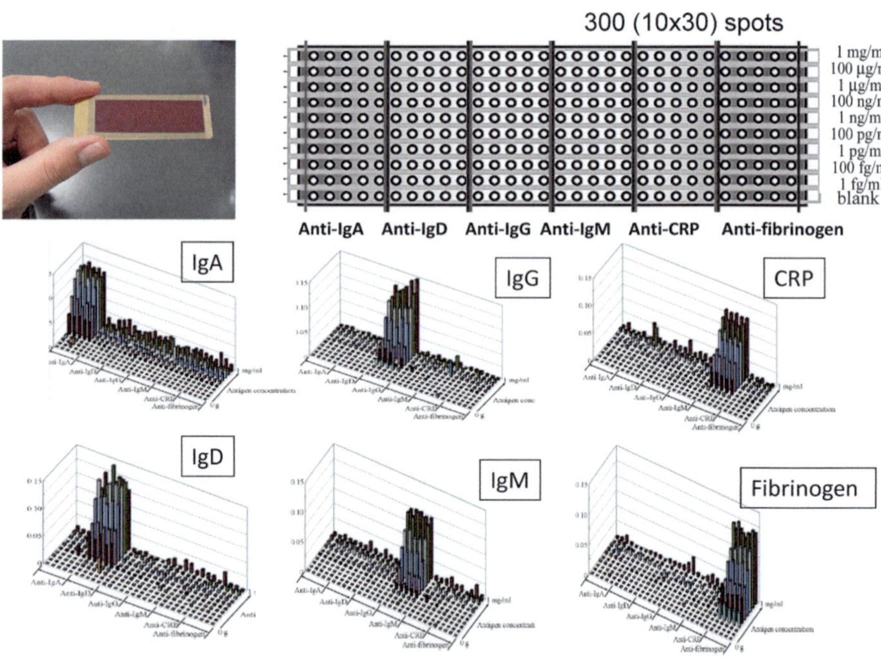

Fig. 1.4 A microarray technology for detection of multiple targets. Reprinted with permission from Endo et al. (2005). Copyright 2005, American Chemical Society

multiplexed PCR as a typical example of this technology; (2) microarray 2D and 3D, a technology first driven by the need to process vast amounts of target molecules (typically proteins and genes). Basically, this is a platform on which target labeled molecules are precisely immobilized at sub-microlitre volumes at designated locations on surfaces, where they can be identified and any interrogation can be easily detected. The popularity of this technology is exemplified by its successful commercialization. By fabricating microstructures onto planar solid surfaces, the 3D microarray system overcomes some of the drawbacks of the 2D systems, thus improving the signal-to-noise ratio, sensitivity and specificity; and (3) bead microarray systems. Like the 3D microarray, this system also improves the signal-to-noise ratio and improves sensitivity and specificity. It differs from the 3D system in that the beads are labeled prior to being introduced onto the microarrays (Ng et al. 2008; Ng and Chong 2011). A biosensor for the detection of beta-thalassemia mutations was successfully developed using this technology (Ng et al. 2010).

Furthermore, hybrid biosensing architectures involving integration with microfluidics (Kumar et al. 2013), and at least two different transduction tools for multi- and simultaneous analysis (Scognamiglio et al. 2012), are gaining ground. A few years back, we reported on biosensor development based on electrochemical and localized surface plasmon resonance (LSPR) detection platforms for the detection of melittin's interaction with a lipid membrane (Hiep et al. 2008). LPSR and electrochemistry measurements connected to core-shell structure nanoparticles were successfully exploited in a simultaneous detection scheme. The surface plasmon band characterizations of this type of nanostructure were initially examined by controlling the core size of the silica nanoparticles and shell thickness of the deposited gold. The results showed that when the shell thickness was increased, keeping the core size constant, the peak wavelength of the LSPR spectra shifted to a shorter wavelength and the maximum peak intensity was achieved at a particular shell thickness. Using this structure and its characteristics, a membrane-based nanobiosensor was developed for optically detecting the binding of the peptide toxin melittin to a hybrid bilayer membrane (HBM) and electrochemically assessing its membrane-disturbing properties.

The versatility of nanobiosensors is not limited to our overview. Although not covered here, miniaturization is similarly extreme. Miniaturization can be at a global (whole biosensor) or local scale. The latter includes the fabrication of screen-printed electrodes.

1.3 Nanobioanalyses

Nanobioanalysis is a broad subject including sample pre-treatment and detection. For the latter, it covers the type of instrumentation used and the fabrication or pre-treatment of each associated analytical component. In nanobiosensing, this may pertain to (1) modification and/or treatment of the substrate surface prior to introduction of the biological recognition element; (2) modification and/or pre-treatment of the recognition elements; and (3) the target analyte.

1.3.1 Nanomaterials in Sample Pre-treatment: Molecular Probes

Nanomaterials are utilized in sample preparation and concentration. The most common analytical procedures are solid-phase extraction, microextraction and filtration techniques (Pyrzynska 2013; Silva et al. 2013; Vuckovic et al. 2009). We will not cover this subject and we refer the interested reader to an excellent recent review on the subject (Pyrzynska 2013). Nanomaterials are also used as molecular probes for target analytes and/or the associated biological recognition elements. In section 1.2.2, we broadly classified nanomaterials as (1) metal-based NPs including quantum dots (QDs); nanowires, nanorods, oxide nanoparticles and noble metals, (2) carbon-based NPs (including carbon nanotubes, graphene, fullerene and carbon dots); and (3) dendrimers.

From the metal-based NPs, noble metals, in particular Au, has been used in molecular probes in a variety of nanobiosensing architectures because of its unique thermal, electrical and optical properties. What has made Au NPs very attractive is that they do not disrupt the activity of the biomolecules onto which they are conjugated. They have been used to tag a variety of biomolecules including antibodies, proteins and nucleotides (Kerman et al. 2004; Möller et al. 2005; Kerman and Kraatz 2009; Wen et al. 2013). Given the vast amount of research conducted, it would be futile to attempt to cover it all. Here, we restrict our discussion to electrical properties of Au NPs in the development of electrochemical nanobiosensors, by proving a couple of examples from our group.

Gold NPs possess a catalytic activity resulting from their large surface-area-to-volume ratio and their interface-dominated properties. They can decrease the over-potentials of many electroanalytical reactions and maintain the reversibility of redox reactions (Thompson 2007; Valden et al. 1998; Turner et al. 2008; Li et al. 2010). They have, therefore, been used as probes in electrochemical biosensor architectures. The pioneering work on the use of Au NPs as electrochemical labels for the voltammetric detection of proteins were performed by González-Garcia, Dequaire and colleagues (González-Garcı́a et al. 2000; Dequaire et al. 2000). We have exploited the ability of Au NPs for detection of protein phosphorylation by labeling of a specific phosphorylation event with Au NPs, followed by electrochemical detection. The phosphorylation reaction is coupled with biotinylation of a kinase substrate using a biotin-modified adenosine 5′-triphosphate [γ]-biotinyl-3,6,9-trioxaundecanediamine (ATP) as a co-substrate. When the phosphorylated and biotinylated kinase substrate is exposed to streptavidin-coated Au NPs, the high affinity between the streptavidin and biotin led to attachment of the metal NPs onto the kinase substrate. An electrochemical response obtained from the NPs enabled monitoring of the activity of the kinase and its substrate. Tyrosine kinase reactions were also performed in the presence of a well-defined inhibitor, 4-amino-5-(4-chlorophenyl)-7-(*t*-butyl) pyrazolo[3,4-d]pyrimidine (PP2) (Kerman et al. 2007). In the same year, we designed a disposable immunosensor protocol with Au NPs, as illustrated in Fig. 1.5 (Idegami et al. 2008). A high affinity between streptavidin and

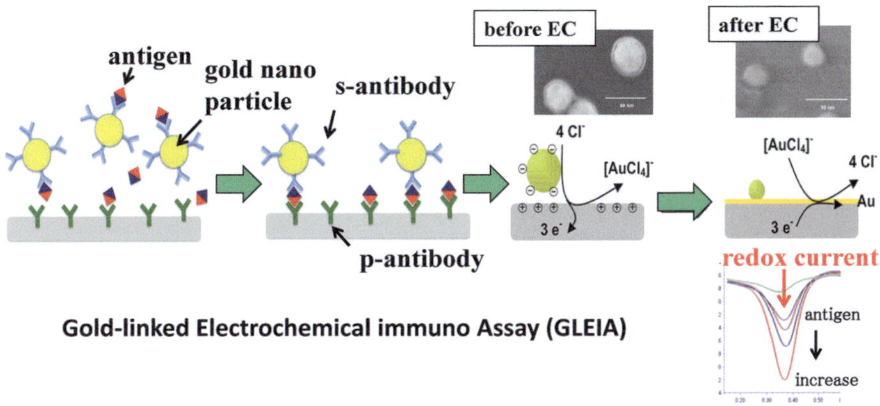

Fig. 1.5 Electrochemical nanobiosensing and bio/electrode interfaces. An illustration of the disposable immunosensor system. The primary antibody was immobilized directly on the screen-printed carbon strip (SPCS) surface, and a series of sandwich-type immunoreactions were performed on a single strip. A high potential is applied in 0.1 M HCl for the oxidation of Au NPs followed by the voltammetric measurement. Part of this figure is used in our original work (Idegami et al. 2008)

biotin was detected using the adsorptive stripping voltammetric signal of Au NPs down to the sub-nanometre streptavidin level. The heterogeneous sandwich immunoassay based on the stripping voltammetric detection of the NPs involved the binding of the Au-labeled secondary antibody, followed by acid dissolution and anodic stripping electrochemical measurement of the solubilized Au tracer on a bare screen-printed carbon strip. First, a primary antibody was immobilized directly on the working electrode surface, and a series of sandwich-type immuno-reactions was performed. A high potential, 1.2 V, was applied to the electrode in 0.1 M HCl for the oxidation of Au NPs, called the pre-oxidation process, and then the voltammetric measurement was performed (Idegami et al. 2008).

The robustness and non-destructive nature of Au NPs have proven very important in in vivo analysis. Recently, we studied differentiation of mouse embryonic stem (mES) cells, including undifferentiated single cells, embryoid bodies (EBs), and terminally differentiated cardiomyocytes using Au NP-based surface-enhanced Raman scattering (SERS). Gold NPs were delivered into all three mES cell differentiation stages without affecting cell viability or proliferation. Localization of Au NPs inside the following cell organelles, mitochondria, secondary lysosome and endoplasmic reticulum, was confirmed using transmission electron microscopy. Using bright- and dark-field imaging, the bright scattering of Au NPs and nanoaggregates in all three ES cell differentiation stages could be visualized. EB (an early differentiation stage) and terminally differentiated cardiomyocytes both showed SERS peaks specific to metabolic activity in the mitochondria and to protein translation (amide I, amide II and amide III peaks). Spatiotemporal changes observed in the SERS spectra from terminally differentiated cardiomyocyte tissues revealed local and dynamic molecular interactions as well as transformations during ES cell differentiation (Fig. 1.6; Sathuluri et al. 2011).

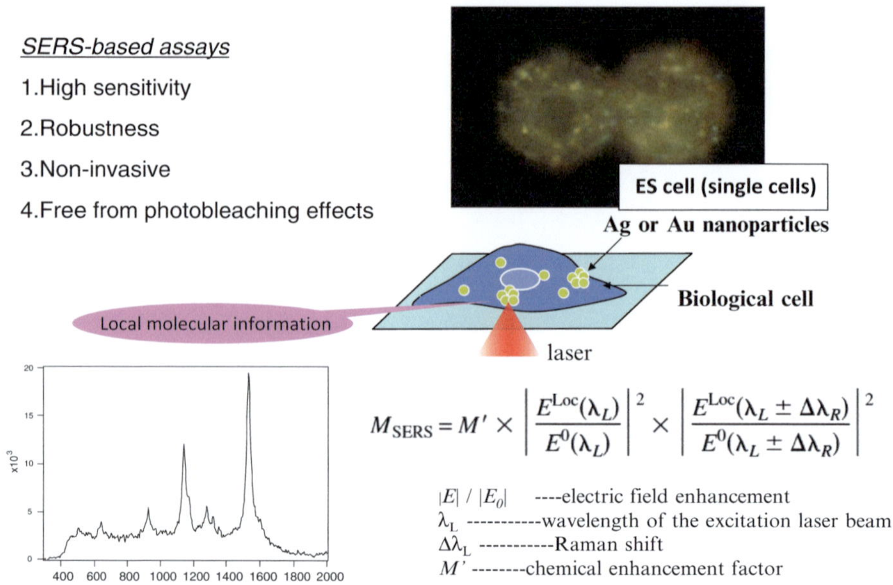

Fig. 1.6 Real-time imaging of live single cells using Au NPs

1.3.2 Nanomaterials in Surface Modification

In this section, we will limit our discussion to the use of Au NPs in LSPR-based biosensors. The principle is similar in Ag NPs.

LSPR biosensors exploit the collective oscillations of free electrons in metal NPs surrounded by dielectric media (Park et al. 2008; Shoute et al. 2009). Specifically, the phenomena involve resonant coupling of surface plasmon resonance and photon(s). Briefly, as the size of a metal structure decreases from the bulk scale (m to μm) to the nano-scale (<100 nm), the movement of electrons through the internal metal framework becomes restricted. Consequently, metal NPs display extinction bands in their UV–Vis spectra when the incident light resonates with the conduction band electrons at their surfaces. These charge density oscillations are called LSPR. The excitation of LSPR by light at an incident wavelength, where resonance would occur, results in the appearance of intense surface plasmon (SP) absorption bands. The type, size and shape of the NPs, and their distribution, affect the intensity and position of the SP absorption (Fig. 1.7). LSPR is also highly sensitive to changes in the surrounding environment (Su et al. 2009; Hutter and Fendler 2004; Haes et al. 2004; Nath and Chilkoti 2002).

In our group, among other applications, we have used Au NPs in the fabrication of LSPR chips for sensitive and selective detection of Alzheimer's tau protein (Vestergaard et al. 2008). From the basic design of an LSPR chip, we further advanced our fabrication procedure by immobilizing Au NPs on a silicon dioxide/

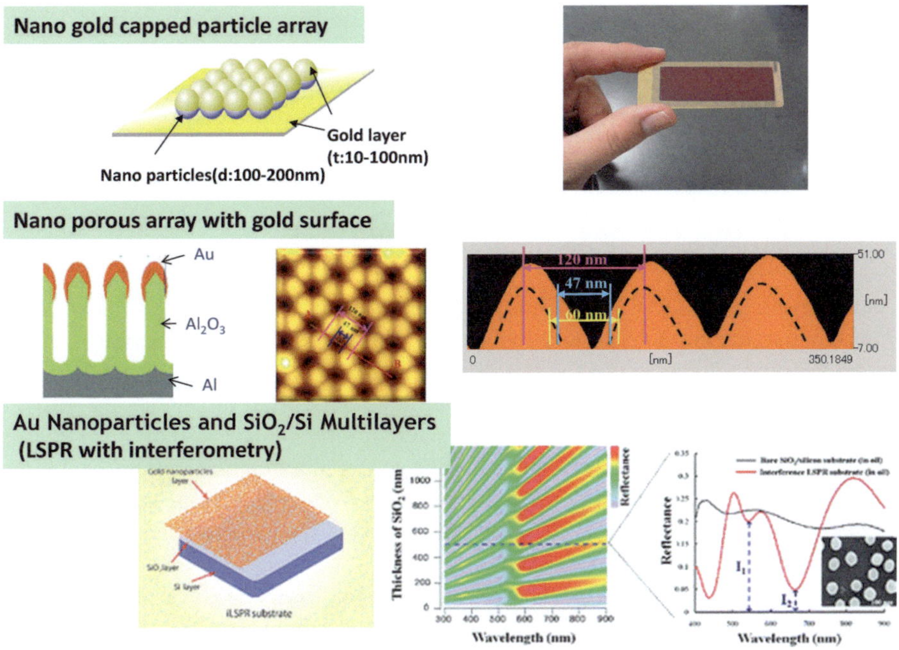

Fig. 1.7 Plasmonic nanostructures and performance. Various Au nanostructures for LSPR-based biosensor chips. Reprinted with permission from Hiep et al. (2009). Copyright 2009, American Chemical Society

silicon (SiO_2/Si) substrate. It is known that thin-film multilayers can be formed on a SiO_2/Si substrate. Since the Fabry–Perot fringe pattern undergoes wavelength shifts upon molecule binding, such structures have been used for biosensors (Lin et al. 1997). A two-dimensional assembly of Au NPs, which could be regarded as a thin film, still retained the optical characteristics of the original NPs. From a combination of these principles, we developed a novel interference LSPR (iLSPR) biosensor (Hiep et al. 2009). First, iLSPR substrates were fabricated by immobilizing Au NPs on a SiO_2/Si substrate: the SiO_2/Si substrate was cut into the desired size, and soaked in a neutral pH detergent solution for 12 h. After cleaning the substrates in solutions of hydrogen peroxide (30 %), ammonia (28 %) and MilliQ water for 30 min at 80 °C, the substrates were immersed in a 10 % (v/v) solution of 3-aminopropyltrimethoxysilane for 15 min. Afterwards, they were rinsed with ethanol and dried at 120 °C for 2 h. Subsequently, Au NPs were dispersed as a monolayer on the modified SiO_2/Si substrates through adsorption for 24 h, and the monolayer was heated under vacuum conditions for 2 h. Bonding of the NPs to the surface was strong enough to resist detachment from the surface because of later chemical modifications. The developed chip was functionalized using alkane-thiol molecules and the addition of the molecules resulted in an increase in relative reflectance. The success motivated us to integrate the developed chip with microfluids

(Hiep et al. 2009), and also inspired development of another nanobiosensor for specific biomolecular interactions. The nanobiosensor chip was fabricated by immobilizing Au NPs onto a porous aluminum oxide (Al_2O_3) layer formed on an aluminum (Al) substrate (Hiep et al. 2010).

1.3.3 *Nanospatial Interface Interactions*

As briefly discussed and shown in Fig. 1.3, there are many types of nanobiosensors. The key parameters determining the type of nanobiosensor to develop are (1) the target analyte; (2) the level of sensitivity needed; and (3) the medium in which the target analyte is present. This may pose some challenges if, for example, the analyte is a low-abundant molecule in a complex environment such as biological fluids. Some high-abundant proteins such as human serum albumin and immunoglubulins have been reported to affect detection sensitivity (Vestergaard and Tamiya 2007). Other secondary but equally important issues are the type of instrumentation available, as well as the trained personnel.

Let us consider these two related biosensing platforms: surface plasmon resonance (SPR) and localized SPR with spatial interfaces of at least an order of magnitude difference. SPR sensing is based on the fact that the refractive index is altered by the medium conditions, which consequently changes the requirements for coupling resonance. The binding of an analyte to a recognition layer immobilized on the metal surface would, for example, result in a shifted resonance dip/peak in the spectrum, which can then be correlated to an analyte concentration. A commonly suggested mechanism of improvement in Au NP-based SPR sensing devices is that the coupling between the LSPR of Au NPs and the propagating SPR of Au films results in a greater SPR incident angle shift. The shape of the peak and size of the shift vary with Au NP size (Su et al. 2009; Hutter and Fendler 2004; Haes et al. 2004; Nath and Chilkoti 2002). Metal nanoparticles exhibit localized surface plasmon resonance (LSPR). This can be observed as a strong UV–Vis absorption band that is absent from the spectrum of the bulk metal.

Some knowledge and understanding of nanospatial interface interactions in the different types of biosensors is important since biorecognition occurs at the interface. In Fig. 1.8, we show four types of nanobiosensors with differing nanospatial interfaces. We show that the sensitive regions are different among types of nanodevices [for example, carbon nanotube field effect transistor (CNT-FET), LSPR, electrochemical devices]. The FET-sensitive region is based on a Debye length of less than 10 nm from the CNT surface. The LSPR region is 10–100 nm from the gold surface. Electrochemical responses (in the case of enzyme sensors) are controlled by substrate diffusion processes, and this is at a micrometre scale. The electron transfer process through the electric double layer is at a nanometre scale. For CNT-FETs, a small gap is therefore important because the effect is then much stronger. This understanding is important in choosing the appropriate technology for nanobiosensing.

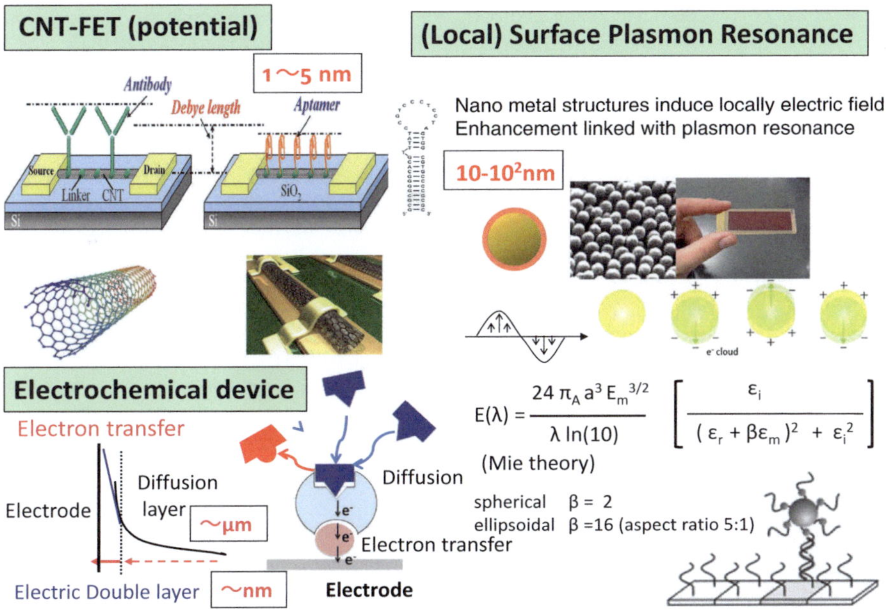

Fig. 1.8 Nanospatial interface interactions in nanobiosensors. Reprinted with permission from Maehashi et al. (2007). Copyright 2007, American Chemical Society

1.4 Concluding Remarks

In the introductory review chapter to this book, we have provided an overview of nanobiosensing and nanobioanalyses, paving way for the subsequent chapters to cover specific topics under this general theme. The chapter is conveniently split into two main parts: we discussed nanobiosensors in the first half and the second presented nanobioanalyses. Throughout, we used examples of applications from our group and other research groups. Both nanobiosensing and nanobioanalyses are born out of the tremendous advances made within the disciplines of materials development, technology, engineering and application areas including medical, environmental, food, pharmaceutical, energy and security. We have provided a brief history of the advances made in the field, for the benefit of the reader new to the field. The more advanced reader will still benefit from being introduced to some of the new developments, in particular the applications and challenges of developed nanobiosensors.

References

Ambrosi A, Castañeda MT, Killard AJ (2007) Double-codified gold nanolabels for enhanced immunoanalysis. Anal Chem 79(14):5232–5240. doi:10.1021/ac070357m

Bailey RE, Nie S (2003) Alloyed semiconductor quantum dots: tuning the optical properties without changing the particle size. J Am Chem Soc 125(23):7100–7106. doi:10.1021/ja035000o

Banin U, Millo O (2003) Tunneling and optical sepectroscopy of semiconductor nanocrystals. Annu Rev Phys Chem 54:465–492. doi:10.1146/annurev.physchem.54.011002.103838
Bertok T, Sediva A, Katrlik J et al (2013) Label-free detection of glycoproteins by the lectin biosensor down to attomolar level using gold nanoparticles. Talanta 108:11–18
Binnig G, Rohrer H, Gerber CH et al (1982) Tunneling through a controllable vacuum gap. Appl Phys Lett 40(2):178–180
Buhleier E, Wehner W, Vögtle F (1978) "Cascade" and "nonskid-chain-like" syntheses of molecular cavity topologies. Synthesis 1978(2):155–158. doi:10.1055/s-1978-24702
Cha C, Shin SR, Annabi N et al (2013) Carbon-based nanomaterials: multifunctional materials for biomedical engineering. ACS Nano 7(4):2891–2897. doi:10.1021/nn401196a
Chen Z-P, Peng Z-F, Zhang P (2007) A sensitive immunosensor using colloidal gold as electrochemical label. Talanta 72(5):1800–1804. doi:10.1016/j.talanta.2007.02.020
Dequaire M, Degrand C, Limoges B (2000) An electrochemical metalloimmunoassay based on a colloidal gold label. Anal Chem 72(22):5521–5528. doi:10.1021/ac000781m
Dreyer DR, Park S, Bielawski CW et al (2010) The chemistry of graphene oxide. Chem Soc Rev 39(1):228–240. doi:10.1039/B917103G
Du D, Chen W, Cai J et al (2009) Acetylcholinesterase biosensor based on gold nanoparticles and cysteamine self assembled monolayer for determination of monocrotophos. J Nanosci Nanotechnol 9(4):2368–2373
Endo T, Kerman K, Nagatani N et al (2005) Label-free detection of peptide nucleic acid–DNA hybridization using localized surface plasmon resonance based optical biosensor. Anal Chem 77:6976–6984
Endo T, Kerman K, Nagatani N et al (2006) Multiple label-free detection of antigen-antibody reaction using localized surface plasmon resonance-based core-shell structured nanoparticle layer nanochip. Anal Chem 78(18):6465–6475
Fiorino DJ (2010) Voluntary initiatives, regulation, and nanotechnology oversight: charting a path. http://www.nanotechproject.org. Accessed 17 Jan 2014.
Gaponik N, Radtchenko IL, Sukhorukov GB et al (2002) Toward encoding combinatorial libraries: charge-driven microencapsulation of semiconductor nanocrystals luminescing in the visible and near IR. Adv Mater 14(12):879–882. doi:10.1002/1521-4095(20020618)14:12<879::AID-ADMA879>3.0.CO;2-A
González-García MB, Fernández-Sánchez C, Costa-García A (2000) Colloidal gold as an electrochemical label of streptavidin–biotin interaction. Biosens Bioelectron 15(5–6):315–321
Gooding JJ, Ciampi S (2011) The molecular level modification of surfaces: from self-assembled monolayers to complex molecular assemblies. Chem Soc Rev 40:2704–2718
Haes AJ, Hall WP, Chang L et al (2004) A localized surface plasmon resonance biosensor: first steps toward an assay for Alzheimer's disease. Nano Lett 4(6):1029–1034. doi:10.1021/nl049670j
Häkkinen H (2012) The gold-sulfur interface at the nanoscale. Nat Chem 4:443–455
Hasanzadeh M, Shadjou N, Eskandani M et al (2014) Dendrimer-encapsulated and cored metal nanoparticles for electrochemical nanobiosensing. Trends Anal Chem 53:137–149
Hiep HM, Endo T, Saito M et al (2008) Label-free detection of melittin binding to a membrane using electrochemical-localized surface plasmon resonance. Anal Chem 80(6):1859–1864. doi:10.1021/ac800087u
Hiep HM, Yoshikawa H, Saito M et al (2009) An interference localized surface plasmon resonance biosensor based on the photonic structure of Au nanoparticles and SiO_2/Si multilayers. ACS Nano 3(2):446–452. doi:10.1021/nn800831a
Hiep HM, Yoshikawa H, Tamiya E (2010) Interference localized surface plasmon resonance nanosensor tailored for the detection of specific biomolecular interactions. Anal Chem 82(4):1221–1227. doi:10.1021/ac902008x
Huang X, Jain PK, El-Sayed IH et al (2007) Gold nanoparticles: interesting optical properties and recent applications in cancer diagnostics and therapy. Nanomedicine 2(5):681–693
Huang H, Huang S, Yuan S et al (2011) High-sensitivity biosensors fabricated by tailoring the localized surface plasmon resonance property of core–shell gold nanorods. Anal Chim Acta 683(2):242–247. doi:10.1016/j.aca.2010.10.033

Hutter E, Fendler JH (2004) Exploitation of localized surface plasmon resonance. Adv Mater 16(19):1685–1706. doi:10.1002/adma.200400271

Idegami K, Chikae M, Kerman K et al (2008) Gold nanoparticle-based redox signal enhancement for sensitive detection of human chorionic gonadotropin hormone. Electroanalysis 20:14–21

Iijima S (1991) Helical microtubules of graphitic carbon. Nature 354:56–58. doi:10.1038/354056a0

Jadzinsky PD, Calero G, Ackerson CJ, Bushnell DA, Kornberg RD (2007) Structure of a thiol monolayer-protected gold nanoparticle at 1.1 Å resolution. Science 318:430–433

Kerman K, Kraatz H-B (2009) Electrochemical detection of protein tyrosine kinase-catalysed phosphorylation using gold nanoparticles. Biosens Bioelectron 24(5):1484–1489. doi:10.1016/j.bios.2008.10.024

Kerman K, Saito M, Morita Y et al (2004) Electrochemical coding of single-nucleotide polymorphisms by monobase-modified gold nanoparticles. Anal Chem 76(7):1877–1884. doi:10.1021/ac0351872

Kerman K, Chikae M, Yamamura S et al (2007) Gold nanoparticle-based electrochemical detection of protein phosphorylation. Anal Chim Acta 588(1):26–33. doi:10.1016/j.aca.2007.02.001

Kumar S, Kumar S, Ali MA et al (2013) Microfluidic-integrated biosensors: prospects for point-of-care diagnostics. Biotechnol J 8(11):1267–1279. doi:10.1002/biot.201200386

Lan J, Zhou X, Liu G et al (2011) Enhancing photocatalytic activity of one-dimensional $KNbO_3$ nanowires by Au nanoparticles under ultraviolet and visible-light. Nanoscale 3(12):5161–5167. doi:10.1039/C1NR10953G

Li Y, Schluesener HJ, Xu S (2010) Gold nanoparticle-based biosensors. Gold Bull 43(1):29–41

Ligler FS, Taitt CR, Shriver-Lake LC et al (2003) Array biosensor for detection of toxins. Anal Bioanal Chem 377(3):469–477. doi:10.1007/s00216-003-1992-0

Lin VSY, Motesharei K, Dancil KPS et al (1997) A porous silicon-based optical interferometric biosensor. Science 278:840–843

Love JC, Estroff LA, Kriebel JK, Nuzzo RG, Whitesides GM (2005) Self-assembled monolayers of thiolates on metals as a form of technology. Chem Rev 105:1103–1169

Maehashi K, Katsuya T, Kerman K et al (2007) Label-free Protein Biosensor Based on Aptamer-Modified Carbon Nanotube Field-Effect Transistors. Anal Chem 79:782–787

Möller R, Powell RD, Hainfeld JF et al (2005) Enzymatic control of metal deposition as key step for a low-background electrical detection for DNA chips. Nano Lett 5(7):1475–1482. doi:10.1021/nl050824k

Nath N, Chilkoti A (2002) A colorimetric gold nanoparticle sensor to interrogate biomolecular interactions in real time on a surface. Anal Chem 74(3):504–509. doi:10.1021/ac015657x

Ng JK-K, Chong SS (2011) Multiplexing capabilities of biosensors for clinical diagnostics. In: Serra PA (ed) Biosensors for health, environment and biosecurity. InTech, Rijeka, pp 241–256

Ng JK-K, Selamat ES, Liu W-T (2008) A spatially addressable bead-based biosensor for simple and rapid DNA detection. Biosens Bioelectron 23(6):803–810. doi:10.1016/j.bios.2007.08.026

Ng JK-K, Wang W, Liu W-T et al (2010) Spatially addressable bead-based biosensor for rapid detection of beta-thalassemia mutations. Anal Chim Acta 658(2):193–196. doi:10.1016/j.aca.2009.11.020

Novoselov KS, Geim AK, Morozov SV et al (2004) Electric field effect in atomically thin carbon films. Science 306(5696):666–669. doi:10.1126/science.1102896

Palchetti I, Mascini M (2010) Sensors and microsystems. Lecture notes in Electrical Engineering 54:15–23

Park T-H, Mirin N, Lassiter JB et al (2008) Optical properties of a nanosized hole in a thin metallic film. ACS Nano 2(1):25–32. doi:10.1021/nn700292y

Pyrzynska K (2013) Use of nanomaterials in sample preparation. Trends Anal Chem 43:100–108

Sathuluri RR, Yoshikawa H, Shimizu E et al (2011) Gold nanoparticle-based surface-enhanced Raman scattering for noninvasive molecular probing of embryonic stem cell differentiation. PLoS One 6(8):e22802. doi:10.1371/journal.pone.0022802

Scognamiglio V, Pezzotti I, Pezzotti G et al (2012) Towards an integrated biosensor array for simultaneous and rapid multi-analysis of endocrine disrupting chemicals. Anal Chim Acta 751:161–170

Selvaraju T, Das J, Jo K et al (2008) Nanocatalyst-based assay using DNA-conjugated Au nanoparticles for electrochemical DNA detection. Langmuir 24(17):9883–9888. doi:10.1021/la801828a
Shoute L, Bergren AJ, Mahmoud AM et al (2009) Optical interference effects in the design of substrates for surface-enhanced Raman spectroscopy. Appl Spectrosc 63(2):133–140. doi:10.1366/000370209787392102
Souza Silva EA, Risticevic S, Pawliszyn J (2013) Recent trends in SPME concerning sorbent materials, configurations and in vivo applications. Trends Anal Chem 43:24–36
Su H, Li Y, Li X-Y et al (2009) Optical and electrical properties of Au nanoparticles in two-dimensional networks: an effective cluster model. Opt Express 17(24):22223–22234
Taniguchi N (ed) (1996) Nanotechnology: integrated processing systems for ultra-precision and ultra-fine products. Oxford University Press, New York
Thompson DT (2007) Using gold nanoparticles for catalysis. Nano Today 2(4):40–43
Turner M, Golovko VB, Vaughan OPH et al (2008) Selective oxidation with dioxygen by gold nanoparticle catalysts derived from 55-atom clusters. Nature 454(7207):981–983. doi:10.1038/nature07194
Ulman A (1996) Formation and structure of self-assembled monolayers. Chem Rev 96(4):1533–1554. doi:10.1021/cr9502357
Valden M, Lai X, Goodman DW (1998) Onset of catalytic activity of gold clusters on titania with the appearance of nonmetallic properties. Science 281(5383):1647–1650. doi:10.1126/science.281.5383.1647
Vericat C, Vela ME, Benitez G, Carro P, Salvarezza RC (2010) Self-assembled monolayers of thiols and dithiols on gold: new challenges for a well-known system. Chem Soc Rev 39:1805–1834
Vestergaard M, Tamiya E (2007) A rapid sample pretreatment protocol: improved sensitivity in the detection of a low-abundant serum biomarker for prostate cancer. Anal Sci 23(12):1443–1446
Vestergaard M, Kerman K, Kim D-K et al (2008) Detection of Alzheimer's tau protein using localised surface plasmon resonance-based immunochip. Talanta 74:1038–1042
Vuckovic D, Shirey R, Chen Y et al (2009) In vitro evaluation of new biocompatible coatings for solid-phase microextraction: Implications for drug analysis and in vivo sampling applications. Anal Chim Acta 638(2):175–185. doi:10.1016/j.aca.2009.02.049
Wen Q, Gu Y, Tang L-J et al (2013) Peptide-templated gold nanocluster beacon as a sensitive, label-free sensor for protein post-translational modification enzymes. Anal Chem 85(24):11681–11685. doi:10.1021/ac403308b
Whatmore RW (2006) Nanotechnology-what is it? Should we be worried? Occup Med 56(5):295–299
Wu S, Lan X, Zhao W et al (2011) Controlled immobilization of acetylcholinesterase on improved hydrophobic gold nanoparticle/Prussian blue modified surface for ultra-trace organophosphate pesticide detection. Biosens Bioelectron 27(1):82–87. doi:10.1016/j.bios.2011.06.020
Xiang Y, Deng K, Xia H et al (2013) Isothermal detection of multiple point mutations by a surface plasmon resonance biosensor with Au nanoparticles enhanced surface-anchored rolling circle amplification. Biosens Bioelectron 49:442–449

Part II
Nanobiosensing Architectures

Chapter 2
Vertically Aligned Nanowire Array-Based Sensors and Their Catalytic Applications

Kafil M. Razeeb, Mamun Jamal, Maksudul Hasan, and Alan Mathewson

Abstract This chapter investigates the fabrication of metallic nanowire arrays and their application as sensors or biosensors to determine glucose, glutamate and H_2O_2. Nanowire-based material provides a very high electrochemically active surface area, thereby leading to very high detection sensitivity. To date, noble metals, such as platinum (Pt), gold (Au), silver (Ag) and their alloys, have been extensively investigated as anodic materials in the design of non-enzymatic sensor surfaces. Vertically aligned structures such as free-standing nanowire arrays are particularly useful for sensing and biosensing applications due to their high sensitivity to surface interactions. Although nanowires have been fabricated by various methods, a simple fabrication technique which incurs reasonable costs for practical applications is highly desired. In this chapter, a template-based fabrication of nanowire arrays of Au, Pt and nickel (Ni) will be described. These materials have been characterized using X-ray diffraction, scanning electron microscopy and transmission electron microscopy. The catalytic activity and the suitability of these materials as electrochemical sensors have been investigated by cyclic voltammetry and amperometry.

Keywords Biosensor • Electrochemistry • Nanowire array • Sensor

2.1 Introduction

Fabrication of reliable and cost-effective catalysts for the precise monitoring of electro-active species is of significant interest in the development of sensing devices for point-of-care, environmental control and industrial systems. From the application point of view, the final goal of sensor or biosensor technology lies in designing high-performance sensors with appropriate characteristics such as sensitivity, selectivity, response time, linearity, stability and reproducibility. In designing such high-performance sensor devices, the key factors are the strategic technologies for on-electrode fabrication and enzyme immobilization. Regarding electrode

K.M. Razeeb (✉) • M. Jamal • M. Hasan • A. Mathewson
Tyndall National Institute, University College Cork,
Dyke Parade, Maltings, Cork, Ireland
e-mail: kafil.mahmood@tyndall.ie

M.C. Vestergaard et al. (eds.), *Nanobiosensors and Nanobioanalyses*,
DOI 10.1007/978-4-431-55190-4_2

fabrication, vertically aligned nanowire array or three-dimensional (3D) electrodes (Kang et al. 2008; Pang et al. 2009; Wen et al. 2009) are becoming popular due to their large active surface area in a small footprint. The term "biosensor" is used where biological substances (such as enzymes, antibodies, etc.) are used as bio-recognition elements. A great amount of work has been done on the immobilization of bio-recognition elements on the surface of electrodes to improve biosensor performance. However, the distribution of enzymes in random 3D matrices such as the sol–gel matrix (Yang et al. 2006) and the random porous matrix (You et al. 2009) are usually not uniform, and they increase steric hindrance for the analytes, resulting in relatively low electron transfer efficiency. Moreover, in this situation, a smaller percentage of immobilized enzyme is electroactive because of the haphazard orientation of the enzyme used. Thus, ordered 3D matrices, such as nanotube arrays (Cui et al. 2007; Lan et al. 2008), ordered porous materials (Zhu et al. 2009) and nanowire (Xu et al. 2010; Jamal et al. 2011), nanorod and nanopillar (Yu et al. 2003; Anandan et al. 2006; Gu et al. 2009) electrode ensembles have all been used for increasing the active surface area for enzyme immobilization. The ordered 3D nanowire arrays with high aspect ratios improve the signal-to-noise ratio significantly, resulting in significantly higher sensitivity and a lower detection limit (De Leo et al. 2007); (Xu et al. 2010) in both sensor and biosensor applications.

Although noble metals are found to be highly sensitive as sensing surfaces, various other metals and metal oxides, such as CuO, RuO_2 and MnO_2, have also been used as efficient materials for direct oxidation of glucose, glutamate and H_2O_2, where there is no enzyme even needed (Chen et al. 2008; Umar et al. 2009). Among them, Ni-based nanomaterials have received special attention owing to their low cost and high catalytic activity towards glucose oxidation. Liu et al. reported glucose biosensors based on Ni nanoparticle (NP) decorated carbon nanofibers (Liu et al. 2009). He et al. reported glucose sensors based on Ni NP modified TiO_2 tube arrays (Yu et al. 2012). In this chapter, we describe the fabrication of various types of vertically aligned nanowire arrays using simple template-based methods and their application in the detection of various analytes such as glutamate, glucose and H_2O_2.

2.2 Synthesis of Pt Nanoparticle-Modified Gold Nanowire Arrays

Vertically standing gold nanowire arrays (Fig. 2.1a) were fabricated using a template electrodeposition technique. A thin layer of Au (300 nm thick) was first sputtered onto one side of the anodic alumina oxide (AAO). Electrodeposition of Au was performed at a constant current density of 1.0 mA cm^{-2} for 9,000 s in the Au bath (PURAMET 402). After deposition, the AAO template was etched away by submerging the sample in 6.0 M KOH solution for 45 min. The resulting gold NAE film was rinsed in deionised water and cut into a circular electrode 5 mm in diameter. This was pasted to a 3 M transparent gold sputtered polymer film for the electrical connection. The resulting nanowire-based electrode was dipped into a Pt

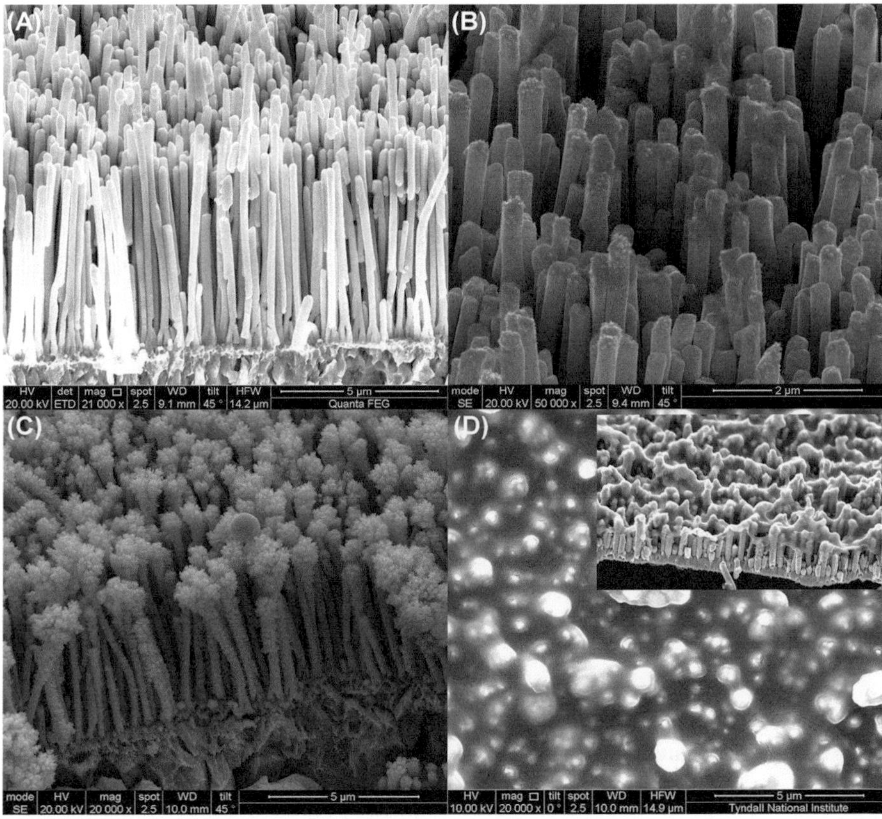

Fig. 2.1 SEM images of the (**a**) AuNAE; (**b**) 100 s PtNP-deposited AuNAE; (**c**) 600 s PtNP-modified AuNAE and (**d**) GlutOx/PtNP/AuNAE (Jamal et al. 2010)

solution made of 5 mM potassium hexachloroplatinate and 10 mM HCl and electrodeposition was performed with a current density of 1.0 mA cm^{-2} from 60 to 900 s to fabricate a Pt nanoparticle-modified gold nanowire array electrode (PtNP/NAE) (Fig. 2.1b, c).

2.2.1 Enzyme Immobilization for Glutamate Detection

Prior to enzyme immobilisation, the PtNP/NAE was rinsed with acetone and deionised water, followed by running cyclic voltammetry (CV) in 0.5 M H_2SO_4 with a potential range of −0.2 to +1.5 V at a scan rate of 0.1 V s^{-1} until a stable CV was obtained. 20 µL of glutamate oxidase (GlutOx) [25 U in 250 µL of 0.01 M phosphate-buffered saline (PBS), pH 7.4] was mixed with 2 mg of bovine serum albumin (BSA), 20 µL of glutaraldehyde (2.5 % w/v) and 10 µL of Nafion (0.5 %). 10 µL of this mixture was then dispensed onto the surface of a PtNP/NAE and allowed to dry for 60 min under ambient conditions. The electrodes were then stored at 4 °C overnight in 0.01 M PBS buffer prior to use (Fig. 2.1d).

2.2.2 Morphology of the NAEs and PtNP/NAEs

Figure 2.1 shows SEM images of a gold NAE, PtNP/NAE and enzyme-immobilised PtNP/NAE. The average length of the nanowires is approximately 6 μm with a diameter of ~200 nm. Figure 2.1a shows the fabricated nanowires after removing the anodic alumina oxide (AAO) template. Figure 2.1b clearly shows that PtNPs were formed on the nanowire surface after 100 s of deposition at a current density of 1 mA cm^{-2}. With increasing deposition time, more PtNPs were deposited on the nanowire, and they coated the surface area of the nanowires and were concentrated at the top of the nanowires like a flower (Fig. 2.1c). Immobilization of GlutOx was found to be uniform, where effective cross-linking and protein immobilization within the Nafion polymer layer (Fig. 2.1d) can be observed.

2.2.3 Electrochemical Characterization of PtNP/NAEs

Figure 2.2a illustrates the stepwise fabrication process for the GlutOx/PtNP/AuNAEs. Figure 2.2b shows the cyclic voltammetry of a flat gold electrode (AuE) (a circular Au-coated polymer 5 mm in diameter) and a PtNP-modified flat gold electrode (PtNP/AuE). Figure 2.2c shows the same for an AuNAE and PtNP-modified AuNAEs (PtNP/AuNAE). All these experiments were performed in 0.5 M H_2SO_4 at a scan rate of 0.1 V s^{-1}, where typical voltammogram characteristics of a metallic Pt electrode in 0.5 M H_2SO_4 have been observed. The electrochemically active surface area of PtNP was determined from the charge consumed during the hydrogen adsorption/desorption voltammogram (Chakraborty and Retna Raj 2009). The loading of nanoparticles was optimized on the surface of the NAE, and 600–800 s was found to be the optimum deposition time at a current density of 5 mA cm^{-2}. This created a surface area of deposited PtNPs of 0.067 cm^2. A longer deposition time of PtNP on NAE has a negative effect on the active surface area, and this may be attributed to the blocking effect by the nanoparticles, due to the larger amount of deposition at the top of the NAEs, as evident from Fig. 2.1c. Moreover, a large amount of PtNPs reduces the inter-particle distance, which effectively reduces the active surface area (Karam and Halaoui 2008).

The roughness ratio of the flat Au, NAE and PtNP/NAE was calculated by integrating the voltammogram from +0.72 to +1.2 V for flat Au and NAE (Fig. 2.2b), and from +0.20 to +0.67 V for PtNP/AuE and PtNP/NAE (Fig. 2.2c) electrodes (Xu et al. 2010). Roughness was increased by 26.6 times for NAE electrodes compared to the flat Au electrode and increased by 4.3 times for PtNP/AuNAE compared to the flat Au/PtNP electrode.

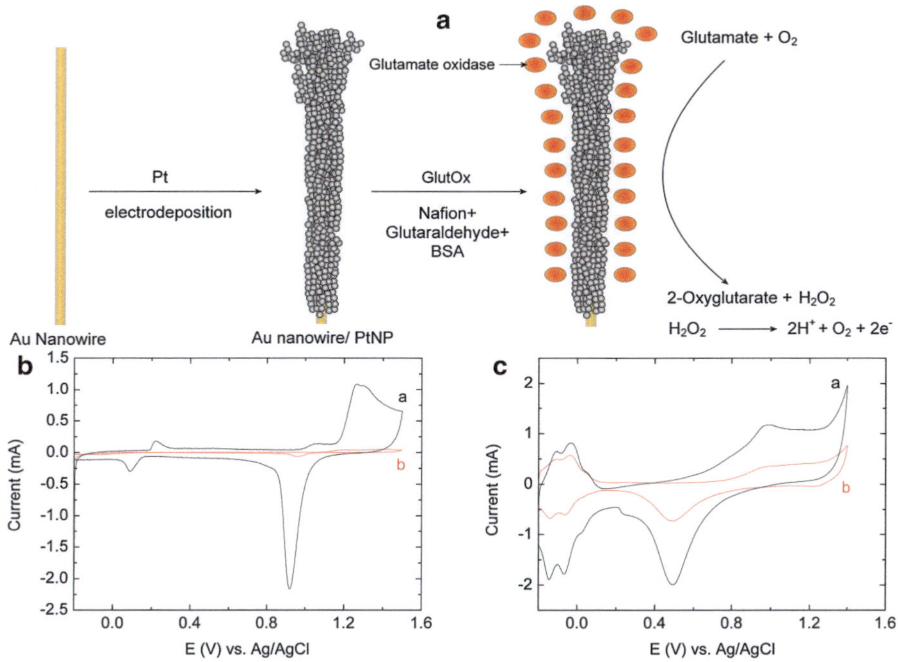

Fig. 2.2 (**A**) Schematic illustration of stepwise fabrication of the GlutOx/PtNP/AuNAE. (**B**) Cyclic voltammograms of the AuNAE (*a*) and flat Au electrode (*b*). (**C**) Cyclic voltammograms of the PtNP/AuNAE (*a*) and PtNP on flat Au electrode (*b*) in 0.5 M H_2SO_4 at 0.1 V s^{-1} (Jamal et al. 2010)

2.2.4 Amperometric Behaviour Towards H_2O_2

Figure 2.3 shows the amperogram of direct electro-oxidation of H_2O_2 on a PtNP/NAE electrode at $E_{app}=0.65$ V vs. Ag/AgCl. To obtain the amperogram, 200 μM of H_2O_2 was added at regular intervals. A reproducible response was obtained upon repeated addition of H_2O_2, with a sensitivity of 194.6 μA mM^{-1} cm^{-2} at 20 °C and a linear range of up to 20 mM. This confirms that the electrode does not undergo deactivation during the experiments. Without using any redox mediator or enzyme, PtNP/NAE can detect H_2O_2 concentrations as low as 1 μM at ambient. Moreover, PtNP/NAE-modified electrodes showed 24 times higher sensitivity toward H_2O_2 than a AuNAE electrode and 82 times more than a flat Au electrode of similar footprint. However, the sensitivity of a PtNP-modified flat Au electrode was found to be 166.4 μA mM^{-1} cm^{-2}, which is closer to that obtained for PtNP/NAE (194.6 μA mM^{-1} cm^{-2}); this suggests that nanoparticles play a major role in H_2O_2 oxidation. The overall superiority of PtNP/NAE compared to other electrode platforms prepared in this work is based on sensitivity, detection limit, t_{90} and r^2 values (Table 2.1).

A comparison of the performance of some efficient sensor platforms using PtNPs on nano-array electrodes for H_2O_2 detection is presented in Table 2.2. It can be seen

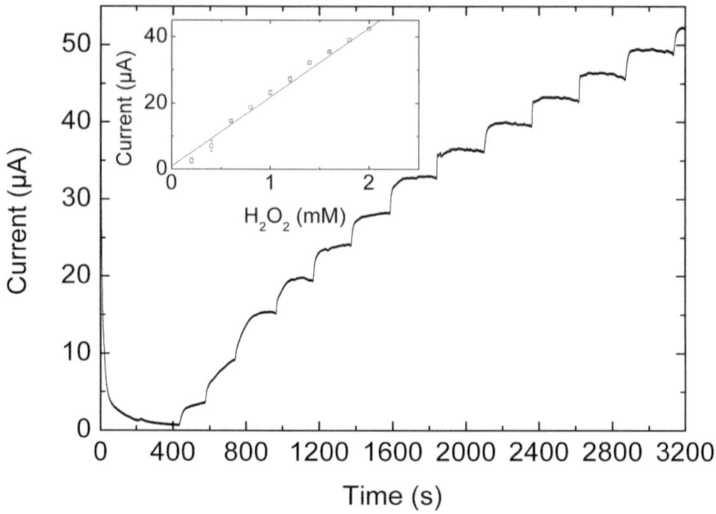

Fig. 2.3 Amperometric response of PtNP/NAE on successive addition of 0.2 mM H_2O_2 (E_{app} = 0.65 V vs. Ag/AgCl) and corresponding concentration plot (*inset*). Background electrolyte is 0.01 M PBS (pH 7.4) (Jamal et al. 2010)

Table 2.1 Analytical performance parameters of AuE, PtNP/AuE, NAE and PtNP/NAE using no enzyme for detection of H_2O_2

Electrodes	r^2	t_{90} (s)	Linearity (mM)	Detection limit (µM)	Sensitivity (µA mM^{-1} cm^{-2})
AuE	0.9972	55	2.0–20	47	2.35 ± 0.11
PtNP/AuE	0.9627	18	0.02–18	23	166.4 ± 7.9
NAE	0.9628	16	1.0–16	5	8.0 ± 0.38
PtNP/NAE	0.9847	11	0.02–20	1	194.6 ± 9.2

Table 2.2 A comparison of the performance of some sensor platforms using PtNPs on nano-array electrodes for H_2O_2 detection

Electrodes	Sensitivity (µA mM^{-1} cm^{-2})	Linear range (mM)	Limit of detection (µM)	References
PtNP/NAE	194.6	0.02–20	1.0	This paper
PtNP-CNT arrays	140	5×10^{-3} to 25	1.5	Wen et al. (2009)
PtNP-TiO_2 nanotube arrays	1.68	4×10^{-3} to 1.25	4.0	Cui et al. (2008)
PtNP-CNT TiO_2 nanotube arrays	0.134	1×10^{-3} to 2	1.0	Pang et al. 2009)
Pt-AuNP TiO_2 nanotube arrays	2.92	10–80 × 10^{-3}	10	Kang et al. (2008)

that the proposed PtNP/NAE sensor template shows a good performance in terms of sensitivity, linear range and limit of detection (LOD). However, the calculated sensitivity for all the electrodes in Table 2.1 was determined by considering a surface area of 0.2 cm^2.

2.2.5 *Analytical Performance of the Glutamate Biosensor*

The analytical performance of the PtNP/NAE platform was examined in the detection of glutamate. Immobilization of glutamate oxidase (GlutOx) was achieved by cross-linking with glutaraldehyde (1 % v/v) and BSA (2 % w/v) onto the surface of PtNP/NAEs. The optimum enzyme loading was determined by examining the activity levels over the range of 0.02–0.8 U GlutOx per electrode . An enzyme loading of 0.58 U GlutOx per electrode was found to be optimal and higher loadings resulted in poor substrate response. On the other hand, lower enzyme loadings also resulted in poor response sensitivity. To obtain the best cross-linking, the ratio of Nafion, glutaraldehyde and BSA was used following the article published by Jamal et al. (2009). The potential of the GlutOx-immobilized electrode was held at 0.65 V and aliquots of glutamate was injected into oxygen-saturated PBS at regular intervals (Fig. 2.4). This electrode could successfully detect enzymatically generated H_2O_2. The sensitivity of the electrode towards glutamate obtained from the calibration plot was 10.76 μA mM^{-1} cm^{-2}. The limit of detection was found to be 14 μM with a linear range of 50 μM to 2 mM and a t_{90} of 4.6 s.

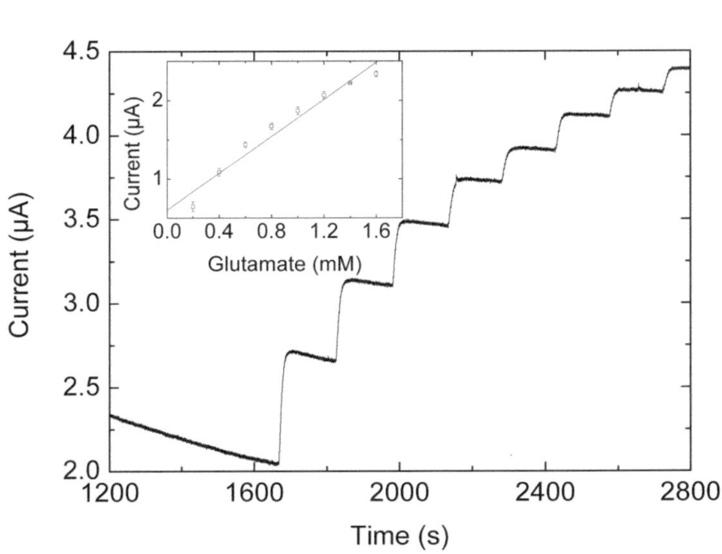

Fig. 2.4 Amperometric response of GlutOx-modified PtNP/NAE on successive addition of 0.2 mM glutamate (E_{app}=0.65 V vs. Ag/AgCl) and corresponding concentration plot (*inset*). Background electrolyte is 0.01 M PBS (pH 7.4) (Jamal et al. 2010)

2.3 Synthesis of Ni Nanowire Arrays

Nickel nanowires were prepared inside the anodic alumina oxide (AAO) template having an average pore diameter of 250 nm. One side of the template was evaporated with a 400 nm layer of Ni film (deposited using Temescal FC-2000), which served as the working electrode. The ohmic contact to the substrate was made using Radionics silver conductive paint, which has a volume resistivity of 0.001 Ω cm when fully hardened. A low stress nickel sulfamate bath was prepared using nickel sulfamate (0.37 mol dm^{-3}), boric acid (0.64 mol dm^{-3}), nickel bromide (0.18 mol.dm^{-3}) and wetting agent ANKOR® F (10 mL L^{-1}). The pH of the solution was adjusted to 3.8 by adding 1 mol dm^{-3} sulfamic acid at room temperature. Deionized water with resistivity ~18 MΩ was used to prepare the solution. The electrolyte temperature was maintained at 50 ± 1 °C, and the solution was stirred slowly at a rate of 100 rpm. A two-electrode cell was used with Ni pellets in a Ti-basket as an anode and Ni/porous AAO template as a working electrode (cathode). A constant current of 50 mA cm^{-2} (as the ratio of the total template area) was applied to the working electrode and the deposition rate was found to be 0.6 μm min^{-1}. After deposition, the free-standing nanowires were released by dissolving the template in 6.0 M KOH solution for 45 min, washed with plenty of de-ionized water and dried in air.

2.3.1 Synthesis of Shell@Core NiO@Ni Nanowire Array Electrode (NAE)

The nickel nanowire arrays were annealed in plasma using a March Plasmod GCM 200 system to fabricate shell@core NiO@NiNAE. The as-prepared film of NiNAEs were attached onto a Si substrate by adhesive tape (Kapton tapes, DuPont™, USA) at the edge, so that the back side was not exposed to the plasma and thereby only the surface of the nanowires was converted to oxide. The Ni nanowire arrays were annealed for different times at a power input of 50 W with a constant oxygen flow in order to achieve different thicknesses of the oxide layer on the nanowires.

2.3.2 Elemental Characterization of Shell@Core NiO@NiNAE

Figure 2.5a, b shows SEM images of the NiNAE and NiO@NiNAE with an average diameter of each nanowire of 250 nm. The images were taken after removing the AAO template. The average height of the nanowire was found to be around 10 μm. The surface morphologies of the as-deposited and plasma-annealed NAEs are different; e.g. the surface roughness of the annealed samples is increased due to oxide formation (Fig. 2.5c). A thin oxide layer is formed around the circumference of NiNAEs.

Fig. 2.5 (**a**) SEM images at 45° perspective view of the NiNAE; (**b**) NiO@NiNAE obtained from metallic NiNAE after oxygen plasma annealing at 50 W for 600 s; (**c**) High-magnification TEM image of a NiO@Ni nanowire; (**d**) STEM-EDS elemental mapping shows the presence of a thin oxide layer around the Ni circumference annealed at 50 W for 600 s: dark field (DF) images of coaxial NiO@Ni nanowire; (**e**) nickel at the inner core; (**f**) oxygen at the outer shell (Jamal et al. 2013)

A high-magnification HR-TEM (STEM-EDS elemental mapping on dark field) image (Fig. 2.5d) clearly shows the presence of a thin oxide layer at the outer surface of the Ni. Elemental mapping by STEM-EDS reveals the presence of Ni in the inner core (Fig. 2.5e) and O in the outer shell (Fig. 2.5f) of the annealed Ni nanowire. The constitution of NiNAE and NiO@NiNAE was further characterized using X-ray diffraction (XRD). The XRD patterns of the as-deposited virgin Ni nanowires and those from after they were annealed at 50 W for 300 s in oxygen plasma are shown in Fig. 2.6a. The XRD patterns of NiO@NiNAE show three well-resolved characteristic peaks of NiO that can be readily indexed at 37.3°, 43.3° and 63° to (111), (200), and (220), respectively. The peaks designated by asterisks are associated with the metallic Ni phases and show no diffraction peaks for NiO in the XRD patterns of the as-deposited sample. The characteristic peak intensities of NiO are lower than that of Ni, because Ni peaks have also been counted from the seed layer and thin film (3–4 μm) deposited onto the back side of the seed layer during the electrodeposition of NiNAE. Furthermore, the selected area diffraction image of a single nanowire (Fig. 2.6b) shows crystal orientation for both Ni and NiO phases.

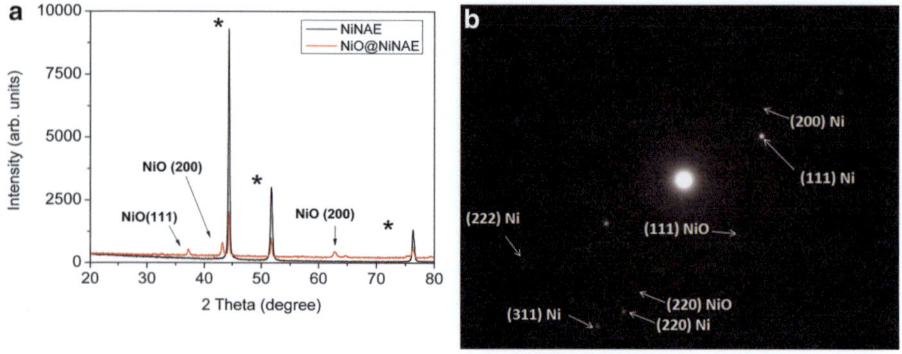

Fig. 2.6 (**a**) XRD of NiNAE and NiO@NiNAE; (**b**) electron diffraction pattern taken from the NiO@Ni nanowire showing the presence of both Ni metal and oxide d-spacing (Jamal et al. 2013)

2.3.3 Electrochemical Characterization of Shell@ Core NiO@NiNAE

Prior to the implementation of the as-prepared NiNAE and NiO@NiNAE as non-enzymatic sensors, their electrochemical behaviours were investigated using cyclic voltammograms (CVs). Figure 2.7a, b presents the CVs obtained for the NiNAE and NiO@NiNAE, respectively, in 1 M NaOH solution at various scan rates. The CV curve of the as-deposited NiNAE shows an anodic peak at 0.46 V and a cathodic peak at 0.34 V at a scan rate of 20 mV s^{-1}. The anodic and cathodic peaks are attributed to the formation and reduction of a few monolayers of NiOOH by the surface corrosion of metallic Ni (Amjad et al. 1977; Nelson et al. 2002). Similar behaviour was obtained for shell@core NiO@NiNAEs (5 min annealing), where anodic and cathodic peaks were obtained at 0.50 and 0.38 V, respectively. In contrast to the electrochemical behaviour of metallic Ni nanowires in alkaline solution, plasma-annealed NiO@NiNAEs show a redox couple resulting from the pseudocapacitive behaviour of faradaic oxidation and reduction reactions of NiO↔NiOOH. The thin oxide shell (NiO) also prevents the metallic core Ni from coming into contact with the alkaline electrolyte, and therefore the measured capacitance is completely attributable to the surface faradaic redox reaction of NiO. The reactions which occur on the surface can therefore be described as follows:

For metallic Ni:

$$Ni + 2OH^- \rightarrow Ni(OH)_2 + 2e^- \tag{2.1}$$

$$Ni(OH)_2 + OH^- \leftrightarrow NiOOH + H_2O + e^- \tag{2.2}$$

For NiO:

$$NiO + OH^- \rightarrow NiOOH + e^- \tag{2.3}$$

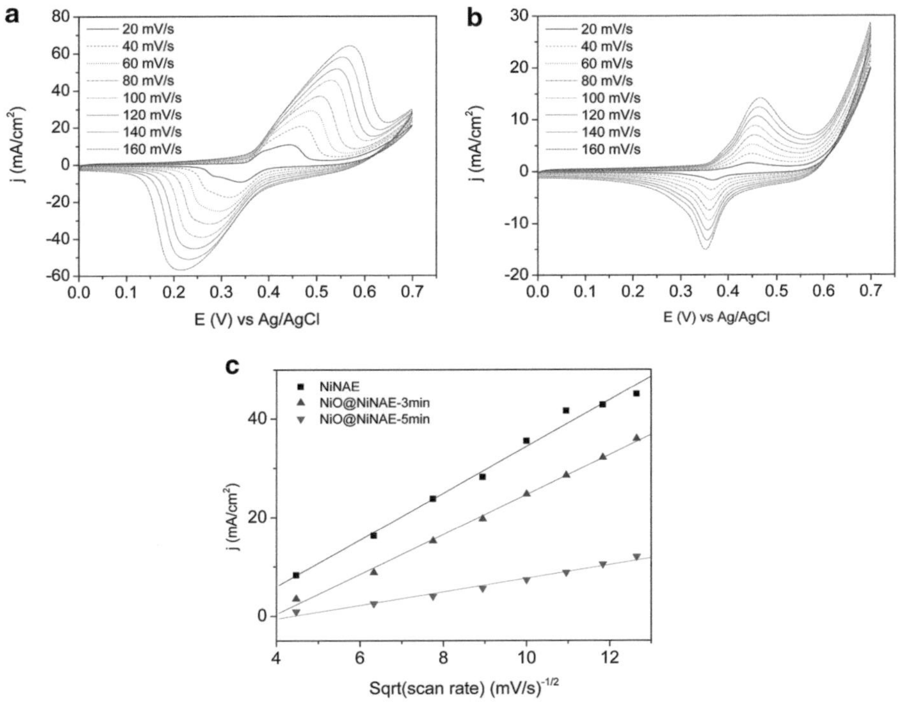

Fig. 2.7 Cyclic voltammograms of (**a**) NiNAE and (**b**) NiO@NiNAE, at different scan rates, (**c**) plot of current density vs. scan rate; inner to outer: 20, 40, 60, 80, 100, 120, 140 and 160 mV s^{-1}; electrolyte: 1 M NaOH (Jamal et al. 2013)

The capacitive current of NiNAEs in alkaline solution is higher than that of NiO, due to the fact that Ni is more corrosive in alkaline solution, and this contributes significantly to the current response. Conversely, NiO is a semiconductor in nature that prevents the corrosion of the metallic Ni core, and the current response is due totally to the surface faradaic oxidation and reduction reaction of the NiO shell. In addition, the reduced current response of annealed Ni nanowires can also be partially attributed to the slight depletion in interspaces between wires (see Fig. 2.7c) through agglomeration of nanowires. In any case, NiNAEs show quasi-reversible behaviour in NaOH, where peak potential separation increases with the scan rate. On the other hand, NiO@NiNAEs show completely reversible characteristics with insignificant peak separation as a function of the scan rate, which is attributed to electrode diffusion-controlled behaviour towards the electrolyte. This phenomenon is significant for the application of NiO@NiNAE as an efficient catalyst in the basic media. The effect of the thickness of the NiO shell (i.e. the effect of annealing) on electrochemical activities has been discussed in previous work (Hasan et al. 2012). At higher annealing times, more Ni turns into NiO and this thins the conductive core of Ni, thereby increasing the electrical resistance of each individual nanowire. Therefore, at higher annealing times (>5 min), thick and poorly conductive NiO in the outer shell of the nanowires becomes practically in electrical isolation from the conductive inner Ni core. On the other hand, shell@core coaxial NiO@Ni nanowire arrays made with less than 3 min annealing time do not show high catalytic response in glucose oxidation due to the absence of an efficient NiO@Ni interface.

2.3.4 *Electrocatalytic Behaviour and Analytical Performance of the NiO@NiNAE Sensor*

In order to address the analytical applicability of these nanowire electrodes, the electrocatalytic activity towards glucose was investigated. Figure 2.8a, b presents the CV responses of NiNAEs and NiO@NiNAEs (5 min annealing) in 1 M NaOH without and with 30 mM glucose. In the absence of glucose, two redox peaks of Ni^{2+}/Ni^{3+} were observed. Upon the addition of glucose, an increase of the anodic peak current can be observed for both NiNAE and NiO@NiNAE (Fig. 2.8), indicating that both electrodes exhibit excellent electrocatalytic activity towards the oxidation of glucose without using any enzyme. It also can be seen that with the addition of glucose, the anodic peak shifts to the higher potential in both electrodes, which may be attributed to the diffusion limitation of glucose at the electrode surface. Similar behaviour was obtained for Ni electrodes in other publications (Lu et al. 2009; Nie et al. 2011). As illustrated in the literature, the catalytic effect of the Ni^{2+}/Ni^{3+} redox couple for oxidation of glucose to gluconolactone was according to the following reactions:

$$\mathrm{Ni(OH)_2 + OH \leftrightarrow NiO(OH) + H_2O + e^-} \tag{2.4}$$

$$\mathrm{NiO(OH) + glucose \rightarrow Ni(OH)_2 + gluconolactone} \tag{2.5}$$

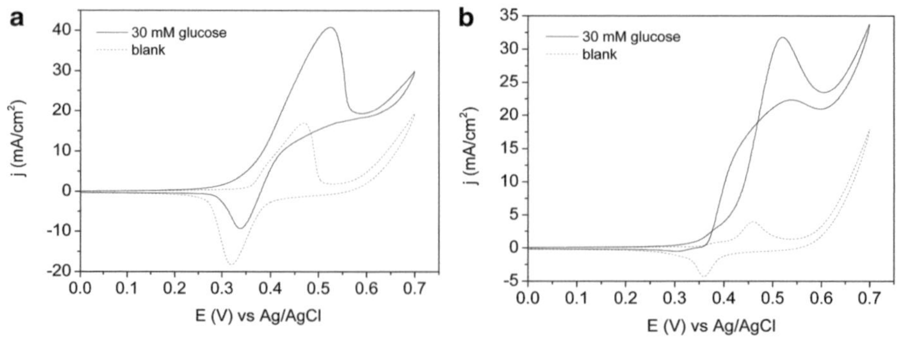

Fig. 2.8 Electrocatalytic response of (**a**) NiNAE, (**b**) NiO@NiNAE (5 min annealing) in the presence of 0 and 30 mM glucose. Scan rate: 20 mV s^{-1}; electrolyte: 1 M NaOH (Jamal et al. 2013)

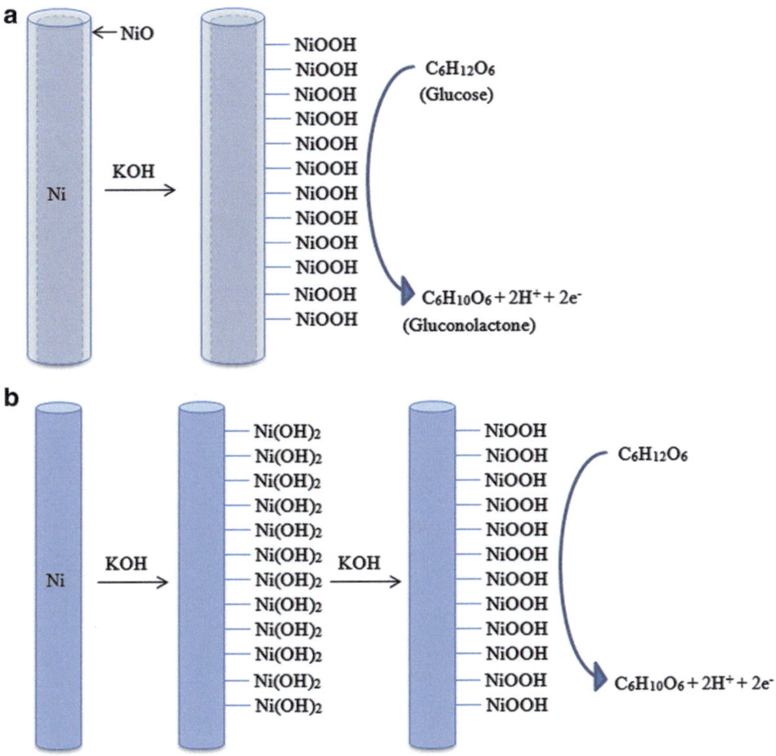

Fig. 2.9 A schematic diagram of the reaction mechanism on the surface of (**a**) NiNAE; (**b**) NiO@NiNAE (Jamal et al. 2013)

The reaction mechanism is described in more detail in Fig. 2.9. The catalytic behaviour of the NiO@NiNAE differs significantly from that of the NiNAE owing to the presence of the thin NiO shell around the exposed surface of each Ni nanowire. Upon the addition of glucose to the electrolyte system, the electro-oxidation current increases dramatically for NiO@NiNAEs compared with NiNAEs and therefore NiO is considered a good electrocatalyst. Moreover, the electro-oxidation current of glucose (calculated from CV before and after addition of glucose to the electrolyte of Fig. 2.8) for NiNAEs is low (23.8 mA cm^{-2}) compared to NiO@NiNAEs (27.8 and 25.4 mA cm^{-2} for 5 and 3 min annealing time, respectively). This is an indication that the current response for NiO@NiNAEs can be interpreted as the oxidation of glucose in the electrolyte system. In addition to the lower background current, the electrocatalytic current of NiO@NiNAEs increases upon the addition of glucose by 15–20 % compared to NiNAEs, but the capacitive behaviour is eliminated. This is due to the fact that in the case of NiO@NiNAEs, the outer shell (NiO) has converted into NiOOH in a single step (Eq. 2.3 and Fig. 2.9b) whereas in NiNAEs it occurs in two steps (Eqs. 2.1 and 2.2 and Fig. 2.9a) during the diffusion of glucose. The formation of the reaction intermediate NiOOH in the case of NiO is one step less

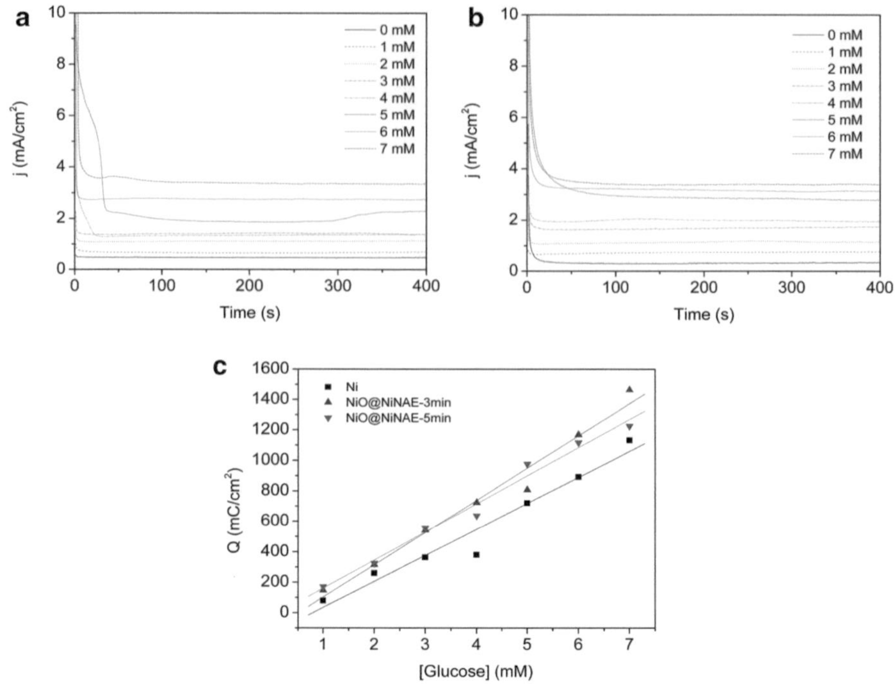

Fig. 2.10 Chronoamperometric responses of (**a**) NiNAE, (**b**) NiO@NiNAE, 5 min upon successive addition of 1 mM glucose. (**c**) Corresponding concentration plot; conditions: electrolyte 1 M NaOH and applied potential $E_{app}=0.52$ V (Jamal et al. 2013)

compared to Ni, which is the rate-determining step for glucose oxidation. This is also evident from the sharp rise of the oxidation current for the NiO@NiNAE compared to the NiNAE in Fig. 2.8b, a, respectively. Moreover, NiO is a good catalyst and thus the electrocatalytic behaviour completely dominates the capacitive behaviour in the presence of glucose and the current response is completely due to the electro-oxidation of glucose. A thin oxide-coated Ni nanowire can therefore be seen to be an excellent electrocatalytic material for the detection of glucose.

For the chronoamperometric sensing application, NiNAEs and NiO@NiNAEs are evaluated by measuring the current response for glucose under optimal conditions. Figure 2.10a, b shows the chronoamperometric responses for the NiNAE and NiO@NiNAE during the successive addition of 1.0 mM glucose in 1 M NaOH at an applied potential of +0.52 V. These electrodes could successfully detect glucose catalytically through electron transition from Ni^{2+} to Ni^{3+}. The sensitivity to glucose obtained from the calibration plot was 80 mC mM^{-1} cm^{-2} for the NiNAE and 170.4 mC mM^{-1} cm^{-2} for the NiO@NiNAE. Glucose sensitivity for the NiO@Ni nanowire array was found to be significantly higher than the NiNAE (Fig. 2.10c). Figure 2.7c shows the plot of current density vs. scan rate for three different electrodes. At higher annealing times, more Ni turned into NiO and that resulted in thinning of the conductive core of Ni, thereby increasing the electrical resistance of the individual nanowire. Therefore, at higher annealing times (>5 min), thick and poorly conductive NiO in the outer shell of the nanowires becomes practically in electrical isolation from the conductive inner Ni core. On the other hand, shell@core coaxial NiO@Ni nanowire arrays made with less than 5 min annealing time, which is not enough to cover the core Ni wire, do not show the highest catalytic response because of the absence of an adequate NiO@Ni interface.

Moreover, glucose sensitivity of the NiO@NiNAE was significantly higher than the majority of the Ni-based glucose sensors reported in the literature, where the sensitivity varies between 0.04 and 0.42 mA mM^{-1} cm^{-2} (You et al. 2003; Liu et al. 2009; Safavi et al. 2009; Wang et al. 2012). This is attributed to the excellent electrocatalytic properties of the NiO@NiNAE in the alkaline medium, as well as its nanostructured shape and vertical orientation. Figure 2.10c shows that both NiNAEs and NiO@NiNAEs are linear for glucose detection up to 7 mM, with a limit of detection of 33 and 14 μM (signal-to-noise ratio = 3), respectively. The stability of the NiNAE and NiO@NiNAE was determined over a period of 4 weeks, with an analysis carried out every 2 days with five assays each day with a 1.0 mM glucose addition. In between the testing, the electrodes were stored in air at ambient and activity remained at 96 % of its initial activity after 4 weeks for the NiO@NiNAEs. This was not the case for the NiNAEs, which needed extra cleaning to obtain a stable current response.

2.3.5 *Interferents Studies*

One of the major challenges in non-enzymatic glucose detection can be electrochemically active interferents, which generate electrochemical signals at the same potential as glucose. In our previous work, we have used the ground subtraction method to eliminate signals generated due to ascorbic acid and uric acid, which are common electrochemical interferents in physiological fluids (Jamal et al. 2009). Therefore, we examined the amperometric responses of the NiNAE and NiO@NiNAE at an applied potential of 0.52 V in 1 M NaOH solution with continuous additions of 1 mM glucose, 100 μM uric acid, 100 μM ascorbic acid, 140 μM acetaminophen and 1 μM dopamine. From the current response in Fig. 2.11, a significant signal was obtained for glucose compared to uric acid, ascorbic acid, dopamine and acetaminophen. Compared to glucose, the interfering species (uric and ascorbic acids) yielded current responses ranging from 5 to 7 %.

2.4 Summary

In this chapter, a new cost-effective and sensitive vertically aligned sensor platform to detect hydrogen peroxide, glutamate and glucose based on PtNP modified AuNAE and coaxial NiO@NiNAE has been reported. Excellent catalytic activity

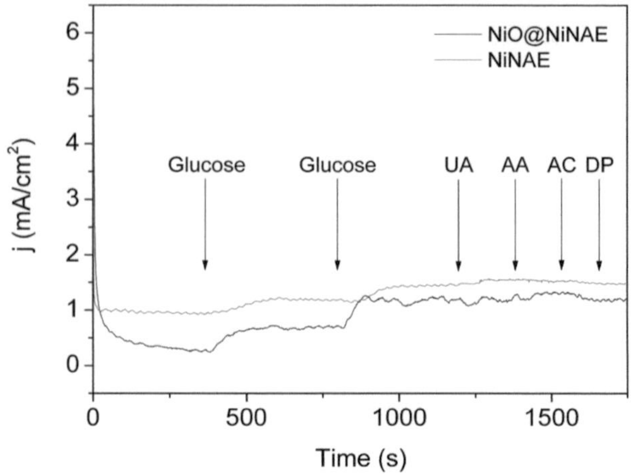

Fig. 2.11 Amperometric response of NiNAE and NiO@NiNAE upon successive additions of 1 mM glucose, 100 μM uric acid, 100 μM ascorbic acid, 140 μM acetaminophen and 1 μM dopamine at −0.52 V vs. Ag/AgCl; background electrolyte 1 M NaOH (Jamal et al. 2013)

has been found for both glutamate and glucose, with a sensitivity of 10.76 μA mM^{-1} cm^{-2} and 170 mC mM^{-1} cm^{-2}, respectively and with stability retained up to a level of 98 % after 2 weeks of regular use. NiO@Ni nanowires with 300 s annealing time shows better catalytic performance for glucose detection in 1 M NaOH solution compare to NiNAEs or NiO@NiNAEs with various annealing times. Selectivity for both glutamate and glucose was found to be promising in the presence of physiological interferents.

Acknowledgments This work was financially supported by The European Union FP7 scheme (e-BRAINS-FP7-ICT-257488).

References

Amjad M, Pletcher D, Smith C (1977) The oxidation of alcohols at a nickel anode in alkaline t-butanol/water mixtures. J Electrochem Soc 124:203–206

Anandan V, Rao YL, Zhang GG (2006) Nanopillar array structures for enhancing biosensing performance. Int J Nanomedicine 1(1):73–79

Chakraborty S, Retna Raj C (2009) Pt nanoparticle-based highly sensitive platform for the enzyme-free amperometric sensing of H_2O_2. Biosens Bioelectron 24(11):3264–3268

Chen J, Zhang W-D, Ye J-S (2008) Nonenzymatic electrochemical glucose sensor based on MnO_2/MWNTs nanocomposite. Electrochem Commun 10(9):1268–1271

Cui HF, Ye JS, Zhang WD, Li CM, Luong JHT, Sheu FS (2007) Selective and sensitive electrochemical detection of glucose in neutral solution using platinum-lead alloy nanoparticle/carbon nanotube nanocomposites. Anal Chim Acta 594(2):175–183

Cui XL, Li ZZ, Yang YC, Zhang W, Wang QF (2008) Low-potential sensitive hydrogen peroxide detection based on nanotubular TiO_2 and platinum composite electrode. Electroanalysis 20(9):970–975

De Leo M, Kuhn A, Ugo P (2007) 3D-ensembles of gold nanowires: preparation, characterization and electroanalytical peculiarities. Electroanalysis 19(2–3):227–236

Gu BX, Xu CX, Zhu GP, Liu SQ, Chen LY, Wang ML, Zhu JJ (2009) Layer by layer immobilized horseradish peroxidase on zinc oxide nanorods for biosensing. J Phys Chem B 113(18): 6553–6557

Hasan M, Jamal M, Razeeb KM (2012) Coaxial NiO/Ni nanowire arrays for high performance pseudocapacitor applications. Electrochim Acta 60:193–200

Jamal M, Worsfold O, McCormac T, Dempsey E (2009) Biosens Bioelectron 24(9):2926–2930

Jamal M, Xu J, Razeeb KM (2010) Disposable biosensor based on immobilisation of glutamate oxidase on Pt nanoparticles modified Au nanowire array electrode. Biosens Bioelectron 26(4):1420–1424

Jamal M, Hasan M, Mischieri J, Mathewson A, Razeeb KM (2011) Fabrication of horseradish peroxidase modified 3D Pt nanowire array electrode and its electro-catalytical activity towards H2O2. ECS Trans 35(26):53–59

Jamal M, Hasan M, Schmidt M, Petkov N, Mathewson A, Razeeb KM (2013) Shell@Core coaxial NiO@Ni nanowire arrays as high performance enzymeless glucose sensor. J Electrochem Soc 160(11):B207–B212

Kang Q, Yang LX, Cai QY (2008) An electro-catalytic biosensor fabricated with Pt-Au nanoparticle-decorated titania nanotube array. Bioelectrochemistry 74(1):62–65

Karam P, Halaoui LI (2008) Sensing of H_2O_2 at low surface density assemblies of Pt nanoparticles in polyelectrolyte. Anal Chem 80(14):5441–5448

Lan D, Li BX, Zhang ZJ (2008) Chemiluminescence flow biosensor for glucose based on gold nano particle-enhanced activities of glucose oxidase and horseradish peroxidase. Biosens Bioelectron 24(4):934–938

Liu Y, Teng H, Hou H, You T (2009) Nonenzymatic glucose sensor based on renewable electrospun Ni nanoparticle-loaded carbon nanofiber paste electrode. Biosens Bioelectron 24(11): 3329–3334

Lu L-M, Zhang L, Qu F-L, Lu H-X, Zhang X-B, Wu Z-S, Huan S-Y, Wang Q-A, Shen G-L, Yu R-Q (2009) A nano-Ni based ultrasensitive nonenzymatic electrochemical sensor for glucose: enhancing sensitivity through a nanowire array strategy. Biosens Bioelectron 25(1):218–223

Nelson PA, Elliott JM, Attard GS, Owen JR (2002) Mesoporous nickel/nickel oxidea nanoarchitectured electrode. Chem Mater 14:524–529

Nie H, Yao Z, Zhou X, Yang Z, Huang S (2011) Nonenzymatic electrochemical detection of glucose using well-distributed nickel nanoparticles on straight multi-walled carbon nanotubes. Biosens Bioelectron 30(1):28–34

Pang XY, He DM, Luo SL, Cai QY (2009) An amperometric glucose biosensor fabricated with Pt nanoparticle-decorated carbon nanotubes/TiO2 nanotube arrays composite. Sens Actuators B Chem 137(1):134–138

Safavi A, Maleki N, Farjami E (2009) Fabrication of a glucose sensor based on a novel nanocomposite electrode. Biosens Bioelectron 24(6):1655–1660

Umar A, Rahman MM, Al-Hajry A, Hahn YB (2009) Enzymatic glucose biosensor based on flower-shaped copper oxide nanostructures composed of thin nanosheets. Electrochem Commun 11(2):278–281

Wang J, Bao W, Zhang L (2012) A nonenzymatic glucose sensing platform based on Ni nanowire modified electrode. Anal Methods 4(12):4009–4013

Wen ZH, Ci SQ, Li JH (2009) Pt nanoparticles inserting in carbon nanotube arrays: nanocomposites for glucose biosensors. J Phys Chem C 113(31):13482–13487

Xu J, Shang F, Luong JHT, Razeeb KM, Glennon JD (2010) Direct electrochemistry of horseradish peroxidase immobilized on a monolayer modified nanowire array electrode. Biosens Bioelectron 25(6):1313–1318

Yang MH, Yang YH, Liu YL, Shen GL, Yu RQ (2006) Platinum nanoparticles-doped sol-gel/carbon nanotubes composite electrochemical sensors and biosensors. Biosens Bioelectron 21(7):1125–1131

You T, Niwa O, Chen Z, Hayashi K, Tomita M, Hirono S (2003) An amperometric detector formed of highly dispersed Ni nanoparticles embedded in a graphite-like carbon film electrode for sugar determination. Anal Chem 75(19):5191–5196

You CP, Li X, Zhang S, Kong JL, Zhao DY, Liu BH (2009) Electrochemistry and biosensing of glucose oxidase immobilized on Pt-dispersed mesoporous carbon. Microchim Acta 167(1–2): 109–116

Yu SF, Li NC, Wharton J, Martin CR (2003) Nano wheat fields prepared by plasma-etching gold nanowire-containing membranes. Nano Lett 3(6):815–818

Yu S, Peng X, Cao G, Zhou M, Qiao L, Yao J, He H (2012) Ni nanoparticles decorated titania nanotube arrays as efficient nonenzymatic glucose sensor. Electrochim Acta 76:512–517

Zhu YH, Cao HM, Tang LH, Yang XL, Li CZ (2009) Immobilization of horseradish peroxidase in three-dimensional macroporous TiO_2 matrices for biosensor applications. Electrochim Acta 54(10):2823–2827

Chapter 3
Device Architecture and Biosensing Applications for Attractive One- and Two-Dimensional Nanostructures

Chun-Da Liao, Tien-Chun Tsai, Yi-Ying Lu, and Yit-Tsong Chen

Abstract In the past decade, a myriad of novel nanoscale materials have been discovered and/or synthesized. New device architectures configured with these novel nanomaterials have been fabricated for innovative experiments and practical applications in electronics, optoelectronics, biosensors, and more. In this chapter, we introduce attractive one-dimensional (1D) and two-dimensional (2D) nanostructures and their fabricated device architectures for biosensing applications. In particular, this chapter focuses on 1D silicon nanowires and 2D graphene. We first describe the preparation and synthesis methods that have been used to generate these nanomaterials. The unique characteristics of these nanostructures and their physical, chemical, mechanical, and electrical properties are discussed. Current techniques for device fabrication are introduced. Examples of various biological and cellular applications are also included. Field-effect transistor devices constructed from 1D silicon nanowires and 2D graphene sheets are described, and we discuss biosensors for the investigation of protein–protein interactions, neural signal transmission, viral infection diagnosis, biomolecular detection, nucleic acid screening and sequencing, and real-time cellular recording.

Keywords Biosensor • Field-effect transistor • Graphene • Silicon nanowire

C.-D. Liao • T.-C. Tsai • Y.-Y. Lu • Y.-T. Chen (✉)
Department of Chemistry, National Taiwan University, Taipei 106, Taiwan

Institute of Atomic and Molecular Sciences, Academia Sinica,
P. O. Box 23-166, Taipei 106, Taiwan
e-mail: ytcchem@ntu.edu.tw

M.C. Vestergaard et al. (eds.), *Nanobiosensors and Nanobioanalyses*,
DOI 10.1007/978-4-431-55190-4_3

3.1 One-Dimensional Device Architecture and Biosensing Applications

3.1.1 Introduction

Recent developments in fabrication techniques for a variety of nanoscale materials, such as nanoparticles, nanowires (NWs), and nanotubes (NTs), have allowed these materials to be extensively investigated and used in novel biosensor devices (Rosi and Mirkin 2005; Hu et al. 1999). Several desired features, such as specificity, ultra-sensitivity, high-speed sample delivery, and low cost, have to be considered when designing and fabricating nanoscale biosensors. For example, quantum dot-based sensing devices have been shown to have high detection sensitivity, high target selectivity, and a short response time. These cost-effective sensing devices are usually integrated with optical instruments, which can translate a successful binding event into a readable signal (Raymo and Yildiz 2007). In applications of nanomaterials to study biological entities, such as nucleic acids, proteins, and viruses, the dimension of nanomaterials of ~1–100 nm provides a perfect feature for these studies because this feature size (comparable to biological entities) can maximize the interactions with biological entities, thus enhancing the sensitivity of detection (Curreli et al. 2008). In addition, nanomaterials with high surface-to-volume ratios have a large proportion of their constituent atoms located at or adjacent to the surface. Therefore, the surface atoms of nanomaterials play an important role in determining their physical, chemical, and electronic properties. In particular, if nanomaterial surfaces can be easily modified, such chemical surface modifications can facilitate the anchoring of a specific receptor for highly selective sensing applications.

In recent years, field-effect transistors (FETs) have been developed as powerful sensing devices that translate target–receptor interactions on the FET surface directly into readable electrical signals (Poghossian et al. 2007). 1D semiconducting nanomaterials, such as silicon nanowires (SiNWs) and carbon nanotubes (CNTs), have attracted great attention for use in FETs because of their high selectivity and sensitivity, real-time responsiveness, and label-free detection capabilities (Patolsky et al. 2006b; Park et al. 2007; Chen et al. 2004b). A multitude of highly sensitive SiNW-FETs have been utilized for biosensors (Cui et al. 2001; Stern et al. 2007; Li et al. 2004) and chemical sensors (McAlpine et al. 2007; Talin et al. 2006), and recently they have also been used to interface with natural neuronal networks (Qing et al. 2010; Kwiat et al. 2012). In this chapter, we first focus on the device fabrication and biosensing applications of SiNW-FETs and then focus on graphene-based FETs in the second section.

3.1.2 Fabrication of SiNW-FETs

"Bottom-up" and "top-down" techniques are the two approaches generally used to fabricate SiNW-FETs. "Bottom-up" methods are based on the catalytically assisted growth of SiNWs (with diameters as small as 5 nm or less) (Shao et al. 2010), which are assembled for device fabrication through photolithography and/or electron beam lithography (EBL). "Top-down" approaches mainly rely on lithographic methods, such as EBL (Trivedi et al. 2011), nanoimprinting (Morton et al. 2008), and scanning probe lithography (Martínez et al. 2008, 2010), to fabricate SiNW-FETs with lateral sizes of around ten to several hundred nanometers.

3.1.2.1 Top-Down Fabrication Methods

- **CMOS-Compatible Fabrication Techniques**
  In a "top-down" fabrication process, SiNWs are built on a silicon-on-insulator (SOI) wafer. The normal structure of a three-layered SOI wafer contains buried silicon dioxide (SiO_2, ~200–400 nm thick) sandwiched between a bottom Si substrate and a top Si device layer (~50–100 nm thick). Through the multiple procedures of photolithography, reactive ion etching (RIE), ion implantation, and thermal/e-beam evaporation, the connecting electrodes and the SiNWs that are pre-defined by EBL can be joined together, forming SiNW-FET devices in which the width of the SiNWs can reach ~100 nm. Figure 3.1 illustrates a standard complementary metal-oxide-semiconductor (CMOS)-compatible fabrication process (Chen et al. 2011), in which the Si layer is first doped with low-density boron or phosphorus at a concentration of $\sim 10^{15}$ cm^{-3} via ion implantation (Fig. 3.1a). The semiconducting properties (*n*- or *p*-type) and the doping ratio of SiNWs are determined in this step. The highly dense doping area ($\sim 10^{19}$ cm^{-3}) is prepared for source and drain leads (Fig. 3.1b). Photolithography is a preferred technique to define the pattern of leads, followed by transferring the drawn photoresist pattern onto the SOI wafer via RIE (Fig. 3.1c). In the last stage (Fig. 3.1d), RIE (or tetramethylammonium hydroxide etching) is executed again to transfer the EBL-defined electric-resist pattern onto the SOI wafer, resulting in nanometer-sized SiNWs. Thermal or e-beam evaporation is subsequently used to produce the contact leads and back gate. Finally, an insulator

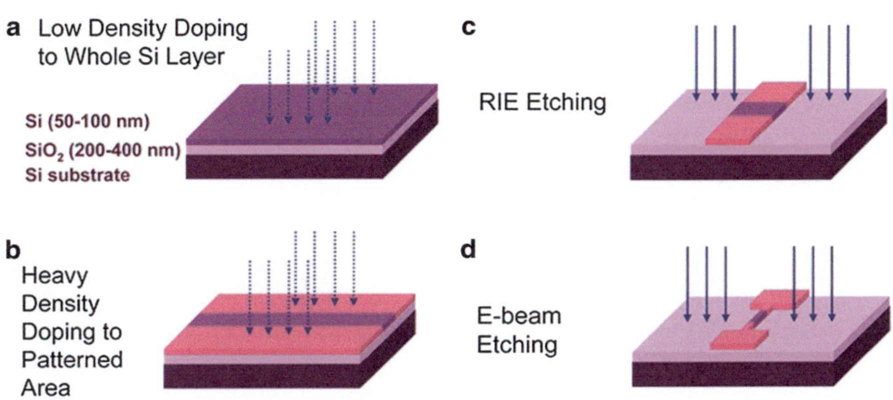

Fig. 3.1 Schematic illustration of a typical "top-down" process for fabricating SiNW-FETs Reprinted with permission from Chen et al. (2011). Copyright Elsevier

layer, made of Al_2O_3 (Stern et al. 2007; Knopfmacher et al. 2010), SiO_2 (Lin et al. 2009), or Si_3N_4 (Vu et al. 2009), is coated onto the SiNW-FET device.

The conventional "top-down" patterning process relies heavily on high-resolution lithography, typically requiring the use of deep UV (DUV)-photolithography, EBL, and focused ion beam (FIB). These advanced nanolithography techniques can scale the feature sizes down to a few nanometers (Vieu et al. 2000; Craighead et al. 1983). Although the "top-down" approach may require expensive instrumentation, it has the advantages of using standard semiconductor techniques to precisely design a desired device-array pattern without the problems associated with positioning SiNWs.

- **Improved CMOS-Compatible Fabrication Techniques**
  In this section, "top-down" approaches without using refined nanolithography techniques, such as metallic nanowire pattern transfer and sidewall method, will be introduced as alternatives for SiNW-FET fabrication.

  1. **Metallic Nanowire Pattern Transfer**
     The fabrication of SiNWs with a "top-down" technique is difficult to accomplish without employing EBL or high-resolution DUV lithography. To overcome this problem, transferring a hard metallic NW mask, built from a $GaAs/Al_{0.8}Ga_{0.2}As$ superlattice nanowire pattern (SNAP), was proposed as shown in Fig. 3.2a (Melosh et al. 2003). A $GaAs/Al_{0.8}Ga_{0.2}As$ superlattice was deposited on the (100) surface of a GaAs substrate through molecular beam epitaxy (MBE). Subsequently, superlattice voids of ~20–30 nm in depth were formed via selective etching of the AlGaAs layers with a dilute mixture of buffered hydrofluoric acid (HF). After creating voids, the GaAs/AlGaAs superlattice was metal-coated by orienting the superlattice at an angle of 36° with respect to the evaporative flux (step (iii) of Fig. 3.2a), followed by putting the metal-coated superlattice onto a Si wafer in contact with a heat-curable epoxy film that serves as an adhesion layer (step (iv) of Fig. 3.2a). The superlattice-attached wafer was then suspended upside down in a KI etchant (4 g)/I_2 (1 g)/H_2O (100 mL) to remove the GaAs layer, leading to well-aligned metallic NWs remaining on the Si wafer (step (v) of Fig. 3.2a). Through an O_2 plasma etching process (step (vi) of Fig. 3.2a), the pattern of metallic NWs can be transferred to an underlying semiconductor substrate, such as an SOI wafer, thereby generalizing both metal (Fig. 3.2b, c) and semiconductor NWs (Fig. 3.2d). The SNAP-based FETs integrated with microfluidic channels can function as highly sensitive biodetectors as shown in Fig. 3.2e, f. Instead of employing photolithography or EBL, the SNAP process relies upon MBE to precisely control the thickness of film, i.e., to define the width of NWs, thus resulting in the production of thinner wires with much smaller pitches than those fabricated by photolithography or EBL. It is also possible to make such metallic NWs without using MBE. A technique named deposition and etching under angles (DEA) (Tong et al. 2009) was further presented as another means for fabricating SiNWs, and is based on standard microfabrication

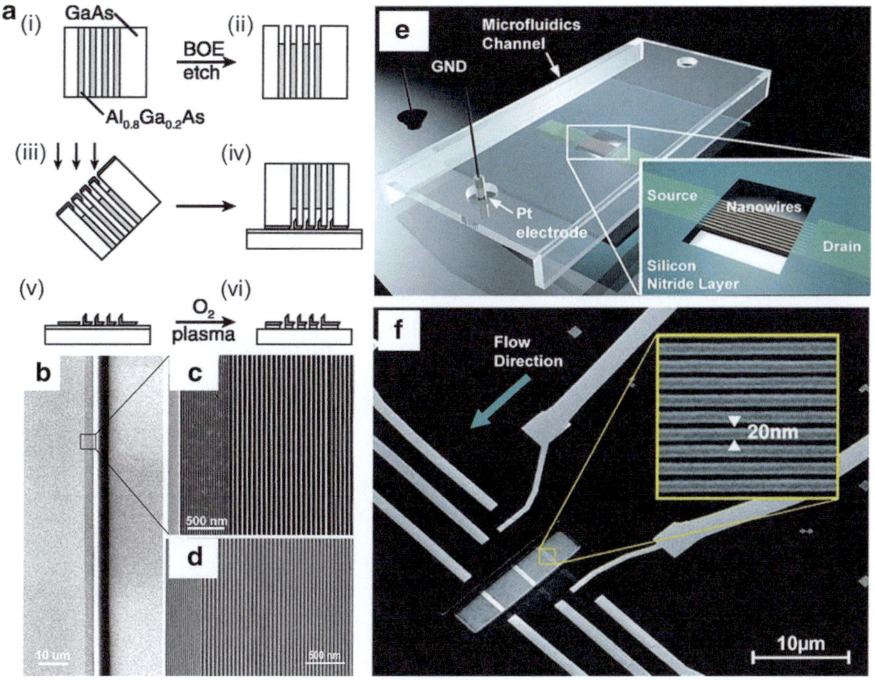

Fig. 3.2 (**a**) The SNAP process for creating small-pitch metallic NWs via a GaAs/AlGaAs superlattice. (**b**) Scanning electron microscopy (SEM) image showing an array of metallic NWs. The *bright line* adjacent to the NWs results from residual epoxy. (**c**) An enlarged section from (**b**) reveals 10-nm-diameter platinum (Pt) NWs with different pitch lengths of 30 and 60 nm. (**d**) SEM image of SiNWs. These wires are straight and continuous for more than 100 µm. (**e**) and (**f**) show that SiNW-FETs made by the SNAP technique contain three groups of SiNWs within a microfluidics channel. The wafer is coated with Si_3N_4, except for an exposed active region containing SiNWs. Reprinted with permission from Melosh et al. (2003) and Bunimovich et al. (2006). Copyright American Association for the Advancement of Science and American Chemical Society

processes including photolithography, thin film deposition, and ion beam etching. The metallic NWs created by DEA can reach ~30 nm in width and up to several millimeters in length. SiNW-FET devices built with the DEA metallic mask can achieve a lateral dimension down to <100 nm.

2. **Sidewall Method**
   To reduce the lateral size of SiNWs, Choi et al. used a lithographically defined sacrificial sidewall as a nano-pattern to produce submicron-scale SiNWs (Choi et al. 2003). The fabrication process is illustrated in Fig. 3.3a. The thermal oxide, serving as a hard mask for the subsequent Si etching process, was grown on a (100) Si wafer. The poly-Si layer, functioning as a sacrificial layer, was then deposited on the top of the thermal oxide shown in step (ii) of Fig. 3.3a. Photoresist was spin-coated on the top of the poly-Si prior to applying

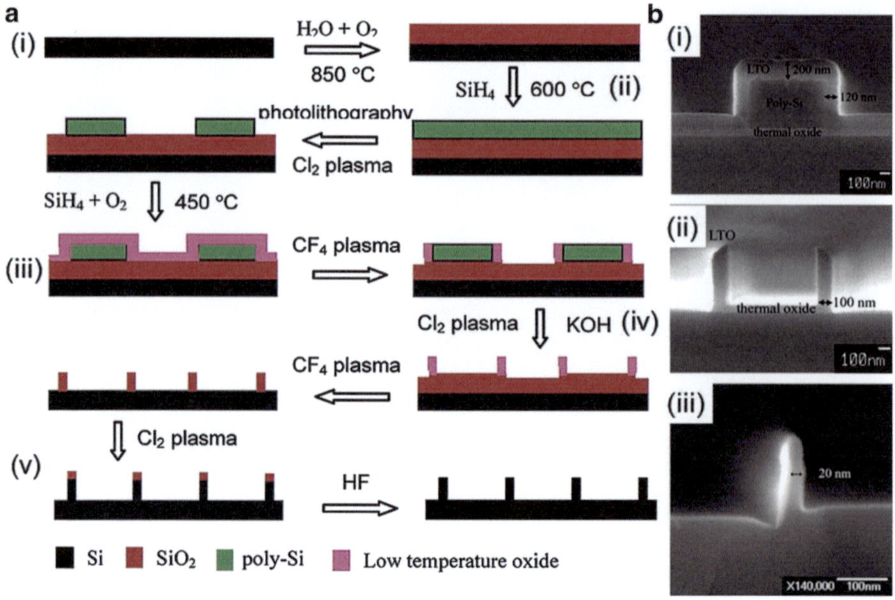

Fig. 3.3 (**a**) The size reduction lithography fabrication process. (**b**) Cross-sectional views: (*i*) after 200-nm-thick LTO deposition and (*ii*) after etching of the top LTO layer and poly-Si layer. (*iii*) An SEM image of the 20-nm SiNW fabricated by size reduction lithography. Reprinted with permission from Choi et al. (2003). Copyright American Chemical Society

UV-lithography to define a starting pattern. The poly-Si, without the protection of the photoresist, was subsequently plasma-etched (step (ii) of Fig. 3.3a). A low temperature oxide (LTO), determining the final pattern size of the SiNWs, was deposited on the top of the poly-Si (step (iii) of Fig. 3.3a). In order to expose the poly-Si and its LTO sidewalls, the top-covered LTO layer was treated in a CF_4 plasma-etching process (step (iii) of Fig. 3.3a) and the exposed poly-Si, i.e., the starting pattern, was plasma-etched again. After this step, the sacrificial poly-Si was entirely removed, leaving LTO narrow wires that are less than 200 nm in width. CF_4 plasma was used again to etch away the SiO_2 layer (thermal oxide), which was not protected by the LTO wires (step (iv) of Fig. 3.3a). The exposed Si wafer was finally etched away by using RIE until the desired SiNW height was obtained (step (v) of Fig. 3.3a). The minimum pattern pitch has to be limited by the photolithographically defined pattern. However, the fabricated feature size, i.e., the NWs, is only determined by the thickness of the deposited material rather than by photolithography. Because the thickness of the deposited film can be controlled down to 10 nm or less with high precision, this method is capable of generating nano-patterns that are far smaller than those created by optical lithography (Fig. 3.3b). In addition to the LTO sidewalls, properly annealing amorphous silicon (a-Si) sidewalls deposited by low pressure chemical vapor deposition (LPCVD) has also been proposed to directly form poly-SiNWs (Hsiao et al. 2009; Hakim et al. 2012; Demami et al. 2012; Lin et al. 2005).

3.1.2.2 Bottom-up Fabrication Methods

- **SiNWs Synthesis and Their Device Fabrication Process**
  Figure 3.4 illustrates a "bottom-up" process starting with the growth of SiNWs and ending with the final stage of device fabrication. Chemical vapor deposition (CVD, Fig. 3.4a) plays a vital role in the SiNW synthesis, and it is governed by a vapor–liquid–solid (VLS) growth mechanism, in which catalytic metal nanoparticles facilitate the formation of SiNWs (Tian et al. 2009; Wagner and Ellis 1964) and control the size of the as-synthesized SiNWs (Wu et al. 2004; Ma et al. 2003). In addition, the dopants of the as-grown SiNWs can be precisely controlled during the CVD reaction (Cui et al. 2000).

  Because the as-synthesized SiNWs are well controlled in the CVD reaction, the "bottom-up" method can exhibit high reproducibility. Compared to the "top-down" technique, the "bottom-up" method has the advantages of synthesizing SiNWs with high single crystallinity, a designated dopant density, thin SiO_2 sheaths, and easily controlled diameters in a cost-effective preparation. However, without deliberate alignment of the randomly orientated SiNWs grown on the Si substrate, the device fabrication would suffer from inefficient yields, which could limit industrial applications. Therefore, the practical use of "bottom-up" SiNW-FETs will require the development of uniform assembly techniques for aligning the as-synthesized SiNWs on a support substrate. Several effective assembly techniques will be discussed in the following section, entitled "Nanowire Assembly Techniques".

  After assembling SiNWs (Fig. 3.4b), device fabrication can be carried out by photolithography or EBL. In the photolithographic steps, a two-layer photoresist consisting of LOR3A and S1805 is deposited onto a Si substrate by spin coating (Fig. 3.4c). Subsequently, the electrodes are defined with a pre-designed photomask, followed by thermal or e-beam evaporation for the deposition of metal

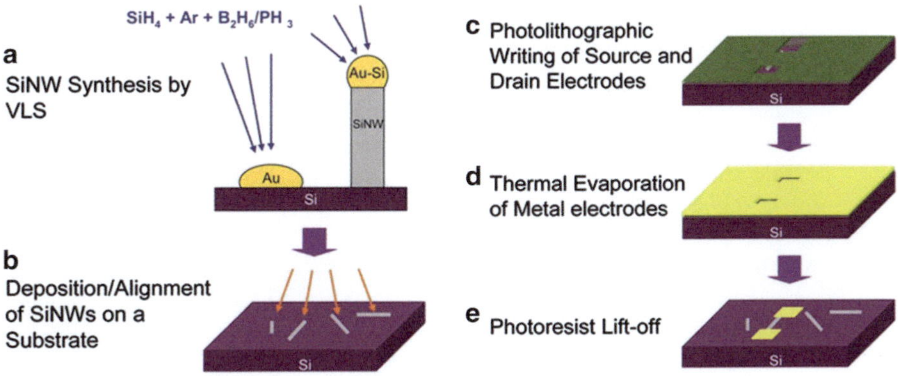

Fig. 3.4 Schematic illustration of a typical "bottom-up" process for fabricating SiNW-FETs. Reprinted with permission from Chen et al. (2011). Copyright Elsevier

contacts (Fig. 3.4d). Finally, the remaining photoresist layer is lifted off by Remover PG to yield the as-fabricated SiNW-FETs (Fig. 3.4e).

- **Nanowire Assembly Techniques**
  Significant efforts have been invested in developing generic methods to align NWs for the fabrication of NW-FETs. Several techniques for the assembly of NWs have been developed, including flow-assisted alignment, the Langmuir–Blodgett technique, the bubble-blown technique, electric-field-directed assembly, the smearing-transfer method, roll-to-roll printing assembly, and the polydimethylsiloxane (PDMS) transfer method.

1. **Flow-Assisted Alignment**
   In the flow-assisted alignment method (Huang et al. 2001), the SiO_2/Si substrate is pre-modified with 3-aminopropyltriethoxysilane (APTES), of which one end is anchored to the SiO_2 surface and the other end forms an NH_2-terminated surface. When the suspended SiNWs pass through a PDMS microfluidic channel covered onto the SiO_2/Si substrate, the NH_2-terminated surface assists the alignment of SiNWs through electrostatic interactions. The angular spread of the SiNWs in the flow direction is flow-rate-dependent, but the density of the SiNWs assembled on the SiO_2/Si substrate is time-dependent. Furthermore, multiple crossed arrays of SiNWs can be produced through a crossed flow involving multiple consecutive assembly processes.
2. **Langmuir–Blodgett Technique**
   The Langmuir–Blodgett technique can be used for the alignment of NWs/NTs (Whang et al. 2003; Li et al. 2007). This solution-based method assembles SiNWs in a monolayer of surfactant at the air–water interface and then compresses the SiNWs on a Langmuir–Blodgett trough to a specified pitch. The aligned SiNWs are then transferred to the surface of a substrate to make a uniform parallel array. Crossed SiNW structures can be formed by uniform transfer of a second layer of aligned parallel SiNWs perpendicular to the first layer. Compared to other methods, the Langmuir–Blodgett technique is capable of producing a SiNW alignment with much higher density.
3. **Bubble-Blown Technique**
   The bubble-blown technique uses nitrogen flow to form SiNW blown-bubble films in a physical assembly method. The SiNWs are suspended in tetrahydrofuran solution and then blown into a single bubble. The unique feature of this method is that blown-bubble films can be transferred to both rigid and flexible substrates during the expansion process. This method can also be scaled to large wafers and non-rigid substrates (Yu et al. 2007).
4. **Electric-Field-Directed Assembly**
   The electric-field-directed assembly of SiNWs, developed by Freer et al. (Freer et al. 2010), is an intriguing and desirable method. Taking advantage of the appropriate electrode design and adjustment of the applied gate voltage (V_g) vs. source-drain voltage (V_{sd}), single SiNWs can be assembled at over

98.5 % of 16,000 pre-patterned electrode sites by controlling the balance of surface, hydrodynamic, and dielectrophoretic forces.

5. **Smearing-Transfer Method**
   The smearing-transfer method (or contact-printing method), developed by Ali Javey's group (Fan et al. 2008), is based on a direct contact printing process that enables the direct transfer and positioning of SiNWs from a donor substrate to a receiver chip. This simple method can efficiently transfer a variety of NWs (such as SiNWs and germanium NWs (GeNW)) to a wide range of receiver substrates, including Si and flexible plastics. The technique, which uses a chemical "lawn" and "lubricant" to increase the density and alignment quality, is regarded as a rapid, efficient, and economical method.
6. **Roll-Printing Assembly**
   Ali Javey's group has also developed an approach based on contact printing for scalable and large-area printing. Roll-to-roll assembly has made it possible to produce highly ordered, dense, aligned, and regular arrays of NWs with high uniformity and reproducibility. The differential-roll-printing process is compatible with the smearing-transfer method and can also be implemented on a wide range of rigid and flexible substrates (Yerushalmi et al. 2007).
7. **PDMS-Transfer Method**
   An NW alignment method using a PDMS stamp was proposed by Chang et al. (Chang and Hong 2009). In their report, a high-speed roller (20–80 cm/min) was used to assist the transfer of zinc oxide (ZnO) NWs from the growth substrate to a PDMS stamp. The NWs were then re-transferred from the PDMS stamp to another receiver substrate. Taking advantage of this method, NWs can be aligned with high density, providing a convenient and efficient approach for the fabrication of NW-FETs.

3.1.3 Applications of SiNW-FET Biosensors

SiNW-FET biosensors have been investigated for their potential use in the biological sciences and healthcare. Possible applications include examining protein–DNA interactions, monitoring neural exocytosis, probing viral infection, and detecting small-molecule biomarkers. Although the traditional optical and electrochemical biosensors are very sensitive, they have several disadvantages, such as being limited to analyzing electro-active analytes and using dyes or fluorescent probes that interfere with the function of the biosystem under study. These traditional biosensors also involve time-consuming procedures and restrictive device constraints. In contrast, SiNW-FETs have high detection sensitivity and target selectivity and can be used for label-free assays involving real-time recording. Unlike traditional 2D FET-based biosensors for detecting analyte binding, 1D SiNW-FETs can deplete or accumulate charge carriers throughout the entire cross-section of the NWs, leading to a greater conductance change and higher detection sensitivity. Additionally, an insulated silica layer intrinsically forms on the SiNW surface, which is beneficial because it allows easy surface modification. Linkers or receptors can be easily attached, and electrical leakage is prevented, which is important because biosensing applications normally take place in a liquid environment.

3.1.3.1 Protein–Protein and Protein–Small Molecule Interactions

SiNW-FETs have been used to explore the mechanisms of biomolecular interactions, such as protein–protein, protein–small molecule, and nucleic acid interactions. The simultaneous detections of different biomarkers with a multiplexed platform could enable precise diagnosis of single or multiple diseases. Zheng et al. successfully conducted label-free multiplexed detection of various cancer biomarkers by using an antibody-functionalized SiNW-FET (Zheng et al. 2005). Antibodies for prostate-specific antigen, carcinoembryonic antigen, and mucin-1 were immobilized separately on three different SiNW-FETs and used for clinical diagnosis. These multiplexed SiNW-FETs have been successfully used for simultaneous detections of multiple proteins at femtomolar (fM) concentrations with excellent target selectivity.

Understanding how proteins recognize other proteins is important for the study of physiological regulation in processes such as muscle contraction and neurotransmitter release. Lin et al. have developed a SiNW-FET biosensor for the rapid screening of protein–protein interactions. In their study to examine the proteins interacting with calmodulin (CaM), a reusable CaM-modified SiNW-FET (referred to as CaM/SiNW-FET) biosensor was fabricated by immobilizing glutathione *S*-transferase (GST)-fused CaM (represented by CaM-GST) on a glutathione (GSH)-modified SiNW-FET (GSH/SiNW-FET) via the reversible GSH-GST association (Lin et al. 2010). The results of real-time measurement revealed a dramatic decrease in conductance of the CaM/SiNW-FET after the binding of positively charged troponin I (TnI, pI ~9.3) to the *p*-type CaM/SiNW-FET in phosphate buffer solution at pH 7.4 containing 100 μM Ca^{2+}. However, with a phosphate buffer solution containing dilute Ca^{2+} ions (10^{-8}–10^{-7} M), the presence of TnI did not induce significant conductance change in the CaM/SiNW-FET until 10^{-6} M Ca^{2+} was added. Therefore, this CaM/SiNW-FET was applied to reveal that the minimal concentration of Ca^{2+} ions required to trigger the association between CaM and TnI was at the micromolar level. In addition, Lin et al. used the GSH/SiNW-FETs to explore molecular interactions in cell lysates containing expressed GST-CaM or green fluorescent protein (GFP)-tagged CaMKIIα (Lin et al. 2013). The conductance of a GSH/SiNW-FET increased by ~3 % after adding a cell lysate expressing GST-CaM, whereas no significant elevation was observed when control cell lysates were added. Moreover, the conductance change in the CaM/SiNW-FET was elevated by ~4 % after the addition of a cell lysate containing GFP-CaMKIIα in the presence of Ca^{2+} ions. This was substantially different from the addition of lysates without Ca^{2+} ions or lysates containing only GFP. These results are consistent with those of a previous study, which demonstrated that GFP-CaMKII can be activated by the Ca^{2+}–CaM complex (Soderling et al. 2001).

When a neuron fires, an initial efflux of K^+ ions across the cell membrane can modulate the membrane potential and subsequently induce the action potential to achieve signal propagation. The abnormal permeability of K^+ ions was suggested to be related to neural degeneration, diabetes mellitus, and epilepsy (MacDonald et al. 2001; Jentsch 2000; Singh et al. 1998). Thus, a practical platform for real-time monitoring of K^+ flux is essential for understanding neural excitability. To detect K^+ ions, a polyvinyl chloride (PVC) membrane, embedded with valinomycin (VAL), coated on a SiNW-FET (referred to as VAL-PVC/SiNW-FET) was used to monitor the efflux of K^+ ions from chromaffin cells (Chang et al. 2012). By using the VAL-PVC/SiNW-FET, Chang et al. were able to determine the apparent association constant (K_a) of VAL for K^+ ions. K_a was determined to be $5{,}974 \pm 115$ M^{-1}, which is significantly higher than the apparent association constants for VAL binding to Li^+, Na^+, and Cs^+ ions (Chang et al. 2012). The VAL-PVC/SiNW-FET was further used to monitor the efflux of K^+ ions from chromaffin cells stimulated with nicotine. The conductance of a VAL-PVC/SiNW-FET seeded with chromaffin cells was gradually elevated in response to increasing amounts of nicotine stimulus. The observations suggest that nicotinic-type acetylcholine receptors (nAChRs) on the cell membranes of chromaffin cells could be activated by nicotine to allow the efflux of K^+ ions through the conjugated ion channels (Sala et al. 2008). As demonstrated, SiNW-FETs can be used for real-time, label-free monitoring of ion efflux. SiNW-FETs are also faster and interfere less with the system being studied compared to other techniques that involve electrophysiological recording or the use of ion-sensitive fluorescent dyes.

3.1.3.2 Neural Signal Transmission

Nanoelectronic 1D SiNW-FETs (or CNT-FETs) could potentially be used for more detailed studies in neuroscience. These devices have excellent detection sensitivity, a fast signal response, and high spatial resolution for sub-cellular and multi-site recordings. An example of extracellular action potential recording from single neurons has been provided by Patolsky et al. using a multiplexed platform of SiNW-FETs (Patolsky et al. 2006a). They aligned a cortical neuron across a series of SiNW-FET device arrays, with each active junction for the axon/SiNW interface having an area of approximately 0.01–0.02 μm^2. The ultra-small width (~20 nm) of the junction area, which is similar to that of a natural synapse, represents a major advantage over other electrophysiological techniques. More significantly, SiNW-FETs could be constructed as multiplexed sensing elements on different regions of a dendrite and axon in a single neuron. Under local electrical stimulation, various architecture designs of SiNW-FET devices were applied to characterize the propagation of an action potential, including measurements of the back-propagation effect, propagation speed, and hyperpolarization. Although 1D SiNW-FET platforms can be used to detect the propagation of extracellular electrical signals, these devices have so far only been constructed on planar substrates, making it difficult to record the desired signals from certain target areas. For example, it is impossible to maneuver the sensing element to contact a cell at a specific point on its surface or to record inside the cell.

Recently developed three-dimensional (3D) FET devices, such as kinked SiNW-FETs (Tian et al. 2010; Jiang et al. 2012), branched intracellular nanotube FETs (BIT-FETs) (Duan et al. 2012), and active Si nanotube transistors (ANTTs) (Gao et al. 2012), have demonstrated the flexible recording of intracellular electrical signals (Duan et al. 2013). Kinked SiNWs can be synthesized in a CVD reaction by adjusting the concentrations of reactant gases and the overall chamber pressure during SiNW growth to control the length, growth direction, and doping components. This type of doping modulation was used to introduce the heavily doped parts as source and drain electrodes, and a short, lightly doped FET channel could be defined around the probe tip. The SiNW probe, synthesized with a tip angle of 60°, was finally fabricated in 3D geometry, allowing the tip to insert into a cell for intracellular recording. Another example of a 3D FET design, BIT-FET, was realized by the integration of bottom-up synthesis and the top-down technique (Duan et al. 2012). The GeNW branches were grown on the top of a SiNW channel, and the surface of the GeNW was coated with a SiO_2 layer. Finally, a hollow SiO_2 nanotube was constructed on the SiNW-FET device by selectively etching the GeNW core. In another approach to fabricating an ANTT device (Gao et al. 2012), the insulated source and drain electrodes were fabricated at two separate contact points on a semiconductor Si nanotube. In measurements by BIT-FET and ANTT devices, after the nanotube penetrated the cell membrane, the cytosol filled the active channel of the nanotube, resulting in highly sensitive recording of intracellular signals. With the BIT-FET and ANTT devices, their innovative designs provide unique advantages for cellular studies, including the ability to conduct real-time monitoring of sub-cellular structure by adjusting the nanotube dimensions. Spatially resolved detection can also be achieved by constructing high-density arrays.

Neurotransmitters released from vesicles play an important role in signal transduction between neurons, and precise quantification of neurotransmitters is essential for the diagnosis of neurological diseases. In addition to conventional electrochemical techniques, 1D single-walled carbon nanotube (SWCNT)-FETs have been used to measure exocytotic chromogranin A (Wang et al. 2007) and dopamine (Sudibya et al. 2009), both of which are important biomarkers for neuroendocrine cancer and neurodegenerative diseases, such as Alzheimer's and Parkinson's diseases (Tsai et al. 2013). In our recent study, a SiNW-FET with selective surface modification (SSM) by boronic acid could be used to sensitively detect the release of dopamine from PC12 cells under K^+ stimulation (Li et al. 2013). As illustrated schematically in Fig. 3.5a, the conventional method of modifying receptors on a SiNW-FET biosensor covers not only the SiNW surface but also the surrounding substrate; this is called all-area modification (AAM). In biosensing measurements with AAM SiNW-FETs, the ratio of target–receptor binding on the SiNW surface, relative to the widespread area of the surrounding substrate, is very small (typically ~10^{-6}). This limits the detection sensitivity because a large proportion

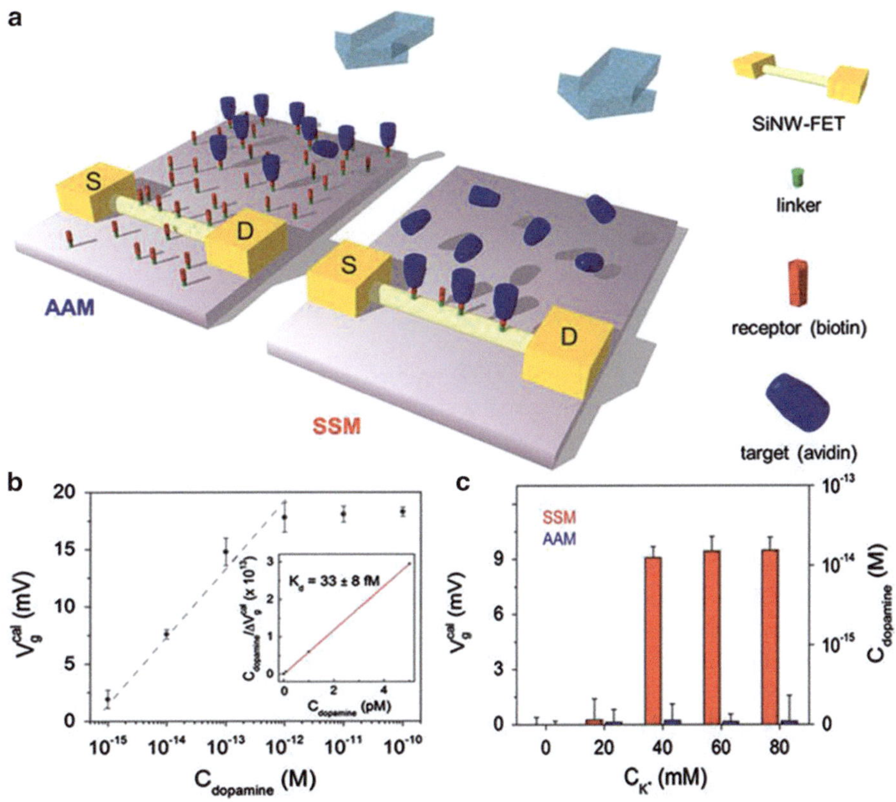

Fig. 3.5 (**a**) Schematic diagrams of the binding competition involved in biomolecular detection by AAM or SSM SiNW-FET devices. (**b**) The plot shows the detection limit and linear working range of a boronic acid-modified SSM SiNW-FET for dopamine detection. (**c**) A comparison of sensing performances for boronic acid-modified AAM and SSM SiNW-FET devices applied to the monitoring of dopamine released from PC12 cells stimulated with various K^+ concentrations. Reprinted with permission from Li et al. (2013). Copyright Elsevier

of target molecules are likely to be captured by the receptors immobilized on the surrounding substrate before arriving at the sensing surface of the SiNW-FET. Thus, a bottom-up technique to build SSM SiNW-FETs was used to improve detection sensitivity. The strategy for manufacturing SSM SiNW-FETs involves modifying the SiNW with a chemical linker of 3-aminopropyltrimethoxysilane (APTMS) prior to photolithographic fabrication. After the device fabrication of SiNW-FETs, the receptors (e.g., boronic acid for probing dopamine) can be selectively immobilized on the APTMS-modified SiNW surface rather than on the surrounding (SiO_2/Si) substrate (Nakajima and Ikada 1995). As shown in Fig. 3.5b, the linear response range of the boronic acid-modified SSM SiNW-FET for dopamine detection ranged from 1 fM to 1 pM. Figure 3.5c compares measurements made by boronic acid-modified AAM and SSM SiNW-FETs of the amounts of dopamine released from living PC12 cells stimulated by various K^+ concentrations. Under stimulation with a concentration of K^+ greater than 40 mM, sizable releases of dopamine (>10 fM) could be detected by the SSM SiNW-FET but not by the AAM SiNW-FET. These results show that the selective modification of receptors on the SiNW surface, without contamination of the surrounding substrate, can substantially increase the detection sensitivity of molecular biosensors.

3.1.3.3 Viral Infection Diagnosis

Swine-derived influenza A (H1N1, 2009) represented a major worldwide epidemic of influenza that spread from Mexico within a short time. Real-time reverse transcription polymerase chain reaction (RT-PCR) was used as a powerful tool for identifying H1N1 viruses, and it served as a highly sensitive assay for the qualitative detection of influenza viruses (Zhang and Ning 2012). However, real-time RT-PCR requires highly skilled personnel and expensive laboratory instruments, which makes it a difficult technique to use for first-line screening. Several commercial kits are available for the rapid screening of seasonal influenza A (FluA), including Binax Now, BD Directigen EZ, and Quidel QuickVue, but the challenges of quantitative screening lie in overcoming the low sensitivity of these tests and their inability to distinguish the subtypes of FluA (Hurt et al. 2009). Kao et al. constructed an integrated platform of SiNW-FETs for the simultaneous screening of H1N1 and FluA viruses (Kao et al. 2011). Peptide nucleic acid (PNA) probes have strong binding affinity for target viral single-stranded DNA (ssDNA). In one application, two individual SiNW clusters were immobilized with different PNA probes to achieve specific recognition of the H1N1 and FluA viruses. Another cluster served as a baseline reference to detect environmental disturbances. It would be expected that when the PNA/DNA hybridization occurred on the SiNW surface, the negative charge on the ssDNA surface backbone created a charge depletion layer in the *n*-type SiNW, consequently increasing the resistance of the SiNW. The PNA probe-modified SiNW-FET could specifically recognize the target virus in a fast screening application (<20 min). For the H5N2 avian influenza virus (AIV), a SiNW-FET biosensor with reversible surface functionalization was used for fast screening and ultrasensitive detection (Chiang et al. 2012). The reversible surface functionalization of a SiNW-FET device was accomplished via the linkage and cleavage of disulfide bonds. First, 3-mercaptopropyltrimethoxysilane (MPTMS) was designed to modify the SiNW-FET surface. A monoclonal antibody against the H5N2 virus (mAb_{H5}) was then anchored to the MPTMS-modified SiNW-FET through the formation of a disulfide bond. The advantage of choosing a disulfide linker is that the immobilization of receptor proteins (e.g., mAb_{H5} in this study) through the disulfide bonds can be carried out in a reaction at room temperature to avoid spoiling the normal functions of proteins or damaging the electrical properties of SiNW-FET device. Finally, after the detection of H5N2 AIVs, the mAb_{H5}–H5N2 complex could be removed efficiently by cleaving the disulfide bonds with dithiothreitol (DTT). This mAb_{H5}-modified SiNW-FET provides an example of a biosensing platform that is reusable and has ultrahigh sensitivity (with a detection limit of 10^4 AIV mL^{-1}, which is equal to 10^{-17} M). This platform also shows high target specificity and can be used for fast screening in the diagnosis of H5N2 AIVs.

3.2 Two-Dimensional Device Architecture and Biosensing Applications

3.2.1 Graphene Structure

Since the discovery of graphene, the newest member in the family of carbon allotropes, by Novoselov, Geim, and co-workers at Manchester University in 2004, this one-atom-thick wonder material has drawn the attention of scientists from various disciplines (Novoselov et al. 2004). Graphene has many extraordinary physicochemical properties. It displays linear dispersion of Dirac electrons, a room temperature quantum Hall effect, and ballistic transportation up to room temperature. These superb characteristics come from the confinement of electron motion to a two-dimensional plane. The properties of graphene make it a promising candidate for use in a variety of applications, ranging from (opto)electronics to biosensing. Here, we will focus on some interesting properties of graphene, its preparation for use in FET devices, and biosensing applications.

Graphene is a single layer of carbon atoms packed in a honeycomb structure with a carbon–carbon distance of 0.142 nm. The crystal structure of graphene can be modeled as a triangular Bravais lattice with a two-atom basis (A and B), as depicted in Fig. 3.6. The lattice vector can be written as follows:

$$a_1 = \frac{a}{2}\left(3,\sqrt{3}\right) \text{ and } a_2 = \frac{a}{2}\left(3,-\sqrt{3}\right). \tag{3.1}$$

The three vectors that connect a site on the A sublattice with its nearest neighbor on the B sublattice are given by the following equations:

$$\delta_1 = \frac{a}{2}\left(1,\sqrt{3}\right),\ \delta_2 = \frac{a}{2}\left(1,-\sqrt{3}\right), \text{ and } \delta_3 = -a\left(1,0\right). \tag{3.2}$$

The reciprocal lattice vectors are given by the following equations:

$$b_1 = \frac{2\pi}{3a}\left(1,\sqrt{3}\right) \text{ and } b_2 = \frac{2\pi}{3a}\left(1,-\sqrt{3}\right). \tag{3.3}$$

Each carbon atom, with sp^2 hybridization to form three sigma bonds, results in deep-lying valence bands, leaving the unhybridized $2p_z$ orbital perpendicular to the plane of graphene to form π bonds. The planar orbitals formed by the sigma bonds are responsible for most of the binding energy and for the excellent mechanical properties of the graphene sheet. The remaining unhybridized $2p_z$ orbitals overlap each other between neighboring carbon atoms to form π bonds, which play key roles in the optical and electronic properties of graphene.

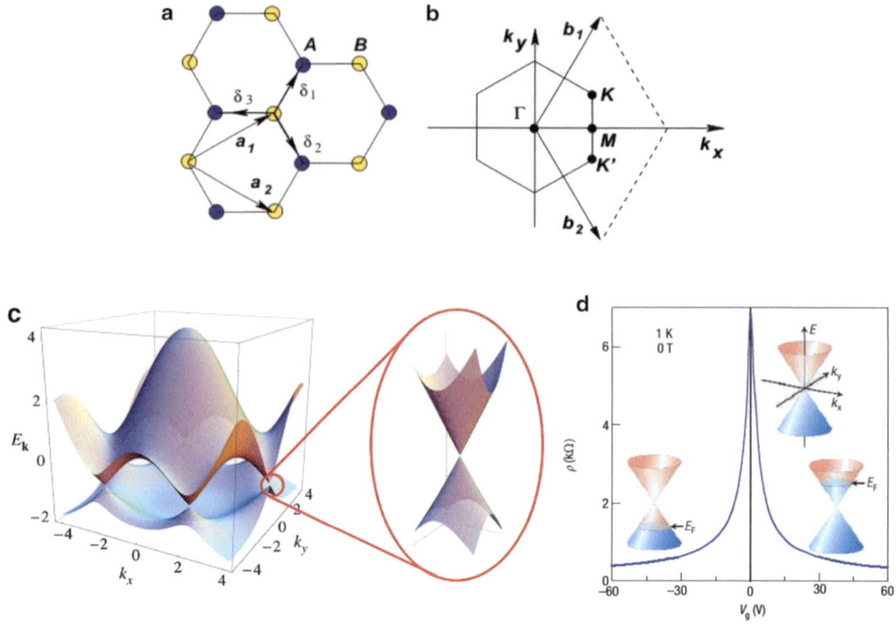

Fig. 3.6 (**a**) The lattice structure of graphene consists of two triangular lattices, *A* and *B* (a_1 and a_2 are the lattice unit vectors, and δ_i, $i = 1, 2, 3$, are the nearest neighbor vectors). (**b**) The first Brillouin zone of a graphene lattice (Castro Neto et al. 2009) (**c**) The dispersion relation of graphene in the first Brillouin zone. The Dirac cones are located at the *K* and *K′* points. (**d**) Ambipolar electric field effect in graphene. The cones represent the low-energy spectrum, indicating the changes in the position of the Fermi energy E_F with increasing or decreasing gate voltage V_g. Negative V_g induces holes; positive V_g induces electrons. Reprinted with permission from Castro Neto et al. (2009) and Geim and Novoselov (2007). Copyright American Physical Society and Nature Publishing Group

Graphene has a unique band structure (linear energy dispersion), with its valence bands and conduction bands meeting at the corners of the Brillouin zone, as shown in Fig. 3.6c, d. This structure makes it possible to continuously drive the Fermi level from the valence to the conduction band simply by applying an electric field. The application of an electric field produces a pronounced ambipolar electric field effect (Geim and Novoselov 2007).

3.2.2 *Doping and Characterization*

Doping is an effective way of tailoring the electronic properties of graphene. Doping of graphene may be broadly categorized into two types: electrical doping (Das et al. 2008), which is performed by changing gate voltage, and chemical doping (Hwang et al. 2007), which involves using chemicals. Chemical doping can further be classified as surface transfer doping or substitutional doping. In this section, we concentrate on surface transfer doping and substitutional doping.

3.2.2.1 Surface Transfer Doping

Surface transfer doping is achieved by electron exchange between a semiconductor and its dopants, which adsorb on the surface of the semiconductor. Surface transfer doping generally preserves the unique band structure of graphene without decreasing the mobility of the charge carriers. The types of doping of graphene are determined by the relative positions of the Fermi level of the dopant and the Fermi level of graphene. If the Fermi level of the dopant is higher (lower) than that of graphene, electrons are transferred from the dopant (graphene) to graphene (the dopant) with the dopant acting as a donor (acceptor). Surface transfer doping of graphene can be accomplished by coating the graphene with dopants or exposing the graphene to dopant vapor. In 2007, Geim and co-workers first demonstrated the possibility of inducing charges in graphene by the adsorption of various gases, including NH_3, H_2O, and NO_2 (Schedin et al. 2007). Hall effect measurements proved that NH_3 is a donor, whereas the latter two types of adsorbates act as acceptors. Doping can also be performed on a selected area. Cheng et al. demonstrated the effectiveness of noncovalent functionalization of graphene by using a completely resist-free approach and a spatially selective chemical modification process (Cheng et al. 2011).

3.2.2.2 Substitutional Doping

Substitutional doping, which has long been used as a strategy to tailor the electronic properties of semiconductors, is achieved by substituting appropriate foreign atoms for atoms in the host lattice. Based on theoretical studies, we know that substitutional doping can modulate the band structure of graphene (Chae et al. 2009), because the incorporation of foreign atoms would disrupt the sp^2 hybridization of carbon atoms. Boron and nitrogen atoms are both suitable dopants because their atomic sizes are similar to that of a carbon atom. Nitrogen-doped graphene on a copper surface has been obtained from a CVD reaction by using NH_3 as the nitrogen source and CH_4 as a carbon source (Wei et al. 2009). The doping concentration can be controlled by tuning the ratio of NH_3 to CH_4 in the reaction. Such nitrogen-doped graphene exhibits *n*-type doping behavior and has a larger on/off current ratio than that of pristine graphene. However, it possesses lower conductivity and lower mobility (approximately 200–450 $cm^2\ V^{-1}\ s^{-1}$) due to the scattering at defects introduced by dopants.

3.2.2.3 Characterization

The number of graphene layers can be identified by optical microscopy when graphene is on a Si substrate of suitable SiO_2 thickness. This is possible because of the optical interference effect. The contrast can be maximized by adjusting the SiO_2 thickness or the incident light wavelength. This method, however, is not able to provide other structural or electronic information about graphene. In contrast, Raman spectroscopy allows us to characterize not only the number of layers but also the amount of defects and impurities, the doping, and the graphene edges.

The Raman spectrum of pristine graphene shows three prominent peaks: a G band at ~1,580 cm^{-1}, a D band at ~1,350 cm^{-1}, and a G′(2D) band at ~2,680 cm^{-1} (Malard et al. 2009). The G band results from the high-frequency E_{2g} phonon at the Brillouin zone center Γ. The D band is caused by the breathing mode of six-atom rings. The G′(2D) band corresponds to a two-phonon band and is approximately twice the D band frequency. The Raman G band is sensitive to chemical doping, and there is an empirical rule that can be used to determine the doping type for the surface transfer doping of graphene. For *n*-type (*p*-typed) doped graphene, G band stiffens (softens) and downshifts (upshifts). In the case of electrically doped graphene, the G band stiffens and upshifts for both electron and hole dopings (Das et al. 2008).

3.2.3 Materials Preparation and Device Fabrication

3.2.3.1 Preparation Methods

Graphene was first isolated by Novoselov, Geim, and co-workers in 2004, and since then many researchers have attempted to produce graphene at a large scale while controlling the number of layers. Here, we briefly overview the six primary ways to produce single- and few-layered graphene sheets.

- **Micromechanical Cleavage**
  This method, also known as micromechanical exfoliation, is a promising way of obtaining high-quality graphene sheets that are suitable for fundamental studies of transport physics and other properties. In this procedure, one- to few-layered graphene sheets are peeled off from highly ordered pyrolytic graphite (HOPG) by using adhesive tape. This is possible due to the weak interlayer attraction (van der Waals forces) in graphite. Currently, this micromechanical cleavage method is still the most reliable way of obtaining high-mobility graphene sheets; however, it is a time-consuming process and not readily scalable.
- **Liquid-Phase Exfoliation of Graphite**
  This method generally involves three steps: (a) dispersion of graphite in a solvent, (b) exfoliation, and (c) purification (Hernandez et al. 2008). The third step is required due to the separation of exfoliated and un-exfoliated flakes, and it is usually carried out via ultrasonication in water and organic solvents. The types of dispersion solvents and the parameters for sonication are key factors that must be optimized to obtain large-scale, defect-free graphene with good electrical performance.
- **Reduction of Graphene Oxide**
  In general, graphene oxide is produced as a precursor for making graphene, and it can be prepared by one of three principal methods developed by Brodie, Staudenmaier, and Hummers (Marcano et al. 2010). The three methods all involve the oxidation of graphite to different levels. Brodie was the first to synthesize graphite oxide in 1859, by treating graphite with a solution of potassium chlorate

($KClO_3$) and nitric acid (HNO_3). In 1958, Hummers introduced the method that is now often used, in which graphite is treated with potassium permanganate ($KMnO_4$) and sodium nitrate ($NaNO_3$) in concentrated sulfuric acid (H_2SO_4) for several hours, followed by the addition of water. In this process, the reaction is safer and faster, and because of the absence of $KClO_3$, it does not produce explosive by products.

- **Epitaxial Growth on SiC**
  This technique starts with heating the SiC substrate to temperatures between 1,000 and 1,500 °C in ultrahigh vacuum. Graphene is produced by the thermal desorption of Si (Emtsev et al. 2009). Thermal decomposition is not self-limiting, and areas of different thicknesses of graphene may exist on the same SiC crystal. The carrier mobility of graphene synthesized by this method is approximately 1,100 $cm^2\ V^{-1}\ s^{-1}$ (Berger et al. 2004).
- **Chemical Vapor Deposition**
  During the CVD reaction, reactant precursors (e.g., gaseous hydrocarbons) are introduced into the reaction chamber. At elevated temperature, hydrocarbon precursors decompose to form carbon radicals on the catalytic metal substrate surface. They then form single- and/or few-layered graphene sheets. The metal substrate not only serves as a catalyst that reduces the energy required for the reaction to occur but also determines the growth mechanism of the graphene, which ultimately affects the number of layers and the quality of the as-grown graphene. Here, we focus on the catalytic metals nickel (Ni) and copper (Cu), which are the two commonly used substrates for CVD-grown graphene.

  - **Growth on Nickel:**
    Synthesis of graphene on Ni mainly involves the annealing of Ni film in an Ar/H_2 atmosphere at 900–1,000 °C. This is followed by the decomposition of hydrocarbon gas and the dissolution of carbon atoms into the Ni film to form a solid solution. Ni has high carbon solubility (>0.1 atomic percentage) (Mattevi et al. 2011) which decreases with decreasing temperature. At a rapid cooling rate, carbon atoms diffuse out from the Ni–C solid solution and precipitate on the Ni surface to form graphene layers. Graphene films grown on Ni substrates are usually continuous, with monolayer and multilayer regions.

    One can also obtain CVD-grown graphene from solid carbon sources (e.g., polymer films or small molecules) instead of gaseous hydrocarbons (Sun et al. 2010). Bilayer graphene films can be formed between the catalytic Ni layer and SiO_2/Si substrate by spin-coating a polymer film, such as poly(methyl methacrylate) (PMMA), on the top of a Ni layer pre-coated on a Si wafer, resulting in the diffusion of carbon atoms from the top to the bottom of the Ni layer during the CVD reaction (Huang et al. 2011b). Recently, Liao et al. grew large-area multilayer graphene sheets directly on a Si wafer by sandwiching hexamethyldisilazane (HMDS) as a carbon source between Ni/Cu films and the Si wafer (Liao et al. 2013).

- **Growth on Copper:**
  The growth of graphene on Cu is carried out by annealing a Cu film in an Ar/H_2 atmosphere at 900–1,000 °C, followed by the decomposition of gaseous hydrocarbons over the Cu substrate at ~1,000 °C. Cu has almost zero carbon solubility (<0.001 atomic percentage), even at 1,000 °C (Mattevi et al. 2011); therefore, the amount of carbon precipitating just outside the Cu surface is not even sufficient to form a single layer graphene. The carbon for graphene formation comes mainly from the hydrocarbons that are catalytically decomposed on the Cu surface. Thus, the growth of graphene on a Cu substrate is considered the more promising way of obtaining single-layered graphene sheets.

3.2.3.2 Device Fabrication

The standard fabrication procedures for graphene-based FETs involve (a) the transfer of graphene to a substrate, (b) the deposition of a gate dielectric on graphene (if preparing top-gated FETs), and (c) the deposition of metal electrodes on graphene. Accordingly, the performance of graphene-FETs is determined by three interfaces: (a) graphene–substrate, (b) graphene–dielectric, and (c) graphene–metal.

- **Transfer and Cleaning**
  Transferring graphene sheets to a substrate, on which some designed device patterns will later be fabricated, is a challenging task in the field of graphene electronics. Currently, various transfer methods have been employed for fabricating graphene-FETs, including wet transfer and dry transfer of exfoliated graphene flakes.

  - **Wet Transfer of Exfoliated Graphene**
    In wet transfer, graphene is in direct contact with liquid throughout the process. A PMMA thin film serving as a sacrificial layer is first spin-coated on top of the graphene, which has been prepared via mechanical exfoliation from bulk graphite and deposited on a SiO_2/Si substrate. The sample substrate containing graphene is then placed in a NaOH solution. The PMMA film that adheres to graphene is separated from the SiO_2/Si substrate by NaOH etching and water intercalation. The graphene can be transferred to a desired position on a receiver substrate by placing the PMMA film onto the substrate surface. Finally, the PMMA is dissolved with acetone, leaving a graphene flake that remains on the substrate (Reina et al. 2008).
  - **Dry Transfer of Exfoliated Graphene**
    In dry transfer, the upper graphene surface is kept out of contact with the liquid solvent/solution. An exfoliated graphene flake adheres to a combined bilayer consisting of a protective polymer layer (PMMA) on the top and a sacrificial releasing layer (water-dissolvable) on the bottom. The sacrificial layer is located above the substrate. Placing this sample structure into deionized water

causes the PMMA film containing the graphene to detach from the substrate. The graphene flake is now available for transfer to a target substrate. A manipulator is used to align the graphene flake to the desired position, and the PMMA film is stuck to the graphene flake on the substrate. Finally, the PMMA film is removed with an acetone rinse. In this way, the graphene flake is transferred without touching a liquid solvent or aqueous solution (Dean et al. 2010).

– **Cleaning Graphene**
Cleaning the graphene surface is an important part of making graphene-FETs, because contaminations can severely alter the electronic properties of graphene. The most common method of eliminating the resist residuals from the fabrication process and other contaminants from tape glue (if exfoliation is used) is thermal annealing. Generally, cleaning involves annealing graphene at high temperature (400 °C) in an Ar/H_2 environment (Ishigami et al. 2007), or under ultrahigh vacuum ($<1.5\times10^{-10}$ Torr), to remove contamination (Stolyarova et al. 2007).

- **Electrodes**
A high-quality metal–graphene interface is also critical for the performance of a graphene-FET. The typical contact metals used as electrodes are Ti/Au, Cr/Au, Cr/Pt, Ni, and Co. Normally, metals in contact with the graphene channel are fabricated by conventional vacuum deposition and photolithographic pattern methods. The contact resistance is generally measured by four-probe or transmission line methods. So far, the best reported contact resistance for lithographically defined contacts deposited onto exfoliated graphene is in the range of 200–500 Ω μm (Xia et al. 2011). However, for CVD-grown graphene, the contact resistance can range from 500 to 1,000 Ω μm (Chan et al. 2012).
- **Dielectrics**
The gate dielectric is an essential component of a transistor, and it significantly affects important features of the device such as the transconductance, subthreshold swing, and frequency response. A typical graphene-FET is fabricated on a Si wafer (SiO_2/Si substrate). The Si substrate serves as a global back gate, and a SiO_2 layer serves as the gate dielectric. However, such a configuration cannot independently drive multiple devices on the same chip. Moreover, the SiO_2 layer has a small gate capacitance due to its relatively small dielectric constant. It therefore requires a large voltage to enable gate control of channel conductivity. As an alternative, a top-gate structure can be used to overcome the limitations of a back-gate configuration for practical applications of graphene-FETs. Physical vapor deposition (PVD) and atomic layer deposition (ALD) are two common methods used to deposit top-gate dielectrics on graphene. ALD is the preferred technique for producing high-quality, homogeneous, and high-κ dielectric films (e.g., ZrO_2, TiO_2, Al_2O_3, and HfO_2) with precise control of thickness at the Ångstrom or monolayer level.

3.2.4 Biosensing Applications of Graphene-Based FETs

Graphene, a single-atom-thick sheet of carbon atoms in a 2D honeycomb lattice, has been used to configure FET devices that exhibit great potential for the electrical detection of biological molecules. These graphene-based FETs have unique electrical and physical properties (Liu et al. 2012). High-quality graphene can display remarkably high electron mobility at room temperature (20,000 $cm^2\ V^{-1}\ s^{-1}$), high carrier density ($10^{13}\ cm^{-2}$), ambipolar field-effect characteristics, and low intrinsic noise (Lin and Avouris 2008; Geim and Novoselov 2007; Schedin et al. 2007). Moreover, its extraordinary structure provides a large detection area and enables homogenous surface functionalization (Huang et al. 2011b). In biosensing applications, graphene-FETs have been used to detect metal ions and various biomolecules, including neurotransmitters, glucose, proteins, and nucleic acids.

3.2.4.1 Biomolecular Detection

Graphene-FET biosensors have been used to detect glucose and glutamate with a detection limit of ~100 and ~5 μM, respectively (He et al. 2010). In these cases, detection was mediated by the immobilization of glucose oxidase (GOD) and glutamate dehydrogenase (GluD) on graphene, respectively. Catalytic reactions involving glucose and glutamate were carried out by the enzymes (GOD and GluD, respectively) to produce H_2O_2, which served as a *p*-dopant to graphene, consequently increasing the conductance of graphene in the *p*-type regime. The current vs. elapsed time ($I-t$) curves show that an obvious increment in source-drain current (I_{sd}) was observed with successive additions of glucose and glutamate to phosphate-buffered saline (PBS). The current responses of GOD-immobilized graphene-FETs tended to saturate at glucose concentrations greater than 10 mM. Moreover, two linear detection ranges of glutamate (5–50 μM and 50 μM–1.2 mM) were displayed when using GluD-functionalized graphene-FET biosensors.

To ensure the specificity of protein detection, an electronic immunoglobulin E (IgE) biosensor was fabricated by functionalizing IgE-specific aptamers onto the surface of monolayer graphene via a chemical linker of 1-pyrenebutanoic acid succinimidyl ester (Ohno et al. 2010). When the positively charged IgE was introduced, it bound to the negatively charged IgE-specific aptamers, and the conductance in the graphene-FET decreased dramatically due to the gating effect. A linear detection range of IgE was measured from 290 pM to 1.60 nM. More importantly, the aptamer-modified graphene-FET could detect IgE specifically. There was no interference when bovine serum albumin or streptavidin were introduced. Another study was carried out by using an antibody as the recognition element on the graphene-FET surface to achieve specific immunoglobulin G (IgG) detection (Mao et al. 2010). Prior to the modification of anti-IgG antibody, gold nanoparticles (AuNPs) were anchored to the graphene sheet to mediate the immobilization of the anti-IgG antibody. During the electrical measurements, the relative resistance of the graphene-FET increased in a stepwise fashion in response to a series of IgG additions. This anti-IgG/AuNP-modified graphene-FET could sensitively detect IgG with a detection limit of ~13 pM and a wide linear sensing range of 2×10^{-9}–2×10^{-5} g mL^{-1}. Additionally, this IgG-sensing device based on a graphene-FET exhibited high target selectivity, avoiding non-specific interactions with other proteins, such as immunoglobulin M and horseradish peroxidase.

3.2.4.2 Nucleic Acid Screening and Sequencing

A molecule-scale nanopore (several nanometers in diameter) placed within an insulating membrane can be used as a molecular detector with excellent sensitivity and exquisite spatial resolution. When a biomolecule passes through a narrow nanopore connecting two separate electrolyte solutions, the ionic current flowing through the nanopore will be partially blocked. The ionic current can be modulated significantly by the subtle structure of the passing biomolecule (Liu et al. 2012). Compared with lipid bilayers (Wendell et al. 2009; Faller et al. 2004) and other artificial solid-state membranes (e.g., Si_3N_4 or SiO_2) (Chen et al. 2004a; Dekker 2007), single- or few-layered graphene sheets are considered ideal candidates for the nanopore-sequencing of DNA or RNA due to their superior structural stiffness and atomic thickness. As shown in Fig. 3.7, nanopores of 8–22 nm in diameter within the graphene sheets have been constructed in several studies (Merchant et al. 2010; Schneider et al. 2010; Venkatesan et al. 2011). The translocation events of unfolded, partially folded, and fully folded double-stranded DNAs (dsDNAs) could be precisely detected as the DNA molecules traveled through the nanopore. Additionally, a nanopore in the center of a graphene ribbon could be created to sensitively distinguish different types of nucleotides by measuring the distinct lateral tunneling currents across the nanopore caused by individual nucleobases and the electrostatic interactions of those nucleotides with the graphene nanoribbon (Nelson et al. 2010). A nanopore device can also be designed using four graphene nanoelectrodes that are thin enough to achieve single-base resolution for the passing DNAs (He et al. 2011); however, with this design, the intrinsic electrical conductivity of the graphene nanoelectrode and the coupling strength between graphene and DNA are small in comparison with the gold nanoelectrode, meaning that the signal-to-noise ratio will be negatively affected. Nevertheless, the detection sensitivity can be improved by hydrogenation of the edges of each graphene nanoelectrode, so that the edges can couple with the nucleobases of DNA molecules through hydrogen bonding and slow the translocation of the DNA. This increases the electron tunneling rate and enhances the average transverse conductivity by about three orders of magnitude (Wuttke et al. 1992). In the future, this graphene nanopore technique, assisted by a novel pore architecture and further functionalization, will make rapid DNA sequencing feasible.

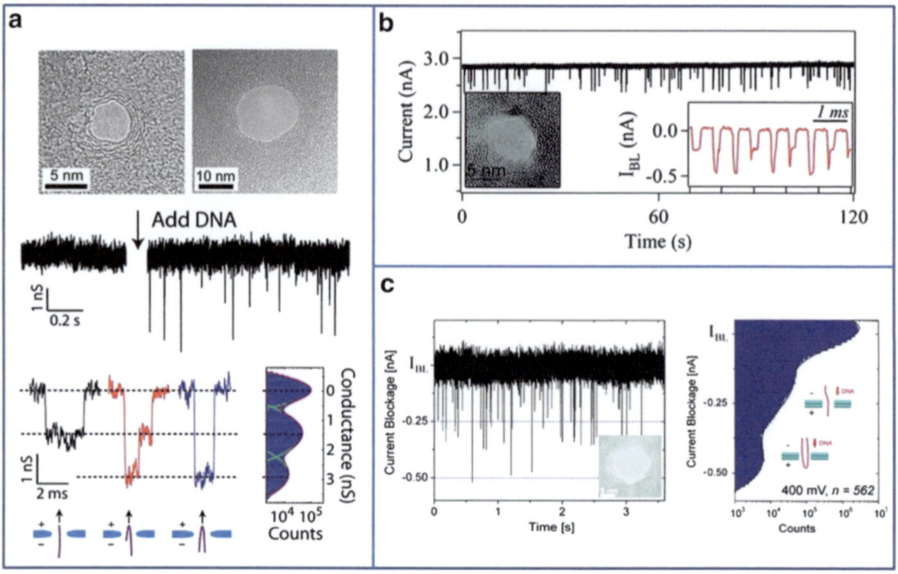

Fig. 3.7 (**a**) (*Top graph*) TEM images showing a 22-nm nanopore within a graphene monolayer. (*Middle graph*) Real-time electrical detection measures the baseline conductance (*left*) and detects blockade events upon the translocation of a 48 kb double-stranded λ-DNA molecule through the nanopore (*right*). (*Bottom graph*) Examples of translocation events of nonfolded (*black*), partially folded (*red*), and fully folded (*blue*) DNA molecules, recorded at 200 mV in the 22-nm pore (*left*), and a conductance histogram representing 1,222 translocation events, including the open-pore conductance before and after the event (*right*). (**b**) Time traces of ionic current reveal DNA translocations for a graphene-TiO_2 nanopore with 15 kb of dsDNA. *Left inset* is a TEM image of the actual nanopore within the graphene-TiO_2 sheet, with a pore diameter of 7.5 nm. *Right inset* shows a concatenated sequence of sample events, with the open pore current excluded. (**c**) Current traces show the translocation of a 48.5 kb λ-DNA molecule through an 11.3-nm graphene-Al_2O_3 nanopore (*left*). Each downward spike from the baseline current level corresponds to the transport of a single λ-DNA molecule. *Inset* shows a TEM image of the nanopore. Current blockade histogram shows two distinct current peaks, corresponding to the transport of unfolded and folded λ-DNA (*right*). The histogram represents $n = 562$ separate translocation events. Reprinted with permission from Schneider et al. (2010), Merchant et al. (2010), and Venkatesan et al. (2011). Copyright American Chemical Society

3.2.4.3 Real-Time Cellular Recording

Unlike the FETs based on 1D nanomaterials (e.g., SiNW, ZnO NW, or CNT), graphene-FETs have a flat structure and unique electrical and physiochemical properties that make them better for use in electrical cellular detection applications (Nguyen and Berry 2012; Huang et al. 2011a). Because of the tight, homogeneous interaction between 2D cell membranes and flat graphene, cellular dynamic activity and chemical fluctuations can sensitively modulate the conductance inside the graphene-FET to yield optimal detection signals (Huang et al. 2011b; Liu et al. 2012). So far, graphene-FETs have been used to record the rapid release of exocytotic catecholamines from living PC12 cells under the stimulation of cell membrane depolarization and voltage-induced Ca^{2+} influx (He et al. 2010). A series of in vitro tests revealed that the conductance of the graphene-FET in the *n*-type regime was sequentially increased in response to increasing catecholamine concentration. This was caused by the strong attachment of positively charged, aromatic catecholamines on the graphene surface via a π–π interaction. These results from in vitro studies could be used to postulate that each observed current spike is likely to be caused by the vesicular secretion of catecholamines from PC12 cells exposed to high K^+ solution. Graphene-FET devices have also been used to characterize the presence and activities of pathogens. An *E. coli* bacteria biosensor was fabricated from CVD-grown graphene, and in this biosensor, anti-*E. coli* antibodies were immobilized as the recognition elements via a chemical linker of 1-pyrenebutanoic acid succinimidyl ester (Huang et al. 2011b). To prevent non-specific binding of *E. coli* bacteria, ethanolamine was used to quench the non-reacted linker molecules, and Tween 20 was applied to protect the exposed graphene surface. The observed percentage change in graphene conductance corresponded to the amount of *E. coli* present, and the biosensor had a low detection limit of 10 cfu mL^{-1} and excellent specificity for the *E. coli* species. These studies demonstrate that graphene-based FET biosensors have great potential for use in temporally resolved monitoring of dynamic cellular functions and in rapid detection of pathogenic microbes.

3.3 Summary

SiNW-FETs and graphene-FETs have been successfully used in biological and cellular studies, such as studies of protein–protein interactions, neural signal transmission, viral infection diagnosis, biomolecular detection, nucleic acid screening and sequencing, and real-time cellular recording. While these SiNW-FETs and graphene-FETs have important performance advantages in various biosensing applications, challenges remain for the future advancement of FET-based biosensors. For example, there is a demand for new designs for device fabrication and signal acquisition systems, which could be used for high-throughput multiplexed sensing and the mass production of robust nanoscale FET devices for point-of-care diagnosis. Notwithstanding these challenges, we believe that the continuing discovery/synthesis of new 1D and 2D nanostructures and the continuing study of their novel device architectures will continue to play a significant role in the field of (opto)electronics and biosensors.

References

Berger C, Song Z, Li T, Li X, Ogbazghi AY, Feng R, Dai Z, Marchenkov AN, Conrad EH, First PN, de Heer WA (2004) Ultrathin epitaxial graphite: 2D electron gas properties and a route toward graphene-based nanoelectronics. J Phys Chem B 108(52):19912–19916

Bunimovich YL, Shin YS, Yeo WS, Amori M, Kwong G, Heath JR (2006) Quantitative real-time measurements of DNA hybridization with alkylated nonoxidized silicon nanowires in electrolyte solution. J Am Chem Soc 128(50):16323–16331

Castro Neto AH, Guinea F, Peres NMR, Novoselov KS, Geim AK (2009) The electronic properties of graphene. Rev Mod Phys 81(1):109–162

Chae SJ, Gunes F, Kim KK, Kim ES, Han GH, Kim SM, Shin HJ, Yoon SM, Choi JY, Park MH, Yang CW, Pribat D, Lee YH (2009) Synthesis of large-area graphene layers on poly-nickel substrate by chemical vapor deposition: wrinkle formation. Adv Mater 21(22):2328

Chan J, Venugopal A, Pirkle A, McDonnell S, Hinojos D, Magnuson CW, Ruoff RS, Colombo L, Wallace RM, Vogel EM (2012) Reducing extrinsic performance-limiting factors in graphene grown by chemical vapor deposition. ACS Nano 6(4):3224–3229

Chang YK, Hong FCN (2009) The fabrication of ZnO nanowire field-effect transistors by roll-transfer printing. Nanotechnology 20(19)

Chang KS, Sun CJ, Chiang PL, Chou AC, Lin MC, Liang C, Hung HH, Yeh YH, Chen CD, Pan CY, Chen YT (2012) Monitoring extracellular K^+ flux with a valinomycin-coated silicon nanowire field-effect transistor. Biosens Bioelectron 31(1):137–143

Chen P, Gu JJ, Brandin E, Kim YR, Wang Q, Branton D (2004a) Probing single DNA molecule transport using fabricated nanopores. Nano Lett 4(11):2293–2298

Chen RJ, Choi HC, Bangsaruntip S, Yenilmez E, Tang X, Wang Q, Chang Y-L, Dai H (2004b) An investigation of the mechanisms of electronic sensing of protein adsorption on carbon nanotube devices. J Am Chem Soc 126(5):1563–1568

Chen KI, Li BR, Chen YT (2011) Silicon nanowire field-effect transistor-based biosensors for biomedical diagnosis and cellular recording investigation. Nano Today 6(2):131–154

Cheng H-C, Shiue R-J, Tsai C-C, Wang W-H, Chen Y-T (2011) High-quality graphene p-n junctions via resist-free fabrication and solution-based noncovalent functionalization. ACS Nano 5(3):2051–2059

Chiang PL, Chou TC, Wu TH, Li CC, Liao CD, Lin JY, Tsai MH, Tsai CC, Sun CJ, Wang CH, Fang JM, Chen YT (2012) Nanowire transistor-based ultrasensitive virus detection with reversible surface functionalization. Chem Asian J 7(9):2073–2079

Choi YK, Zhu J, Grunes J, Bokor J, Somorjai GA (2003) Fabrication of sub-10-nm silicon nanowire arrays by size reduction lithography. J Phys Chem B 107(15):3340–3343

Craighead HG, Howard RE, Jackel LD, Mankiewich PM (1983) 10-nm line width electron-beam lithography on gaas. Appl Phys Lett 42(1):38–40

Cui Y, Duan X, Hu J, Lieber CM (2000) Doping and electrical transport in silicon nanowires. J Phys Chem B 104(22):5213–5216

Cui Y, Wei QQ, Park HK, Lieber CM (2001) Nanowire nanosensors for highly sensitive and selective detection of biological and chemical species. Science 293(5533):1289–1292. doi:10.1126/science.1062711

Curreli M, Zhang R, Ishikawa FN, Chang HK, Cote RJ, Zhou C, Thompson ME (2008) Real-time, label-free detection of biological entities using nanowire-based FETs. IEEE Trans Nanotechnol 7(6):651–667

Das A, Pisana S, Chakraborty B, Piscanec S, Saha SK, Waghmare UV, Novoselov KS, Krishnamurthy HR, Geim AK, Ferrari AC, Sood AK (2008) Monitoring dopants by Raman scattering in an electrochemically top-gated graphene transistor. Nat Nanotechnol 3(4):210–215

Dean CR, Young AF, Meric I, Lee C, Wang L, Sorgenfrei S, Watanabe K, Taniguchi T, Kim P, Shepard KL, Hone J (2010) Boron nitride substrates for high-quality graphene electronics. Nat Nanotechnol 5(10):722–726

Dekker C (2007) Solid-state nanopores. Nat Nanotechnol 2(4):209–215

Demami F, Ni L, Rogel R, Salaun AC, Pichon L (2012) Silicon nanowires based resistors as gas sensors. Sensor Actuat B Chem 170:158–162

Duan XJ, Gao RX, Xie P, Cohen-Karni T, Qing Q, Choe HS, Tian BZ, Jiang XC, Lieber CM (2012) Intracellular recordings of action potentials by an extracellular nanoscale field-effect transistor. Nat Nanotechnol 7(3):174–179

Duan XJ, Fu TM, Liu J, Lieber CM (2013) Nanoelectronics-biology frontier: from nanoscopic probes for action potential recording in live cells to three-dimensional cyborg tissues. Nano Today 8(4):351–373

Emtsev KV, Bostwick A, Horn K, Jobst J, Kellogg GL, Ley L, McChesney JL, Ohta T, Reshanov SA, Rohrl J, Rotenberg E, Schmid AK, Waldmann D, Weber HB, Seyller T (2009) Towards wafer-size graphene layers by atmospheric pressure graphitization of silicon carbide. Nat Mater 8(3):203–207

Faller M, Niederweis M, Schulz GE (2004) The structure of a mycobacterial outer-membrane channel. Science 303(5661):1189–1192

Fan ZY, Ho JC, Jacobson ZA, Yerushalmi R, Alley RL, Razavi H, Javey A (2008) Wafer-scale assembly of highly ordered semiconductor nanowire arrays by contact printing. Nano Lett 8(1):20–25

Freer EM, Grachev O, Duan XF, Martin S, Stumbo DP (2010) High-yield self-limiting single-nanowire assembly with dielectrophoresis. Nat Nanotechnol 5(7):525–530

Gao RX, Strehle S, Tian BZ, Cohen-Karni T, Xie P, Duan XJ, Qing Q, Lieber CM (2012) Outside looking in: nanotube transistor intracellular sensors. Nano Lett 12(6):3329–3333

Geim AK, Novoselov KS (2007) The rise of graphene. Nat Mater 6(3):183–191

Hakim MMA, Lombardini M, Sun K, Giustiniano F, Roach PL, Davies DE, Howarth PH, de Planque MRR, Morgan H, Ashburn P (2012) Thin film polycrystalline silicon nanowire biosensors. Nano Lett 12(4):1868–1872

He QY, Sudibya HG, Yin ZY, Wu SX, Li H, Boey F, Huang W, Chen P, Zhang H (2010) Centimeter-long and large-scale micropatterns of reduced graphene oxide films: fabrication and sensing applications. ACS Nano 4(6):3201–3208

He YH, Scheicher RH, Grigoriev A, Ahuja R, Long SB, Huo ZL, Liu M (2011) Enhanced DNA sequencing performance through edge-hydrogenation of graphene electrodes. Adv Funct Mater 21(14):2674–2679

Hernandez Y, Nicolosi V, Lotya M, Blighe FM, Sun Z, De S, McGovern IT, Holland B, Byrne M, Gun'Ko YK, Boland JJ, Niraj P, Duesberg G, Krishnamurthy S, Goodhue R, Hutchison J, Scardaci V, Ferrari AC, Coleman JN (2008) High-yield production of graphene by liquid-phase exfoliation of graphite. Nat Nanotechnol 3(9):563–568

Hsiao CY, Lin CH, Hung CH, Su CJ, Lo YR, Lee CC, Lin HC, Ko FH, Huang TY, Yang YS (2009) Novel poly-silicon nanowire field effect transistor for biosensing application. Biosens Bioelectron 24(5):1223–1229

Hu J, Odom TW, Lieber CM (1999) Chemistry and physics in one dimension: synthesis and properties of nanowires and nanotubes. Acc Chem Res 32(5):435–445

Huang Y, Duan XF, Wei QQ, Lieber CM (2001) Directed assembly of one-dimensional nanostructures into functional networks. Science 291(5504):630–633

Huang YX, Cai D, Chen P (2011a) Micro- and nanotechnologies for study of cell secretion. Anal Chem 83(12):4393–4406

Huang YX, Dong XC, Liu YX, Li LJ, Chen P (2011b) Graphene-based biosensors for detection of bacteria and their metabolic activities. J Mater Chem 21(33):12358–12362

Hurt AC, Baas C, Deng Y-M, Roberts S, Kelso A, Barr IG (2009) Performance of influenza rapid point-of-care tests in the detection of swine lineage a(H1N1) influenza viruses. Influenza Other Respir Viruses 3(4):171–176

Hwang EH, Adam S, Das Sarma S (2007) Transport in chemically doped graphene in the presence of adsorbed molecules. Phys Rev B 76(19):195421

Ishigami M, Chen JH, Cullen WG, Fuhrer MS, Williams ED (2007) Atomic structure of graphene on SiO_2. Nano Lett 7(6):1643–1648

Jentsch TJ (2000) Neuronal KCNQ potassium channels: physiology and role in disease. Nat Rev Neurosci 1(1):21–30

Jiang Z, Qing Q, Xie P, Gao RX, Lieber CM (2012) Kinked p–n junction nanowire probes for high spatial resolution sensing and intracellular recording. Nano Lett 12(3):1711–1716

Kao LTH, Shankar L, Kang TG, Zhang GJ, Tay GKI, Rafei SRM, Lee CWH (2011) Multiplexed detection and differentiation of the DNA strains for influenza A (H1N1 2009) using a silicon-based microfluidic system. Biosens Bioelectron 26(5):2006–2011

Knopfmacher O, Tarasov A, Fu WY, Wipf M, Niesen B, Calame M, Schonenberger C (2010) Nernst limit in dual-gated Si-nanowire FET sensors. Nano Lett 10(6):2268–2274

Kwiat M, Elnathan R, Pevzner A, Peretz A, Barak B, Peretz H, Ducobni T, Stein D, Mittelman L, Ashery U (2012) Highly ordered large-scale neuronal networks of individual cells-toward single cell to 3D nanowire intracellular interfaces. ACS Appl Mater Inter 4(7):3542–3549

Li Z, Chen Y, Li X, Kamins T, Nauka K, Williams RS (2004) Sequence-specific label-free DNA sensors based on silicon nanowires. Nano Lett 4(2):245–247

Li XL, Zhang L, Wang XR, Shimoyama I, Sun XM, Seo WS, Dai HJ (2007) Langmuir-Blodgett assembly of densely aligned single-walled carbon nanotubes from bulk materials. J Am Chem Soc 129(16):4890–4891

Li BR, Chen CW, Yang WL, Lin TY, Pan CY, Chen YT (2013) Biomolecular recognition with a sensitivity-enhanced nanowire transistor biosensor. Biosens Bioelectron 45:252–259

Liao C-D, Lu Y-Y, Tamalampudi SR, Cheng H-C, Chen Y-T (2013) Chemical vapor deposition synthesis and Raman spectroscopic characterization of large-area graphene sheets. J Phys Chem A 117(39):9454–9461

Lin YM, Avouris P (2008) Strong suppression of electrical noise in bilayer graphene nanodevices. Nano Lett 8(8):2119–2125

Lin HC, Lee MH, Su CJ, Huang TY, Lee CC, Yang YS (2005) A simple and low-cost method to fabricate TFTs with poly-Si nanowire channel. IEEE Trans Nanotechnol 26(9):643–645

Lin SP, Pan CY, Tseng KC, Lin MC, Chen CD, Tsai CC, Yu SH, Sun YC, Lin TW, Chen YT (2009) A reversible surface functionalized nanowire transistor to study protein-protein interactions. Nano Today 4(3):235–243

Lin TW, Hsieh PJ, Lin CL, Fang YY, Yang JX, Tsai CC, Chiang PL, Pan CY, Chen YT (2010) Label-free detection of protein-protein interactions using a calmodulin-modified nanowire transistor. Proc Natl Acad Sci U S A 107(3):1047–1052

Lin TY, Li BR, Tsai ST, Chen CW, Chen CH, Chen YT, Pan CY (2013) Improved silicon nanowire field-effect transistors for fast protein-protein interaction screening. Lab Chip 13(4):676–684

Liu YX, Dong XC, Chen P (2012) Biological and chemical sensors based on graphene materials. Chem Soc Rev 41(6):2283–2307

Ma D, Lee C, Au F, Tong S, Lee S (2003) Small-diameter silicon nanowire surfaces. Science 299(5614):1874–1877

MacDonald PE, Ha XF, Wang J, Smukler SR, Sun AM, Gaisano HY, Salapatek AMF, Backx PH, Wheeler MB (2001) Members of the Kv1 and Kv2 voltage-dependent K^+ channel families regulate insulin secretion. Mol Endocrinol 15(8):1423–1435

Malard LM, Pimenta MA, Dresselhaus G, Dresselhaus MS (2009) Raman spectroscopy in graphene. Phys Rep 473(5–6):51–87

Mao S, Lu GH, Yu KH, Bo Z, Chen JH (2010) Specific protein detection using thermally reduced graphene oxide sheet decorated with gold nanoparticle-antibody conjugates. Adv Mater 22(32): 3521–3526

Marcano DC, Kosynkin DV, Berlin JM, Sinitskii A, Sun Z, Slesarev A, Alemany LB, Lu W, Tour JM (2010) Improved synthesis of graphene oxide. ACS Nano 4(8):4806–4814

Martínez J, Martínez RV, Garcia R (2008) Silicon nanowire transistors with a channel width of 4 nm fabricated by atomic force microscope nanolithography. Nano Lett 8(11):3636–3639

Martínez RV, Martínez J, Garcia R (2010) Silicon nanowire circuits fabricated by AFM oxidation nanolithography. Nanotechnology 21(24):245301

Mattevi C, Kim H, Chhowalla M (2011) A review of chemical vapour deposition of graphene on copper. J Mater Chem 21(10):3324–3334

McAlpine MC, Ahmad H, Wang DW, Heath JR (2007) Highly ordered nanowire arrays on plastic substrates for ultrasensitive flexible chemical sensors. Nat Mater 6(5):379–384

Melosh NA, Boukai A, Diana F, Gerardot B, Badolato A, Petroff PM, Heath JR (2003) Ultrahigh-density nanowire lattices and circuits. Science 300(5616):112–115

Merchant CA, Healy K, Wanunu M, Ray V, Peterman N, Bartel J, Fischbein MD, Venta K, Luo Z, Johnson ATC, Drndić M (2010) DNA translocation through graphene nanopores. Nano Lett 10(8):2915–2921

Morton KJ, Nieberg G, Bai S, Chou SY (2008) Wafer-scale patterning of sub-40 nm diameter and high aspect ratio (>50: 1) silicon pillar arrays by nanoimprint and etching. Nanotechnology 19(34):345301

Nakajima N, Ikada Y (1995) Mechanism of amide formation by carbodiimide for bioconjugation in aqueous media. Bioconj Chem 6(1):123–130

Nelson T, Zhang B, Prezhdo OV (2010) Detection of nucleic acids with graphene nanopores: ab initio characterization of a novel sequencing device. Nano Lett 10(9):3237–3242

Nguyen P, Berry V (2012) Graphene interfaced with biological cells: opportunities and challenges. J Phys Chem Lett 3(8):1024–1029

Novoselov KS, Geim AK, Morozov SV, Jiang D, Zhang Y, Dubonos SV, Grigorieva IV, Firsov AA (2004) Electric field effect in atomically thin carbon films. Science 306(5696):666–669

Ohno Y, Maehashi K, Matsumoto K (2010) Label-free biosensors based on aptamer-modified graphene field-effect transistors. J Am Chem Soc 132(51):18012–18013

Park IY, Li ZY, Li XM, Pisano AP, Williams RS (2007) Towards the silicon nanowire-based sensor for intracellular biochemical detection. Biosens Bioelectron 22(9–10):2065–2070

Patolsky F, Timko BP, Yu G, Fang Y, Greytak AB, Zheng G, Lieber CM (2006a) Detection, stimulation, and inhibition of neuronal signals with high-density nanowire transistor arrays. Science 313(5790):1100–1104

Patolsky F, Zheng G, Lieber CM (2006b) Nanowire sensors for medicine and the life sciences. Nanomedicine 1(1):51–65

Poghossian A, Ingebrandt S, Abouzar M, Schöning M (2007) Label-free detection of charged macromolecules by using a field-effect-based sensor platform: experiments and possible mechanisms of signal generation. Appl Phys A Mater 87(3):517–524

Qing Q, Pal SK, Tian B, Duan X, Timko BP, Cohen-Karni T, Murthy VN, Lieber CM (2010) Nanowire transistor arrays for mapping neural circuits in acute brain slices. Proc Natl Acad Sci U S A 107(5):1882–1887

Raymo FM, Yildiz I (2007) Luminescent chemosensors based on semiconductor quantum dots. Phys Chem Chem Phys 9(17):2036–2043

Reina A, Son H, Jiao L, Fan B, Dresselhaus MS, Liu Z, Kong J (2008) Transferring and identification of single- and few-layer graphene on arbitrary substrates. J Phys Chem C 112(46): 17741–17744

Rosi NL, Mirkin CA (2005) Nanostructures in biodiagnostics. Chem Rev 105(4):1547–1562

Sala F, Nistri A, Criado M (2008) Nicotinic acetylcholine receptors of adrenal chromaffin cells. Acta Physiol 192(2):203–212

Schedin F, Geim AK, Morozov SV, Hill EW, Blake P, Katsnelson MI, Novoselov KS (2007) Detection of individual gas molecules adsorbed on graphene. Nat Mater 6(9):652–655

Schneider GF, Kowalczyk SW, Calado VE, Pandraud G, Zandbergen HW, Vandersypen LMK, Dekker C (2010) DNA translocation through graphene nanopores. Nano Lett 10(8): 3163–3167

Shao M, Ma DDD, Lee ST (2010) Silicon nanowires– synthesis, properties, and applications. Eur J Inorg Chem 2010(27):4264–4278

Singh NA, Charlier C, Stauffer D, DuPont BR, Leach RJ, Melis R, Ronen GM, Bjerre I, Quattlebaum T, Murphy JV, McHarg ML, Gagnon D, Rosales TO, Peiffer A, Anderson VE, Leppert M (1998) A novel potassium channel gene, KCNQ2, is mutated in an inherited epilepsy of newborns. Nat Genet 18:25–29

Soderling TR, Chang B, Brickey D (2001) Cellular signaling through multifunctional Ca^{2+}/calmodulin-dependent protein kinase II. J Biol Chem 276:3719–3722

Stern E, Klemic JF, Routenberg DA, Wyrembak PN, Turner-Evans DB, Hamilton AD, LaVan DA, Fahmy TM, Reed MA (2007) Label-free immunodetection with CMOS-compatible semiconducting nanowires. Nature 445(7127):519–522

Stolyarova E, Rim KT, Ryu S, Maultzsch J, Kim P, Brus LE, Heinz TF, Hybertsen MS, Flynn GW (2007) High-resolution scanning tunneling microscopy imaging of mesoscopic graphene sheets on an insulating surface. Proc Natl Acad Sci U S A 104(22):9209–9212

Sudibya HG, Ma J, Dong XC, Ng S, Li LJ, Liu XW, Chen P (2009) Interfacing glycosylated carbon-nanotube-network devices with living cells to detect dynamic secretion of biomolecules. Angew Chem Int Ed 48(15):2723–2726

Sun Z, Yan Z, Yao J, Beitler E, Zhu Y, Tour JM (2010) Growth of graphene from solid carbon sources. Nature 468(7323):549–552

Talin AA, Hunter LL, Léonard F, Rokad B (2006) Large area, dense silicon nanowire array chemical sensors. Appl Phys Lett 89(15):153102

Tian B, Xie P, Kempa TJ, Bell DC, Lieber CM (2009) Single-crystalline kinked semiconductor nanowire superstructures. Nat Nanotechnol 4(12):824–829

Tian BZ, Cohen-Karni T, Qing Q, Duan XJ, Xie P, Lieber CM (2010) Three-dimensional, flexible nanoscale field-effect transistors as localized bioprobes. Science 329(5993):830–834

Tong HD, Chen S, van der Wiel WG, Carlen ET, van den Berg A (2009) Novel top-down wafer-scale fabrication of single crystal silicon nanowires. Nano Lett 9(3):1015–1022

Trivedi K, Yuk H, Floresca HC, Kim MJ, Hu W (2011) Quantum confinement induced performance enhancement in sub-5-nm lithographic Si nanowire transistors. Nano Lett 11(4):1412–1417

Tsai TC, Huang FH, Chen JJJ (2013) Selective detection of dopamine in urine with electrodes modified by gold nanodendrite and anionic self-assembled monolayer. Sensor Actuat B Chem 181:179–186

Venkatesan BM, Estrada D, Banerjee S, Jin XZ, Dorgan VE, Bae MH, Aluru NR, Pop E, Bashir R (2011) Stacked graphene-Al_2O_3 nanopore sensors for sensitive detection of DNA and DNA-protein complexes. ACS Nano 6(1):441–450

Vieu C, Carcenac F, Pepin A, Chen Y, Mejias M, Lebib A, Manin-Ferlazzo L, Couraud L, Launois H (2000) Electron beam lithography: resolution limits and applications. Appl Surf Sci 164: 111–117

Vu XT, Eschermann JF, Stockmann R, GhoshMoulick R, Offenhausser A, Ingebrandt S (2009) Top-down processed silicon nanowire transistor arrays for biosensing. Phys Status Solid A 206(3):426–434

Wagner RS, Ellis WC (1964) Vapor–liquid-solid mechanism of single crystal growth. Appl Phys Lett 4(5):89–90

Wang CW, Pan CY, Wu HC, Shih PY, Tsai CC, Liao KT, Lu LL, Hsieh WH, Chen CD, Chen YT (2007) In situ detection of chromogranin A released from living neurons with a single-walled carbon-nanotube field-effect transistor. Small 3(8):1350–1355

Wei D, Liu Y, Wang Y, Zhang H, Huang L, Yu G (2009) Synthesis of N-doped graphene by chemical vapor deposition and its electrical properties. Nano Lett 9(5):1752–1758

Wendell D, Jing P, Geng J, Subramaniam V, Lee TJ, Montemagno C, Guo PX (2009) Translocation of double-stranded DNA through membrane-adapted phi29 motor protein nanopores. Nat Nanotechnol 4(11):765–772

Whang D, Jin S, Wu Y, Lieber CM (2003) Large-scale hierarchical organization of nanowire arrays for integrated nanosystems. Nano Lett 3(9):1255–1259

Wu Y, Cui Y, Huynh L, Barrelet CJ, Bell DC, Lieber CM (2004) Controlled growth and structures of molecular-scale silicon nanowires. Nano Lett 4(3):433–436

Wuttke DS, Bjerrum MJ, Winkler JR, Gray HB (1992) Electron-tunneling pathways in cytochrome c. Science 256(5059):1007–1009

Xia F, Perebeinos V, Lin Y-M, Wu Y, Avouris P (2011) The origins and limits of metal-graphene junction resistance. Nat Nanotechnol 6(3):179–184

Yerushalmi R, Jacobson ZA, Ho JC, Fan Z, Javey A (2007) Large scale, highly ordered assembly of nanowire parallel arrays by differential roll printing. Appl Phys Lett 91(20):203104

Yu GH, Cao AY, Lieber CM (2007) Large-area blown bubble films of aligned nanowires and carbon nanotubes. Nat Nanotechnol 2(6):372–377

Zhang GJ, Ning Y (2012) Silicon nanowire biosensor and its applications in disease diagnostics: a review. Anal Chem Acta 749:1–15

Zheng GF, Patolsky F, Cui Y, Wang WU, Lieber CM (2005) Multiplexed electrical detection of cancer markers with nanowire sensor arrays. Nat Biotechnol 23(10):1294–1301

Chapter 4
Nanoimprinted Plasmonic Biosensors and Biochips

Keiichiro Yamanaka and Masato Saito

Abstract In this chapter, the novel Au-capped nanopillar localized surface plasmon resonance (LSPR) chip is introduced. In the nanobiosensing field, LSPR-based sensors have attracted attention because they make it possible to perform label-free detection with high sensitivity. Moreover, a simplified optical system can be employed due to the LSPR signal being obtained at long wavelengths. Generally, electron beam exposure and etching techniques are used to make the nano-structured mold of the LSPR chip. However, these methods are very costly and time-consuming. Thus, novel fabrication methods based on the thermal nanoimprinting technique and use of porous alumina molds are introduced here. Porous alumina molds which have small or large pore size and pitch were successfully fabricated by controlling the anodizing conditions. By using thermal nanoimprinting, the nanopillar structure was transferred onto the cyclo-olefin polymer substrate from porous alumina molds. To obtain the plasmonic surface, a thin layer of gold was generated on the nanopillar structure. After optimization of the size of the alumina pores, the Au cap thickness, the diameter of the nanopillars and bovine serum albumin blocking conditions for high sensitivity detection, immunoglobulin (Ig) G/anti-IgG reactions were measured. As result, the sensitivity achieved using the Au-capped LSPR sensor was 6.7 pM antigen IgG. A noteworthy achievement of our study is the mass production of a high-density gold-based plasmon flexible chip for biosensing applications, paving the way to the commercialization of low-cost, high-sensitivity biosensors.

Keywords Biosensor • Localized surface plasmon resonance • Nanoimprint lithography • Porous alumina

---

K. Yamanaka
Photonics Advanced Research Center (PARC), Graduate School of Engineering, Osaka University, 2-1 Yamadaoka, Suita, Osaka 565-0871, Japan

M. Saito (✉)
Graduate School of Engineering, Osaka University, 2-1 Yamadaoka, Suita, Osaka 565-0871, Japan
e-mail: saitomasato@ap.eng.osaka-u.ac.jp

M.C. Vestergaard et al. (eds.), *Nanobiosensors and Nanobioanalyses*,
DOI 10.1007/978-4-431-55190-4_4

4.1 Introduction

4.1.1 Localized Surface Plasmon Resonance (LSPR)

Nowadays, LSPR-based sensors are studied for development as highly sensitive biosensing platforms. The collective oscillation of free electrons on the surface of noble metals is excited using total reflected light; this is called surface plasmon resonance (SPR). When nanometer-sized structures (not thin film) of noble metals are used, the polarization occurs by collective oscillation and results in a highly enhanced near-field amplitude. This is called local surface plasmon resonance (LSPR). LSPR is generated by the interactions between the incident light and surface electrons of conductive nanoparticles without using, and the adsorption spectrum changes immediately in the interfacial refractive index (RI) of the surrounding medium. The excitation of LSPR by the incident wavelength, where resonance occurs, results in the appearance of intense surface plasmon adsorption bands. This phenomena is strongly dependent on the type of nanomaterial, size, composition and particle–particle distance of the nanoparticles.

The merits of biosensing utilizing LSPR are label-free measurement and the highly sensitive signal obtained at long wavelengths (visible light region). Therefore, a simplified optical system can be employed for highly sensitive detection of biomolecules. Generally, such nanoparticles of noble metal are fabricated by the reduction of silver or gold ion solutions (Mayer et al. 2008; Nath and Chilkoti 2004; Frederix et al. 2003). The reason for using silver or gold is that their absorbance response is in the visible region. In our previous work, a LSPR chip for biosensing was fabricated by immobilizing gold nanoparticles on the surface of an anodic alumina oxide layer on an aluminum substrate (Hiep et al. 2010). However, gold nanoparticles could not be uniformly distributed on the substrate by simply depositing the colloidal Au solution. Therefore, we examined a new fabrication process for making uniformly distributed gold nanoparticles on a substrate by utilizing the thermal nanoimprinting technique, and an enhancement of sensitivity can be expected because of reducing the nonspecific binding area and making a high density of gold on the measurement spot.

4.1.2 Nanoimprint Lithography (NIL)

NIL is one method for fabrication of nano- to micrometer-scale patterns (Buyukserin et al. 2009). By controlling the temperature and pressure, the patterns are translated onto the surface via deformation of the thermoplastic substrate (Chou et al. 1995). The advantages of using NIL for nanoscale patterning are that it is a simple technique with low cost, high throughput and excellent reproducibility (Chou et al. 1997). This makes NIL a promising technique for mass production of nanostructured devices, such as single electron transistors (Guo et al. 1997), highly integrated magnetic memory disks (Wu et al. 1998) and optical devices (wang et al. 1999; Li et al. 2000). Moreover, nanopatterning on polymer film is also possible by NIL

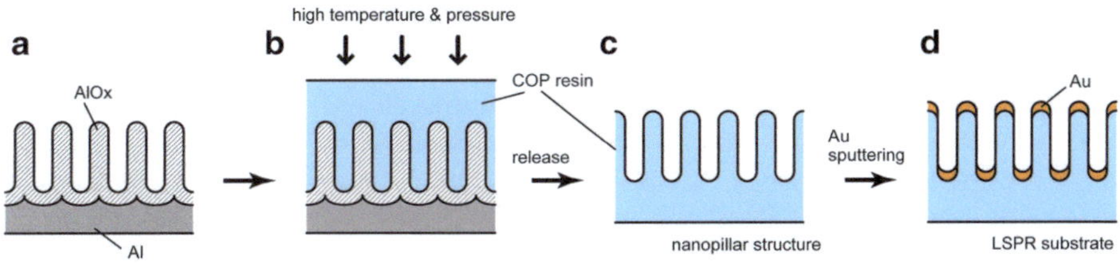

Fig. 4.1 Schematic diagram of the fabrication of LSPR substrates by nanoimprinting. (**a**) Fabrication of porous alumina nanoimprinting mold. (**b**) Thermal nanoimprinting with pressure applied to the COP resin under vacuum. (**c**) Formation of the nanopillar structure by releasing from the alumina porous mold. (**d**) Au-capped nanopillar structure for the LSPR substrate

techniques, and thus it can be applied in the flexible electronics field. There are two major kind of NIL techniques: thermoplastic nanoimprint lithography (T-NIL) and photo nanoimprint lithography (P-NIL) (Fig. 4.1). In T-NIL, the mold is pressed onto the thermoplastic resin substrate, and heated above its glass transition temperature. After being cooled below the glass transition temperature, the mold is separated from the substrate, and thus the pattern is transferred onto the surface. In a P-NIL, the mold is pressed onto the photo-curable liquid resist. During pressing, UV light irradiation is used to cure the resist and it becomes solid, and thus the pattern is transferred onto the surface.

NIL as mentioned above is a very simple technique; therefore, it has great potential for high-throughput nanofabrication. However, the common mold fabrication method such as using electron beam lithography and dry etching is a time-consuming technique; moreover, the cost of mold fabrication using these methods is very high. Therefore, the improvement of the mold fabrication process is important to ensure a high throughput of devices using NIL techniques. Here, we propose self-ordered anodic aluminum oxide (AAO) as a nanostructure array mold for NIL to achieve the NIL mold fabrication process. The porous alumina nanostructure with high density like a honeycomb is able to form by anodization of aluminum; its diameter is from 5 to 450 nm and the hole interval is from 10 to 500 nm. We believe that the low cost, large area, size tuning capability and ease of the anodizing process may enhance the production and marketability of NIL technology.

In this chapter, novel fabrication methods of LSPR biosensors based on the thermal nanoimprinting technique and the use of porous alumina molds as described above are introduced (Saito et al. 2012).

4.2 LSPR Chip Fabrication

First, two types of porous alumina (PA) molds which had small or large pore size and pitch were fabricated by controlling the anodizing conditions. The PA was formed by a two-step anodizing procedure which we developed (Kim et al. 2007). The first anodizing step was to generate an aluminum oxide layer; the polished aluminum plates were held at 40 V (for small-sized pores) or 80 V (for large-sized pores) in a 0.3 M aqueous oxalic acid solution for 60 min. The aluminum oxide layer generated was removed to obtain a concave pattern on the surface by immersing the substrate in an aqueous solution containing phosphoric acid (1.16 % w/v) and chromic acid (5 % w/v) at 333–353 K. The second anodizing step was to generate a PA layer on the concave surface. For small-sized pores, a constant voltage of 40 V was applied to the aluminum substrate for 4 min, and immersed in a 0.23 M H_3PO_4 etching solution for 7.5 min. For large-sized pore fabrication, a voltage of 80 V was applied for 60 min and immersed in the etching solution for 12.5 min. These molds were denoted PA-mold40 (small-sized) or PA-mold80 (large-sized), respectively. To confirm the size and pitch of the fabricated PA, the surface of the molds was observed by using atomic force microscopy (AFM). The pore pitch of PA-mold40 and PA-mold80 was approximately 116.7 and 152 nm, respectively, from the AFM line profiles. Thus, it was demonstrated that the size of PA was successfully controlled by changing the applied voltage and times.

To fabricate the nanopillar structure on the polymer film, T-NIL was carried out. In this study, cyclo-olefin polymer (COP) was selected for the chip material. The material features of COP are glass-like optical clarity, low birefringence, high glass-transition temperature and low water adsorption; therefore, COP is a suitable material for LSPR and T-NIL substrates. At first, the PA molds were immersed in a mold release agent (Optool, Daikin Industries Ltd.). The PA mold and COP film were set onto the T-NIL machine (X-300H, Scivax Corp.), and a pressure of 0.83 MPa was applied immediately at 100 °C. Next, the temperature was increased to 160 °C (the glass transition temperature of COP), and when the temperature achieved 160 °C the pressure was increased to 2 MPa. After 10 min, the temperature was reduced to 100 °C and the pressure was released. The fabricated nanopillar structured COP films were denoted by NP40 (from PA-mold40) and NP80 (from PA-mold80).

For LSPR, Au sputtering on the imprinted COP films was carried out to form the Au-capped nanopillar structure. One cycle of Au sputtering was 10 min and the thickness of the Au layer deposited was 24 nm. To examine the relationship between the absorbance spectra and the thickness of the Au layer, four different thicknesses of 24, 48, 72 and 96 nm were formed by increasing the sputtering cycles.

4.3 Structural Characterization of Au-Capped Nanopillars

First, the Au-capped nanopillar structure fabricated on the COP film was observed using scanning electron microscopy (SEM) to confirm the size and density of the deposited Au cap. As shown in Fig. 4.2, the pillar structure was successfully formed on the COP films using T-NIL utilizing our PA molds, and the density of Au caps on NP80 was higher than those of NP40, because the pore size of NP80 was larger than that of NP40; therefore, the density was also higher in the measurement area. Additionally, the diameter distribution of Au-capped nanopillars was quantified by image analysis based on an ellipsoidal fitting (Fig. 4.3). In the case of NP40, the

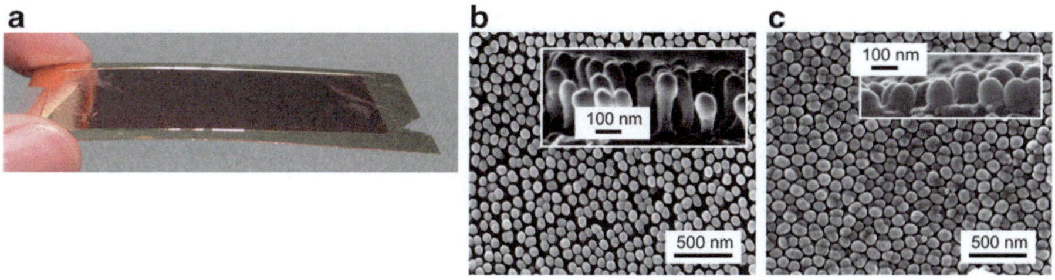

Fig. 4.2 Photo and SEM images of nanopillar chips which were nanoimprinted on COP substrate. (**a**) Nanoimprinted chip after Au sputtering. (**b**) Nanopillar structure NP40 transformed from the PA-mold40. (**c**) Nanopillar structure NP80 transformed from the PA-mold80

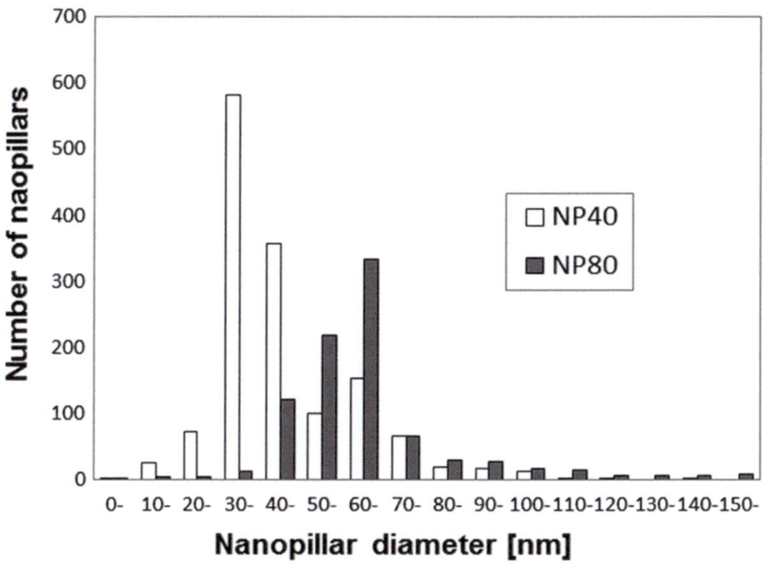

Fig. 4.3 Diameter distribution of Au-capped nanopillar structures

diameter distribution of Au caps was centered around 30.0–39.9 nm, and the fraction of Au-capped surface area over the whole geometric area was 54.3 %. In the case of NP80, the diameter was centered around 60.0–69.9 nm and the fraction was 67.0 %. The gold nanoparticle density reflects the adsorption intensity (Fujiwara et al. 2009), and therefore our fabricated high-density Au-capped structure is expected to be suitable for high-sensitivity detection via LSPR. Moreover, the high-density structure is considered to be able to contribute to development of miniaturized plasmonic biosensors.

4.4 Optical Characterization of Au-Capped Nanopillars

Firstly, the optical properties with various thicknesses of Au cap were examined by measuring the absorbance spectra. The absorbance peak position appeared at 637.7 nm with 24 nm Au caps and shifted to 576.7 nm on increasing the thickness of the Au caps to 48 nm. This result corresponds to the computational model of Zheng et al. in which they observed the dimension-dependent LSPR characteristics of an Au nanodisk (Zheng et al. 2008). As mentioned above, the LSPR peak position was able to be controlled by tuning the sputtering cycles.

Secondly, the effect of nanopillar size for LSPR was analyzed. The peak positions of Au-capped NP40 and NP80 with a constant Au cap thickness were observed in air, as shown in Fig. 4.4a, b, respectively. There was a significant difference in the peak position, that of NP40 being 556.5 nm and NP80 575 nm. This result indicated that the nanopillar size is also a key factor for improving the sensitivity of the LSPR sensor. Additionally, the sensitivity of LSPR to the RI of surrounding environments was

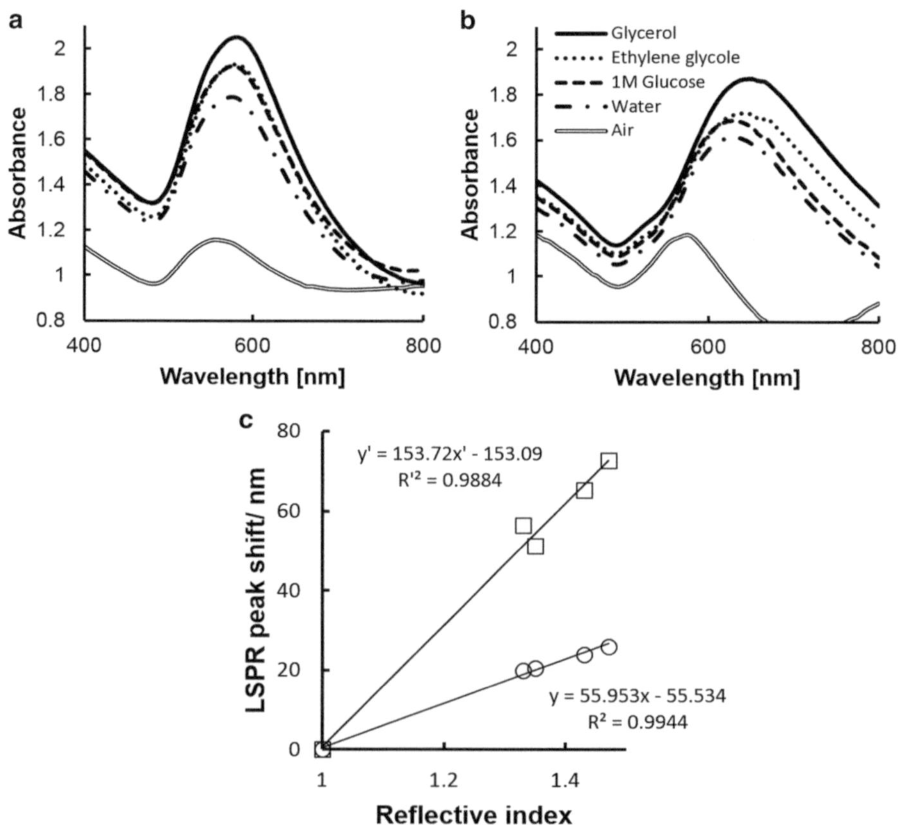

Fig. 4.4 Absorption spectrum of Au-capped nanopillars of (**a**) NP40 and (**b**) NP80, when surrounded by air ($n = 1.0$), water ($n = 1.33$), 1 M glucose ($n = 1.35$), ethylene glycol ($n = 1.43$) and glycerol ($n = 1.47$). (**c**) The RI value dependence of peak shifts for NP40 (*open circle*) and NP80 (*open square*)

evaluated using five solutions with different RIs. As shown in Fig. 4.4, the LSPR peak positions of both Au-capped NP40 and NP80 were red-shifted and the peak intensity was linearly increased depending on the surrounding RI. The slope of the line using NP80 was greater than that using NP40, and therefore the sensitivity of Au-capped NP80 is higher due to its larger nanopillar diameter (Sekhon and Verma 2011).

4.5 LSPR Biosensing on Au-Capped Nanopillar Structures

An antibody–antigen reaction was measured on the Au-capped NP80 for evaluation of the LSPR biosensing applicability of the fabricated Au-capped nanopillar structure. In this study, the immunoglobulin (Ig) G/anti-IgG reaction was measured as a model bioreaction. First, the self-assembled monolayer (SAM) was formed on the Au-capped nanopillar by using 10-carboxyl-1-decanethiol. Next, NHS and EDC were introduced onto the formed SAM to activate the carboxyl groups. After washing with phosphate-buffered saline (PBS) and drying, anti-IgG in PBS was added to the functionalized surface and incubated, followed by washing with PBS containing Tween-20 (PBST) and drying. After immobilization, bovine serum albumin (BSA) was added to inhibit non-specific adsorption (Lahav et al. 2004; Huang et al. 2003; Peterfi and Kocsis 2000; Paulsson et al. 1993). Consequently, the IgG solution was introduced onto the substrate from which the excess BSA was already washed away using PBST and drying. After drying, the absorbance spectrum was measured in air using UV-3600 (Shimadzu). For negative control, C-reactive protein (CRP) was used.

The blocking treatment is an important step in LSPR detection, because the BSA concentration affected the differentiation between positive and negative samples. To optimize the blocking condition for our LSPR sensor, various BSA concentrations (~0.01–1 %) were examined, as shown in Fig. 4.5. As a result, the concentration of 0.1 % BSA most clearly distinguished the positive and negative samples in the LSPR peak shift. When the BSA concentration was lower than 0.1 %, the differences in the peak shift values between the positive and negative sample were decreased due to the nonspecific adsorption which occurred. In the case of 1 % BSA concentration, the error bar became larger in comparison with those of lower concentrations. Due to the overloading of BSA, the PBS was not able to wash away the excess BSA completely, leading to unstable data for each measurement due to the errors resulting from the different RIs of PBS and BSA. Therefore, the concentration of 0.1 % BSA was deemed to be the suitable condition for blocking with our Au-capped nanopillar for detection of the IgG/anti-IgG reaction.

To evaluate the detection limit of the IgG/anti-IgG reaction using our Au-capped nanopillar system, various concentrations (~0–100 μg/mL) of antigen IgG solutions were introduced, and a standard curve generated. As shown in Fig. 4.6, the amount of the LSPR peak shift reflects the concentration of antigen IgG, and its dynamic range was wide. Moreover, the minimum detection limit was 1.0 ng/mL, which is equal to 6.7 pM. This result indicated that biosensing using our novel Au-capped nanopillar structured LSPR chip has a high sensitivity in comparison with the previous similar report (Mayer et al. 2008).

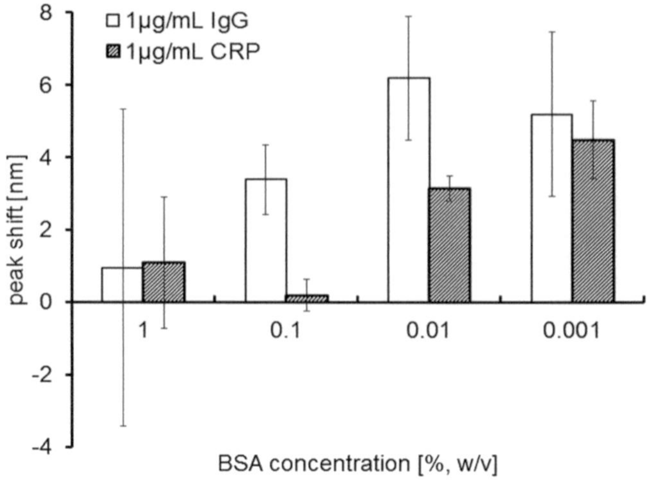

Fig. 4.5 Optimum BSA blocking concentration for antigen detection

Fig. 4.6 Standard curve for IgG/anti-IgG measurement using Au-capped nanopillar LSPR chip. *N* is 1 μg/mL of CRP as a negative control

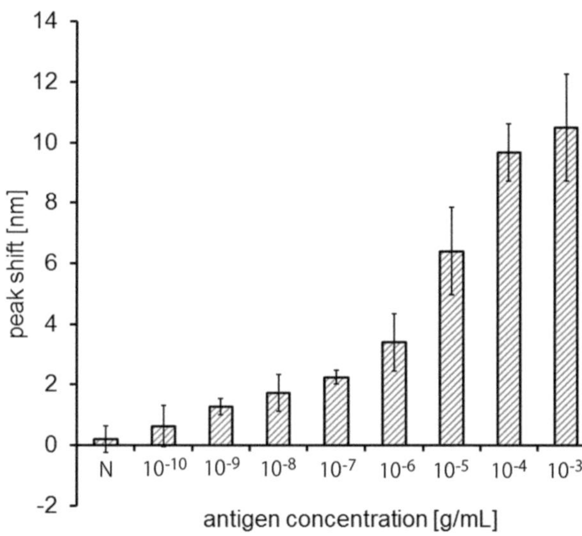

4.6 Conclusions

In this study, we have demonstrated the novel Au-capped nanopillar LSPR chip for detection of anti-antigen reactions. Those results showed the high sensitivity and the wide dynamic range detection possible from our LSPR chip which was optimized for Au cap thickness, nanopillar size and blocking condition. Label-free detection has an advantage for point-of-care testing as there is no need for any treatment of the sample such as tagging using fluorescence; therefore, rapid measurement becomes possible. Moreover, our sensor fabrication process is simplified and low-cost as it is based on the combination of thermal nanoimprinting techniques and porous alumina molds; therefore, it is suitable for mass production. With these advantages, our novel Au-capped nanopillar plasmonic sensor is ready for the transition from academic interest to a commercialized system for practical use in diagnostic applications. Future work will involve combination with a micro-TAS device for development of automated detection systems including sample injection and measurement of target biomolecules.

References

Buyukserin F, Aryal M, Gao J et al (2009) Fabrication of polymeric nanorods using bilayer nanoimprint lithography. Small 5:1632–1634

Chou SY, Krauss PR, Renstrom PJ (1995) Imprint of sub-25 nm vias and trenches in polymers. Appl Phys Lett 67:3114–3116

Chou SY, Krauss PR, Zhang W et al (1997) Sub-10 nm imprint lithography and applications. J Vac Sci Technol B 15:2897–2904

Frederix F, Friedt JM, Choi KH et al (2003) Biosensing based on light absorption of nanoscaled gold and silver particles. Anal Chem 75:6894–6900

Fujiwara K, Kasaya H, Ogawa N (2009) Gold nanoparticle monolayer formation on a chemically modified glass surface. Anal Sci 25:241–248

Guo L, Leobandung E, Zhuang L et al (1997) Fabrication and characterization of room temperature silicon single electron memory. J Vac Sci Technol B 15:2840–2843

Hiep HM, Yoshikawa H, Tamiya E (2010) Interference localized surface plasmon resonance nanosensor tailored for the detection of specific biomolecular interactions. Anal Chem 82:1221–1227

Huang TT, Sturgis J, Gomez R et al (2003) Composite surface for blocking bacterial adsorption on protein biochips. Biotechnol Bioeng 81:618–624

Kim DK, Kerman K, Saito M et al (2007) Label-free DNA biosensor based on localized surface plasmon resonance coupled with interferometry. Anal Chem 79(5):1855–1864

Lahav M, Vaskevich A, Rubinstein I (2004) Biological sensing using transmission surface plasmon resonance spectroscopy. Langmuir 20:7365–7367

Li M, Wang J, Zhuang L, Chou SY (2000) Fabrication of circular optical structures with a 20 nm minimum feature size using nanoimprint lithography. Appl Phys Lett 76:673–675

Mayer KM, Lee S, Liao H et al (2008) A label-free immunoassay based upon localized surface plasmon resonance of gold nanorods. ACS Nano 2:687–692

Nath N, Chilkoti A (2004) Label-free biosensing by surface plasmon resonance of nanoparticles on glass: optimization of nanoparticle size. Anal Chem 76:5370–5378

Paulsson M, Kober M, Freij-Larsson C et al (1993) Adhesion of staphylococci to chemically modified and native polymers, and the influence of preadsorbed fibronectin, vitronectin and fibrinogen. Biomaterials 14:845–853

Peterfi Z, Kocsis B (2000) Comparison of blocking agents for an ELISA for LPS. J Immunoassay 21:341–354

Saito M, Kitamura A, Murahashi M et al (2012) Novel gold-capped nanopillars imprinted on a polymer film for highly sensitive plasmonic biosensing. Anal Chem 84:5494–5500

Sekhon JS, Verma SS (2011) Refractive index sensitivity analysis of Ag, Au, and Cu nanoparticles. Plasmonics 6:311–317

Wang J, Schablitsky S, Yu Z et al (1999) Fabrication of a new broadband waveguide polarizer with a double-layer 190 nm period metal-gratings using nanoimprint lithography. J Vac Sci Technol B 17:2957–2960
Wu W, Cui B, Sun XY et al (1998) Large area high density quantized magnetic disks fabricated using nanoimprint lithography. J Vac Sci Technol B 16:3825–3829
Zheng YB, Juluri BK, Mao X et al (2008) Systematic investigation of localized surface plasmon resonance of long-range orded Au nanodisk arrays. J Appl Phys 103:014308

Chapter 5
Nanoparticle Biosensing with Interferometric Reflectance Imaging

Derin Sevenler, Neşe Lortlar Ünlü, and M. Selim Ünlü

Abstract Interferometric reflectance imaging is a biosensing modality that performs digital detection of individual synthetic and biological nanoparticles in a multiplexed microarray format. The size of nanoparticles of known optical properties is measured indirectly, by calculating the particle scattering cross-section at specific wavelengths. This has enabled the development of label-free, digital detection of whole virions in a multiplexed assay with unprecedented sensitivity and specificity. DNA and protein microarrays using small metallic nanoparticle labels instead of fluorescence have also demonstrated high sensitivity to dilute analytes in biological solutions such as serum and unprocessed whole blood. The optical design is based on off-the-shelf components and the resulting instrument is inexpensive and compact. Thus the sensing platform has many potential clinical applications in screening, diagnostics and monitoring. In this chapter we discuss the principles that have led to the development of this technique and its applications in assay development and biophysical analysis, and our progress in developing this technology into a rapid and cost-effective point-of-care instrument.

Keywords Microarray • Optical biosensors • Single nanoparticle detection

D. Sevenler • N. Lortlar Ünlü
Department of Biomedical Engineering, Boston University, Boston, USA

M.S. Ünlü (✉)
Department of Biomedical Engineering, Boston University, Boston, USA

Department of Electrical and Computer Engineering, Boston University, Boston, USA
e-mail: selim@bu.edu

M.C. Vestergaard et al. (eds.), *Nanobiosensors and Nanobioanalyses*,
DOI 10.1007/978-4-431-55190-4_5

5.1 Introduction

This chapter discusses two different methods by which optical interference can be used to improve the performance of microarray assays. In some respects, the gains are afforded by such margins as to allow microarray assays to be used to address qualitatively different problems in biological research and medical diagnostics. We will discuss briefly how a thin film interference reflectance microarray substrate can be used to measure biomolecule accumulation as a change in its spectral reflectance curve, thereby enabling calibration of fluorescence readouts or replacing fluorescent labels entirely. As a label-free technique, this has enabled experiments measuring the native binding kinetics of DNA and proteins in a massively multiplexed fashion. We shall also discuss in detail how such a thin film substrate can be used to implement an entirely different physical principle—homodyne interferometry—to perform detection of individual biological or metallic nanoparticles, enabling the detection readout to be limited only by Poisson shot noise. This detector has sensitivity and selectivity limited in principle only by the thermodynamic binding interaction between the various molecular species of interest, allowing measurements to be performed at the theoretical performance limit of solid-phase, affinity-based biosensors. We will discuss how these techniques have been used to measure scarce protein or DNA detection for cancer diagnostics, bacterial drug resistance screening or allergy testing, or detection of deadly contagious virions at presymptomatic concentrations. Together, these two optical interference technologies provide a comprehensive platform that is based on simple physical principles, and yet is capable of a wide variety of microarray assays for diagnostics and biomedical research.

5.2 Principles

5.2.1 The Microarray

A microarray consists of a solid surface that is patterned with a microscale array of molecular probes. These probes are typically designed to specifically bind to a single biomolecule of interest, and are commonly either single-stranded DNA or antibody. These probes are deposited in "spots" usually ~100 μm across, so hundreds to thousands of different spots can be placed within a single square-millimeter sensor area. Microarrays exist that allow massively multiplexed assays to be performed on small volumes. Traditionally, target molecules bound to the spots are measured with fluorescently labeled reporter molecules, which also specifically bind to targets of interest. In the case of DNA microarrays, this usually consists of a DNA oligonucleotide functionalized to a small fluorescent molecule that is complementary to an adjacent region of a particular captured nucleic acid.

5.2.2 Microarray Readout with Thin Film Interference Reflectance Imaging

The idea of using thin film interference in a biochemical sensor has been around for decades (Brecht et al. 1993). Consider a transparent film thinner than the wavelength of light with known optical properties that has been deposited on a transparent surface with a different refractive index, all submerged in a surrounding medium of likewise known optical properties. Light rays traversing this layered structure in the perpendicular direction will reflect off of the two interfaces, at ratios that are proportional to the index mismatch at each interface in a manner described by the Fresnel equations. The deposition of biomolecules onto the film increases the optical path length difference between the reference reflection at the substrate–film interface and the reflection at the medium–biofilm interface. A given optical path length difference has a characteristic spectral curve, as some wavelengths interfere constructively and others destructively. Changing the optical path length shifts this spectral reflectance curve in a predictable manner. Thin film interference reflectance spectroscopy is the technique by which sub-nanometer scale changes in the optical path length due to biomolecule accumulation can be measured by inspecting the reflectance spectra of the microarray.

The interferometric reflectance imaging sensor (IRIS) layered substrate consists of a polished silicon wafer with 100 nm of thermally grown silicon dioxide. Traditionally, the spectral reflectance of a surface is measured by illuminating with a broadband light source and collecting all the reflected light using a spectrometer. This method permits monitoring of the entire spectral reflectance simultaneously, but excludes any spatial information about the sample. Instead, a tunable laser may be used in conjunction with a monochrome CCD, whereby the illumination is limited to a narrow band of wavelengths and spectroscopic information is collected serially with high spatial resolution. This second method is the working principle behind the spectral reflectance imaging biosensor (Özkumur et al. 2009). Imaging laser spectral reflectance is challenging, due to spatial artifacts in the laser illumination due to its high coherence (speckle) and temporal fluctuations in laser output (necessitating external monitoring of laser power). Four narrowband LEDs may be used instead to interrogate the reflectance spectra with similar precision (Fig. 5.1) (Daaboul et al. 2011). Knowledge of the illumination spectra of the LEDs allows the spectrum at each pixel in the image to be fitted with picometer resolution.

IRIS can perform detection of biomolecules directly, without any molecular tagging. One application of this is in calibrating microarray spots before they are used. The density of probe molecules at each spot is often inconsistent between spots on the same chip, as well as between chips from different batch spotting runs, and this source of variability leads to uncertainty in quantifying target concentrations from a fluorescence measurement. This uncertainty is not acceptable in applications such as gene expression profiling, where quantification is of vital interest. This challenge is traditionally addressed by spotting several replicate spots on each chip and including on-chip positive controls. Using IRIS, the molecular probe densities of each spot

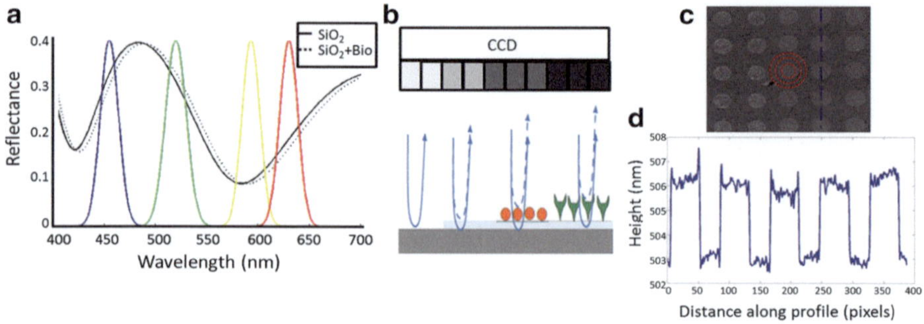

Fig. 5.1 Label-free microarray measurement with IRIS. (**a–b**) The spectral response of the Si–SiO_2 surface is sampled by four LEDs, and the shift in the spectral reflectance curves due to biomolecule adsorption and increased optical path length difference is measured and converted to a "height" of deposited biomolecules. (**c–d**) An IRIS image, with the calculated surface height (corresponding to probe immobilization) along the *dashed line*. Adapted with permission from Daaboul et al. (2011)

may be measured before the chip is used, improving repeatability of fluorescence measurements (Reddington et al. 2013a). Another application involves removing fluorescence reporting entirely, and quantifying the amount of biomolecule accumulation label-free (Daaboul et al. 2011) in a dynamic context. IRIS has been used to measure the hybridization and dehybridization kinetics of spotted DNA, for example, and the association and dissociation rates of proteins bound to antibody spots, by performing measurements in real time. This was used to perform sensitive detection of single nucleotide polymorphisms in DNA (Özkumur et al. 2010a, b). Such dynamic experiments cannot be performed with fluorescence-based systems since unbound fluorophores in the flow cell will saturate the detector. Finally, only label-free detection can directly quantify the surface capture rate of a target molecule, which depends on not only the affinity but also the diffusion rate of that molecule. IRIS can perform all of these measurements in a highly multiplexed format, allowing many different probe–target interactions to be investigated in parallel.

5.2.3 Metallic and Biological Nanoparticle Detection on Microarrays

Fluorescent reporters have the disadvantage of being susceptible to photobleaching and therefore are difficult to detect in low numbers. Alternatives for overcoming this lower sensitivity limit include using other labels such as radioactive nanoparticles (Posner et al. 2007) or gold nanoparticles, which can be enhanced before detection with a silver ion solution (Taton et al. 2000). In this technique, dissolved silver ions reduce onto the gold nanoparticles, forming nanoparticle-mediated silver deposition that can be quantified with a simple colorimetric measurement using a flatbed scanner. In contrast, the Single Particle IRIS (SP-IRIS) instrument is capable of detecting individual small gold nanoparticles (of order ~40 nm) without a silver enhancement step. Additionally, it may also be used to detect biological nanoparticles such as whole virus particles.

Such small nanoparticles are usually only detectable with specialized sensitive optical microscopy techniques. Nanoparticles much smaller than the microscope point spread function can be approximated as point scatterers, and therefore the measured intensity of light scattered from the particle back towards the detector is proportional to the optical properties and size of the particle, as well as the optical properties of the medium surrounding the particle:

$$I_{\text{det}} \propto \sigma_{\text{sca}} \propto \left| \frac{\epsilon_{\text{p}} - \epsilon_{\text{m}}}{\epsilon_{\text{p}} + 2\epsilon_{\text{m}}} \right|^2 R^6$$

Here, E_{sca} corresponds to the scattered light electric field and σ_{sca} corresponds to the particle's scattering cross-section. The scattering cross-section depends strongly on the dielectric index of the nanoparticle ϵ_p and of the medium ϵ_m and even more strongly on the particle size indicated by the radius R. The trouble is that the strong size dependence puts a rather hard limit on the size of the smallest particle that can be detected due to scattering alone. However, this limit may be overcome by mixing the scattered electric field with a stronger reference field before it reaches the detector. Then, the intensity of the detector depends on both field intensities, which interfere according to their phase difference θ_{rs}:

$$I_{\text{det}} \propto \left| E_{\text{ref}} + E_{\text{sca}} \right|^2$$

$$I_{\text{det}} \propto \left| E_{\text{ref}} \right|^2 + \left| E_{\text{sca}} \right|^2 + 2\left| E_{\text{ref}} \right| \left| E_{\text{sca}} \right| \cos \theta_{\text{rs}}$$

When the scattering-only term $|E_{sca}|^2$ is small enough for this technique to be applicable, it is negligibly small compared to the constant background term $|E_{ref}|^2$ and the phase term, which is proportional to R^3. The phase difference between the two channels θ_{rs} is a function of the path length difference between the scattered light from the nanoparticle and the reference light.

Optical heterodyning has been used to push the sensitivity limits of nanoparticle detection and characterization with light. For example, Hong and colleagues combined a modified Mach–Zehnder interferometer with cross-polarization optics and heterodyne detection, and were able to demonstrate the detection of gold nanoparticles immobilized on a glass substrate with a diameter of just 5 nm (Hong et al. 2011). This impressive measurement exploits the fact that light scattered by small nanoparticles has a slightly different polarization than the incident light, which is exacerbated by partial polarization axis conversion of the high numerical aperture (NA) illumination (Juskaitis et al. 1997). By combining polarization optics to isolate polarization-shifted light scattered from the nanoparticles, they achieved a shot noise limited optical signal from the 5 nm nanoparticles. This low-intensity signal was then detected using a heterodyne interferometry scheme, where the weak optical signal was interfered with a modulated reference beam, and then monitored with lock-in amplification. While the sensitivity of heterodyne detection is unsurpassed, homodyne detection does not require such complex components as a phase modulator and lock-in amplifier. Homodyne detection of individual dielectric nanoparticles using a Michelson interferometer has been demonstrated for detecting particles as they flowed through a microchannel, and was capable of measuring their size in real time (Mitra and Novotny 2013).

SP-IRIS utilizes the same thin film Si–SiO_2 substrate described earlier as a common-path interferometer capable of homodyne detection of single nanoparticles across a large field of view. The reflection off the Si–SiO_2 interface serves as the reference arm, which interferes with scattered light from particles on the SiO_2 surface (Fig. 5.2). In order to have a higher spatial resolution, a high-NA objective (0.8) is used. The inclusion of both low-NA and high-NA angles in this system is a potential pitfall, since the path length difference for those different angles will be different and will therefore have different strengths, since the scattered light all has the same phase. To minimize this effect, the oxide film thickness was reduced to 100 nm.

Notably, the illumination source does not need to have a long coherence length because the optical path length difference is so small. This allows a narrowband LED to be used instead of a laser, sidestepping the challenges associated with using coherent sources discussed earlier. The coherence length of an illumination source in the visible band is given by the equation

$$L = \frac{2\ln(2)}{n\pi}\frac{\lambda^2}{\Delta\lambda}$$

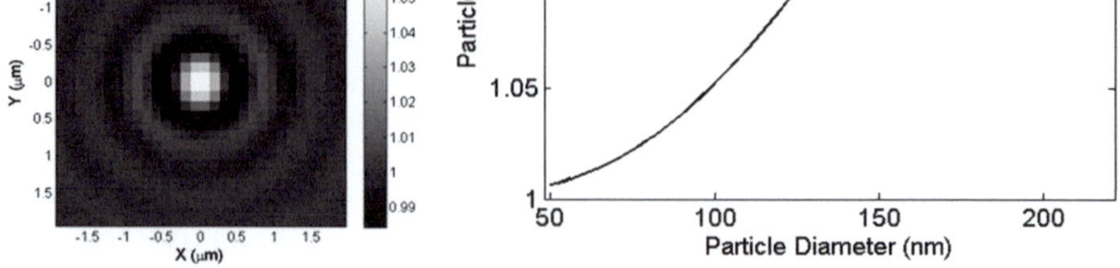

Fig. 5.2 Optical response of a nanoparticle on the SP-IRIS substrate. (**a**) The Si–SiO_2 substrate behaves as a thin film interfering substrate. (**b**) Light scattered from a nanoparticle interferes with the reflection, forming a diffraction-limited response resembling an Airy disk. (**c**) Simulated particle contrast (peak intensity divided by the reference intensity) as a function of particle diameter, for a dielectric nanoparticle on a 100 nm SiO_2 film with a 530 nm (*green*) peak wavelength illumination source

Here, λ is the center wavelength of the illumination (530 nm), $\Delta\lambda$ is the full width half max (FWHM) bandwidth of the illumination spectrum (about 40 nm), and n is the refractive index of the medium (SiO_2, about $n = 1.5$). The coherence length L of such an illumination source is approximately 2.0 μm—much greater than the optical path length difference. This demonstrates that a narrowband high-power LED with these properties is suitable for interference reflectance imaging.

5.3 Applications I: Metallic Nanoparticle Labels for Multiplexed Digital Molecular Detection

For biosensing applications which require greater sensitivity than that afforded currently with fluorescent molecule labels, nanoparticle labels have several advantages—their optical and chemical stability gives them greater signal-to-noise ratio and efficiency in the single-molecule (or single-particle) detection regime (Seydack 2005). Digital nanoparticle detectors provide a method for microarray assays to perform measurements in the single-molecule regime. Techniques for the effective functionalization of nanoparticles with proteins or nucleic acids have been well-studied (Hill and Mirkin 2006), so functionalization of nanoparticles with any specific probe molecule is now routine.

Metallic nanoparticles have nano-optical properties that are especially advantageous for this application, as compared to dielectric nanoparticles. The spectral response of metallic nanoparticles is characterized by strong scattering and absorption at a particular wavelength (or wavelengths) that has been shown to be dependent on their shape and size and on the dielectric material properties of both the nanoparticle and the surrounding medium (Mock et al. 2002; Eustis and El-Sayed 2006; Willets and Van Duyne 2007). This effect is called local surface plasmon resonance, or LSPR, and occurs when the oscillating electric field of the incident light causes coherent oscillations in the free conducting electrons on the surface of the material, thereby causing a resonant oscillating electric dipole (Fig. 5.3).

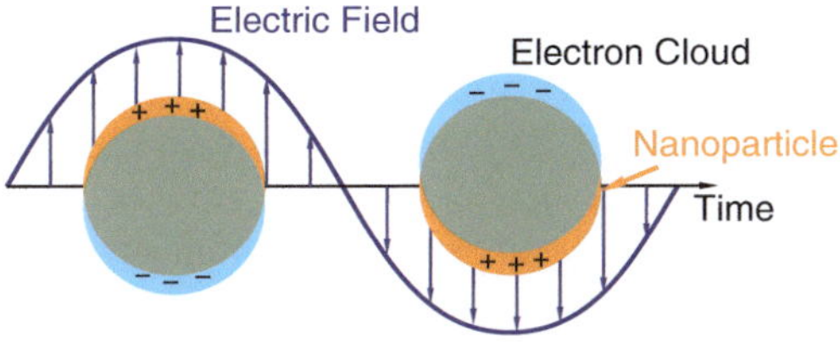

Fig. 5.3 Schematic of localized surface plasmon resonance (LSPR). Coherent oscillations in the free electrons of the material may be induced by the oscillating electric field of the incident light. This induced oscillating electric dipole has greatly increased scattering and absorption of its resonant wavelength as compared to dielectric nanoparticles of similar size, making it a superior label for use with single-nanoparticle detection assays

Due to LSPR, the non-propagating electromagnetic field amplitude (near-field) can be orders of magnitude greater than that of the incident illumination. At resonance, the optical scattering cross-section is greatly enhanced:

$$\left|\frac{\epsilon_{\mathrm{p}}\left(\lambda_{\mathrm{SPR}}\right)-\epsilon_{\mathrm{m}}}{\epsilon_{\mathrm{p}}\left(\lambda_{\mathrm{SPR}}\right)+2\epsilon_{\mathrm{m}}}\right| \gg 1$$

The LSPR spectrum of a particular nanoparticle depends strongly on the dielectric properties of the local surrounding medium, relative to the strength of the non-propagating electric field strength (which decays exponentially with distance from the interface). Changes in the dielectric properties within ~150 nm of the surface of the particle cause changes in the LSPR spectrum, which may be detected via spectroscopy—this is the working principle of surface plasmon resonance (SPR) biosensors. When using metallic nanoparticles as molecular reporters, however, this extra information is superfluous—it is sufficient to sensitively detect surface-bound nanoparticles without characterizing their near-field optical properties.

For detection to occur, the nanoparticle must be tethered to the sensor surface. Therefore, the analyte must have two binding domains: one to specifically bind the nanoparticle label to the analyte, and another to tether the analyte to the sensor surface. An analyte-specific probe molecule is required for each binding step. In the case where the analyte is a specific single-stranded nucleic acid sequence, primer sequences complementary to opposite ends of the target sequence are usually chosen. In the case where the analyte is a protein with multiple independent epitopes, two different antibodies are typically used.

In theory, the order in which these binding steps occur would not have an effect on the overall sensitivity or selectivity of the assay if the epitopes were sufficiently independent and stable and steric effects were negligible. In practice, however, this is often not the case, and the order of binding steps is therefore usually optimized. When the analyte is first bound to the sensor surface and then labeled, it is said to be a *heterogeneous detection* format. In contrast, *homogeneous detection* is performed by first mixing the sample solution with the labels and then binding the labeled analyte to the sensor surface (Fig. 5.4). The diffusivity of the label, which scales inversely with radius, is an important factor that can limit assay performance, especially when homogeneous detection is used with a surface-capture modality. This puts a lower limit on the optical cross-section of the labels to particles visible with conventional methods (Nicewarner-Peña et al. 2001).

Monroe and colleagues detail the development of a multiplexed protein assay that incorporates SP-IRIS in order to enable the use of gold nanoparticle labels with a radius $\leq$40 nm, of the same order of size as the protein itself (hydrodynamic radius ~5 nm) natively, without silver enhancement (Monroe et al. 2013). The authors demonstrated a limit of detection of allergen-specific immunoglobulin E (IgE) of about 60 attomolar (10^{-18} M) in human serum samples, and 500 attomolar in human whole blood samples (Fig. 5.5). They showed that in principle an even lower limit of detection is possible by simply increasing the surface area of the analyzed probes. Nevertheless, clinically relevant concentrations of most protein biomarkers for a

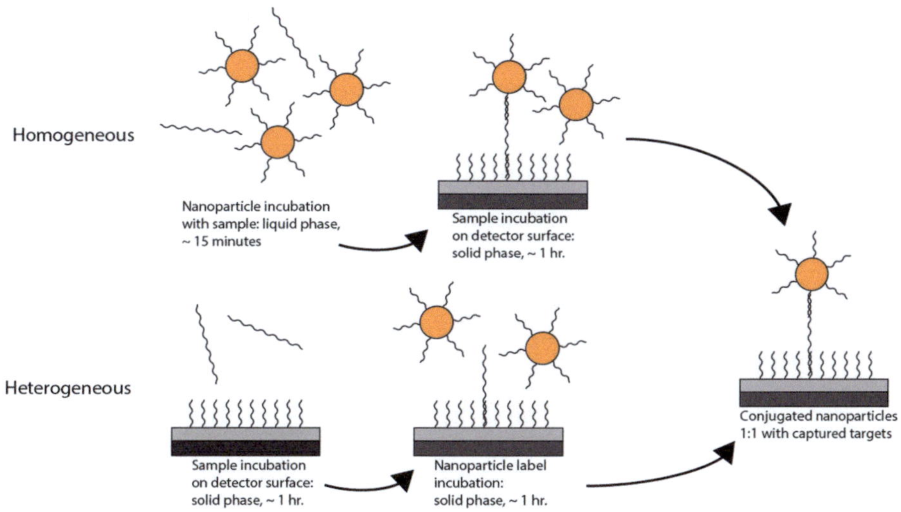

Fig. 5.4 Flow chart of homogeneous and heterogeneous detection of a target single-stranded nucleic acid. Homogeneous detection begins with hybridization of the target with the functionalized nanoparticle probes, while heterogeneous detection begins with hybridization of the target molecule to the microarray surface. The final result is the same in both processes, so detection using IRIS is identical. The terms solid phase and liquid phase refer to the regimes of diffusion under consideration—solid phase diffusion means simply that the molecule or complex must diffuse to a solid surface rather than interact with another solvent (liquid phase). Solid phase diffusion follows slower kinetics so is generally slower than liquid-phase diffusion, which has implications for assay design

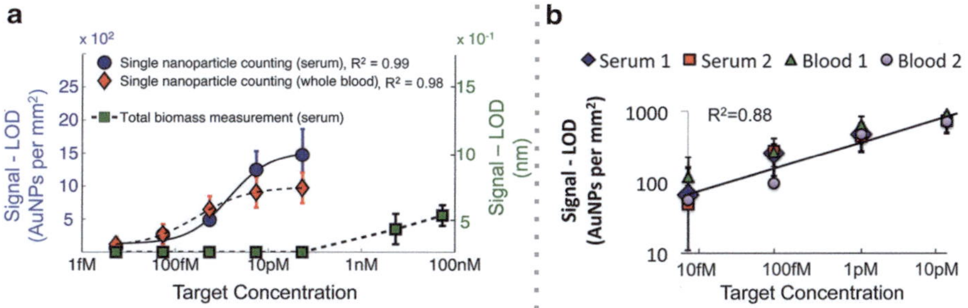

Fig. 5.5 (**a**) The signal above the limit of detection for IgE is shown for detection in both unprocessed whole blood and serum using nanoparticle labels (*left vertical axis*), and is compared to that of label-free overall biomass measurement using IRIS (*right vertical axis*). The dynamic range of SP-IRIS spans from the limit of detection of IRIS to 1 femtomolar, demonstrating that they may be used easily together, and the two techniques have a very high dynamic range as well as very high sensitivity. (**b**) Signal minus background for two serum and two whole blood experiments correlate well, demonstrating assay repeatability. Adapted with permission from Monroe et al. (2013)

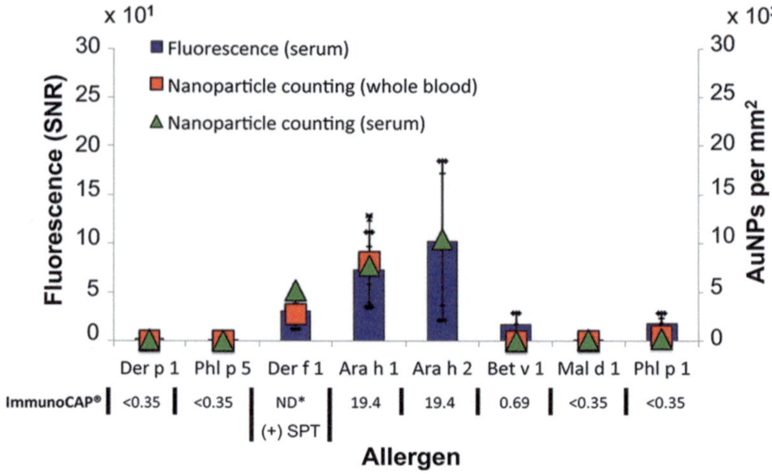

Fig. 5.6 Cross-validation of the SP-IRIS multiplexed allergen immunoassay with Phadia ImmunoCAP (fluorescence), the current state-of-the-art allergen immunoassay. 50 μL serum and unprocessed whole blood were tested with SP-IRIS, and found to correlate well with the fluorescence ImmunoCap reading of serum samples. Adapted with permission from Monroe et al. (2013)

range of diseases fall within the range of 0.01–100 picomolar (10^{-12} M), but commonly exist at the femtomolar range (10^{-15} M) as well (Rusling et al. 2010; Serra 2011). The authors demonstrate assay multiplexing and cross-validation with a current gold standard technique by testing fingerprick volumes (50 μL) of whole blood and serum from four patients across a panel of eight common allergens (the target molecules are in this case antibodies specific for those allergens) (Fig. 5.6).

5.4 Applications II: Label-Free, Digital Detection of Whole Virus Particles

SP-IRIS may also be used to detect single dielectric (i.e., nonmetallic) nanoparticles of biological or synthetic origin. One important biological nanoparticle that can be detected with SP-IRIS is a virus. Demand for rapid and simple virus sensing technologies for providing effective and specific viral diagnostics has been driven by events involving influenza pandemics, development and intentional dissemination of highly virulent biowarfare agents, and deadly global outbreaks of viral hemorrhagic fevers. In order to be effective, rapid tests that can provide an answer at the point of need are required. The gold standard technologies for virus detection and identification consist of immunoassay-based protein detection such as ELISA or polymerase chain reaction-based detection of specific nucleic acids (Amano and Cheng 2005)—however, sample preparation requirements and protocol complexity are still barriers to their application in many areas. Daaboul and colleagues detail label-free detection of individual whole intact virions with SP-IRIS (Daaboul et al. 2014). Previous experiments had demonstrated that measurements of the total biomass accumulation on a multiplexed label-free virus immunoassay could be performed using an IRIS substrate and low-NA objective (Lopez et al. 2011). The authors demonstrate detection of virus from complex solutions of fetal bovine serum containing *E. coli*, without the need for sample preparation.

Biological nanoparticles are difficult to detect optically in comparison to metallic nanoparticles because of the two-fold lack of a plasmon resonance and low index contrast to the surrounding medium. In quasi-static theory, the interaction of a dielectric nanoparticle with a light wave causes an induced dipole, whose strength is proportional to the particle's shape, composition and size, as discussed earlier. Using SP-IRIS, surface-bound individual virions ~100 nm in diameter are detected by identifying local maxima that fall within the point spread function of the optical system and determining their size from the local maxima peak intensity and a forward model (Fig. 5.7). Daaboul and colleagues demonstrate a sensitivity limit of detection of about 8×10^4 plaque-forming units/mL, which is similar to or better than the reported sensitivity of other antibody-based viral diagnostic techniques (Marzi et al. 2011). To demonstrate the robustness and specificity of this technique, the authors performed experiments in samples contaminated with *E. coli*, or in samples of virus-spiked whole blood or serum. Remarkably, the detection limit remained the same for all of these conditions at 8×10^4 pfu/mL, despite the presence of large amounts of various macromolecules and biological nanoparticles generally found in those conditions.

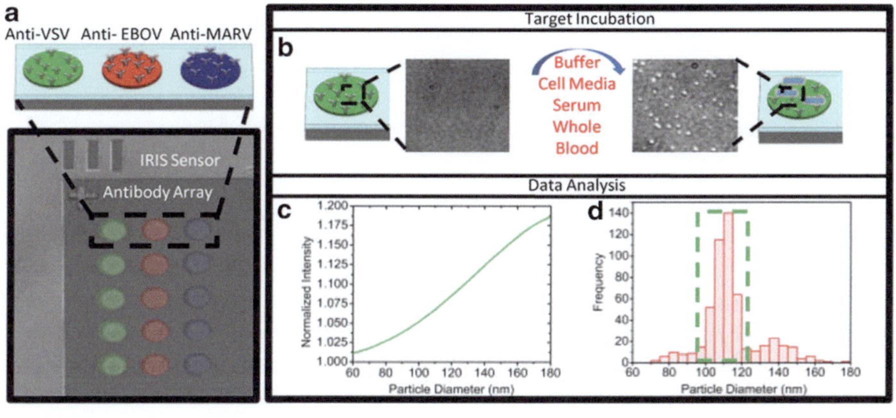

Fig. 5.7 Detection and identification of single virions using a multiplexed microarray format. (**a**) Representative image of anti-VSV, anti-Ebola-VSV, and anti-Marburg-VSV antibody microarray spots. (**b**) Pre- and post-incubation SP-IRIS images showing virus accumulation. This assay may be performed in various complex media. (**c**) A forward model of peak particle intensity as a function of particle diameter was verified using polystyrene beads of known size and is used to determine the size of unknown particles. (**d**) A histogram of particle sizes, calculated using the forward model, allows particles outside the expected range of VSV (*green rectangle*) to be ignored. Virus capture is calculated as the number of particles found within this size window in the post-incubation image minus the number of particles of that size observed in the pre-incubation image. Adapted with permission from Daaboul et al. (2014)

The ability to perform sensitive and specific detection of deadly viruses with a simple and robust platform has clinical significance today. SP-IRIS can perform such measurements robustly in complex media by utilizing both affinity-based immobilization (providing molecular specificity) and size-based identification and discrimination of virions (providing structural specificity), to directly measure the presence of the clinically relevant disease vector.

5.5 Translation from the Lab to the Clinic

Chief among the advantages afforded by the IRIS substrate is that it allows both probe quantification and highly sensitive nanoparticle detection on the same substrate, which is readily transferable to clinical use. Furthermore, the ability to calibrate IRIS microarray slides before use has important implications for fluorescence-based assays, since spot morphology and probe density must be taken into account to avoid large inter-chip variability (Peterson et al. 2001). Reddington and colleagues detail recent efforts to develop an automated, label-free virus detection system utilizing SP-IRIS (Reddington et al. 2013a, b). While significant advances have been made for the translation of interferometric biosensing based on diffraction gratings or photonic crystals to clinical use (Cunningham 2010), single-particle modalities are decidedly more difficult, due to increased tolerances required by a higher numerical aperture objective. Nevertheless, the authors describe a prototype system that is capable of automated scanning, spot identification, focusing, particle detection and analysis (Reddington et al. 2013a, b).

5.5.1 Image Processing for Single Nanoparticle Detection

When metallic nanoparticles are used as specific labels for sensitive molecular assays, nanoparticle detection and classification do not necessitate their exhaustive characterization. Since the optical properties of the detection label are known, data analysis requires only that these particles be identified and counted, while ignoring particles whose optical response does not match what is expected. The nanoparticle population present in a particular SP-IRIS image is analyzed in the following manner: (1) key points are detected and localized using the scale-invariant feature transform, (2) anomalous regions unlikely to contain nanoparticles of interest are detected using morphological operations and key points within those regions are discarded, (3) an appropriate Gaussian kernel is cross-correlated with image region of each key point, and key points with a low correlation coefficient (i.e., those which are not well approximated by a 2D Gaussian) are discarded, (4) the approximate size of each particle is calculated from its image intensity with a forward model that assumes the material properties of the nanoparticles of interest, and (5) nanoparticles whose approximated size is outside of the range of expected particles are discarded (Fig. 5.7). These steps are discussed below.

Scale-invariant feature transform (SIFT) is a feature detection algorithm used in applications like object recognition and panorama image stitching, the principle of which involves obtaining so-called "key points" in an image from the extrema of an assembled difference-of-Gaussians pyramid (Burt and Adelson 1983; Crowley and Stern 1984; Lowe 1999). Although not originally developed for this purpose, this algorithm performs exceptionally well at rapidly detecting and localizing Gaussian-like point spread functions in a microscope image with sub-pixel precision.

SIFT inclusively captures not only local maxima but other salient features in the image as well—for example, the high-contrast edges of the microarray spot. Furthermore, poor spot morphology may sometimes preclude nanoparticle detection yet result in an increased number of key points. Spot morphology may deviate from optimal during preparation (the probe molecules may distribute into an oblong, spatially heterogeneous or coffee-ring pattern) or later during the assay (salt precipitate or other contaminants can deposit onto the spot). Often, these defects obscure only a small portion of the spot's useable surface. However, they are often feature-rich and therefore contain a disproportionally high density of key points and are therefore an undesirable source of noise. This is mitigated in two steps. First, morphological operations are used to detect regions of the spot that are unexpectedly bright or dark or contain unexpected edges, and key points within those regions are discarded. Second, the intensity distribution of the image domains about each key point are cross-correlated with a two-dimensional Gaussian kernel, in order to distinguish key points which correspond to point sources and which ones do not. The Gaussian kernel size is chosen as an approximation of the point spread function of the optical system. Key points with a sufficiently high correlation coefficient are identified as nanoparticles—those with small correlation coefficients are discarded. One limitation of this technique is that two nanoparticles closer than the diffraction limit of the optical system will provide an optical measurement that has a low Gaussian correlation coefficient, and both key points will be discarded. This is the principal limitation of the dynamic range.

The cross-correlation function is amplitude-invariant. However, a nanoparticle of known size and material has a predictable local contrast, i.e., peak intensity with respect to the immediate local background. Therefore, nanoparticles are finally filtered based on their size and material properties (or, equivalently, local contrast) such that nanoparticles smaller or larger than a specified size window are discarded. The number of particles within this window is the detected signal from this image.

5.6 Conclusion

In this chapter, we presented unique properties of the interferometric reflectance imaging sensor (IRIS) and its applications in assay development and biophysical analysis. We also discussed the prototyping efforts in developing the IRIS technology into a rapid and cost-effective point-of-care instrument. In high-resolution optical or single-particle detection modality, IRIS is utilized for multiplexed digital molecular detection using metallic nanoparticle labels. We believe that the ability to detect and study individual nanoparticles will have a major impact on the advancement of many fields in biosciences and biotechnology. For instance, studying how morphology affects binding affinity and specific targeting could further optimize nanoparticles being designed for drug delivery. Single nanoparticle techniques can also serve as very sensitive biosensors for the detection of natural nanoparticles such as viruses that are a serious threat to human health. Thus, this chapter emphasized label-free, digital detection of whole virus particles. Detection with IRIS was rapid, repeatable and quantifiable, and demonstrates the potential of this system for inexpensive clinical and field-deployable pathogen detection and diagnostics of infectious diseases.

As a conclusion, IRIS is a promising platform for diverse clinical purposes. With the ongoing research efforts, we expect that the performance of the IRIS technique will further improve and will lead to a robust and widely used platform that will find practical use in real-world sensing and diagnostic applications.

References

Amano Y, Cheng Q (2005) Detection of influenza virus: traditional approaches and development of biosensors. Anal Bioanal Chem 381:156–164. doi:10.1007/s00216-004-2927-0

Brecht A, Gauglitz G, Polster J (1993) Interferometric immunoassay in a FIA-system: a sensitive and rapid approach in label-free immunosensing. Biosens Bioelectron 8:387–392. doi:10.1016/0956-5663(93)80078-4

Burt PJ, Adelson EH (1983) The laplacian pyramid as a compact image code. IEEE Trans Commun 31:532–540. doi:10.1109/TCOM.1983.1095851

Crowley JL, Stern RM (1984) Fast computation of the difference of low-pass transform. IEEE Trans Pattern Anal Mach Intell PAMI-6:212–222. doi:10.1109/TPAMI.1984.4767504

Cunningham BT (2010) Photonic crystal surfaces as a general purpose platform for label-free and fluorescent assays. J Assoc Lab Autom 15:120–135. doi:10.1016/j.jala.2009.10.009

Daaboul GG, Vedula RS, Ahn S, Lopez CA, Reddington A, Ozkumur E, Ünlü MS (2011) LED-based interferometric reflectance imaging sensor for quantitative dynamic monitoring of biomolecular interactions. Biosens Bioelectron 26:2221–2227. doi:10.1016/j.bios.2010.09.038

Daaboul GG, Lopez CA, Chinnala J, Goldberg B, Connor JH, Ünlu MS (2014) Digital sensing and sizing of vesicular stomatitis virus pseudotypes in complex media; a model for Ebola and Marburg detection. ACS Nano. doi:10.1021/nn501312q

Eustis S, El-Sayed MA (2006) Why gold nanoparticles are more precious than pretty gold: noble metal surface plasmon resonance and its enhancement of the radiative and nonradiative properties of nanocrystals of different shapes. Chem Soc Rev 35:209–217. doi:10.1039/B514191E

Hill HD, Mirkin CA (2006) The bio-barcode assay for the detection of protein and nucleic acid targets using DTT-induced ligand exchange. Nat Protoc 1:324–336. doi:10.1038/nprot.2006.51

Hong X, van Dijk EMPH, Hall SR, Götte JB, van Hulst NF, Gersen H (2011) Background-free detection of single 5 nm nanoparticles through interferometric cross-polarization microscopy. Nano Lett 11:541–547. doi:10.1021/nl1034489

Juskaitis R, Higdon PD, Wilson T (1997) Extinction coefficient in confocal polarization microscopy. *Proc. SPIE* 2984. Three-Dimensional microscopy: image acquisition and processing IV, pp 82–84. doi:10.1117/12.271253

Lopez CA, Daaboul GG, Vedula RS, Özkumur E, Bergstein DA, Geisbert TW, Fawcett HE, Goldberg BB, Connor JH, Ünlü MS (2011) Label-free multiplexed virus detection using spectral reflectance imaging. Biosens Bioelectron 26:3432–3437. doi:10.1016/j.bios.2011.01.019

Lowe DG (1999) Object recognition from local scale-invariant features, vol 2. In: Proceedings of the seventh IEEE international conference on computer vision, 1999. Presented at the proceedings of the seventh IEEE international conference on computer vision, 1999, pp. 1150–1157. doi:10.1109/ICCV.1999.790410

Marzi A, Feldmann H, Geisbert TW, Falzarano D (2011) Vesicular stomatitis virus-based vaccines for prophylaxis and treatment of filovirus infections. J Bioterror Biodef S1. doi:10.4172/2157-2526.S1-004

Mitra A, Novotny L (2013) Real-time optical detection of single nanoparticles and viruses using heterodyne interferometry. In: Bartolo BD, Collins J (eds) Nano-optics for enhancing light-matter interactions on a molecular scale, NATO science for peace and security series B: physics and biophysics. Springer, The Netherlands, pp. 3–22

Mock JJ, Barbic M, Smith DR, Schultz DA, Schultz S (2002) Shape effects in plasmon resonance of individual colloidal silver nanoparticles. J Chem Phys 116:6755–6759. doi:10.1063/1.1462610

Monroe MR, Daaboul GG, Tuysuzoglu A, Lopez CA, Little FF, Ünlü MS (2013) Single nanoparticle detection for multiplexed protein diagnostics with attomolar sensitivity in serum and unprocessed whole blood. Anal Chem 85:3698–3706. doi:10.1021/ac4000514

Nicewarner-Peña SR, Freeman RG, Reiss BD, He L, Peña DJ, Walton ID, Cromer R, Keating CD, Natan MJ (2001) Submicrometer metallic barcodes. Science 294:137–141. doi:10.1126/science.294.5540.137

Özkumur E, Yalçın A, Cretich M, Lopez CA, Bergstein DA, Goldberg BB, Chiari M, Ünlü MS (2009) Quantification of DNA and protein adsorption by optical phase shift. Biosens Bioelectron 25:167–172. doi:10.1016/j.bios.2009.06.033

Özkumur E, Ahn S, Yalçın A, Lopez CA, Çevik E, Irani RJ, DeLisi C, Chiari M, Selim Ünlü M (2010a) Label-free microarray imaging for direct detection of DNA hybridization and single-nucleotide mismatches. Biosens Bioelectron 25:1789–1795. doi:10.1016/j.bios.2009.12.032

Özkumur E, Lopez CA, Yalçın A, Connor JH, Chiari M, Unlu MS (2010b) Spectral reflectance imaging for a multiplexed, high-throughput, label-free, and dynamic biosensing platform. IEEE J Sel Top Quantum Electron 16:635–646. doi:10.1109/JSTQE.2009.2037438

Peterson AW, Heaton RJ, Georgiadis RM (2001) The effect of surface probe density on DNA hybridization. Nucleic Acids Res 29:5163–5168. doi:10.1093/nar/29.24.5163

Posner B, Hong Y, Benvenuti E, Potchoiba M, Nettleton D, Lui L, Ferrie A, Lai F, Fang Y, Miret J, Wielis C, Webb B (2007) Multiplexing G protein-coupled receptors in microarrays: a radioligand-binding assay. Anal Biochem 365:266–273. doi:10.1016/j.ab.2007.03.014

Reddington AP, Monroe MR, Ünlü MS (2013a) Integrated imaging instrument for self-calibrated fluorescence protein microarrays. Rev Sci Instrum 84:103702. doi:10.1063/1.4823790

Reddington AP, Trueb JT, Freedman DS, Tuysuzoglu A, Daaboul GG, Lopez CA, Karl WC, Connor JH, Fawcett H, Unlu MS (2013b) An interferometric reflectance imaging sensor for point of care viral diagnostics. IEEE Trans Biomed Eng 60:3276–3283. doi:10.1109/TBME.2013.2272666

Rusling JF, Kumar CV, Gutkind JS, Patel V (2010) Measurement of biomarker proteins for point-of-care early detection and monitoring of cancer. Analyst 135:2496. doi:10.1039/c0an00204f

Serra P (ed) (2011) Biosensors – emerging materials and applications. InTech, Croatia

Seydack M (2005) Nanoparticle labels in immunosensing using optical detection methods. Biosens Bioelectron 20:2454–2469. doi:10.1016/j.bios.2004.11.003

Taton TA, Mirkin CA, Letsinger RL (2000) Scanometric DNA array detection with nanoparticle probes. Science 289:1757–1760. doi:10.1126/science.289.5485.1757

Willets KA, Van Duyne RP (2007) Localized surface plasmon resonance spectroscopy and sensing. Annu Rev Phys Chem 58:267–297. doi:10.1146/annurev.physchem.58.032806.104607

Part III
Nanomaterial Functionalization and Nanobioelectronics

Chapter 6
Nanomaterial-Based Dual Detection Platforms: Optics Meets Electrochemistry

Nan Li and Kagan Kerman

Abstract Nanoparticles are becoming popular tools for optical and electrochemical biosensors; in this chapter we aim to provide a basic understanding of gold nanoparticles and quantum dots. We begin with the intrinsic properties of gold nanoparticles and quantum dots, and then extend into their novel applications in biosensing. The first half of the chapter focuses on gold nanoparticles and then shifts to quantum dots in the second half. In the first part, a short history of gold nanoparticles is followed by different gold nanoparticle synthesis methods. Different modification methods have facilitated the application of gold nanoparticles in biosensing; their variation in functions, characteristics, and advantages allow gold nanoparticles to take different roles in the qualitative and quantitative analysis associated with biosensors. In the second half of the chapter, we start with the current understanding of quantum dots, and then introduce the structural model and common modification techniques for biosensing applications. Different optical and electrochemical techniques are used in quantum dot-based DNA sensors and immunosensors.

Keywords Biosensors • DNA sensors • Gold nanoparticles • Immunosensors • Quantum dots

6.1 Introduction

Nanotechnology has been recognized as an emerging and enabling technology for the twenty-first century. Nanoparticles are essential in nanotechnology and possess advantageous intrinsic properties over bulk materials. There is no international definition of a nanoparticle, but typical nanoparticles have diameters smaller than 100 nm in at least one dimension. The advantageous properties of nanoparticles can be summarized into four major categories. First, as the nanoparticles get smaller, the

N. Li • K. Kerman (✉)
Department of Physical and Environmental Sciences, University of Toronto Scarborough, 1265 Military Trail, Toronto, ON M1C 1A4, Canada
e-mail: Kagan.kerman@utoronto.ca

M.C. Vestergaard et al. (eds.), *Nanobiosensors and Nanobioanalyses*,
DOI 10.1007/978-4-431-55190-4_6

interfacial area-to-volume ratio increases to allow information to be observed at the particles' interfaces. Second, the wavelike properties of electrons in nanoparticles are affected by the variation in dimensions of nanoparticles. Third, quantum electromagnetic interactions and size confinement effects become significant for organizational structures under 50 nm, and they manifest even at room temperature if their size is less than 10 nm. Fourth, nanoparticles possess size-dependent physical, chemical, or biological properties, such as the change in color of suspensions due to nanoparticle aggregation. New atomic, molecular, and macromolecular structures made of nanoparticles have been generated by various methods in nanofabrication and biotechnology. The smaller size of nanoparticles allows a significant increase in the degree of complexity and speed of processes in particulate systems. There are various types of nanoparticles, and this part of the chapter will focus on the physical theories, chemical structures, and biological applications of gold nanoparticles (AuNPs) and quantum dots (QDs) in electrochemical sensing.

6.2 Gold Nanoparticle-Based Biosensors

The history of the medical application of gold dates back to 2,500 BC with the Egyptians and the Chinese. Later, gold was used by Europeans in the medieval period to treat fever, and then the Germans in the 1890s used gold for bacteriostatic purposes. In modern studies, the biochemical application of gold has never been greater, with gold nanostructures including nanoparticles, nanoshells, nanoclusters, nanostars, nanocubes, and nanorods, all having unique structural, electronic, magnetic, optical and catalytic properties (Thakor et al. 2011). Various sizes and shapes of gold nanostructures have been reported to couple with nucleic acids, proteins, and cells for transportation, separation, and detection of biomolecules.

6.2.1 Synthesis of Gold Nanoparticles

Of all the reported metal nanostructures, AuNPs have been most intensely applied for biochemical analysis. Research and applications have been advancing in recent decades and numerous approaches to the preparation of these particles are well documented. AuNP suspensions can have an intense red colour for particles in the range of less than 100 nm, and most nanoparticles applied in biosensing are less than 100 nm in diameter (Cheng et al. 2011). Numerous approaches are available for the production of AuNPs, starting in the 1850s when Michael Faraday initiated the scientific investigation of colloidal gold (Gentry 1995).

The synthesis of AuNPs usually involves the reduction of gold salts in either organic, aqueous or two phases. The simplistic approach to AuNP fabrication reduces gold salts ($HAuCl_4$) with a reducing agent to initiate nucleation of gold ions. The weakness of the synthesis protocol is aggregation caused by the extreme reactivity of AuNPs and high surface energy, which was often a reccurring issue if the surfaces of the nanoparticles were not passivated. As a result, stabilizers are added to avoid aggregation or precipitation. The stabilizers bind to the AuNP surface to protect the particles and functionalize the AuNPs with the desired charge or surface chemistry. Protection can be also obtained with the use of self-assembled monolayers (SAM), such as encapsulation in the pools of reverse microemulsions and dispersion in polymeric matrixes (Arvizo et al. 2010). In citrate reduction, $HAuCl_4$ is first reduced from Au^{3+} to Au^0, and then sodium citrate is added, through which the negative citrate ions impose a charge around the nanoparticle surface to minimize the aggregation. The size of AuNPs can be monitored by color change, a shift from faint yellow to red, and the red color suggests AuNPs with diameters ranging from 10 to 20 nm (Blakey et al. 2010). Another method requires $HAuCl_4$ to be mixed with a phase transfer agent, such as tetraoctylammonium bromide (TOAB) in toluene. Sodium borohydride ($NaBH_4$) is then added to reduce $HAuCl_4$, followed by thiol ligand capping to prevent AuNP aggregation. The organic phase undergoes a color change almost instantaneously upon the addition of $NaBH_4$, from orange to deep brown, and particles with diameters in the range of 1.5–6 nm are synthesized (Dey et al. 2013; Hostetler et al. 1999). The synthesized AuNPs can be modified further to interact with biomolecules for various purposes.

6.2.2 *Biomolecule–Gold Nanoparticle Conjugation*

AuNPs have been employed in developing novel electrochemical biosensors due to their small size and unique chemical, physical, and electronic properties. The metal nanoparticles can participate in various tasks in electrochemical biosensors, one objective of which was to enhance the sensitivity of the device. Some of the basic functions of AuNPs in an electrochemical biosensor include immobilization of biomolecules, catalysis of electrochemical processes, enhancement of electron transfer, and labeling of biomolecules (Luo et al. 2006).

Table 6.1 summarizes a few of the characteristics of AuNPs and their functions. Incorporation of biomolecules is mandatory for the creation of biosensors.

Table 6.1 Utilization of gold nanoparticles in electrochemical biosensors

Function	Relevant characteristics	Sensor advantages
Biomolecule immobilization	Biocompatibility and large surface area	Improved stability
Catalysis of reactions	High surface energy of AuNP	Enhance sensitivity and selectivity
Efficient electron transfer	Conductivity and small dimensions of particles	Direct electrochemistry with redox proteins
Labeling of biomolecules	Small dimensions; facile surface modification	Improve sensitivity and indirect detection
Increasing surface area	Large surface-to-volume ratio	Improve sensitivity

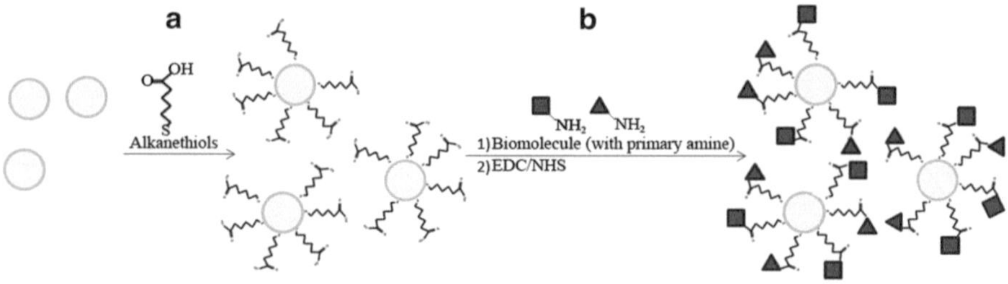

Fig. 6.1 AuNP modification techniques. (**a**) AuNPs are coated with alkanethiols by Au–S interaction, and then (**b**) the carboxylic acid functional groups on alkanethiols are activated with EDC/NHS to form covalent bonds with primary amine groups on biomolecules such as DNA, enzymes or antibodies

As such, much research attention has been directed towards synthesizing AuNPs with biomolecular recognition motifs on their surface.

As shown in Fig. 6.1, these water-soluble AuNPs can then be modified with peptides, enzymes, DNA, and antibodies for their further use in electrochemical biosensors. Electrochemical sensors allow miniaturization, low cost, high sensitivity, and fast response time. Various electrochemical sensors have been demonstrated involving conjugation of AuNPs as biomolecule labels or immobilization on the electrode surface. Common electrochemical methods previously reported to monitor AuNPs include electrochemical impedance spectroscopy, amperometry, potentiometry, cyclic voltammetry, conductometry, anodic stripping voltammetry, and differential pulse voltammetry (Pingarrón et al. 2008). These methods are carried out on conventional gold electrodes, miniaturized or inter-digitized micro-gold electrodes (Li et al. 2013).

6.2.3 *Immobilization of Biomolecules onto Gold Nanoparticles*

AuNPs have been reported to overcome the shortcomings of bulk materials to minimize denaturation of molecules and bioactivity loss. During the synthesis process, AuNPs are often charged, which allows for electrostatic interaction between their surface and the biomolecules. It is possible to attach biomolecules by other means such as covalent bonding, as in many studies discussing the attachment of enzymes to AuNPs including, but not limited to, horseradish peroxidase (Jia et al. 2002), microperoxidase-11 (Patolsky et al. 1999), tyrosinase (Liu et al. 2004), and hemoglobin (Gu et al. 2001). AuNPs can enhance redox processes because the detection of enzymatic products is directly proportional to the concentration of the original substrate. AuNPs have also been used in electrochemical immunosensors, and the mechanism for attaching antibodies is similar to that of enzymes. Yuan et al. was able to bind alpha-1-fetoprotein antibody onto AuNPs for an immunosensor (Ansari et al. 2010). AuNPs have been used to develop immunoassays for immunoglobulin

G (Dequaire et al. 2000), interleukin-6 (Liang and Mu 2006), carcinoma antigen (Tang et al. 2006), and many more. A common use of AuNPs was for the immobilization of oligonucleotide probes for the detection of DNA targets. One of the methods to immobilize DNA molecules on the nanoparticle surface was by modifying either end of the DNA with functionalities that can interact with the AuNP surface, such as thiolated DNA. Taking this approach one step further, modifying the electrode surface with AuNPs has also been shown to increase the saturation quantities of DNA probes immobilized on the electrode by about ten times (Kerman et al. 2007a, b). Although this method can increase the stability and maintain the activity of the biomolecule, one of the pitfalls of using AuNPs is the aggregations of AuNPs.

6.2.4 Catalysis of Electrochemical Reactions

The catalytic properties of AuNPs can be used for the purpose of decreasing overpotentials of redox reactions. Ohsaka et al. designed a sensor for the selective detection of dopamine in the presence of ascorbic acid. This was done by utilizing the catalytic properties AuNPs towards the oxidation of ascorbic acid and reducing its overpotential. This allowed for the distinction between the oxidation potentials originating from ascorbic acid and dopamine (Raj et al. 2003). In addition to selective detection of analytes, the use of nanoparticles also lowers the detection limit of electrochemical biosensors. This was demonstrated by the detection of streptavidin in as low as 2 fM concentrations (Begoña González-García and Costa-García 2000).

As shown in Fig. 6.2, Fiorella et al. (2011) demonstrated the induction of gold nanostructures on the surface of disposable electrochemical printed (DEP) gold chips using acetylcholinesterase. Growth of the gold nanostructures was confirmed via differential pulse voltammetry (DPV) and colorimetric assay. The formation of the gold nanostructures demonstrated the ability to produce high-intensity current signals and low reduction potentials. It is a sensitive technique using only 20 μL of sample volume. Another interesting function of AuNPs is the enhancement of electron transfer.

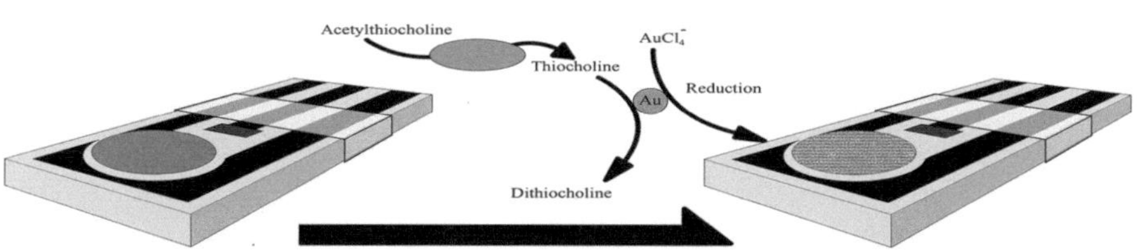

Fig. 6.2 Detection of Au nanostructures on disposable electrochemical printed (DEP) chips. Au nanostructure growth is facilitated by acetylcholinesterase (AChE), and shows physically visible growth, build up, and color change (*right diagram*). Adapted from Fiorella et al. (2011)

6.2.5 *Facilitating Electron Transfer*

In electrochemical biosensors, redox proteins placed directly on the electrode surface require the transfer of electrons from the active center of the enzyme to the electrode surface, as shown in Fig. 6.3, but the active center of the enzyme is often enclosed by the protein shell which acts as an insulator. The conductive properties of AuNPs can assist in the transfer of electrons from the active center to the electrode surface. This has been shown in the work of Willner et al., where the AuNPs were modified with N6-(2-aminoethyl)-flavin adenine dinucleotide and then reconstituted with apo-glucose oxidase (Xiao et al. 2003). The conjugate was placed on a gold electrode containing thiolated monolayer. They also assembled AuNPs directly on the electrode before binding the enzyme. This work reported an electron transfer rate of ~5,000/s, which is seven times the rate of electron transfer between the enzyme and its oxygen substrate. Cheng et al. (2011) discussed the novel modification of screen-printed carbon electrodes with electrodeposition of gold nanostructures. A significant increase in electroactivity suggested that the gold nanoparticles facilitated charge transfer from the redox probe. Cyclic voltammetry was used for electrodeposition of gold nanostructures, and deposition was characterized using electrochemical impedance spectroscopy, UV–Vis spectroscopy, and scanning electron microscopy. Both optical and electrochemical techniques have been demonstrated to characterize a surface.

6.2.6 *Labeling Biomolecules with Gold Nanoparticles*

The ability of biomolecules to maintain their activity in the presence of nanoparticles has also shown the way for these particles to be used as labels. They can be used for tagging antibodies, antigens, or DNA. Once labeled, it is the AuNP that is detected in order to eventually deduce the concentration of the analytes. The AuNPs must first be dissolved and their ions detected subsequently. Stripping voltammetry technique is ideal as it allows for the detection of trace amount gold ions. This was demonstrated by Hepel and Zhong (2010), using indirect detection platform with AuNP labels for the detection of picomolar concentrations of goat immunoglobulin G. The antibodies were labeled with AuNPs, and then dissolved later in acidic solutions and the ions detected using anodic stripping voltammetry. The uses of nanoparticles as catalysts, labeling tools, immobilizing platforms, and electrical bridges for electron transfer will be discussed further as they are incorporated within various types of electrochemical sensors.

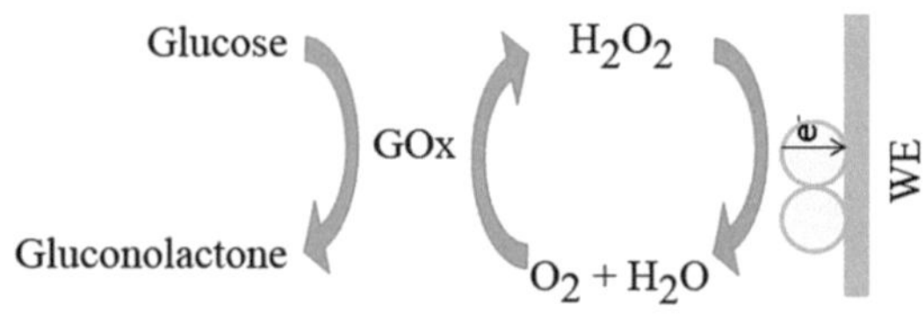

Fig. 6.3 AuNPs act as redox centers to facilitate electron transfer to the electrode surface

6.2.7 Gold Nanoparticle Application in Biosensors

Direct amperometry, cathodic linear sweep voltammetry, and anodic stripping voltammetry are common electrochemical methods used to quantitatively analyze AuNP. Cyclic voltammetry and differential pulse voltammetry of small AuNPs shows multiple peaks of reversible or semi-reversible single electron oxidation/reduction at room temperature, because the AuNPs have an energy gap larger than 0.74 V and behave like molecules (Jackman et al. 1998; Balasubramanian et al. 2005; Kumar et al. 2011). The current peak was determined by molecular-like AuNP diffusion rate; the diffusion coefficient (~10^{-6} cm^2/s) was small enough that the faradic current was proportional to the AuNP population (Bagotsky 2005). AuNPs with diameters larger than 4 nm show bulk-continuum voltammetric behaviour, where the redox currents represent the double layer of the AuNPs (Miles and Murray 2001). A common drawback of direct amperometry analysis of AuNPs was caused by the small population in biochemical applications, limiting current to nano-ampere levels and resulting in low sensitivity of the analysis method (Quinn et al. 2003; Jimenez et al. 2003). Furthermore, the protective monolayer of AuNPs also contributes to the signal, meaning that the currents are not solely from the core of the AuNP. Cathodic linear voltammetry requires AuNP oxidation reaction at 1.25 V vs. Ag/AgCl in HCl electrolyte solution, then electro-deposition to the electrode surface by linear sweeping voltammetry between 1.2 and 0.00 V (Pumera et al. 2005a, b). The reduction current peak depends on the scan rate, oxidation potential, oxidation time, and AuNP size. The peak current was proportional to scan rate between 2 and 100 mV/s; however, the peak current was proportional to the diameter of AuNPs at 2.94×10^{-6} M due to unclear relationships (Garcia and Garcia 1995). Cathodic linear voltammetry is advantageous in terms detection of biomolecule-conjugated AuNP, because conjugation does not affect the redox reaction of AuNPs and various dynamic ranges were reported (Oishi et al. 2010; Hung and Kerman 2011; Li et al. 2012). Anodic stripping voltammetry does not depend on immobilization of AuNPs, but the pre-concentration step is essential for high sensitivity (Fiorella et al. 2011; Fu et al. 2005; Li et al. 2011). The AuNPs can be stripped from the electrode by potentials from 0 to 1 V versus saturated calomel electrodes.

As shown in Fig. 6.4, Hiep et al. (2008) have used core–shell structure nanoparticles to detect melittin binding to a membrane using electrochemistry and localized surface plasmon resonance (LSPR), where the thickness of deposited gold leads to a shift in LSPR signal. Hiep et al. (2010) reported a gold-plated nanostructured biochip that was modified for the label-free detection of polymerase chain reaction by analyzing the relative reflected intensity of the nanostructured substrates. The 5′-thiolated primers were immobilized on the biochip, allowed the label-free detection and amplification of DNA to be performed at the same time. Based on analysis of the relative reflected intensity peak values, observable signals were detected five to six times faster than fluorescent real-time PCR (3–4 cycles versus 23–25 cycles). The results proved gold-coated nanostructured substrates to show promise as a rapid, label-free, and real-time DNA detection method.

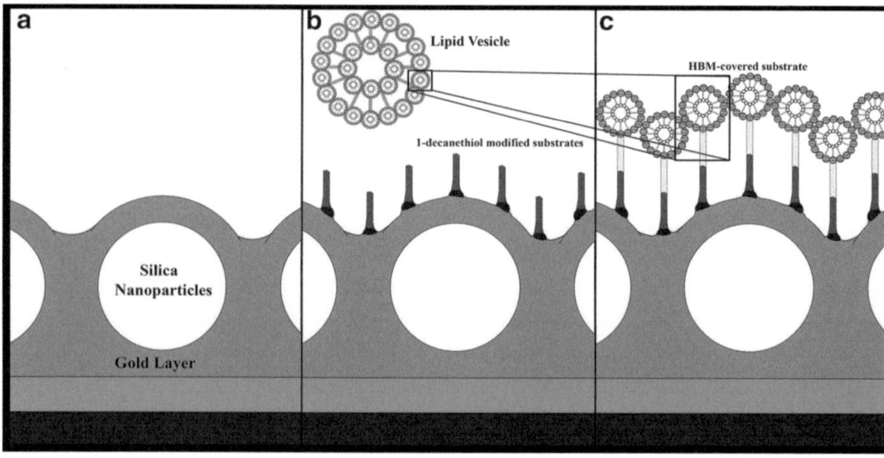

Fig. 6.4 Label-free detection of melittin binding to a membrane using electrochemical-localized surface plasmon resonance. (**a**) Deposition of 5 nm chromium layer and 30 nm gold layer on silicon slides. Silica nanoparticles modified with amino groups are placed on top of previous layers. A final gold-cap layer finishes the core–shell structure nanoparticle substrate. (**b**) Lipid vesicles are prepared by dissolving dimyristoylphosphatidylcholine (DMPC) in pure chloroform. Lipid vesicles are modified for electrochemical, LSPR-based dual detection. 1 mM decanethiol solution on substrate surface led to an alkanethiol layer. (**c**) Lipid vesicles are fused onto alkanethiol-modified surface. Adapted from Hiep et al. (2008)

Kim et al. (2008) detected aptamer–protein interactions via a method which involved LSPR with interferometry using gold-coated porous anodic alumina layer chips. Both the LSPR and interferometric characteristics of the chips indicated reliable detection of the targeted protein, thrombin. The combination of the two techniques allowed Kim et al. (2008) to prove their method to be cost-efficient, simple, and effective.

As shown in Fig. 6.5, Kerman et al. (2004) designed a cost-effective, sensitive method for the detection of single nucleotide polymorphisms (SNPs). The SNPs were determined by monobase-modified colloidal Au nanoparticles. The change in electrochemical signal corresponded to the existence of the SNPs. Label-free detection of cells from mouse thymus was introduced in this study by Endo et al. (2008). A localized surface plasmon resonance biosensor using nanoparticle layer substrates was the proposed detection method. The detection limit was determined to be 10 pg/mL, and the method was also proven to rival the standard detection method, enzyme-linked immunosorbent assay. Based on their results, the method was concluded to show great promise as it was cost-effective, highly specific, and highly sensitive. Vestergaard et al. (2008) developed a new method for the detection of tau protein. The use of Au-capped nanoparticles LSPR chips was the foundation of their method. The detection of tau proteins could be observed at 10 pg/mL, which is much lower than the 195 pg/mL clinical cut-off. The method's robustness was also validated when it was not affected by bovine serum albumin (BSA). Overall, the method showed great promise due to its high-caliber performance and selectivity.

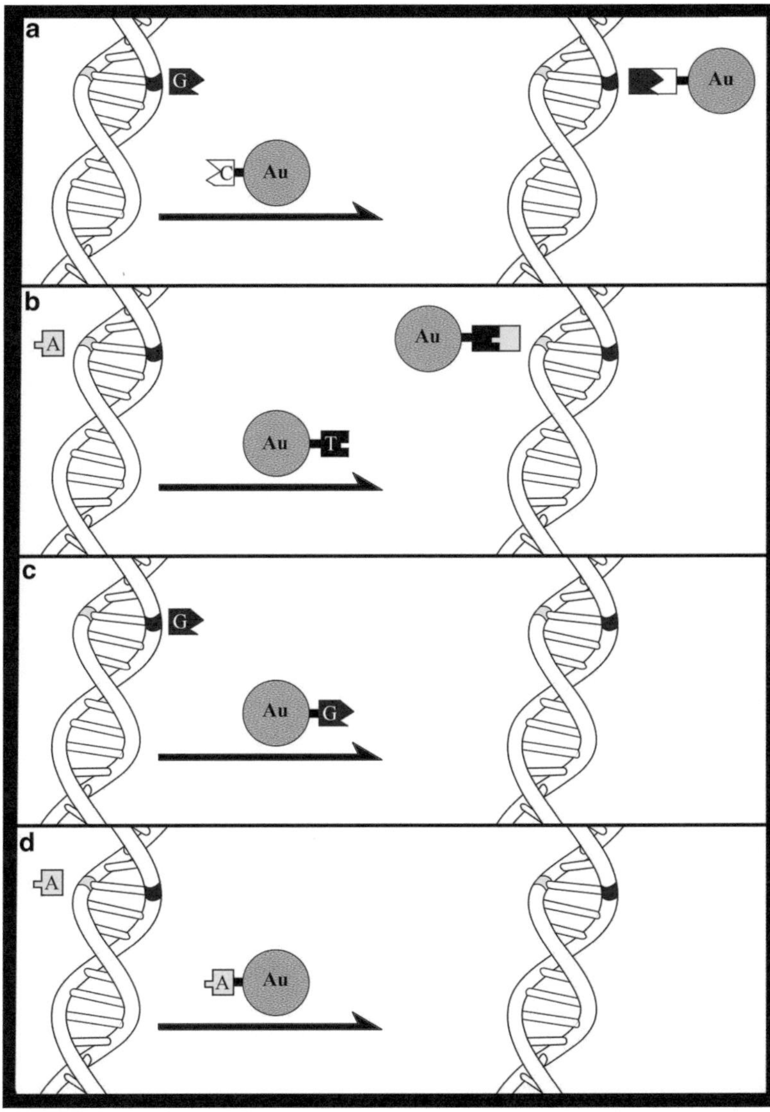

Fig. 6.5 Detection of single nucleotide polymorphisms (SNPs) using monobase-modified AuNPs. Four different monobase-modified AuNPs were used. Identity of the mismatch could be detected from Au redox signal generated by the binding of the appropriate monobase-modified AuNP (**a**, **b**). Adapted from Kerman et al. (2004)

6.3 Quantum Dot-Based Biosensors

In general, quantum dots (QDs) are nano-sized semiconductors that range between 1 and 10 nm in diameter with their central core made of the same materials as other ordinary semiconductors (Kerman et al. 2007a, b). However, strictly speaking, the size of QDs depends on the exciton Bohr radius. In bulk semiconductors, electrons from a lower energy valence band gain enough energy from heat or light to overcome the bandgap energy and get excited to a higher energy band, the conduction band, leaving behind a positive hole. The positive hole and the excited electron are called an exciton, and the distance between the positive hole and the excited electron in the conduction band is called the exciton Bohr radius (Xie 2001). As shown in Fig. 6.6, in bulk semiconductors the dimensions of the metalloid are much larger than the exciton Bohr radius, allowing the exciton to extend to its natural limit and the electron energy levels to be treated as continuous. If one dimension of the bulk material is reduced to a size less than the exciton Bohr radius, then the electrons may only possess freedom of movement in the other two dimensions, and such a system is called a quantum well (Pan and Fonstad Jr 2000). If another dimension of the bulk material is reduced to a size less than the exciton Bohr radius, then the electrons only possess freedom of movement in one dimension, and such a system is called a quantum wire (Zanardi et al. 2002). QDs have such small size that the electron is confined to a size smaller than exciton Bohr radius in all three spatial dimensions. In this case, the electron energy levels can no longer be treated as

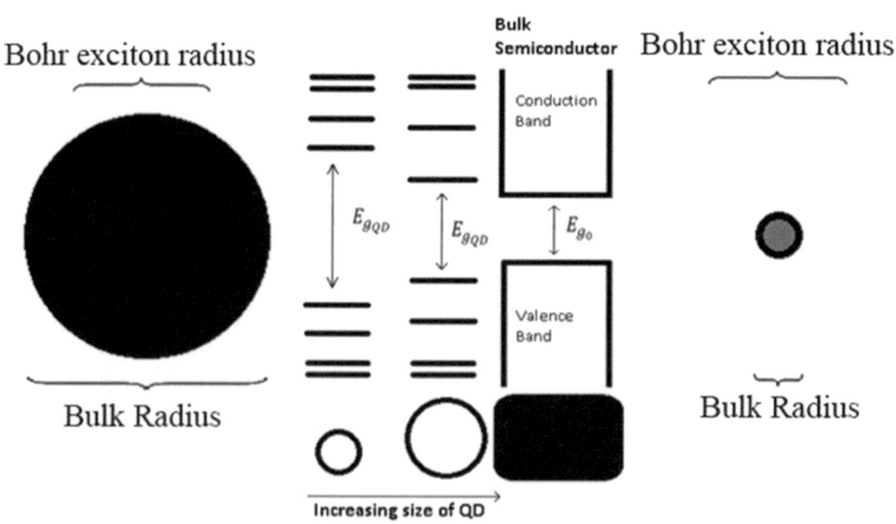

Fig. 6.6 Representative diagram for the bandgap in QDs and bulk semiconductors. The bulk semiconductor has continuous conduction and valence energy bands separated by a "fixed" energy gap, E_{g0} (bulk). QDs are characterized by discrete atomic-like states with energies E_{gQD} that are determined by the QD radius R. Adapted from Xie (2001)

continuous and must be treated as discrete. Such behavior of discrete energy levels is called quantum confinement, which means that there is a small and finite separation between energy levels (Ciurla et al. 2002).

Due to the quantized energy levels of QDs, they are often referred to "artificial atoms". One important feature of the particle-in-a-box model is "confinement energy"; thus, comparisons can be drawn from QDs to the "particle in a box" example from quantum mechanics. The "particle-in-a-box" model cannot be perfectly applied to QDs, but it is a good approximation and can be used to gain some insights into the behavior of QDs (Wojciechowski and Bogdanow 1998).

6.3.1 Quantum Dot Structure

As illustrated in Fig. 6.7, a QD is composed of an inorganic core, inorganic shell, organic capping, and bioconjugates. The chemical nature of the core nanocrystal determines the emission wavelength and the fluorescence color, because the exciton Bohr radius is different for each element. Despite the emergence of other core material, the inorganic compound cadmium selenium (CdSe) is most often used as the core nanocrystal material (Micic et al. 1994; Xie et al. 2009). If the QDs only have a semiconductor core made of CdSe, then only poor photoluminescence and moderate photostability can be obtained, because the surface of the semiconductor core can trap electrons and stop being excited (Loukanov et al. 2004). This drawback can be minimized by growing an inorganic shell over the inorganic core to maintain stability. QDs with a CdSe core are often covered with a layer of zinc sulfide (ZnS), as ZnS inorganic shells can enhance quantum yield from the photophysical aspect and the physiochemical aspect (Hines and Guyot-Sionnest 1996; Fu et al. 2008; Li et al. 2003).

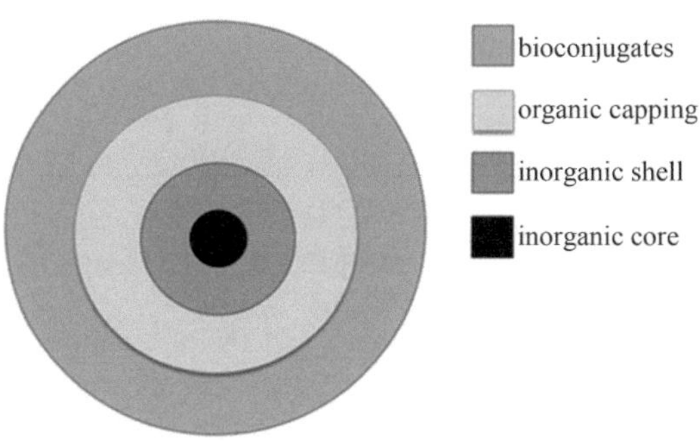

Fig. 6.7 Schematic description for the chemical structure of a QD. The *inner black circle* represents the core nanocrystal of the QD. The next *grey circular band* represents the inorganic shell, the *light grey circular band* represents the organic coating, and the *outermost circular band* represents the conjugated biomolecules

The organic capping surrounding the inorganic shell of QDs prevents uncontrolled growth and agglomeration of the nanoparticles. The organic capping also allows the QD to be manipulated as a large molecule, with solubility and chemical reactivity determined by the identity of the capping organic molecules. Furthermore, the organic capping provides "electronic" passivation of QDs, a process which terminates any dangling bonds that remain on the semiconductor's surface. This "electronic" passivation is absolutely crucial to maintaining emission efficiency, because any un-terminated dangling bonds can result in the trapping of electrons at the surface of QD before they have a chance to emit photons (Kuppayee et al. 2011). There are variety types of suitable organic coating being studied and used. Most commercially available QDs are commonly coated by trioctylphosphine oxide (TOPO) or oleic acid. Although these coatings allow QDs to be stored for years with high quantum yield, they can only dissolve in organic solvents. While water solubility and high stability against oxidation are essential for the application of QDs in biological systems, the essential but water-insoluble inorganic component of QDs hinder them from being water-soluble.

6.3.2 Quantum Dot Modification

As shown in Fig. 6.8, ligand exchange and ligand capping are two common strategies for modifying QDs to increase solubility. The ligand exchange method normally utilizes the natural affinity of the inorganic core nanocrystal for sulfur to exchange the organic coating with various water-soluble thiol ligands. Such modification may result in the aggregation of nanocrystals due to loss of surface shielding, but the QDs remain small in dimension (Comparelli et al. 2003). On the other hand, the ligand capping method surrounds the water-insoluble QDs with amphiphilic polymers and results in large polymer-coated particles. Sometimes more than one nanocrystal is contained in a coating layer (Mansur et al. 2011). Nevertheless, these larger particles obtained by the ligand capping strategy have sufficient chemical stabilization of the surface and a reliable protection against aggregation. In addition, either the thiol ligand or the amphiphilic ligand can contain other functional groups such as carboxylic acid and amine groups to bind with biomolecules for further applications in biology.

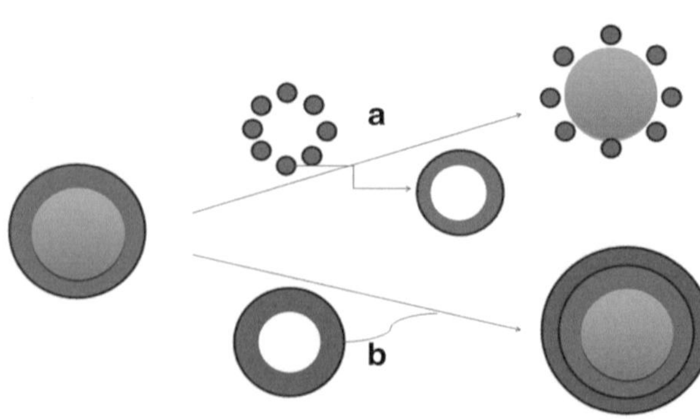

Fig. 6.8 Quantum dot modification techniques. (**a**) Ligand exchange and (**b**) ligand capping strategy

6.3.3 Measurement Techniques

Anodic stripping analysis is widely recognized as a powerful tool for performing trace metal analysis. This is a sensitive technique, because the first step is a preconcentration step through reduction of the metal ions and formation of amalgams, and the second step strips off the metal ions through oxidation, which results in a change in current that correlates to the number of metal ions on the electrode (Huang et al. 2010). Linear sweep voltammetry (LSV), pulsed scan voltammetry, differential pulse voltammetry (DPV), or square wave voltammetry (SWV) can be used during the stripping step to measure current response for various analyte species. However, it has been reported that the differential pulse mode of stripping analysis on mercury film-coated glassy carbon electrodes (GCEs) can detect trace amounts of metal ions. Using anodic DPV, the detection limit was 3–5 times lower than the detection limit measured with LSV (Shiddiky et al. 2012). DPV yields base current slopes 3–5 orders of magnitude lower than LSV (Yin et al. 2010). It should be noted that noise levesl and current sensitivities are similar in the voltammograms for differential pulse and linear sweep modes of anodic stripping voltammetry. Although greater sensitivities can be achieved by using higher pulse amplitudes in the case of DPV, the drawback is the loss in peak resolution after 10 mV amplitude (Zhu et al. 2012). In additional, DPV would be more susceptible to peak height depression than LSV in the presence of surface active agents, e.g. 1-octanol. Moreover, in the case of DPV, adsorption and desorption of organic species can yield an unusually higher faradaic response from the metal ions due to the faradaic response from organic species. However, this problem can be minimized if the organic species is selectively filtered or modified (Kim et al. 2012).

6.3.4 Working Electrodes for Quantum Dot Analysis

6.3.4.1 Mercury Film Electrodes (MFE)

A working electrode (WE) commonly used for anodic stripping voltammetry of heavy metals is the hanging mercury drop electrode (HMDE) and mercury film electrode deposited onto glassy carbon electrode (GCE) or carbon fiber electrode (Hernandez et al. 1988). A major advantage of mercury electrodes is their ability to tolerate large negative potentials from high overpotential reduction of water, which results in the formation of hydrogen gas (Beilby 1972). Another advantage of mercury films on the WE for anodic stripping voltammetric analysis is that mercury can form amalgams with heavy metals (lead, cadmium or copper), which upon oxidation results in a sharp peak, hence improving the resolution between different analyte species (Zhou et al. 2010). The amalgamation also helps in containing the target metal ions in close proximity to the electrode surface (within the diffusion layer) even when the cathodic potential on the GCE is removed. Mercury film can be easily removed from the GCE by cleaning the electrode surface with a moist piece of filter paper followed by a polishing step with a dry filter paper of the same type. For analyte species with oxidation potential higher than mercury, a solid inert metal such as gold, silver or platinum can be employed as a WE. Additionally, in case of HMDEs, a fresh metallic surface can be easily generated by simply producing a new droplet of mercury. This is important considering that in voltammetry the reproducibility of the current response is very sensitive to contamination and irregularities on the electrode surface (Galian and Guardia 2009). However, HMDEs have low sensitivity, poor precision, and poor resolution of neighboring peaks (Gherghi et al. 2003). Although preplated MFEs offer improved sensitivity and excellent resolution, the reproducibility of their response is worse than HMDEs.

6.3.4.2 Bismuth Film Electrode

An alternative working electrode system that can be used for anodic stripping voltammetric analysis of metal ions is the bismuth film electrode (BFE) (Wang et al. 2000). BFEs can be prepared in situ by adding 400 μg/L of bismuth(III) directly to the sample solution and subsequent deposition of bismuth along with other target analyte metal ion species on glassy carbon or carbon fiber electrodes by applying a cathodic potential of −1.2 V (−1.4 V for analyte sample solution containing Zn) (Wang et al. 2000). BFEs possess a wide potential window (−1.2 to −0.2 V), allowing quantification of most metals (except copper, antimony, and bismuth). Additionally, the response from BFEs exhibits high precision and bismuth film can be easily removed from the WE surface by applying a potential of +0.3 V, allowing recycling of the electrode substrate for depositing another layer of bismuth film. Another attractive feature of BFEs is that their performance is less susceptible to oxygen interference, and they are environmentally friendly with a performance that similar that of mercury film electrodes.

6.3.5 Quantum Dot Application in Electrochemical Sensors

6.3.5.1 DNA Sensors

Wang et al. reported on the use of a carbon nanotube (CNT)-based amplification platform along with ultrasensitive electrochemical stripping detection using QD tags for enhanced electrochemical detection of DNA in a sandwich-based assay platform (He et al. 2010). CdS QDs were loaded as monolayer onto activated CNT by hydrophobic interactions and subsequently modified with streptavidin for the immobilization of biotin-labeled reporter DNA sequence, where the reporter sequence hybridized with the overhang portion of the target sequence following the probe target hybridization event. The detection was based on square-wave anodic stripping voltammetry using mercury film GCE as WE (Ag/AgCl as reference electrode and Pt wire as counter electrode) to measure the Cd content after the dissolution of CdS QDs in 1.0 M nitric acid solution. The loading of multiple (~500) CdS tags on a single CNT provided a limit of detection that was 500-fold lower than a conventional assay that used a single CdS tag per reporter sequence (40 pg/mL vs. 20 ng/mL) (He et al. 2010). The assay exhibited excellent selectivity between a fully complementary target and a non-complementary target, where even a 250-fold excess of non-complementary target sequence provided a signal response which was considerably lower than the fully complementary target. Further enhancement of the limit of detection was suggested with the use of a bismuth film-coated carbon electrode and by the catalytic increase in the size of the CdS particles. QDs have also been used for electrochemical coding of single nucleotide polymorphisms (SNPs) (Liu et al. 2005), in which different nucleobases were conjugated to QDs by phosphoramidite chemistry through a cysteamine linker. In particular, adenosine nucleotides were linked to ZnS-based QDs (A-ZnS), cytosine nucleotides were linked to CdS-based QDs (C-CdS), guanine nucleotides were linked to PbS-based QDs (G-PbS), and thymine nucleotides were linked to CuS-based QDs (T-CuS). QD-conjugated nucleobases are then sequentially added, in the order A-ZnS, C-CdS, G-PbS, and finally T-CuS, to the sample solution containing a particular mismatch which was immobilized on a bead. The addition of nucleotide-conjugated QDs allowed specific binding by complementarity of the Watson–Crick base pairing to different mismatched sites as well as to the previously linked conjugates. Depending on the identity of the mismatch, the voltammogram originating from each of the eight possible mismatch combinations resulted in a different signature, thus deciphering the identity of a particular mismatch. In this method, the signal amplification for the SNP detection was provided by the accumulation step of the stripping voltammetry using a Hg film-coated glassy carbon electrode. In addition, the authors also showed that it was possible to electrochemically code two known mismatches within a single DNA target using QDs that are modified with two different nucleobases. The unique feature of SNP coding using QDs is that a single voltammogram was required to encode a particular mismatch, due to the peak potential signal resolution provided by each of the different compositions of the QDs. QDs have also been used as tags for multiplex detection of DNA targets in an electrochemical-based assay. Different compositions of QDs such as ZnS, CdS, and PbS were used for multiplex detection of three different DNA targets using a sandwich-based assay platform, where the metal composition of the QDs provided a potential-resolved stripping voltammetric signal (Wang et al. 2008). For their assay platform, different oligonucleotide probes (maximum three) were immobilized onto magnetic beads, followed by hybridization with the sample solution containing unlabeled DNA targets. This was followed by a second hybridization event using QD-labeled reporter sequences, where each kind of reporter sequence was labeled with a different kind of QD and corresponded to a different probe sequence. Dissolution of QDs following the second hybridization event and subsequent stripping voltammetry provided the analytical signal. An important aspect of this kind of assay is the selection of QDs, in terms of their composition, that provides a well-resolved and non-overlapping oxidation potential signal (Pumera et al. 2005a, b). There are many different kinds of QDs that satisfy this criterion (PbS, ZnS, CdS, CuS, InAs, and GaAs), thus allowing multiplex detection of up to five or six different kinds of targets. All of the previously mentioned metal ions possess well-defined

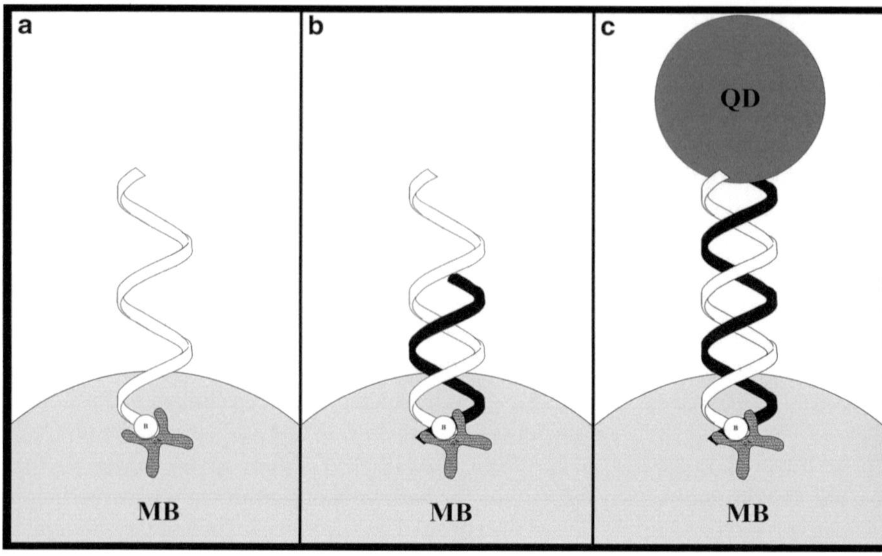

Fig. 6.9 Nanocrystal-based bioelectronic coding of single nucleotide polymorphisms. (**a**) Biotinylated single-stranded DNA (*b* in illustration is biotin) was attached on a conjugated streptavidin magnetic bead (MB) to act as probe. (**b**) First hybridization with target DNA. (**c**) Biotinylated QDs in the form of ZnS, CdS, and PbS (only one is shown in illustration) perform second hybridization. Adapted from Liu et al. (2005)

sharp stripping peaks with some overlap between Cd-In and Pb-In; however, it should be noted that it is possible to resolve the individual contributions of signals from each of the metal ions to the overall signal in the overlapping region of the voltammogram using deconvolution. Another interesting feature of their assay design is the use of a sandwich-based assay platform by employing a reporter sequence, hence the direct labeling of DNA targets is not required. The response for the assay design described above was found to be linear to increasing concentrations of targets, and the detection limit was reported to be about 270 pM. Additionally, lower detection limits were reported (2.7 pM) by incorporating multiple QDs in a polystyrene carrier bead (Wang et al. 2008).

As shown in Fig. 6.9, Liu et al. (2005) used different quantum dots with magnetic beads to code unknown single nucleotide polymorphisms. The mismatch recognition events are amplified by the metal accumulation feature of the stripping voltammetric transduction mode.

6.3.5.2 Immunosensors

QDs are attractive labels for electrochemical detection of target analytes in a competitive-based displacement assay. This is due to the fact that attaching QD labels to a protein or DNA alters its binding association constant for its complementary target to a greater degree than other conventional electrochemical tags due to the size of the QD tags, which is considerably larger than other electrochemical tags. Additionally, with the possibility of attaching multiple QD tags to a single protein or DNA molecule, the thermodynamic association constant is lowered to an even greater extent, thus facilitating the displacement kinetics (Liao et al. 2010). The electrochemical interrogation is based on a competitive (displacement) assay where thiol-modified aptamers are immobilized on a gold substrate and subsequently bound to their corresponding QD-tagged protein targets. The sensing surface was then incubated with the sample solution containing target proteins, where the binding event between an aptamer and its target protein was detected by monitoring the electrochemical signal from the remaining QD-tagged proteins (Bagalkot and Gao 2011). Hence, a larger signal originating from undisplaced QD-tagged proteins could be observed for lower concentrations of target analyte proteins. This assay allowed high sensitivity and selectivity (due to the presence of the blocking action of mixed film of 6-mercapto-1-hexanol) with a limit of detection approaching 0.5 pM, which is 3–4 orders of magnitude lower than a typical aptamer-based assay (~1–6.4 nM) (Kerman et al. 2004). Multiplex detection of thrombin and lysozyme proteins has been demonstrated using CdS- and PbS-based QD tags, respectively. It was suggested that this assay format could be adapted for multiplex detection of up to six different protein targets within the allowed potential window range (Hansen et al. 2006). Multiplex detection of protein markers [human serum albumin (HAS), macroglobulin (MG), and human myoglobin (Mb)] using different inorganic QDs (ZnS, CdS, and PbS) was demonstrated by Kim et al. (Takei et al. 2011). They utilized a sandwich-based assay platform where each of the secondary antibodies was covalently conjugated to a different inorganic nanocrystal by carbonyl diimidazole (CDI) activation of the hydroxyl groups present on the surface of the dithiothreitol (DDT) functionalized QDs. The biotinylated primary antibodies (anti-HSA, anti-MG, and anti-Mb) were immobilized on streptavidin-modified magnetic beads. Following the incubation of primary antibodies in the presence of antigens and subsequent incubation in the presence of QD–secondary antibody conjugates, the QDs were dissolved using 1 M nitric acid solution, followed by analysis of the dissolved metal ions using square-wave anodic stripping voltammetry, where mercury film-GCE electrodes served as a working electrode that was prepared in situ (Deng et al. 2012). The assay provided well-defined highly sensitive peaks in the presence of each of the individual antigen targets (−1.11 V for ZnS anti-MG conjugates, −0.67 V for CdS anti-Mb conjugates, and −0.52 V for PbS anti-HAS conjugates), where the response in terms of peak height was proportional to the concentration of target proteins. The limits of detection were reported to be 10, 11, and 9.5 ng/mL for MG, Mb, and HAS proteins, respectively. The reported limit of detection is well below the concentration that is considered as hazardous for a patient; hence this assay is able to quantify very low concentrations of protein markers. The assay also exhibited excellent selectivity, where the presence of a tenfold excess of non-specific antigens (bovine serum albumin and hemoglobins) provided a response which was 40-fold lower than that in the presence of specific antigens (Rauf et al. 2010). Immunoassays employing potentiometric analysis of trace metal ions using QDs as labels have also been reported in the literature. Potentiometric determination of trace metal ions using ion-selective electrodes (ISEs) is attractive due to their ability to handle extremely small sample volumes (pL), thus allowing limits of detection that can reach 300 amol (Thuer et al. 2007). Additionally, there is a direct relationship between the sample activity and electromotive force as governed by the Nernst equation, allowing concentration-dependent quantification to be made in a predictable manner. Moreover, it is possible to use QDs as labels for potentiometric analysis due to the fact that they can be easily dissolved using hydrogen peroxide, which prevents any changes in the pH of the solution that might interfere with the measurement (Palecek and Bartosik 2012).

QDs are also attractive labels for potentiometric biosensing due to the fact that labels are not generated in situ, when compared with silver plating of AuNPs; this therefore promotes less non-specific labelling (Pfeiffer et al. 2010). In the paper by Thurer et al., the researchers employed a sandwich-based immunoassay platform for the detection of mouse IgG protein using CdSe QDs as labels which were conjugated to secondary antibody. The detection of Cd^{2+} ions was performed using a Cd^{2+}-selective micropipet electrode, where a Na^{+}-selective electrode served as a pseudo-reference electrode. The limit of detection was reported to be less than 10 fmol (Numnuam et al. 2008).

6.4 Conclusions

Biosensors have become a pivotal research area in the search for novel point-of-care systems and diagnostic tools. Since the innovation of the first glucose biosensor, significant improvements have been made to biosensor design to increase sensitivity, reduce signal acquisition time, and allow for simple and portable analysis. Current improvements to biosensors include the use of nano-materials and, taking advantage of their attractive characteristics, further chemical modifications can shape these biosensing platforms to expand their detection capabilities. We envision that optical and electrochemical dual detection methods would use nanoparticles as novel platforms for innovative applications in clinical diagnosis. In the search for novel sensing techniques, the applications of biosensors will continue to flourish and find their place in untapped domains.

References

Ansari AA, Alhoshan M, Alsalhi MS et al (2010) Prospects of nanotechnology in clinical immunodiagnostics. Sensors 10(7):6535–6581

Arvizo R, Bhattacharya R, Mukherjee P (2010) Gold nanoparticles: opportunities and challenges in nanomedicine. Expert Opin Drug Deliv 7(6):753–763

Bagalkot V, Gao X (2011) SiRNA-aptamer chimeras on nanoparticles: preserving targeting functionality for effective gene silencing. ACS Nano 5(10):8131–8139

Bagotsky VS (2005) Electrochemistry and the environment. In: Fundamentals of electrochemistry, 2nd edn. Wiley, Hoboken. doi:10.1002/047174199X.ch24
Balasubramanian R, Guo R, Mills AJ et al (2005) Reaction of $Au_{55}(PPh_3)_{12}Cl_6$ with thiols yields thiolate monolayer protected Au_{75} clusters. J Am Chem Soc 127(22):8126–8132
Begoña González-García M, Costa-García A (2000) Silver electrodeposition catalyzed by colloidal gold on carbon paste electrode: application to biotin–streptavidin interaction monitoring. Biosens Bioelectron 15(11–12):663–670
Beilby AL (1972) Principles of instrumental analysis (Skoog, Douglas A.). J Chem Edu 49(6):A362
Blakey I, Schiller TL, Merican Z et al (2010) Interactions of phenyldithioesters with gold nanoparticles (AuNPs): implications for AuNP functionalization and molecular barcoding of AuNP assemblies. Langmuir 26(2):692–701
Cheng XR, Veloso AJ, Kerman K (2011) Reflection spectroscopy for the determination of Au nanostructure formation on carbon surfaces. Talanta 86(30):421–424
Ciurla M, Adamowski J, Szafran B et al (2002) Modelling of confinement potentials in quantum dots. Physica E 15(4):261–268
Comparelli R, Zezza F, Striccoli M et al (2003) Improved optical properties of CdS quantum dots by ligand exchange. Mater Sci Eng C 23(6–8):1083–1086
Deng Z, Samantz A, Nangreave J et al (2012) Robust DNA-functionalized core/shell quantum dots with fluorescent emission spanning from UV-vis to near-IR and compatible with DNA-directed self-assembly. J Am Chem Soc 134(42):17424–17427
Dequaire M, Degrand C, Limoges B (2000) An electrochemical metalloimmunoassay based on a colloidal gold label. Anal Chem 72(22):5521–5528
Dey P, Blakey I, Thurecht KJ et al (2013) Self-assembled hyperbranched polymer–gold nanoparticle hybrids: understanding the effect of polymer coverage on assembly size and SERS performance. Langmuir 29(2):525–533
Endo T, Yamamura S, Kerman K et al (2008) Label-free cell-based assay using localized surface plasmon resonance biosensor. Anal Chim Acta 614(2):182–189
Fiorella D, Cheng XR, Chan T et al (2011) Electrochemical detection of AChE-induced gold nanostructure growth on disposable chips. Electrochem Commun 13(2):190–193
Fu Y, Yuan R, Xu L et al (2005) Indicator free DNA hybridization detection via EIS based on self-assembled gold nanoparticles and bilayer two-dimensional 3-mercaptopropyltrimethoxysilane onto a gold substrate. Biochem Eng J 23(1):37–44
Fu HK, Chen CW, Wang CH et al (2008) Creating optical anisotropy of CdSe/ZnS quantum dots by coupling to surface plasmon polariton resonance of a metal grating. Opt Express 16(9):6361–6367
Galian RE, Guardia MDL (2009) The use of quantum dots in organic chemistry. TrAC Trends Anal Chem 28(3):279–291
Garcia MBG, Garcia AC (1995) Adsorptive stripping voltammetric behaviour of colloidal gold and immunogold on carbon paste electrode. Bioelectrochem Bioenerg 28(2):389–395
Gentry JW (1995) The aerosol science contributions of Michael Faraday. J Aerosol Sci 26(2):341–349
Gherghi IC, Girousi ST, Voulgaropoulos AN et al (2003) Study of interactions between actinomycin D and DNA on carbon paste electrode (CPE) and on the hanging mercury drop (HMDE) surface. J Pharm Biomed Anal 31(6):1065–1078
Gu HY, Yu AM, Chen HY (2001) Direct electron transfer and characterization of hemoglobin immobilized on a Au colloid–cysteamine-modified gold electrode. J Electroanal Chem 516(1–2):119–126
Hansen JA, Wang J, Kawde A et al (2006) Quantum-dot/aptamer-based ultrasensitive multi-analyte electrochemical biosensor. J Am Chem Soc 128(7):2228–2229
He Y, Zeng K, Zhang X et al (2010) Ultrasensitive electrochemical detection of nucleic acid based on the isothermal strand-displacement polymerase reaction and enzyme dual amplification. Electrochem Commun 12(7):985–988
Hepel M, Zhong CJ (2010) Functional nanoparticles for bioanalysis, nanomedicine, and bioelectronic devices. ACS Symp Ser. doi:10.1021/bk-2012-1113.ch008

Hernandez L, Zapardiel A, Lopez JAP et al (1988) Voltammetric studies of pinazepam and BrTDO accumulation at the HMDE surface. J Electroanal Chem 255(1–2):85–95
Hiep HM, Endo T, Saito M et al (2008) Label-free detection of melittin binding to a membrane using electrochemical-localized surface plasmon resonance. Anal Chem 80(6): 1859–1864
Hiep HM, Kerman K, Endo T et al (2010) Nanostructured biochip for label-free and real-time optical detection of polymerase chain reaction. Anal Chim Acta 661:111–116
Hines MA, Guyot-Sionnest P (1996) Synthesis and characterization of strongly luminescing ZnS-capped CdSe nanocrystals. J Phys Chem 100(2):468–471
Hostetler MJ, Templeton AC, Murray RW (1999) Dynamics of place-exchange reactions on monolayer-protected gold cluster molecules. Langmuir 15(11):3782–3789
Huang H, Li J, Tan Y et al (2010) Quantum dot-based DNA hybridization by electrochemiluminescence and anodic stripping voltammetry. Analyst 135(7):1773–1778
Hung VWS, Kerman K (2011) Gold electrodeposition on carbon nanotubes for the enhanced electrochemical detection of homocysteine. Electrochem Commun 13(4):328–330
Jackman RJ, Brittain ST, Adams A, Prentiss MG et al (1998) Design and fabrication of topologically complex, three-dimensional microstructures. Science 280(5372):2089–2091
Jia J, Wang B, Wu A et al (2002) A method to construct a third-generation horseradish peroxidase biosensor: self-assembling gold nanoparticles to three-dimensional Sol – Gel network. Anal Chem 74(9):2217–2223
Jimenez VL, Leopold MC, Mazzitelli C et al (2003) HPLC of monolayer-protected gold nanoclusters. Anal Chem 75(2):199–206
Kerman K, Saito M, Morita Y et al (2004) Electrochemical coding of single-nucleotide polymorphisms by monobase-modified gold nanoparticles. Anal Chem 76(7):1877–1884
Kerman K, Chikae M, Yamamura S (2007a) Gold nanoparticle-based electrochemical detection of protein phosphorylation. Anal Chim Acta 588(1):26–33
Kerman K, Endo T, Tsukamoto M et al (2007b) Quantum dot-based immunosensor for the detection of prostate-specific antigen using fluorescence microscopy. Talanta 71(4):1494–1499
Kim DK, Kerman K, Hiep HM et al (2008) Label-free optical detection of aptamer-protein interactions using gold-capped oxide nanostructures. Anal Biochem 379(1):1–7
Kim T, El-said WA, Choi J (2012) Highly sensitive electrochemical detection of potential cytotoxicity of CdSe/ZnS quantum dots using neural cell chip. Biosens Bioelectron 32(1):266–272
Kumar SS, Kwak K, Lee D (2011) Electrochemical sensing using quantum-sized gold nanoparticles. Anal Chem 83(9):3244–3247
Kuppayee M, Nachiyar GKV, Ramasamy V (2011) Synthesis and characterization of Cu 2+ doped ZnS nanoparticles using TOPO and SHMP as capping agents. Appl Surf Sci 257(15): 6779–6786
Li JJ, Wang YA, Guo W et al (2003) Large-scale synthesis of nearly monodisperse CdSe/CdS Core/Shell nanocrystals using Air-stable reagents via successive Ion layer adsorption and reaction. J Am Chem Soc 125(41):12567–12575
Li LD, Zhao HT, Chen ZB et al (2011) Aptamer biosensor for label-free impedance spectroscopy detection of thrombin based on gold nanoparticles. Sens Actuator B 157(1):189–194
Li N, Brahmendra A, Veloso AJ et al (2012) Disposable immunochips for the detection of *Legionella pneumophila* using electrochemical impedance spectroscopy. Anal Chem 84(8):3485–3488
Li N, Cheng XR, Brahmendra A et al (2013) Photonic crystals on copolymer film for bacteria detection. Biosens Bioelectron 41(15):354–358
Liang KZ, Mu WJ (2006) Flow-injection immuno-bioassay for interleukin-6 in humans based on gold nanoparticles modified screen-printed graphite electrodes. Anal Chim Acta 580(2): 128–135
Liao S, Liu Y, Zeng J et al (2010) Aptamer-based sensitive detection of target molecules via RT-PCR signal amplification. Bioconjug Chem 21(12):2183–2189

Liu ZM, Wang H, Yang Y (2004) Amperometric tyrosinase biosensor using enzyme-labeled Au colloids immobilized on cystamine/chitosan modified gold surface. Anal Lett 37(6):1079–1091

Liu G, Lee T, Wang J (2005) Nanocrystal-based bioelectronic coding of single nucleotide polymorphisms. J Am Chem Soc 127(1):38–39

Loukanov AR, Dushkin CD, Papazova KI et al (2004) Photoluminescence depending on the ZnS shell thickness of CdS/ZnS core-shell semiconductor nanoparticles. Colloid Surf A 245(1–3): 9–14

Luo X, Morrin A, Killard AJ (2006) Application of nanoparticles in electrochemical sensors and biosensor. Electroanal 18(4):319–326

Mansur HS, Gonzalez JC, Mansur AAP (2011) Biomolecule-quantum dot systems for bioconjugation applications. Colloid Surf B 84(2):360–368

Micic OI, Curtis CJ, Jones KM et al (1994) Synthesis and characterization of InP quantum dots. J Phys Chem 98(19):4966–4969

Miles DT, Murray RW (2001) Redox and double-layer charging of phenothiazine functionalized monolayer-protected clusters. Anal Chem 73(5):921–929

Numnuam A, Chumbimuni-Torres KY, Xiang Y et al (2008) Aptamer-based potentiometric measurements of proteins using ion-selective microelectrodes. Anal Chem 80(3):707–712

Oishi J, Jung J, Tsuchiya A et al (2010) A gene-delivery system specific for hepatoma cells and an intracellular kinase signal based on human liver-specific bionanocapsules and signal-responsive artificial polymer. Int J Pharm 396(1–2):174–178

Palecek E, Bartosik M (2012) Electrochemistry of nucleic acids. Chem Rev 112(6):3427–3481

Pan JL, Fonstad Jr CG (2000) Theory, fabrication and characterization of quantum well infrared photodetectors. Mater Sci Eng R 28(3–4):65–147

Patolsky F, Gabriel T, Willner I (1999) Controlled electrocatalysis by microperoxidase-11 and Au-nanoparticle superstructures on conductive supports. J Electroanal Chem 479(1):69–73

Pfeiffer M, Lindfors K, Wolpert C et al (2010) Enhancing the optical excitation efficiency of a single self-assembled quantum dot with a plasmonic nanoantenna. Nano Lett 10(11): 4555–4558

Pingarrón JM, Yáñez-Sedeño P, González-Cortés A (2008) Gold nanoparticle-based electrochemical biosensors. Electrochim Acta 53(19):5848–5866

Pumera M, Aldavert M, Mills C et al (2005a) Direct voltammetric determination of gold nanoparticles using graphite-epoxy composite electrode. Electrochim Acta 50(18):3702–3707

Pumera M, Castaneda MT, Pividori MI et al (2005b) Magnetically trigged direct electrochemical detection of DNA hybridization using Au67 quantum dot as electrical tracer. Langmuir 21(21):9625–9629

Quinn BM, Liljeroth P, Ruiz V et al (2003) Electrochemical resolution of 15 oxidation states for monolayer protected gold nanoparticles. J Am Chem Soc 125(22):6644–6645

Raj CR, Okajima T, Ohsaka T (2003) Gold nanoparticle arrays for the voltammetric sensing of dopamine. J Electroanal Chem 543(2):127–133

Rauf S, Glidle A, Cooper JM (2010) Layer-by-layer quantum dot constructs using self-assembly methods. Langmuir 26(22):16934–16940

Shiddiky MJA, Rauf S, Kithva PH et al (2012) Graphene/quantum dot bionanoconjugates as signal amplifiers in stripping voltammetric detection of EpCAM biomarkers. Biosens Bioelectron 35(1):251–257

Takei K, Fang H, Kumar S et al (2011) Quantum confinement effects in nanoscale-thickness InAs membranes. Nano Lett 11(11):5008–5012

Tang D, Yuan R, Chai Y (2006) Electrochemical immuno-bioanalysis for carcinoma antigen 125 based on thionine and gold nanoparticles-modified carbon paste interface. Anal Chim Acta 564(2):158–165

Thakor AS, Jokerst J, Zavaleta C et al (2011) Gold nanoparticles: a revival in precious metal administration to patients. Nano Lett 11(10):4029–4036
Thuer R, Vigassy T, Hirayama M et al (2007) Potentiometric immunoassay with quantum dot labels. Anal Chem 79(13):5107–5110
Vestergaard M, Kerman K, Kim DK et al (2008) Detection of Alzheimer's tau protein using localized surface plasmon resonance-based immunochip. Talanta 74:1038–1042
Wang J, Lu J, Hocevar SB et al (2000) Bismuth-coated carbon electrodes for anodic stripping voltammetry. Anal Chem 72(14):3218–3222
Wang H, Tessmer I, Croteau DL et al (2008) Functional characterization and atomic force microscopy of a DNA repair protein conjugated to a quantum dot. Nano Lett 8(6):1631–1637
Wojciechowski KF, Bogdanow H (1998) Quantum size effects of ultrathin simple metal layers on the example of lithium. Surf Sci 397(1–3):53–57
Xiao Y, Patolsky F, Katz E et al (2003) "Plugging into enzymes": nanowiring of redox enzymes by a gold nanoparticle. Science 299(5614):1877–1881
Xie W (2001) Analysis of a positively charged exciton in a parabolic quantum dot. Physica B 302(1–4):112–118
Xie R, Rutherford M, Peng X (2009) Formation of high-quality I-III-VI semiconductor nanocrystals by tuning relative reactivity of cationic precursors. J Am Chem Soc 131(15):5691–5697
Yin H, Zhou Y, Ai S et al (2010) Sensitivity and selectivity determination of BPA in real water samples using PAMAM dendrimer and CoTe quantum dots modified glassy carbon electrode. J Hazard Mater 174(1–3):236–243
Zanardi P, D'Amico I, Ionicioiu R et al (2002) Quantum information processing using semiconductor nanostructures. Physica B 314(1–4):1–9
Zhou J, Huang H, Xuan J et al (2010) Quantum dots electrochemical aptasensor based on three-dimensionally ordered macroporous gold film for the detection of ATP. Biosens Bioelectron 26(2):834–840
Zhu JJ, Huang H, Wang W et al (2012) Preparation and analytical applications of quantum dots. In: Reference module in chemistry, molecular sciences and chemical engineering comprehensive sampling and sample preparation analytical techniques for scientists 3:169–187

Chapter 7
Nanocarbon Film-Based Electrochemical Detectors and Biosensors

Osamu Niwa and Dai Kato

Abstract For electroanalytical or electrochemical biosensor applications, carbon electrodes are more suitable than other electrode materials including gold, platinum and various metal oxide-based electrodes. This is because carbon-based electrodes have a wide potential window and a relatively low background noise level, which is very important as regards the electrochemical detection of certain biomolecules. Recently, new carbon materials including carbon nanotubes (CNTs), graphene and boron-doped diamond (BDD) have been developed and employed as electrode materials. BDD has been applied to electroanalysis because of its extremely wide potential window and stability. In contrast, nanocarbon materials such as CNTs and graphene have been actively developed as the electrodes of energy devices such as batteries and fuel cells. Recently, these materials have been studied as sensing platforms for such biomolecules as DNA and proteins. Compared with these nanocarbons, which have a powder-like or fiber morphology, carbon films are advantageous as electrodes for biochemical detection or biosensor platforms. This is because they are sufficiently conductive without needing to be doped with other atoms, and they have a relatively low background current, which is effective for obtaining a high signal-to-noise ratio or a low detection limit. Film electrodes can be microfabricated into sensor electrodes of various sizes and shapes with excellent reproducibility. Here, we introduce different kinds of carbon-based film electrodes fabricated with a variety of techniques including the pyrolysis of organic films and vacuum processes such as sputtering or chemical vapor deposition (CVD), and we describe their application for electroanalysis and chemical and biosensors.

Keywords Direct electron transfer (DET) • Electrochemical biosensor • Interdigitated array (IDA) • Nanocarbon film electrode • Sputtering

---

O. Niwa (✉) • D. Kato
National Institute of Advanced Industrial Science and Technology,
1-1-1 Higashi, Tsukuba, Ibaraki 305-8566, Japan
e-mail: niwa.o@aist.go.jp

M.C. Vestergaard et al. (eds.), *Nanobiosensors and Nanobioanalyses*,
DOI 10.1007/978-4-431-55190-4_7

7.1 Introduction

Electrochemical detection techniques are widely employed for chemical, biochemical and environmental analysis because they offer the advantages of simplicity, a short measurement time, low cost, versatility and sensitivity as regards the construction of sensing devices. Moreover, they can be miniaturized to deal with very small volume samples. Various kinds of electrode materials have been used for electroanalysis. Carbon materials such as glassy carbon (GC) and highly oriented pyrolytic graphite (HOPG) have been employed in a number of electroanalytical studies, including research on an electrochemical detector for HPLC and on various biosensors, since these electrodes are chemically stable, highly conductive and have low permeability with respect to gases and solvents. A recent review article has well described the electrochemistry of certain carbon-based electrodes (McCreery 2008).

In the last 20 years, electrochemical measurements using boron-doped diamond (BDD) electrodes (Compton et al. 2003; Swain and Ramesham 1993) have become more widespread. A BDD electrode is extremely chemically stable and BDD exhibits a much wider potential window and lower background noise level than other electrode materials. These unique electrochemical characteristics of BDD electrodes are advantageous in terms of detecting various species including heavy metal ions (Pb^{2+}, Cd^{2+}) (Manivannan et al. 2004), histamine and serotonin (Sarada et al. 2000; Singh et al. 2010) and even nonmetal proteins (Chiku et al. 2008). Although BDD electrodes have excellent electroanalytical characteristics, a relatively high temperature of 400–700 °C is needed to obtain high-quality films, and expensive plasma or thermal chemical vapor deposition (CVD) equipment must be utilized, which increases the deposition cost and limits the substrates that can be used for deposition. In addition, the doping of other atoms such as boron is needed to obtain sufficient conductivity for electrochemical measurements, because diamond is a wide bandgap semi-conductor unlike sp^2 carbons.

In the last decade, nanocarbon materials including carbon nanotubes (CNTs) and graphene have been more intensively studied with a view to using them as electrode materials for both biosensors (Pumera et al. 2010; Qureshi et al. 2009; Vashist et al. 2011; Zhou et al. 2009) and biofuel cells (Miyake et al. 2011). However, the above nanocarbon materials have a fiber or flake-like structure and require a substrate electrode for biosensor construction. On the other hand, film electrodes are more suitable for electroanalytical applications. Although the wide surface area of CNTs and graphene can provide a large current, the detection limit cannot be improved due to their large background noise current. On the other hand, film electrodes are needed to improve the electron transfer rate of analytes in order to retain diffusion-limited electrochemical reactions. This is because their smooth surface has fewer active sites than the surfaces of nanocarbon materials, which often induces an electron transfer limited electrochemical reaction with analytes that have relatively low electron transfer rates.

It is very important to be able to fabricate electrodes of any shape and size with high reproducibility for use as platforms for chemical or biochemical sensors (Niwa 2005).

Conventional photolithography or micromachining fabrication processes can be more easily applied to film electrodes than to powder or fiber-like nanomaterials. These processes also allow us to fabricate electrodes with various shapes and excellent throughput. Carbon films are particularly useful for such applications since carbon is a sustainable material compared with noble metals and also has electrochemical properties that can be tuned by changing the structure and surface termination group. This chapter describes various processes for forming pure carbon film electrodes and reports the electrodes' electrochemical performance. We then describe the application of carbon film electrodes for the direct detection of biomolecules.

7.2 Fabrication, Structure and Basic Electrochemical Performance of Pure Carbon Film Electrodes

In the early stages, the pyrolysis of organic films including various polymers was used to form carbon film electrodes. Table 7.1 shows examples of carbon film preparation by pyrolysis. Kaplan et al. deposited 3,4,9,10-perylenetetracarboxylic dianhydride (PTDA) films on the substrate, pyrolyzed them above 700 °C and obtained conducting carbon film (Kaplan et al. 1980). The conductivity was 250 S/cm, which is similar to that of a GC electrode but lower than that of graphite. Rojo et al. obtained carbon film using a similar method to Kaplan et al. and employed it for electrochemical measurements (Rojo et al. 1986). They used the film electrode to measure catechol and catecholamines and obtained sharp current responses. Recently, Morton et al. fabricated carbon film by pyrolyzing parylene C formed by CVD at 900 °C, and reported that the electrochemical properties of their film are similar to those of carbon film electrodes fabricated from pyrolyzed photoresist films (Morton et al. 2011).

Table 7.1 Fabrication of carbon film electrodes by pyrolysis process

Carbon film	Procedures and properties	References
Pyrolysis of PTDA[a]	PTDA deposited in quartz tube and pyrolyzed at 850 °C at 0.01 Torr	Kaplan et al. (1980), Rojo et al. (1986)
	Conductivity: 250 S/cm	
Pyrolysis of parylene C	CVD of parylene C and pyrolyzed at 900 °C	Morton et al. (2011)
	Similar properties to pyrolyzed photoresist	
Pyrolyzed photoresist films (PPF)	Pyrolysis of phenol-formaldehyde resin around 1,000 °C	Lyons et al. (1983)
	Pyrolysis of photoresist AZ4330 from 600 to 1,100 °C.	Kim et al. (1998), Ranganathan and McCreery (2001)
	Near atomic flatness <0.5 nm	
	Pyrolysis of photoresist AZ4620 at 1,100 °C.	Brooksby and Downard (2004)
	Conductivity comparable to GC	
	Pyrolysis of photoresist AZ4562 by rapid	Javier del Campo et al. (2011)
	Thermal process (140 °C/min to 1,000 °C).	

[a]3,4,9,10-Perylenetetracarboxylic dianhydride

A number of groups have fabricated carbon film electrodes by pyrolyzing positive photoresist polymers, which mainly consist of phenol resin. Most of the groups pyrolyzed the film at around 1,100 °C. Lyons et al. spin-coated phenolic resin on silicon and Al_2O_3 substrates and obtained their carbon films by changing the pyrolysis temperature (Lyons et al. 1983). The resistivity was changed from 2×10^{-2} to $2 \times 10^{-3}\,\Omega$ cm depending on the pyrolysis temperature. The electrochemical performance of pyrolyzed photoresist films (PPF) has been intensively studied by McCreery and Madou's groups (Kim et al. 1998; Ranganathan and McCreery 2001). PPF film has a lower O/C ratio than a GC electrode and relatively larger peak separations were observed for $Fe^{3+/2+}$ and dopamine measurements. Since the resist solution was optimized to form homogeneous thin films, the carbon films formed with this method tend to have a relatively smooth surface. Ranganathan et al. reported the atomic force microscopy (AFM) measurement of PPF carbon film and observed less than 0.5 nm flatness (Ranganathan and McCreery 2001). Brooksby et al. reported the surface modification of PPF which they achieved by diazonium reduction and observed with AFM (Brooksby and Downard 2004), which could be important for modifying various molecules for biosensor applications. The fabrication conditions of PPF-based carbon films including types of resists and heating programs, and their properties, such as resistivity and surface roughness, were recently well summarized by Compton's group (Javier del Campo et al. 2011).

On the other hand, carbon film electrodes including diamond-like carbons (DLC) and tetrahedral amorphous carbons have been developed by using various vacuum deposition techniques, which include electron beam evaporation, plasma-assisted chemical vapor deposition (PACVD), radio-frequency plasma enhanced chemical vapor deposition (RF-PECVD) and radio frequency (RF) magnetron sputtering. DLC electrodes are chemically inert and have low surface roughness, which results in a high signal-to-noise ratio and low capacitance (Schnupp et al. 1998). Blackstock et al. reported ultraflat carbon film (~0.1 nm) whose electrochemical response is similar to that of GC (Blackstock et al. 2004). A number of groups have developed a variety of carbon films for biosensor applications, as described later in Sect. 7.3.2. Hirono et al. developed a very hard carbon film using electron cyclotron resonance (ECR) sputtering (Hirono et al. 2002). The film consists of sp^2 and sp^3 hybrid bonds with a nanocrystalline structure. Figure 7.1 shows a transmission electron microscope (TEM) image of ECR carbon films formed by changing the ion acceleration voltage during ECR sputtering. Strong ion irradiation on the substrate surface changed the nanostructure of the carbon films depending on the ion acceleration voltage applied between the target and the substrate. The sp^3 content increased from about 15 to 43 % as the ion acceleration voltage increased from 20 to 85 V. In fact, a parallel layered structure identified as a nano-order graphite crystalline structure can be observed at a low ion acceleration voltage, but a curved and closed nanostructure is dominant at a high ion acceleration voltage. The surface of the film is extremely flat with an average roughness (Ra) of 0.07 nm measured with AFM (Niwa et al. 2006). Figure 7.2a shows the potential window of sputter-deposited ECR nanocarbon film. The potential window is much wider than that of GC and

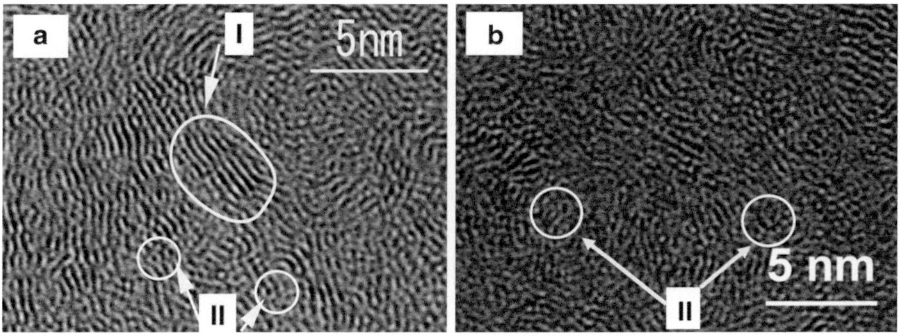

Fig. 7.1 TEM images of ECR nanocarbon films prepared using ion acceleration voltages of 20 V (**a**) and 85 V (**b**). Regions I and II indicate "graphite-layered" structure and "curved and fine-closed" structure, respectively

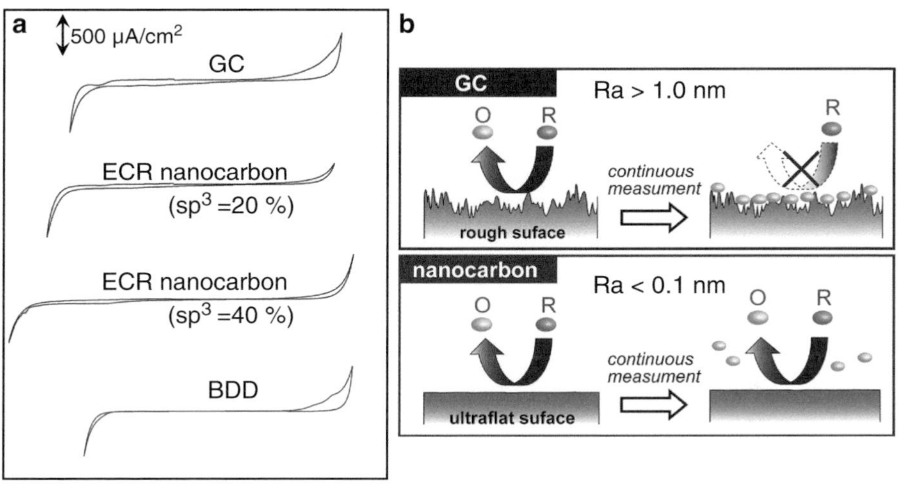

Fig. 7.2 (**a**) Voltammograms for GC, ECR nanocarbon and BDD electrodes in 0.05 M H_2SO_4 deoxygenated with argon. Scan rate = 0.1 V/s. (**b**) Schematic representation of surface fouling of electroactive molecules on GC and ECR nanocarbon film electrode surface

almost the same as that of BDD in the positive potential region because the film contains more than 40 % of sp^3 bond. The flat surface of the film also achieves a low capacitive current and suppresses electrode fouling. Figure 7.2b shows a schematic representation of the mechanism for suppressing fouling. With a conventional electrode such as a GC electrode, the relatively rougher surface adsorbs the molecules after an electrochemical reaction. In contrast, after the electrochemical reaction the molecules easily desorbed from the ECR nanocarbon film electrode surface because of its flat and chemically stable surface.

7.3 Application of Carbon Film Electrodes

This section describes three different applications for carbon film. The first is the direct electrochemical measurement of biomolecules. The second is a biosensor application realized by modification with various proteins and DNAs. The third application is miniaturized carbon film electrodes, which are currently becoming very important for combination with microchips and microfluidic systems.

7.3.1 Direct Electrochemical Detection of Biomolecules Using Nanocarbon Film

As an example of direct electrochemical measurement using a carbon film electrode, this section describes the detection of certain biomolecules using an ECR sputtered nanocarbon film electrode. ECR nanocarbon film has a wide potential window and low surface fouling. These characteristics have been used to study the highly sensitive detection of 8-hydroxy 2′-deoxyguanosine (8-OHdG), a damaged urinary DNA molecule, at the nM level (Kato et al. 2011). Figure 7.3 compares the signals obtained when 8-OHdG was continuously injected onto nanocarbon film installed in a radial flow cell. The 8-OHdG detection limit was 3 nM at the nanocarbon film electrode with a high reproducibility (C.V. value = 0.75 % ($n = 12$)) that

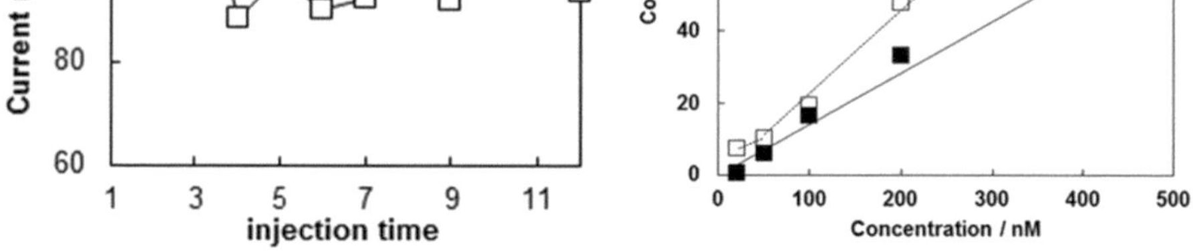

Fig. 7.3 (**a**) Reproducibility of 1 μM 8-OHdG measurement at the nanocarbon film (*black squares*) and the GC (*white squares*) electrode by flow injection analysis (FIA) measurement, showing the ratios of the obtained FIA responses in each measurement on the carbon electrodes to that in the first measurement as a function of time. Conditions: 20 μL injection; run buffer, PB (pH 7.5)/MeOH (99/1, v/v); flow rate, 200 μL/min; detection potential, 1.0 and 0.8 V vs. Ag/AgCl at the nanocarbon film and GC electrode, respectively. (**b**) Calibration curves of 8-OHdG at the nanocarbon film (*black squares*) and the GC (*white squares*) electrode obtained by HPLC-ECD. Conditions: mobile phase, PB (pH 7.5)/MeOH (95/5, v/v); temperature, 40 °C; the other conditions are the same as those in Fig. 7.3a

was superior to that of GC electrodes (low detection limit = 7.2 nM, C.V. value = 9.28 % (n = 12)). This electrode could be successfully used to detect a glial transmitter in combination with high-performance liquid chromatography with electrochemical detection (HPLC-ECD) and microdialysis (Yamamura et al. 2013). These film properties can be employed to measure much larger biomolecules such as oligonucleotides. We have reported extremely simple electrochemical DNA analysis techniques such as DNA methylation (Kato et al. 2008a) and single nucleotide polymorphism (SNP) (Kato et al. 2008b) detection based on the quantitative measurement of all the bases by direct electrochemical oxidation.

Figure 7.4 shows background-subtracted square-wave voltammograms (SWVs) of 3 μM of CpG oligonucleotides (**1**: 5′-CGCGCG-3′, **2**: 5′-(mC)GCGCG-3′, **3**: 5′-(mC)G(mC)GCG-3′, **4**: 5′-(mC)G(mC)G(mC)G-3′) at the nanocarbon film electrode in 50 mM pH 5.0 acetate buffer containing 0.3 M $NaNO_3$. The nanocarbon film electrode could distinguish both 5-methylcytosine (mC) and cytosine (C) bases individually, by measuring the peak potential differences caused by C methylation. These clear results allow us to obtain direct quantitative measurements of both the C and mC bases of oligonucleotides with much longer actual sequences by employing nanocarbon film electrodes. For example, we measured an oligonucleotide (13-mer) including a sequence from the tumor suppressor retinoblastoma (RB1) gene containing hotspot methylation sites (5′-CGAACACCCAGGC-3′), and three kinds of methylated oligonucleotides with different methylation statuses and positions. We were able to quantitatively differentiate the voltammograms for these oligonucleotides (Goto et al. 2010). These results indicate that the influence of the methylation position of the sequences was almost negligible as regards electrochemical DNA methylation detection. Furthermore, we also measured the longer

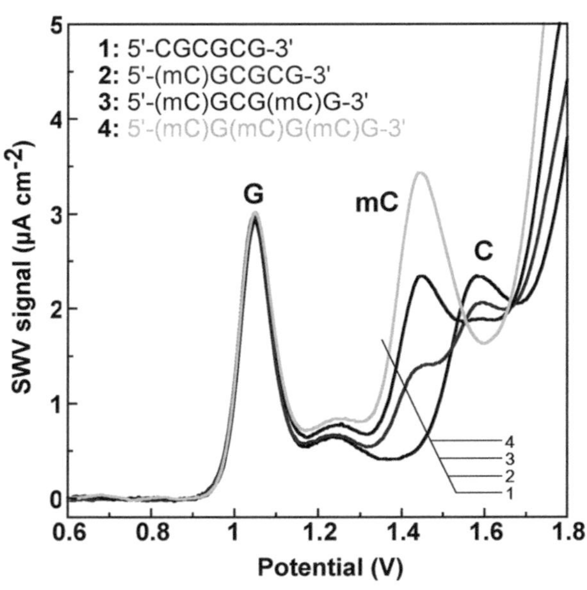

Fig. 7.4 Background-subtracted SWVs of 3 μM of CpG oligonucleotides (**1**–**4**) at ECR nanocarbon film electrode in 50 mM pH 5.0 acetate buffer containing 0.3 M $NaNO_3$. Amplitude = 25 mV, ΔE = 5 mV, f = 10 Hz

oligonucleotides (60-mers) that constitute a non-methylated and a methylated CpG dinucleotide with different methylation ratios. The relationship between the methylation ratio in the oligonucleotides and the current response assigned to the mC of the CpG oligonucleotides was linear ($r=0.9955$) despite the use of the direct DNA oxidation method (Goto et al. 2010). In general, we must use a bisulfite or the restriction enzyme treatment of the DNA samples to distinguish C and mC in DNA since both bases exhibit identical Watson–Crick base-pair behavior. Therefore, this is highly advantageous as regards achieving both mC- and C-positive assays solely by using the electrochemical oxidation of oligonucleotides without a bisulfite or a labeling process.

7.3.2 Enzyme and DNA Modified Carbon Film Electrodes

Carbon film-based electrochemical sensors have been studied by modifying the films with various biomolecules such as DNA and proteins including enzymes and antibodies. Table 7.2 shows a selection of recent publications related to carbon film-based biosensors. Metal films or carbon paste electrode-based biosensors have been used as blood glucose sensors. These electrode materials must be suitable for glucose measurement because of its high concentration in blood. Early studies reported glucose sensors that were fabricated by a DLC electrode modified with glucose oxidase (Higson and Vadgama 1995). However, the detection potential of H_2O_2 is very high at a carbon film surface and the detection limit is even worse than that of a GC electrode (Maalouf et al. 2006) because of the high overpotential of H_2O_2 oxidation on a carbon film electrode. In contrast, Yu et al. functionalized a tetrahedral amorphous carbon thin film surface by covalently attaching peptide nucleic acid to detect target DNA. They successfully measured the impedance change

Table 7.2 Thin carbon film-based biosensors

Carbon film	Procedures and properties	References
DLC	DLC electrode was modified with glucose oxidase (GOD) and detection limit is 50 μM (not as good as GC)	Maalouf et al. (2006)
	GOD was modified by BSA and glutaraldehyde on porous polymer membrane and pressed to DLC	Higson and Vadgama (1995)
	DLC electrode was modified with antibody to detect C-reactive protein (CRP)	Lee et al. (2008)
Boron-doped DLC	Boron-doped DLC microelectrode was applied for electrochemical ELISA of viruses (HIV, HBV, HCV)	Kim et al. (1998)
Tetrahedral amorphous carbon film	Covalent attachment of PNA on the carbon film by detecting impedance change by target DNA hybridization.	Yu et al. (2011)
ECR-sputtered nanocarbon film	ECR nanocarbon film was modified with GABAse and detected γ-aminobutyric acid (GABA) by detecting NADPH (coenzyme)	Sekioka et al. (2008)

caused by the hybridization of target DNA sequences (He et al. 2011). Jo-Il et al. (2009) used a boron-doped DLC microelectrode for an electrochemical enzyme-linked immunoabsorbent assay (ELISA) to detect the human immunodeficiency virus (HIV), human hepatitis B virus (HBV) and human hepatitis C virus (HCV). Both DNA and antigen measurements require a low detection limit since the target concentrations are usually low. For such applications, a carbon film electrode is more suitable because the film can be fabricated reproducibly and has a low noise level, which makes it suitable for achieving a low detection limit.

We used ECR sputtered nanocarbon film to detect γ-aminobutyric acid (GABA), which is a well known inhibitory neurotransmitter whose in vivo concentration is in the low μM region or below. GABA sensors are designed to use an enzymatic GABAse reaction. The electrochemical detection of GABA requires the measurement of β-nicotinamide adenine dinucleotide phosphate (NADPH) as a coenzyme for various enzymatic reactions. However, the high overpotential needed for NAD(P)H oxidation and the electrode surface fouling caused by the irreversible adsorption of $NADP^+$ (oxidized form) make the electrochemical detection of NAD(P)H difficult (Sekioka et al. 2008). We obtained very stable responses for NADPH with a very low background noise level because the nanocarbon film has an extremely flat surface, a wide potential window and suppresses surface fouling by oxidized NADPH ($NADP^+$). As a result, we obtained a low detection limit of 10 nM for NADPH, which was more than one order of magnitude lower than that for a GC electrode (250 nM) (Fig. 7.5a). We then constructed a GABA sensor based on highly sensitive and reproducible NADPH detection with the nanocarbon film electrode. The sensor comprised the nanocarbon film electrode modified with GABAse

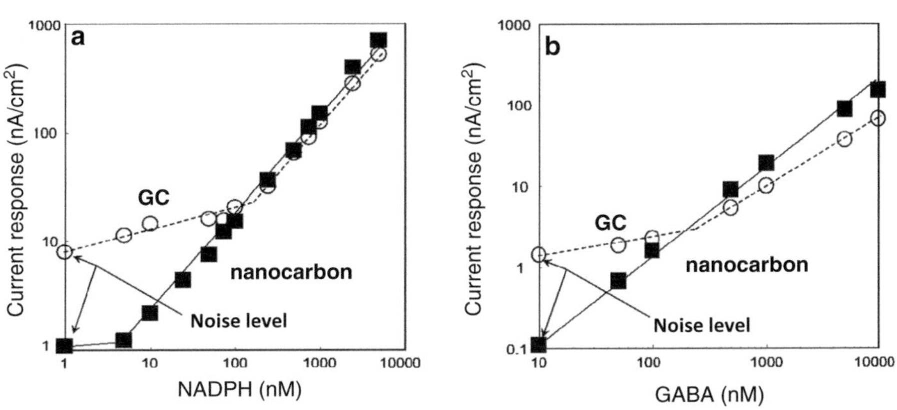

Fig. 7.5 (**a**) Calibration curves of NADPH at the GC and the ECR nanocarbon film electrode. (**b**) Calibration curves of GABA at the BSA-GABAse film-modified GC and the ECR nanocarbon film electrode-based sensor. The α-ketoglutarate and $NADP^+$ concentrations were 50 and 100 μM, respectively. The electrolyte was used for phosphate buffer (pH 8.2). The potential of each electrode was held at 0.9 V vs. Ag/AgCl at a flow rate of 20 μL/min. The current noise of this measurement was determined by the blank response

co-immobilized with bovine serum albumin. The proposed electrochemical GABA sensor achieved a detection limit of 30 nM GABA, which is lower than that of any previously reported GABA sensor (Fig. 7.5b).

Recently, third-generation biosensors have been developed using nanocarbon films such as graphenes (Pumera et al. 2010). As this type of biosensor is based on direct electron transfer (DET) between the electrode surface and the enzymes, an extremely wide surface area and nanocarbon materials with nano-order structures are suitable for reducing the distance between an enzyme active center and the electrode surface. Therefore, the extremely flat surface of carbon film electrodes such as DLC and sputtered nanocarbon film electrodes make it difficult to fabricate third-generation biosensors. Recently, we have developed a fine nanostructure on a sputtered nanocarbon film surface to realize efficient DET with enzymes. The nanostructure was fabricated using a UV/ozone treatment for the nanocarbon film without a mask (Ueda et al. 2011). After the UV/ozone treatment, a nanothorn-like structure, which is typically 2–3.5 nm high, was confirmed with AFM observation (Fig. 7.6a). X-ray photoelectron spectroscopy and a TEM image revealed that these nanostructures could be formed by

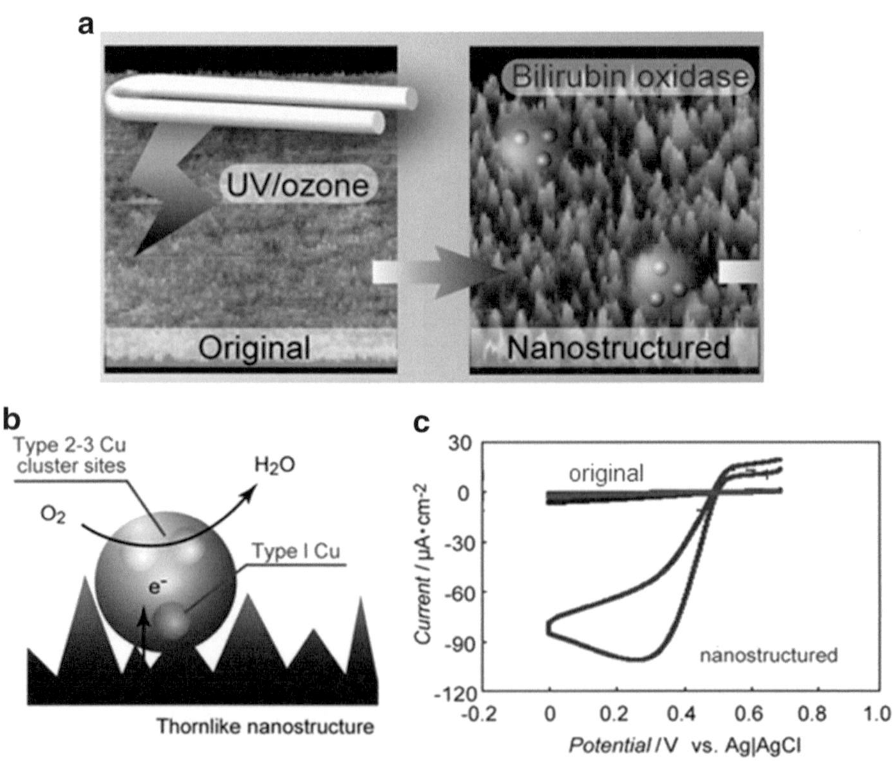

Fig. 7.6 (**a**) Nanothorn-like structure fabricated by UV/ozone treatment of a sputtered nanocarbon film. (**b**) Schematic of DET-type bioelectrocatalysis of BOD on the thorn-like nanostructured carbon film and (**c**) voltammograms of BOD physically adsorbed on original and nanostructured carbon film

employing significantly different etching rates depending on nanometer-order differences in the local sp^3 content of the sputtered nanocarbon film. In fact, the etching rate for the sputtered nanocarbon film was ~7 nm/h, whereas carbon film containing more sp^2 bonds was etched much faster than that with lower sp^2 content. For example, the etching rate for GC is about 10^3 nm/h, but a BDD surface is largely unetched. Therefore, such nanostructures could not be realized using, for example, GC, BDD, DLC or other carbon films formed by the pyrolysis of organic polymers because those carbons have relatively homogeneous structures or micrometer-order crystalline structures. We modified bilirubin oxidase (BOD) on the nanostructured carbon by physical adsorption (Fig. 7.6b). The DET catalytic current amplification was 30 times greater than that obtained with the original flat nanocarbon film, indicating the very efficient electron transfer between our thorn-like carbon film and BOD (Fig. 7.6c). We also compared DET using different carbon materials such as GC, DLC and BDD, but the DET current values of BOD for these materials were much lower than that at our thorn-like carbon film, indicating that the surface nanostructure is very important in terms of improving DET.

7.3.3 Microfabrication of Carbon Film and Applications

One very important advantage of carbon thin film electrodes relates to micro- and nanofabrication because the widely used photolithography or electron beam lithography and etching processes can be easily applied because of the thin and flat structure of carbon films. There are two main methods for fabricating carbon film-based microstructures. One involves patterning the resist by photolithography and pyrolyzing it at high temperature. The other involves forming an electrode pattern on the carbon film by photolithography and then employing an etching process such as oxygen plasma etching. We fabricated carbon-based interdigitated array (IDA) electrodes using photolithography and an oxygen plasma dry-etching process. The IDA electrode was first fabricated by using a Pt film under-layer to increase the electrical conductivity to a level sufficient for voltammetric measurements (Tabei et al. 1992). However, it has been proved that a carbon-based IDA electrode can be applied to conventional measurement and to generation and collection voltammetric measurement without a Pt layer (Niwa and Tabei 1994). When fabricating carbon film by oxygen plasma etching, it is difficult to fabricate the pattern with a resist mask because both carbon film and resist polymer are etched by oxygen plasma. We used a silicon-based positive photoresist to mask the carbon film because silicon-based resist changes into silicon dioxide in oxygen plasma and thus provides an excellent mask. Later, other groups also reported carbon-based IDA electrodes that they fabricated by patterning RF-sputtered carbon film (Fiaccabrino et al. 1996) followed by the pyrolysis of a patterned photoresist (Kostecki et al. 1999). The pyrolysis of electrochemically polymerized film formed on a Pt IDA electrode also provided a carbon-based IDA electrode (Niwa et al. 1994). Since carbon film-based IDA electrodes are suitable for detecting trace level analytes as a result of the current enhancement

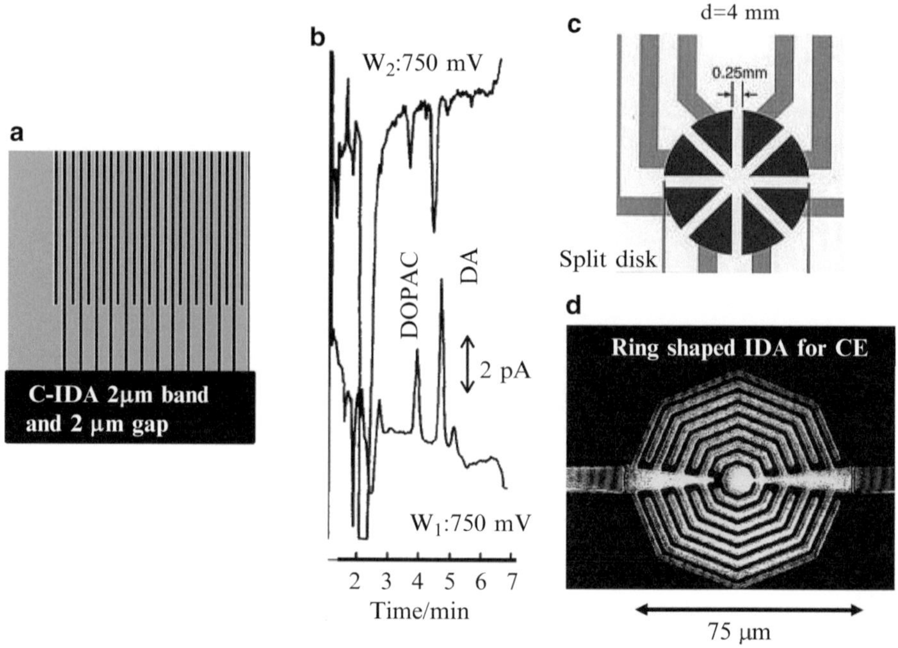

Fig. 7.7 (**a**) SEM image of carbon-based IDA electrode. The band width and gap of the IDA are both 2 μm (250 pairs). (**b**) Typical chromatogram of DA and DOPAC measured with the carbon-based IDA electrode. (**c**) An 8-split disk array electrode for a radial flow cell. (**d**) A ring-shaped IDA electrode for capillary electrophoresis

caused by redox cycling and the lower background current compared with that of metal film electrodes, they have been employed as detectors for liquid chromatography (LC) and capillary electrophoresis (CE). Figure 7.7a shows an image of a typical carbon-based IDA electrode. The band width and gap of the IDA are both 2 μm (250 pairs). We installed this electrode in a channel flow cell used as an amperometric detector for microbore LC. Figure 7.7b shows a typical chromatogram of dopamine (DA) and 3,4-dihydroxyphenylacetic acid (DOPAC) measured with a carbon film IDA electrode. Clear DA and DOPAC peaks were observed by injecting only 100 fg (in 5 μl) of DA and DOPAC, owing to the current enhancement caused by redox cycling and the low background noise level (Tabei et al. 1992). As a result, low detection limits of 32 and 57 amol were obtained. Figure 7.7c shows a split disk array electrode for a radial flow cell. As the buffer solution is introduced into the center of eight channel electrodes, a hydrodynamic-voltammetry-like curve can be obtained by applying different potentials to each electrode simply by injecting the sample solution (Iwasaki et al. 1996). Figure 7.7d is a ring-shaped IDA electrode for CE (Liu et al. 2000). Since fused silica capillaries with diameters between 25 and 75 μm are used for CE, the total electrode diameter is suppressed to less than 100 μm. A microfabricated carbon film-based electrode has been used for biosensors. Lee et al. (2008) fabricated a gold film-based electrochemical cell consisting of working, reference and counter electrodes. Pyrolyzed carbon film was then formed on the working electrode surface and modified with an aptamer. Label-free thrombin detection was demonstrated with an electrochemical impedance measurement. Hayashi et al. (2000) fabricated a 64-channel (8 × 8) rectangular electrode array that they modified with glutamate oxidase and Os-polymer mediator containing horseradish peroxidase (Os-polymer-HRP) (Ohara et al. 1993). A change in the L-glutamate concentration was clearly imaged by using the enzyme-modified array electrode.

Carbon film microelectrodes have been used as detectors in microfluidic devices. A well-known example is chip-based electrophoresis with electrochemical detection. PPF films were incorporated in the CE chip and used to detect 100 nM level DA (Hebert et al. 2003). Luntes' group compared various electrode materials including PPF, carbon fiber, Pd and carbon ink and proved that PPF provides the best sensitivity and lowest detection limit (Fischer et al. 2009). A carbon film microelectrode has also been used as an on-chip detector for the online measurement of cellular release. For example, a PPF electrode was used as an on-chip detector of epinephrine released from PC 12 cells (Larsen et al. 2013). Another application is in vivo measurement. Figure 7.8a shows an image of a microfluidic device consisting of two working carbon film microelectrodes, along with reference and counter electrodes. One of the working electrodes was modified with Os-polymer-HRP and glucose oxidase and the other was modified with Os-polymer-HRP and lactate oxidase. A microdialysis (MD) probe was connected to the upstream port of the chip and the MD probe was inserted in a rat brain. Figure 7.8b shows the variation in the glucose and lactate concentrations in the rat cortex simulated by veratridine at a flow rate of 1.5 μL/min. When veratridine (50 μM) was perfused through the MD probe, a rapid and long-lasting glucose reduction and a lactate increase were observed in real time (Kurita et al. 2002). The effect of L-ascorbic acid, which is a major electroactive interferent in the brain, was removed by L-ascorbate oxidase immobilized upstream of the detection electrodes.

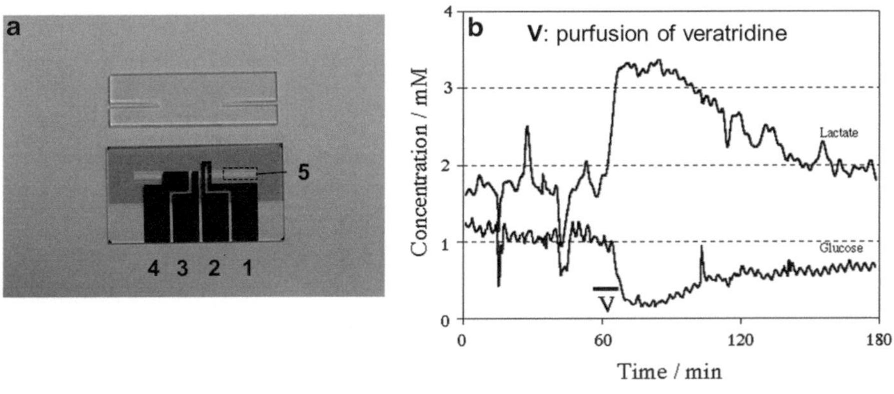

Fig. 7.8 (**a**) Image of microfluidic device consisting of two working carbon film microelectrodes (**1**, **2**), reference (**3**) and counter (**4**) electrodes. The working microelectrodes **1** and **2** were modified with GOD/Os-polymer-HRP and LOD/Os-polymer-HRP, respectively. The portion **5** indicates L-ascorbate oxidase immobilized channel. (**b**) Variation in glucose (*lower trace*) and lactate (*upper trace*) concentration in rat cortex stimulated by veratridine at a flow rate of 1.5 μl/min. *V* indicates that the MD probe was perfused with 50 mM veratridine for 10 min

7.4 Conclusion and Future Trends

This chapter described the fabrication of carbon film electrodes, their electrochemical performance and electroanalytical applications. The pyrolysis of polymeric materials or vacuum deposition processes have been used to form carbon film. The electrochemical performance varied depending on such factors as surface roughness, sp^2/sp^3 ratio and crystallinity. In a direct electrochemical measurement, carbon film has been employed for measuring catecholamines such as DA or L-ascorbic acid. A wide potential window and low background noise level were achieved by increasing the sp^3 content. Nanocarbon films with a higher sp^3 concentration above 40 % formed by the ECR sputtering method improve the detection limit and can even detect large molecules such as DNA by suppressing surface fouling. Carbon films have been applied to enzyme-modified electrochemical sensors and immunosensors. Carbon films have been successfully microfabricated into microarray electrodes of various shapes and applied to LC and CE detectors. On-chip CE devices and microfluidic devices for detecting neurotransmitters and in vivo glucose and lactate concentrations have been fabricated by using carbon films as electrochemical detectors. To improve electrochemical activity, hybrid carbon films have recently been developed, and they include nitrogen-doped carbon films (Kamata et al. 2013; Yang et al. 2012; Zeng et al. 2002; Lagrini et al. 2004) and carbon films doped with metal nanoparticles (Pocard et al. 1992; You et al. 2003). These hybrid-type films are promising as regards extending the area of application of carbon film electrodes to both the electroanalysis of various molecules and biosensors, and energy devices.

References

Blackstock JJ, Rostami AA, Nowak AM et al (2004) Ultraflat carbon film electrodes prepared by electron beam evaporation. Anal Chem 76(9):2544–2552

Brooksby PA, Downard AJ (2004) Electrochemical and atomic force microscopy study of carbon surface modification via diazonium reduction in aqueous and acetonitrile solutions. Langmuir 20(12):5038–5045

Chiku M, Nakamura J, Fujishima A et al (2008) Conformational change detection in nonmetal proteins by direct electrochemical oxidation using diamond electrodes. Anal Chem 80(15): 5783–5787

Compton RG, Foord JS, Marken F (2003) Electroanalysis at diamond-like and doped-diamond electrodes. Electroanalysis 15(17):1349–1363

Fiaccabrino GC, Tang XM, Skinner N et al (1996) Electrochemical characterization of thin-film carbon interdigitated electrode arrays. Anal Chim Acta 326(1–3):155–161

Fischer DJ, Hulvey MK, Regel AR et al (2009) Amperometric detection in microchip electrophoresis devices: effect of electrode material and alignment on analytical performance. Electrophoresis 30(19):3324–3333

Goto K, Kato D, Sekioka N et al (2010) Direct electrochemical detection of DNA methylation for retinoblastoma and CpG fragments using a nanocarbon film. Anal Biochem 405(1):59–66

Hayashi K, Horiuchi T, Kurita R et al (2000) Real-time electrochemical imaging using an individually addressable multi-channel electrode. Biosens Bioelectron 15(9–10):523–529

He Y, Peng L, Robertson J (2011) Fabrication and bio-functionalization of tetrahedral amorphous carbon thin films for biosensor applications. Diam Relat Mater 20(7):1020–1025

Hebert NE, Snyder B, McCreery RL et al (2003) Performance of pyrolyzed photoresist carbon films in a microchip capillary electrophoresis device with sinusoidal voltammetric detection. Anal Chem 75(16):4265–4271

Higson SPJ, Vadgama PM (1995) Diamond-like carbon-coated films for enzyme electrodes – characterization of biocompatibility and substrate diffusion limiting properties. Anal Chim Acta 300(1–3):77–83

Hirono S, Umemura S, Tomita M et al (2002) Superhard conductive carbon nanocrystallite films. Appl Phys Lett 80(3):425–427

Iwasaki Y, Niwa O, Morita M et al (1996) Selective electrochemical detection using a split disk array electrode in a thin-layer radial flow system. Anal Chem 68(21):3797–3800

Javier del Campo F, Godignon P, Aldous L et al (2011) Fabrication of PPF electrodes by a rapid thermal process. J Electrochem Soc 158(1):H63–H68

Jo-Il K, Bordeanu A, Jae-Chul P (2009) Diamond-like carbon (DLC) microelectrode for electrochemical ELISA. Biosens Bioelectron 24(5):1394–1398

Kamata T, Kato D, Hirono S et al (2013) Structure and electrochemical performance of nitrogen-doped carbon film formed by electron cyclotron resonance sputtering. Anal Chem 85(20): 9845–9851

Kaplan ML, Schmidt PH, Chen CH et al (1980) Carbon-films with relatively high conductivity. Appl Phys Lett 36(10):867–869

Kato D, Sekioka N, Ueda A et al (2008a) A nanocarbon film electrode as a platform for exploring DNA methylation. J Am Chem Soc 130(12):3716–3717

Kato D, Sekioka N, Ueda A et al (2008b) Nanohybrid carbon film for electrochemical detection of SNPs without hybridization or labeling. Angew Chem Int Ed 47(35):6681–6684

Kato D, Komoriya M, Nakamoto K et al (2011) Electrochemical determination of oxidative damaged DNA with high sensitivity and stability using a nanocarbon film. Anal Sci 27(7):703–707

Kim J, Song X, Kinoshita K et al (1998) Electrochemical studies of carbon films from pyrolyzed photoresist. J Electrochem Soc 145(7):2314–2319

Kostecki R, Song X, Kinoshita K (1999) Electrochemical analysis of carbon interdigitated microelectrodes. Electrochem Solid State Lett 2(9):465–467

Kurita R, Hayashi K, Fan X et al (2002) Microfluidic device integrated with pre-reactor and dual enzyme-modified microelectrodes for monitoring in vivo glucose and lactate. Sens Actuator B 87(2):296–303

Lagrini A, Deslouis C, Cachet H et al (2004) Elaboration and electrochemical characterization of nitrogenated amorphous carbon films. Electrochem Commun 6(3):245–248

Larsen ST, Argyraki A, Amato L et al (2013) Pyrolyzed photoresist electrodes for integration in microfluidic chips for transmitter detection from biological cells. Electrochem Lett 2(5):B5–B7

Lee JA, Hwang S, Kwak J et al (2008) An electrochemical impedance biosensor with aptamer-modified pyrolyzed carbon electrode for label-free protein detection. Sens Actuator B 129(1): 372–379

Liu ZM, Niwa O, Kurita R et al (2000) Carbon film-based interdigitated array microelectrode used in capillary electrophoresis with electrochemical detection. Anal Chem 72(6):1315–1321

Lyons AM, Wilkins CW, Robbins M (1983) Thin pinhole-free carbon-films. Thin Solid Films 103(4):333–341

Maalouf R, Soldatkin A, Vittoria O et al (2006) Study of different carbon materials for amperometric enzyme biosensor development. Mater Sci Eng C 26(2–3):564–567

Manivannan A, Kawasaki R, Tryk DA et al (2004) Interaction of Pb and Cd during anodic stripping voltammetric analysis at boron-doped diamond electrodes. Electrochim Acta 49(20):3313–3318

McCreery RL (2008) Advanced carbon electrode materials for molecular electrochemistry. Chem Rev 108(7):2646–2687

Miyake T, Yoshino S, Yamada T et al (2011) Self-regulating enzyme-nanotube ensemble films and their application as flexible electrodes for biofuel cells. J Am Chem Soc 133(13):5129–5134

Morton KC, Morris CA, Derylo MA et al (2011) Carbon electrode fabrication from pyrolyzed parylene C. Anal Chem 83(13):5447–5452

Niwa O (2005) Electroanalytical chemistry with carbon film electrodes and micro and nano-structured carbon film-based electrodes. Bull Chem Soc Jpn 78(4):555–571

Niwa O, Tabei H (1994) Voltammetric measurements of reversible and quasi-reversible redox species using carbon-film based interdigitated array microelectrodes. Anal Chem 66(2):285–289

Niwa O, Horiuchi T, Tabei H (1994) Electrochemical properties of carbon-based interdigitated microarray electrodes fabricated by the pyrolysis of electrochemically prepared conducting films. J Electroanal Chem 367(1–2):265–269

Niwa O, Jia J, Sato Y et al (2006) Electrochemical performance of Angstrom level flat sputtered carbon film consisting of sp(2) and sp(3) mixed bonds. J Am Chem Soc 128(22):7144–7145

Ohara TJ, Rajagopalan R, Heller A (1993) Glucose electrodes based on cross-linked Os(bpy)(2) (+/2+) complexed poly(L-vinylimidazole) films. Anal Chem 65(23):3512–3517

Pocard NL, Alsmeyer DC, McCreery RL et al (1992) Nanoscale platinum(0) clusters in glassy-carbon – synthesis, characterization, and uncommon catalytic activity. J Am Chem Soc 114(2):769–771

Pumera M, Ambrosi A, Bonanni A et al (2010) Graphene for electrochemical sensing and biosensing. Trends Anal Chem 29(9):954–965

Qureshi A, Kang WP, Davidson JL et al (2009) Review on carbon-derived, solid-state, micro and nano sensors for electrochemical sensing applications. Diam Relat Mater 18(12):1401–1420

Ranganathan S, McCreery RL (2001) Electroanalytical performance of carbon films with near-atomic flatness. Anal Chem 73(5):893–900

Sarada BV, Rao TN, Tryk DA et al (2000) Electrochemical oxidation of histamine and serotonin at highly boron doped diamond electrodes. Anal Chem 72(7):1632–1638

Schnupp R, Kuhnhold R, Temmel G et al (1998) Thin carbon films as electrodes for bioelectronic applications. Biosens Bioelectron 13(7–8):889–894

Sekioka N, Kato D, Kurita R et al (2008) Improved detection limit for an electrochemical gamma-aminobutyric acid sensor based on stable NADPH detection using an electron cyclotron resonance sputtered carbon film electrode. Sens Actuator B 129(1):442–449

Singh YS, Sawarynski LE, Michael HM et al (2010) Boron-doped diamond microelectrodes reveal reduced serotonin uptake rates in lymphocytes from adult rhesus monkeys carrying the short allele of the 5-HTTLPR. ACS Chem Neurosci 1(1):49–64

Swain GM, Ramesham R (1993) The electrochemical activity of boron-doped polycrystalline diamond thin-film electrodes. Anal Chem 65(4):345–351

Tabei H, Morita M, Niwa O et al (1992) Fabrication and electrochemical features of new carbon based interdigitated array microelectrodes. J Electroanal Chem 334(1–2):25–33

Ueda A, Kato D, Kurita R et al (2011) Efficient direct electron transfer with enzyme on a nano-structured carbon film fabricated with a maskless top-down UV/ozone process. J Am Chem Soc 133(13):4840–4846

Vashist SK, Zheng D, Al-Rubeaan K et al (2011) Advances in carbon nanotube based electrochemical sensors for bioanalytical applications. Biotechnol Adv 29(2):169–188

Yamamura S, Hoshikawa M, Kato D et al (2013) ONO-2506 inhibits spike-wave discharges in a genetic animal model without affecting traditional convulsive tests via gliotransmission regulation. Br J Pharmacol 168:1088–1100

Yang X, Haubold L, DeVivo G et al (2012) Electroanalytical performance of nitrogen-containing tetrahedral amorphous carbon thin-film electrodes. Anal Chem 84(14):6240–6248

You TY, Niwa O, Tomita M et al (2003) Characterization of platinum nanoparticle-embedded carbon film electrode and its detection of hydrogen peroxide. Anal Chem 75(9):2080–2085

Zeng A, Liu E, Tan SN et al (2002) Cyclic voltammetry studies of sputtered nitrogen doped diamond-like carbon film electrodes. Electroanalysis 14(15–16):1110–1115

Zhou M, Zhai YM, Dong SJ (2009) Electrochemical sensing and biosensing platform based on chemically reduced graphene oxide. Anal Chem 81(14):5603–5613

Chapter 8
Hybrid Metallic Nanoparticles: Enhanced Bioanalysis and Biosensing via Carbon Nanotubes, Graphene, and Organic Conjugation

Michael A. Daniele, María Pedrero, Stephanie Burrs, Prachee Chaturvedi, Wan Wardatul Amani Wan Salim, Filiz Kuralay, Susana Campuzano, Eric McLamore, Allison A. Cargill, Shaowei Ding, and Jonathan C. Claussen

Abstract Composite materials, incorporating noble metal and metal oxide nanoparticles, have attracted much interest as active substrates for biosensor electronics. These nanoparticles provide a viable microenvironment for biomolecule immobilization by retaining their biological activity with desired orientation and for facilitating transduction of the biorecognition event. Herein, we discuss various methods for fabrication of metal and metal oxide nanoparticle composite materials

M.A. Daniele
U.S. Naval Research Laboratory, Center for Bio-Molecular Science and Engineering, Code 6900, 4555 Overlook Ave SW, Washington, DC 20375, USA

M. Pedrero • S. Campuzano
Departamento de Química Analítica, Universidad Complutense de Madrid, Avda. Complutense s/n, E-28040 Madrid, Spain

S. Burrs • P. Chaturvedi • E. McLamore
Agricultural and Biological Engineering Department, Institute of Food and Agricultural Sciences, University of Florida, 1741 Museum Rd, Bldg 474, 105 Frazier Rogers Hall, Gainesville, FL 32611, USA

W.W.A. Wan Salim
Department of Biotechnology Engineering (BTE), Kuliyyah of Engineering (KOE), IIUM, Gombak, Kuala Lumpur

F. Kuralay
Department of Chemistry, Faculty of Science and Arts, Ordu University, 52200 Ordu, Turkey

A.A. Cargill • S. Ding • J.C. Claussen (✉)
Department of Mechanical Engineering, Iowa State University, 2025 Black Engineering, Ames, IA 50011, USA
e-mail: aacargill@gmail.com; swding@iastate.edu; jcclauss@iastate.edu

M.C. Vestergaard et al. (eds.), *Nanobiosensors and Nanobioanalyses*, DOI 10.1007/978-4-431-55190-4_8 and their applications in different electrochemical biosensors. The materials are organized by the corresponding component with the nanoparticles, i.e. carbon-based composites, polymers, and DNA. The performance of hybrids is compared and examples of biosensing apparatus are discussed. In all cases, the engineering of morphology, particle size, effective surface area, functionality, adsorption capability, and electron-transfer properties directly impact the resultant biosensing capabilities. Ultimately, these attractive features of metal and metal-oxide hybrid materials are expected to find applications in the next generation of smart biosensors.

Keywords Carbon nanotubes • DNA • Graphene • Metal oxide nanoparticles • Metallic nanoparticles • Polymer

8.1 Introduction

Hybrid metallic nanoparticles offer unique opportunities for designing powerful electrochemical bioassays and biosensors, and add a new dimension to such assays and devices. Metal and semiconductor nanomaterials, one-dimensional nanotubes and nanowires have rapidly become attractive labels for bioaffinity assays, offering unique signal amplification and multiplexing capabilities. The coupling of different metallic nanoparticle-based platforms and amplifications processes have dramatically enhanced the intensity of the analytical signal and led to the development of ultrasensitive bioassays and biosensors (Wang 2007). The orders of magnitude of amplification afforded by such metallic nanoparticle-based schemes opens up the possibility of detecting a plethora of agents and markers more rapidly and at lower detection limits. These highly sensitive biodetection schemes might provide an early detection of biomarkers, toxins, pathogens, and diseases in agricultural, food, and medical systems by using ultrasensitive bioelectroanalytical protocols unachievable with standard electrochemical methods.

Applications of biosensors are developed mainly for environmental and bioprocess monitoring, agriculture, bioterrorism, as well as medical and food biosensor systems. The presence of unsafe levels of chemical compounds, toxins, and pathogens in food constitutes a growing public health problem that necessitates new technology for the detection of these contaminants along the food continuum from production to consumption. The recent attention to food safety and regulatory issues towards consumer welfare is of utmost concern. While traditional techniques that are highly selective and sensitive exist, there is still a need for simpler, more rapid, and cost-effective approaches to food safety evaluation. Biosensors offer advantages over current analytical methods. In addition to their good selectivity, low cost, and portability, they have the ability to measure samples with minimal sample preparation required. Microbial metabolism, antibody, and DNA/RNA-based biosensors display promise for use as electrochemical biosensors used in food safety applications (Arora et al. 2011).

8.2 Carbon–Metal Hybrids

The combination of highly conductive carbon nanomaterials and highly catalytic metallic nanoparticles has led to new leaps in electrochemical biosensor performance. Carbon–metal hybrid carrier-molecules and carbon nanotube (CNT)– or graphene–metallic nanoparticle-based electrodes have all greatly enhanced the sensitivity, linear sensing range, and limit of detection of electrochemical biosensors. These nanomaterials display a high degree of catalytic activity, conductivity, and biocompatibility that act in a synergetic manner to improve biosensor performance. These unique material properties and the design and fabrication of biosensors that incorporate CNTs, graphene, and metallic nanoparticles will be thoroughly discussed. The subsequent functionalization and resultant interaction with biorecognition agents (e.g., peptides, proteins, and nucleic acids) for use as biosensors will be reviewed including those associated with health care, environmental monitoring, security surveillance, food safety, and biodefense.

8.2.1 Carbon–Metal Hybrids as Carrier Labels in Biosensing

Better understanding of synthesis routes for nanostructured materials has opened new avenues for developing biosensing platforms. The use of carbon/inorganic hybrid nanostructures as carrier labels in bioanalysis offers very elegant ways of interfacing biomolecule recognition events with inherent signal amplification. Particular attention is given to hybrid nanostructures involving carbon and inorganic nanomaterials. Elegant advances for tagging in biosensing based on such hybrids of carbon/inorganic-nanomaterial heterostructures are underlined along with future prospects and challenges.

Ultrasensitive detection of biomolecules is required in a variety of societal areas including clinical diagnostics, food safety, and environmental protection. Many efforts have been devoted to accomplishing ultrasensitive and even single-molecule detection by using signal amplification based on polymerase chain reaction (PCR) and mass spectrometric (MS) techniques. Although these methodologies have adequate sensitivity, they are destructive and often suffer from time-consuming derivatization, high cost, and the need for professional operation. With the emergence of nanotechnology and nanoscience, nanomaterial-based signal amplification holds great promise for achieving high sensitivity and selectivity for in situ or online detection of biomolecules, due to the use of rapid analysis procedures and easy miniaturization. The applications of nanomaterials in bioanalysis can be classified into two categories according to their functions: nanomaterial-modified transducers to facilitate bioreceptor immobilization or improvement of transducers' properties, such as low-background signals and high signal-to-noise ratio, and nanomaterial–biomolecule conjugates as labels for signal amplification. In particular, nanomaterial labels are showing the greatest promise for developing ultrasensitive bioanalysis strategies (Liu and Lin 2007). In these approaches, the nanomaterials usually act as catalysts to trigger the detectable signal or as carriers for both large loading of signal molecules and the accumulation of reaction products (Lei and Ju 2012).

Bioconjugates integrating nanomaterials with the catalytic and recognition properties of biomolecules have led to advanced electrochemical biosensors with ultrahigh sensitivity and multiplexed capability (Wang 2003; Pingarrón et al. 2008; Merkoçi 2010; Yáñez-Sedeño et al. 2010). These biofunctional nanomaterials can not only produce a synergic effect among catalytic activity, conductivity, and biocompatibility to accelerate the signal transduction, but also provide amplified recognition events by high loading of signal tags, leading to highly sensitive and specific biosensing (Lei and Ju 2012). The range of nanomaterials used in biosensors is wide and depends on the specific assay and application. Due to the diverse properties of different nanomaterials, in many situations the coupling of two different nanoscale materials offsets the insufficiency of each individual nanomaterial to fulfill the growing requirements of emerging sensing devices. Moreover, it also endows the resultant nanohybrid material with a greatly enhanced performance, superior to that observed when a single nanomaterial is used.

Inorganic nanomaterials, such as metal nanowires, nanoparticles, quantum dots (QDs) or inorganic nanocrystals, and carbon nanomaterials, including carbon nanotubes (CNTs), fullerenes or graphene, have received considerable interest in the field of nanoscience owing to their unique physical and chemical properties (in comparison with bulk materials), which offer excellent prospects for enhancing the performance of chemical, biological, and electrochemical sensors (Storhoff and Mirkin 1999; Caruso 2001; Claussen et al. 2011a; Willner and Willner 2002; Wang 2003; Pingarrón et al. 2008; Merkoçi 2010; Yáñez-Sedeño et al. 2010; Pei et al. 2013). When reduced to the nanoscale, such nanomaterials display new and unique size- and shape-dependent properties compared to those they display on a macroscale. A wide variety of inorganic nanoscale materials of different sizes, shapes, and compositions are now available, leading to tunable electronic and optical properties. Particularly attractive are the heteronanostructures of carbon and inorganic nanomaterials (CNTs modified with metal nanoparticles, QDs, semiconductor nanocrystals and metal oxide nanoparticles and graphene–inorganic nanohybrids), which have shown extremely useful integration of the unique properties of these types of nanomaterials. Thus, they exhibit some new functions and superior properties to those of their individual constituents and impart excellent analytical performance to biosensing (Peng et al. 2009; Eder 2010; Wu et al. 2011; Campuzano and Wang 2011) as well as molecular logic paradigms that are well suited for multiplexed biosensing (Manesh et al. 2011; Claussen et al. 2013, 2014). Often a third component, such as ionic liquids or chitosan, plays a key role in the preparation of these carbon–inorganic nanohybrids, acting as effective binder systems, inducing the solubilization of the corresponding carbon nanomaterial, and facilitating its manipulation and functionalization (Zhang et al. 2004; Shan et al. 2010; Zeng et al. 2011).

The power and scope of these nanoheteromaterials can be greatly enhanced by coupling them with immunoreactions and electrical processes (i.e., nanobioelectronics) (Liu and Lin 2007). Commonly used enzyme immunosensors can greatly benefit from the highly enhanced response of the biocatalytic reaction product at the electrode transducers from nanoscale inorganic/carbon amplification platforms carrying multiple tags. Such hybrid nanoarchitectures open up the possibility of detecting ultralow levels of biomarkers that cannot be measured by conventional methods or in connection to a single nanomaterial. This section summarizes recent significant advances and progress in the use of biofunctional carbon/inorganic nanoheterostructures as excellent electronic signal tags in ultrasensitive bioanalysis and biosensing, particularly using bioelectronic affinity assays and electrochemical detection, highlighting some elegant applications. While the new capabilities offered by nanoscale hybrid materials will be illustrated mainly in connection with ultrasensitive electrochemical immunosensors and immunoassays, a wide range of important biomolecules can benefit from similar improvements.

As an example, Zhong et al. (2010) developed a new amplification core–shell nanolabel based on a chitosan-protected graphene nanosheet core and multi-nanogold particles as the shell. Such a nanogold-enwrapped graphene nanocomposite led to ultrasensitive measurements down to 100 pg mL^{-1} of carcinoembryonic antigen (CEA) and convenient assays of serum samples. QD-functionalized graphene sheets (GS) were prepared and used as labels for the preparation of sandwich-type electrochemical immunosensors for the detection of prostate-specific antigen (PSA). The immunosensor displayed a linear response within a wide concentration range (0.005–10 ng mL^{-1}), a low detection limit (3 pg mL^{-1}), and applicability to detect PSA in patient serum samples (Yang et al. 2011). A novel enzyme-free sandwich electrochemical immunoassay for alpha-fetoprotein (AFP) was developed by Tang et al. (2011) using gold nanoparticle-coated carbon nanotubes (CNTs/AuNPs) as nanolabels/nanocatalysts. This highly sensitive approach is based on the catalytic reduction of *p*-nitrophenol (NP) by the CNT/AuNP labels and the redox cycling of *p*-aminophenol (AP) to *p*-quinone imine (QI) by $NaBH_4$ to offer an extremely low detection limit of 0.8 fg mL^{-1}.

Lai et al. (2011) proposed an ultrasensitive multiplexed immunoassay method for tumor markers based on the use as a trace tag of a novel functional CNT/AgNP nanohybrid functionalized with streptavidin and the corresponding biotinylated signal antibody. Through a sandwich-type immunoreaction on a disposable immunosensor array, the high-content AgNPs can be captured on the immunosensor surface to further induce the silver deposition, which greatly amplifies the detection signal (Fig. 8.1). Based on the electrochemical stripping detection of the AgNPs on the immunosensor surface, the proposed simultaneous multianalyte immunoassay method, using CEA and AFP as model analytes, showed an ultrahigh sensitivity with limits of detection (LODs) of 0.093 and 0.061 pg mL^{-1}, respectively, and wide linear ranges over four orders of magnitude.

Han et al. (2012) developed a novel multiple-label method and dual catalysis amplification strategy for the simultaneous detection of free and total PSA (fPSA and tPSA, respectively). AuNP-modified Prussian blue nanoparticles and AuNP-modified nickel hexacyanoferrate nanoparticles decorated onion-like mesoporous graphene sheets, denoted as Au@PBNPs/O-GS and Au@NiNPs/O-GS, respectively, functionalized with streptavidin and biotinylated alkaline phosphatase (AP) were

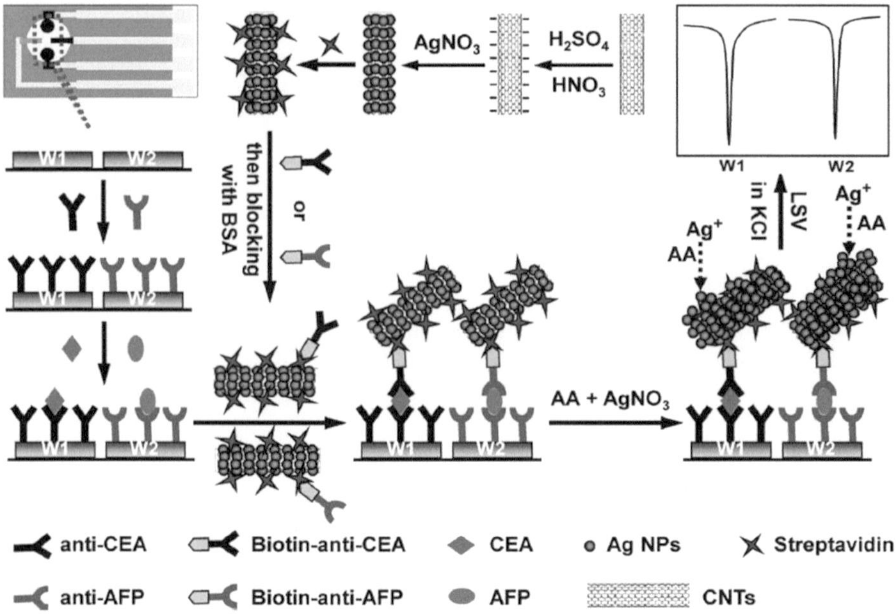

Fig. 8.1 Schematic representation of the multiplexed immunosensor developed based on the use of novel functional CNT/AgNP nanohybrids functionalized with streptavidin and biotinylated signal antibodies as tracer tags, and of the detection strategy by linear-sweep stripping voltammetric (LSV) analysis of AgNPs captured on the immunosensor surface (Reproduced from Lai et al. (2011) with permission). Copyright 2011 Wiley

utilized as distinguishable signal tags to label different detection antibodies. The dual catalysis amplification could be achieved by biocatalysis of AP towards ascorbic acid 2-phosphate (AA-P) to in situ production of ascorbic acid (AA) and then the chemicatalysis of Au@PBNPs/O-GS and Au@NiNPs/O-GS nanohybrids towards AA to generate dehydroascorbic acid (DHA). The linear ranges of the proposed immunosensor were defined from 0.02 to 10 ng mL^{-1} and 0.01 to 50 ng mL^{-1} with LODs of 6.7 pg mL^{-1} and 3.4 pg mL^{-1} for fPSA and tPSA, respectively.

Very recently, an ultrasensitive immunoassay for *Shewanella oneidensis* was presented by employing a novel conjugate featuring gold nanoparticles (AuNPs) and antibodies (Ab) assembled on bovine serum albumin (BSA)-modified GO (Ab/AuNPs/BSA/GO) as carrier and a silver enhancement detection strategy in the presence of hydroquinone. This electrochemical immunoassay offers excellent detectability (LOD of 12 cfu mL^{-1}) and a wide range of linearity (from 7.0×10^{1} to 7.0×10^{7} cfu mL^{-1}) (Wen et al. 2013).

The need for ultrasensitive bioassays and the trend towards miniaturized assays has made the biofunctionalization of nanomaterials one of the more popular fields of research. Although inorganic/carbon nanohybrids are still in an early stage of material science, the use of these nanohybrids as carriers of the signalling molecules for amplification transduction of biorecognition events has taken off rapidly and will surely continue to expand at an accelerated pace. A wide variety of nanoscaled materials with different sizes, shapes, and compositions have been introduced into biosensing for amplification detection. The judicious coupling of two different nanomaterials has been shown to offer greatly enhanced analytical performance, superior to that observed when a single nanomaterial is used, which is extremely attractive for bioelectronic transduction of biomolecular recognition events. Future efforts will certainly aim at guiding and tailoring the synthesis of nanohybrid materials for meeting specific electrochemical biosensing applications and needs.

The rapid recent progress in electrochemical biosensors based on the use of these bioconjugated nanoscale hybrid materials as tracing bionanotags, characterized by enormous signal enhancement and ultralow LODs, suggests the major impact that such nanobioelectronic sensing devices will have in the near future on important fields, including health care, environmental monitoring, security surveillance, food safety, and biodefense. Many exciting opportunities and challenges thus remain in the development and use of bioconjugated hybrid nanoarchitectures for future bioelectronic sensing applications. For example, a better understanding of their nanofabrication process is necessary to improve the properties of these nanohybrid labels. Furthermore, in order to increase their biocompatibility, a significant direction to explore is a mild biofunctional way to fix the biomolecules on the surface of these bioconjugated hybrid heteronanostructures. Another interesting opportunity for achieving exponential signal amplification is the development of hybrid nanomaterials-based autocatalytic systems, in which each step produces a product that acts both as a template or a stoichiometric trigger and a catalyst (or activates a catalyst) to produce more products. With the demand in life sciences and clinical diagnosis, the ultimate goal of this field is the utilization of nanomaterials which not only enhance the biosensing capabilities compared with conventional platforms, but also bring out new approaches such as miniaturization, reagent-less biosensing, and single-molecule detection.

8.2.2 Graphene–Metal Hybrid Biosensors

There is evidence that graphene and its derivatives can exhibit good electrochemical performance compared with other electrodes such as glassy carbon (Shang et al. 2008), graphite (Shan et al. 2009), or even carbon nanotubes (Alwarappan et al. 2009; Wang et al. 2009). When focusing on graphene-based biosensors, it is noteworthy that few sensors were demonstrated to have actually incorporated pure graphene (Ohno et al. 2009; Guo et al. 2011; Lidong et al. 2012; Ruan et al. 2012); the remainder have employed graphene oxide (and its derivatives) (Alwarappan et al. 2009; Shan et al. 2009; Mohanty and Berry 2008; Lu et al. 2009) or multilayer graphene and related structures (Shang et al. 2008; Lu et al. 2007, 2008; Claussen et al. 2012).

Conjugation of reduced graphene oxide with metal nanoparticles has been recently studied for the development of electrocatalytic platforms in amperometric biosensors (see Table 8.1). The hybridization of reduced graphene oxide with metal nanoparticles helps maintain the interplanar spacing between reduced graphene

Table 8.1 Examples of recently developed graphene–nanoparticle biosensors

Biosensor	Analyte	Sensitivity	Response time	Operating range	Reference
GO decorated with AuNR-antibody conjugates	Transferrin	10.2 nm shift in resonant wavelength at 10 mg/mL transferrin	60 min	0.0375–40 μg/mL	Zhang et al. (2013)
MIP/Gr-AuNPs/GCE	Glycoprotein	0.405 μA/log *M*	8 min	1×10^{-11}–1×10^{-5} g/mL	Wang et al. (2013)
GN-AuNRs-ADH	Ethanol	102 μA/mM cm^2	10 s	5–377 μM	Li et al. (2013)
Nafion/Mb-Gr-Pt/CILE	Trichloroacetic acid	30.08 μA/mM	NR	0.9–9 mM	Sun et al. (2013)
C-SPE/Pt-NPs/RGO/lacc/ Nafion	Caffeic acid	2,147.38 nA/μM	60 s	0.2–2 μM	Eremia et al. (2013)
Pt nanoparticles/Graphene Petals/Ti/SiO_2	Glucose	NR	NR	0.01–50 mM	
GS–CS–PtPd/ChOx	Cholesterol	NR	<7 s	2.2–520 μM	Cao et al. (2013)
(HRP-Pd)/f-graphene-Gr	Hydrogen peroxide	92.82 μA/mM cm^2	<2 s	25–3,500 μM	Nandini et al. (2013)
AgNPs-Pdop@Gr/GCE	Catechol	0.087 μA/μM	NR	0.5–240 μM	Huang et al. (2013)
Ni/Nafion/graphene/GCE	Ethanol	NR	< 5 s	0.43–88.15 mM	Jia and Wang (2013)

NR not reported in the manuscript

oxide sheets while connecting the nanomaterial network to the electrode surface. Functionalization of electrodes with graphene–nanoparticle composites has shown promising potential for improvement in the electrochemical performance of the sensing devices (e.g. sensitivity, response time, operating range).

The unique properties of graphene–nanoparticle hybrids have been widely used in the fabrication of biosensor systems for food safety applications, to detect toxins (Gan et al. 2013a; Srivastava et al. 2013; Tang et al. 2012; Yang et al. 2013b), pesticides (Oliveira et al. 2013a), allergens (Eissa et al. 2013), ingredients and bioactives (Labroo and Cui 2013; Zhou et al. 2013; Si et al. 2013; Wang et al. 2011), controlled and prohibited substances (Kong et al. 2013; Xie et al. 2012; Lin et al. 2013; Gan et al. 2013b, c; Zhao et al. 2011; Zhang et al. 2011; Cui et al. 2011; Huang et al. 2012; Wei et al. 2012; Wang et al. 2012b; Ma et al. 2013; Ye et al. 2013), and a variety of foodborne pathogens (Hu et al. 2013; Singh et al. 2013a; Chang et al. 2013; Liu et al. 2011a, b; Wan et al. 2011; Jung et al. 2010).

Perhaps some of the most promising graphene-based biosensors have been displayed from those that combine graphene with noble metal nanoparticles (Taguchi et al. 2014). For example, Hong et al. self-assembled positively charged gold nanoparticles (2–6 nm in diameter) onto the surfaces of 1-pyrene butyric acid functionalized graphene (PFG) sheets via a facile chemical mixing technique (Hong et al. 2010). When immobilized onto a glassy carbon electrode, the graphene/gold nanoparticle composite material showed strong electrocatalytic activity and was able to sense uric acid with high sensitivity and a rapid response. Zheng et al. utilized graphene that was covalently functionalized and decorated with palladium nanoparticles immobilized with the enzyme glucose oxidase on a glassy carbon electrode to sense glucose (Zeng et al. 2011). The biosensor had high glucose sensitivity with a linear range from 1.0 μM to 1.0 mM as well as a low detection limit of 0.2 μM ($S/N = 3$). Claussen and coworkers demonstrated the growth of a three-dimensional matrix of graphene petals on a silicon wafer via a chemical vapor deposition process (Claussen et al. 2012). The petals were decorated with platinum nanoparticles of varying size, density, and morphology through a current-pulse electrochemical deposition technique (Fig. 8.2). Likewise the enzyme glucose oxidase mixed with the conductive polymer PEDOT:PSS was electrodeposited onto the biosensor surface for subsequent glucose biosensing. The results demonstrated a robust sensor design that demonstrated exceptional performance with regards to glucose sensitivity (0.3 μM detection limit, 0.01–50 mM linear sensing range), stability (shelf life >1 month), and low interference from electroactive species typically found endogenously in human serum samples.

8.2.3 Carbon Nanotube–Metal Nanoparticle Biosensors

An important property of carbon nanotubes (CNT) for electrochemical detection is their ability to promote electron transfer in electrochemical reactions (Gooding 2005; McCreery 2008; Katz and Willner 2004). CNTs are commonly referred to as rolled-up graphene sheets, and both allotropes have a meshwork of

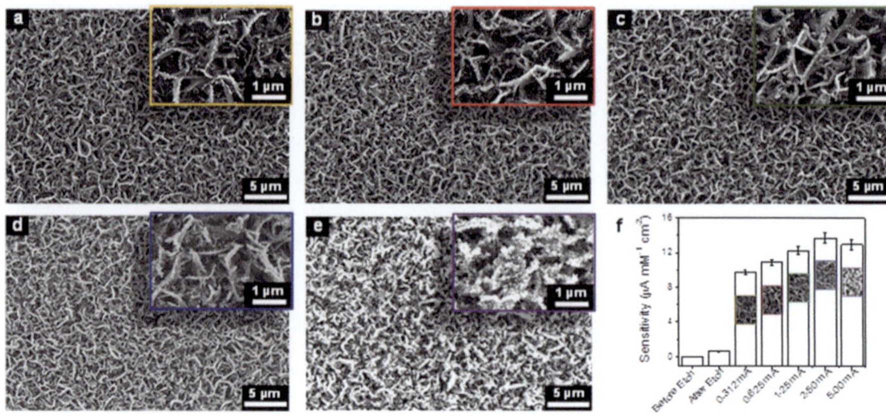

Fig. 8.2 Field emission scanning electron microscopy micrographs of platinum nanoparticles electrodeposited on multilayered graphene petal nanosheets (MPGNs) grown on a silicon wafer. Current pulses (500 ms) of (**a**) 312 μA (*orange*), (**b**) 625 μA (*red*), (**c**) 1.25 mA (*green*), (**d**) 2.5 mA (*blue*), and (**e**) 5.0 mA (*purple*) were used to electrodeposit Pt nanoparticles of distinct size and density onto the MGPNs. (**f**) *Bar graph* displaying the H_2O_2 sensitivity of the MGPN electrode (before and after the oxygen plasma etch) and the PtNP-MGPN electrodes. *Errors bars* show standard deviations for three different experiments. The MPGNs with the most needle-like Pt nanostructures (**d**) had the highest H_2O_2 sensitivity and highest glucose sensitivity in subsequent analysis (Reproduced from Claussen et al. (2012) with permission). Copyright 2012 Wiley

sp^2-hybridized carbon atoms (Yang et al. 2010b). CNTs have high aspect ratios, high mechanical strength, high surface areas, excellent chemical and thermal stability, and rich electronic and optical properties (Ajayan 1999). The properties of CNTs can also be capitalized on by combining them with other functional materials, such as conducting polymers or metal nanoparticles, in order to enhance their electrochemical sensing performance (Dai 2007). Table 8.2 shows some examples of carbon nanotube nanoparticle-mediated biosensors for the quantification of analytes such as glucose, glutamate, xanthine, t-DNA, ethanol, and D-amino acid.

The combination of excellent conductivity, good electrochemical properties, and nanometer dimensions has seen CNTs used frequently for the detection of diverse biological structures such as DNA, viruses, antigens, disease markers, and whole cells. In food systems particularly, there have been many reports of CNT-based electrochemical biosensors for the detection of toxins (Singh et al. 2013b; Palaniappan et al. 2013; Temur et al. 2012; Yang et al. 2010a), pesticides (Oliveira et al. 2013b; Cesarino et al. 2012; Liu et al. 2012), allergens (Liu et al. 2010; Cao et al. 2011), ingredients (Antiochia et al. 2013; Monosik et al. 2012a, 2013; Ziyatdinova et al. 2013; Wang et al. 2012a), controlled and prohibited substances (Batra et al. 2013; Kim et al. 2012a, b; Monosik et al. 2012b), microbial metabolism markers (Lim et al. 2013; Park et al. 2012), and a variety of foodborne pathogens (Yang et al. 2013a; Li et al. 2012; Pandey et al. 2011; Garcia-Aljaro et al. 2010; Zelada-Guillen et al. 2010).

The ability to control the position and density of carbon nanotube arrays and the size, density, and morphology of nanoparticles on a biosensor surface is

Table 8.2 Examples of previously developed carbon nanotube–nanoparticle biosensors

Biosensor	Analyte	Sensitivity	Response time	Operating range	Reference
GOx/Pd-Au-SWCNT/PAA	Glucose	5.2 μA/mM cm^2	5 s	1.3–50 mM	Claussen et al. (2009)
GOx/Pt-SWCNT/PAA	Glucose	70 μA/mM cm^2	8 s	LOD=380 nM	Claussen et al. (2010)
GluOx/Pt-SWCNT/PAA	Glutamate	27.4 μA/mM cm^2	NR	0.05–1,600 μM	Claussen et al. (2011c)
GOx/MPA/Au nanodisks/PAA	Glucose	0.47 nA/mM cm^2	NR	1–21 mM	Claussen et al. (2011a)
Pt/MWNT/Nafion/GO	Glucose	531 pA/mM	0.88 s	0.01–17.5 mM	McLamore et al. (2011)
XO-CD/pAuNP/SWNT/GCE	Xanthine	2.47 A/M cm^2	5 s	0.05–9.5 μM	Villalonga et al. (2012)
GluOx/cMWCNT/AuNP/CHIT	Glutamate	155 nA/μM cm^2	2 s	5–500 μM	Batra and Pundir (2013)
CGi+p-DNA	t-DNA	NR	NR	100 fM to 1 μM	Ko et al. (2011)
CNT-Ni hybrid	Ethanol	14.81 μA/μM cm^2	<10 s	50–600 μM	Chen and Huang (2010)
DAAO/c-MWCNT/CuNPs/PANI/Au	D-amino acid	54.85 μA/mM cm^2	2 s	0.001–0.7 mM	Lata et al. (2013)

NR not reported in the manuscript

challenging. Pang et al. demonstrated how CNT arrays can be grown from titania nanotubes with subsequent uniform decoration of Pt nanoparticles (~3 nm) (Pang et al. 2009). After immbolization of the enzyme GOx, the biosensor sensed glucose with a high sensitivity of 0.24 μA mM^{-1} cm^{-2} and a linear sensing range of 0.006–1.5 mM. Zhao et al. developed aligned CNT arrays by chemical vapor deposition on a silicon wafer. A thin film of gold (200 nm) was subsequently deposited onto the CNTs and the Au-CNTs were peeled from the Si floor with the assistance of 10 % hydrofluoric solution (Zhao et al. 2007). Platinum nanoparticles were electrodeposited onto the CNTs to improve electrocatalytic performance while GOx and Nafion were drop-casted onto the Pt nanoparticles/CNTs to transform the electrode into a glucose biosensor. The biosensor was able to sense glucose within a linear sensing range from 0.010 to 7 mM and a fast response time within 5 s. Fisher and coworkers demonstrated how low-density single-walled CNT arrays can be developed from a porous anodic alumina template (Franklin et al. 2008; Claussen et al. 2009). In particular, the team demonstrated how the spacing of Pt nanospheres electrodeposited concentrically around single-walled CNT arrays and functionalized with the enzyme GOx can dramatically change glucose biosensing from 300 μM to 15 mM with a theoretical glucose detection limit of 74 μM ($S/N=3$) to a linear sensing range of 100 μM to 20 mM and a detection limit of 5.8 μM ($S/N=3$) (Claussen et al. 2011b).

8.3 Polymer–Metal Hybrids

As stated, the integration of bio-recognition and signal transduction elements has been an important area of biosensor research for several decades. Similar to the incorporation of carbon nanostructures, the functional properties (electronic, optical, and magnetic) and relative stability when bound to biological molecules (peptides, proteins, and nucleic acids) of metallic and metal oxide nanoparticles (MNPs) make them attractive candidates for integrated biosensors. Proteins, peptides, and antibodies have been utilized to conjugate MNPs for use in a wide array of biosensors to detect and amplify various small-molecule signals. To conjugate biomacromolecules with MNPs, residual thiol groups are often reacted with MNPs to form metal–sulfide bonds (Dreaden et al. 2012), or electrostatic interactions are exploited for physisorption of biomacromolecules to the surface of the MNP. Both methods result in the random placement of the biomacromolecule on the surface of the MNP and may also result in reduced bioactivity; therefore, MNPs composed with tailored, functional polymer coatings have emerged as a popular substrate for precision bio-conjugation. Encapsulation of MNPs within polymer matrices avoids the deleterious effects of MNP aggregation; moreover, polymers provide a versatile, functional platform for attaching organic moieties and biomacromolecules with various signal readout strategies, e.g. electrochemical, enzymatic, colorimetric, fluorometric, magnetic, and chemiluminescent. Both polymer conjugates of noble metal and metal-oxide MNPs are currently being explored as biosensors, and this section will focus on advances in the synthesis of polymer–MNP hybrids and their use in the bio-recognition component of electrochemical biosensors.

8.3.1 Polymer–Metal Hybrid Synthesis

Firstly, reproducible sensor materials require controlled MNP growth, particle size distribution, and particle–polymer interactions. There are many well accepted ex situ and in situ routes for the synthesis of polymer–MNP hybrids that include both standard engineering polymers and advanced conducting polymers (Sperling and Parak 2010; Shaidarova and Budnikov 2008). The polymer–MNP hybrid can be generated by compounding powders of MNPs with common melt-processed polymers such as poly(styrene), poly(ethylene) or poly(methyl methacrylate) (ex situ). Alternatively, the polymer–MNP hybrid can be formed by precipitation of the MNP from metal or metal-oxide precursors dissolved within a swellable polymer or polymer solution, such as poly(acrylic acid), poly(vinyl alcohol), polyaniline, polypyrrole or poly(vinylpyrrolidone) (in situ) (Ferey 2008; Ramesh et al. 2009; Njagi and Andreescu 2007). A schematic for the synthesis of polymer–nanoparticle hybrids is shown in Fig. 8.3.

The ex situ approach has two steps: (1) bulk synthesis of MNP powder and (2) dispersion of MNP powder throughout a polymer matrix, often by compounding with a polymer melt. The bulk synthesis of noble metal and metal-oxide nanoparticle powders are direct processes and have been reviewed elsewhere (Shi et al. 2013). The more nuanced in situ synthesis of polymer–MNP hybrids can take many forms. In this route, MNPs are synthesized inside a polymer matrix by either decomposition or reduction of precursors dissolved into the polymer film/monolith or polymer solution. Polymer–MNP hybrids prepared by the reduction of metal salts in the presence of stabilizing polymers have tightly controlled size and size distribution; thus, the in situ preparation method has been the route of choice for biosensor materials (Rao 2012; Antonietti et al. 1995; Spatz et al. 1996). For instance, $HAuCl_4 \cdot 4H_2O$ gives stable gold MNPs upon refluxing in methanol/water in the presence of poly(vinylpyrrolidone) and NaOH. In poly(acrylamide), $AuCl_4$ can be directly reduced by $NaBH_4$. Reduction of metal ions in the presence of these polymers results in the conjugation of the metal cations with the ligand, and this dramatically limits the MNP size and controls size distribution (Daniel and Astruc 2004). Synthesis of metal-oxide nanoparticles in solutions of polymer stabilizers also aids to the control of crystallinity and oxidation. For example, the reduction of Fe^{2+} and Fe^{3+} can be driven toward either Fe_2O_3 or Fe_3O_4 by altering poly(acid) concentration and constituents (Daniele et al. 2013; Qi et al. 2013; Zhang et al. 2012). Other common polymers used to stabilize MNPs are hydrophilic and biocompatible polymers, such as poly(ethylene glycol), poly(ethylene oxide), poly(lactic-*co*-glycolic acid), poly(vinyl alcohol) and poly(acrylic acid). These in situ methods and selected polymers for the synthesis of polymer-MNPs have been extensively studied for both optical and electrochemical biosensor applications; therefore, the remainder of this section will concentrate on in situ generated polymer–MNP hybrids with attached bio-recognition agents and their utilization as biosensors.

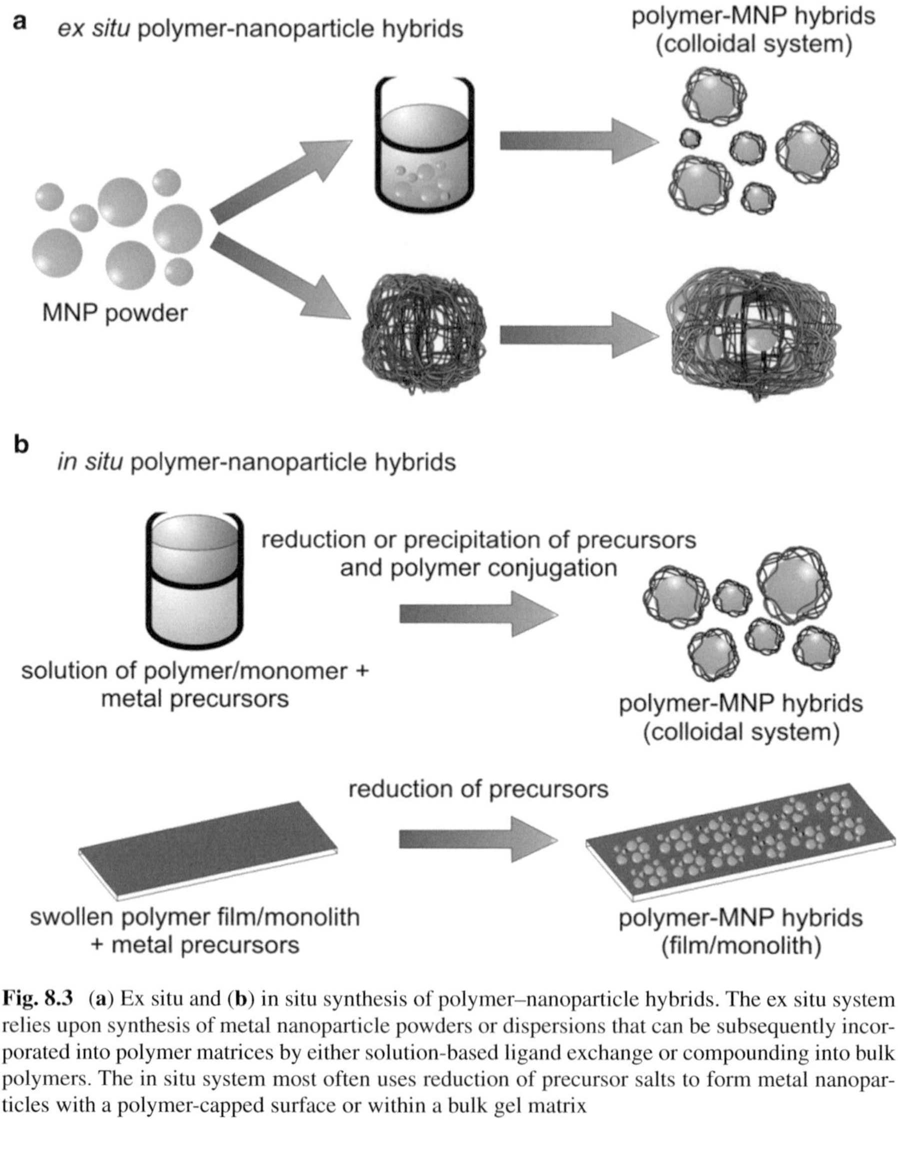

Fig. 8.3 (**a**) Ex situ and (**b**) in situ synthesis of polymer–nanoparticle hybrids. The ex situ system relies upon synthesis of metal nanoparticle powders or dispersions that can be subsequently incorporated into polymer matrices by either solution-based ligand exchange or compounding into bulk polymers. The in situ system most often uses reduction of precursor salts to form metal nanoparticles with a polymer-capped surface or within a bulk gel matrix

8.3.2 Polymer–Noble Metal Hybrids

The unique physicochemical properties of noble metals at the nanoscale have led to the development of a wide variety of biosensors, such as (1) nanobiosensors for point-of-care disease diagnosis, (2) nanoprobes for in vivo sensing/imaging, cell tracking, and monitoring disease pathogenesis or therapy monitoring, and (3) other nanotechnology-based tools that benefit scientific research on basic biology. Gold MNPs are among the most extensively studied nanomaterials and have led to the development of substrates for the conjugation of bio-recognition agents. The earliest gold–MNP hybrid utilized single-strand DNA. In 1996, Mirkin et al. described the first use of gold MNPs as biosensors. Functionalized with thiol-modified ssDNA probes, the gold MNPs would form a cross-linking network upon detection of a complementary ssDNA–MNP hybrid (Mirkin et al. 1996). This cross-linking network lead to the aggregation of the gold NPs causing a bathochromic shift in the surface plasmon resonance. More recent explorations of DNA–MNP hybrids will be assessed in Sect. 8.3.3.

Although similar surface plasmon resonance methods have shown that direct detection of biomolecules and analyte interaction is possible, many have benefited from the electroactive or catalytic properties of NPs as reporters for electrochemical biosensing with unprecedented levels of sensitivity. Noble metal–polymer hybrids have been used as biosensors for detection of analytes including, but not limited to, glucose (Wei et al. 2012; Crespilho et al. 2006), dopamine (Prakash et al. 2013), hydrogen peroxide (Kim et al. 2010; Muraviev et al. 2006), cholesterol (Yang et al. 2006; Ansari et al. 2009), and urea (Kozitsina et al. 2009). Several authors have described the development of amperometric-based biosensors, which are usually more suited for mass production than potentiometric biosensors. In this approach the working electrode is usually a noble metal MNP covered by the bio-recognition component, which enables the amperometric signal. For example, Kim et al. (2000) developed a disposable immuno-chromatographic sensor for on-line quantitative determination of human serum albumin. The polymer–MNP sensor used gold MNPs in a polyaniline matrix (a conducting polymer) for signal generation. The immunoassay was a membrane strip sensor, where the reaction between the conjugate and analyte took place and was carried up into a membrane that contained the immobilized antibody. The secondary antigen–antibody reaction formed a "sandwich complex" at the electrode and the gold–polyaniline polymer–MNP hybrid generated a conductimetric signal. Yin et al. used gold NPs over a surface of poly(styrene-acrylic acid) nanospheres, which served as a matrix to conjugate alkaline phosphatase, to detect the tumor necrosis factor. Omidfar et al. (2012) have also developed a high-sensitivity electrochemical human serum albumin sensor based on human serum albumin MNPs (electrochemical label) within a PVA monolith (polymer matrix) which exhibited high sensitivity and excellent stability.

Ranging from surface plasmon, infrared spectroscopy, and fluorescence to traditional electrochemical methods, noble metal polymer–MNP hybrids are providing a new horizon for biosensing and bioanalyses in clinical diagnostics and biological research; furthermore, the new range of biocompatible polymers systems is providing hybrids and biosensors that can be utilized in complex biosystems.

8.3.3 Polymer–Metal Oxide Hybrids

In contrast to noble metal polymer–MNP hybrids which are commonly utilized as cast films for charge transport purposes, metal-oxide nanoparticles (MONPs) are most often employed as colloidal systems and exhibit varying electrochemical properties, ranging from nanocatalysis to semiconduction. Synthesized by the in situ precipitation of metallic precursors in the presence of stabilizing ligands, MONPs have been prepared with a variety of inorganic chemistries, e.g. Fe_2O_3, Fe_3O_4, TiO_2 (Chen et al. 2001; Lee et al. 2007), SiO_2, and ZnO.

Of the available MONP chemistries, the magnetic properties of Fe_2O_3 and Fe_3O_4 have been exploited across the gamut of biosensors, which include pollution detection (Xu et al. 2012; Horak et al. 2007), disease diagnostics and therapeutics (Sandhu et al. 2010; Veiseh et al. 2010), blood analyses, bioimaging (Lee and Hyeon 2012), and chemical and biological separation. Similar to noble metal hybrids, controlling the particle size and size distribution are important; however, unlike noble metal nanoparticles, the shape, crystal structure, and defect distribution in the polymer–MONP hybrids dramatically affects their electrochemical properties. For example, iron-oxide MNPs have magnetic properties directly correlated to their particle size and crystal structure. To tailor these properties, numerous synthetic approaches have been developed for the generation of Fe_2O_3 and Fe_3O_4 polymer–MONP hybrids (Kievit and Zhang 2011; Netto et al. 2013; Sandhu et al. 2010). Due to the importance of the crystal structure/oxidation state to subsequent properties, the choice of polymer ligand is critical for the stability and functionality of the polymer–MONP hybrids. The methodology used is based on direct precipitation of iron salts inside the pores of the porous polystyrene seed and was pioneered by Ugelstad et al. (1973), similar to the deposition of noble metal MNPs in polymer monoliths. The particles obtained exhibit a narrow particle size distribution with a good magnetic separation, which are critical parameters. Hydrophilic magnetic latexes were first reported by Kawaguchi et al. using acrylamide as the initial monomer, and more recently Lee et al. have modified nanoparticle surfaces with PVA by precipitation of iron salts in PVA aqueous solution to form a stable dispersion. They found that the crystallinity of the particles decreased with increasing PVA concentration, while the morphology and particle size remained almost unchanged. This phenomenon has been shown to be a result of metal–organic chelation and in situ ligand exchange, and it is a critical factor in controlling polymer–MONP hybrid morphology.

Various biomacromolecules, such as antibodies, proteins, and DNA, and bioactive small molecules have been covalently incorporated onto the polymer–MNPs, and the possibilities of the chemistries to do so have been as wide-ranging as the array of ligands. Some interesting ligands with regard to biosensors include 1-ethyl-3-(3-dimethylaminopropyl)carbodi-imide hydrochloride (EDC), *N*-succinimidyl-3-(2-pyridyldithio)-propionate (SPDP), *N*-hydroxysuccinimide (NHS), and methylene bis-acrylamide (MBA). These coupling ligands are readily utilized for the simple and efficient conjugation of proteins as bio-recognition agents. More recently, the incorporation of "clickable" moieties into the polymer–MNP matrix has led to a broader range of conjugation ligands (Ge et al. 2013; Daniele et al. 2013; Zhou et al. 2008; He et al. 2009; Liu et al. 2009).

Enzymes, such as glucose oxidase, hydrolase, horseradish peroxidase, creatinase, lactase, and lactate dehydrogenase have been successfully immobilized on the surface of Fe_3O_4 polymer–MNP hybrids via covalent immobilization (Ge et al. 2013; Zou et al. 2010; Peng et al. 2013; Yang et al. 2009; Zhang et al. 2008a, b; Cevik et al. 2012;

Villalonga et al. 2011). An early enzyme biosensor utilizing polymer–MONP hybrids was reported by Rossi et al. (2004), in which GOx was conjugated with Fe_3O_4 through a poly(ethylene glycol) linker. In order to improve the sensitivity of such enzymatic biosensors, some electron mediators or electron promoters were introduced into the biosensing system. Accordingly, an amperometric glucose biosensor was developed by entrapping GOx in chitosan composite doped with ferrocene monocarboxylic acid-modified Fe_3O_4 nanoparticles. With the aid of a permanent magnet, these polymer–MNP hybrids with incorporated GOx were attached to the surface of an electrode and acted as mediator to transfer electrons between the enzyme and the electrode. The large surface area of Fe_3O_4 nanoparticles and the porous morphology of chitosan lead to a high loading of enzyme and increased sensitivity. Zhuo et al. developed a three-layer composite composed of Fe_3O_4 magnetic core, Prussian blue interlayer, and gold shell to fabricate an electrochemical immunosensor by functionalization with bi-enzyme of horseradish peroxidase and GOx (Zhuo et al. 2009). Besides the metal oxides mentioned above, CuO (Li et al. 2011), Bi_2O_3 (Ding et al. 2010), and CeO_2 (Saha et al. 2009) nanoparticles have been reported to be used for GOx immobilization and biosensor design.

It should be pointed out that, in some MONP-based biosensing systems, the recognition mechanism is not based on the direct reaction between enzyme and analyte. Most biological samples exhibit negligible magnetic susceptibility; therefore, magnetic nanoparticle polymer–MNP hybrids can be used for detection of biomolecules and cells based on magnetic resonance effects. Diagnostic magnetic resonance (DMR) technology encompasses numerous assay configurations and sensing principles, and diamagnetic nanoparticle biosensors have been designed to detect a wide range of targets including DNA/mRNA, proteins, enzymes, drugs, pathogens, and tumor cells. The core principle behind DMR is the use of magnetic nanoparticles as proximity sensors that modulate the spin–spin relaxation time of neighboring water molecules, which can be quantified using clinical MRI scanners or bench-top nuclear magnetic resonance (NMR) relaxometers. DMR biosensor technology holds considerable promise to provide a high-throughput, low-cost, and portable platform for large-scale molecular and cellular screening in clinical and point-of-care settings.

Ultimately, polymer–MNP hybrids display a range of beneficial electrochemical attributes; moreover, the tailored functionality of the polymer matrices provides for unique routes for the precision attachment of bio-recognition molecules. In the last decade, these benefits have been successfully employed for biosensing and bio-analysis applications, and continued development of polymer–MNP hybrids and conjugation chemistries will see gains in both biosensor precision and sensitivity.

8.3.4 Polymer–Metal Hybrids for DNA Sensing

As discussed in Sect. 8.2, DNA–MNP hybrids are a unique class of materials that have generated much interest for biosensing applications. Specifically, in the arena of nucleic acid detection, a DNA ligand provides both possible ligation chemistry and detection elements in one unique package. Detection of nucleic acids, deoxyribonucleic acid (DNA) or ribonucleic acid (RNA), is very important in many life sciences for understanding their basic functions and for identifying certain targets (Wang 2000, 2002; Pyun 2012). DNA is one of the most important molecules of life which encodes the genetic information and instructs the biological synthesis of proteins and enzymes through the process of replication and transcription of genetic information in all known living organisms and many viruses. Since its backbone is resistant to cleavage and, furthermore, the double-stranded structure provides the molecule with a built-in duplicate of the encoded information, it becomes indispensable. A sequence of nucleotides, guanine (G), adenine (A), thymine (T) and cytosine (C), are responsible for encoding the genetic information for further generations. Hence, understanding the structural properties of DNA can lead to understanding the origin of many of diseases, the mutation of genes, and the action mechanism of antitumor/antivirus drugs. To date there have been many scientific and commercial attempts to design and prepare DNA detection systems based on different techniques. These systems have found great interest in medical diagnostics, assessment of gene expression, drug discovery, identification of genetic mutations or single nucleotide polymorphisms, forensic, environmental (pollution, pathogen classification), bioterrorism, and food applications. Classical methods for DNA detection are mostly time-consuming and expensive. Thus, large-scale DNA testing/detection requires the development of small, portable, inexpensive, sensitive, selective, fast, and easy-to-use methods. A biosensor which is an analytical device with a biologically active material (DNA) can offer great promises for achieving this goal. Among different types of biosensors, which are classified according to the transducer that is used, electrochemical biosensors are the most commonly used ones because of their sensitivity, selectivity, compact size, low cost of construction, real-time analysis, and simplicity of use (Gooding 2002; Palecek 2002).

Recent advances in nanotechnology have provided great progress for biosensing purposes. Among different technologies, combining nanoparticles and polymer technology has provided enhanced stability and sensitivity. There has been great interest in terms of research on polymer–nanoparticle hybrids for different types of applications in the sensing area. This sub-section mainly focuses on the use and development of polymer–nanoparticle hybrids for electrochemical DNA biosensing. Electrochemical DNA hybridization detection has advanced a long way with the use of nanoparticle materials and polymer modification was successfully used to stabilize the dispersion of nanoparticles on the electrode surface (Muti et al. 2010; Yumak et al. 2011; Fang et al. 2008; Chang et al. 2008; Zhang et al. 2008a, b, 2009; Yang et al. 2007; Sun et al. 2010; Du et al. 2009; Radhakrishnan et al. 2013; Wang et al. 2003).

Hybridization probe biosensors are one of the most crucial improvements in the field of gene-related biomolecule detection. These kinds of biosensors most commonly rely on the immobilization of an oligonucleotide (ODN) probe onto a transducer surface for hybridization with its complementary target sequence (Fig. 8.3).

Muti et al. (2010) fabricated tin oxide (SnO_2) nanoparticles (SNPs)–poly(vinylferrocenium) (PVF^+)-modified single-use graphite electrodes for electrochemical DNA hybridization detection. SnO_2 is a semiconductor and because of its conductive properties, these nanoparticles can be used in several applications (Wang

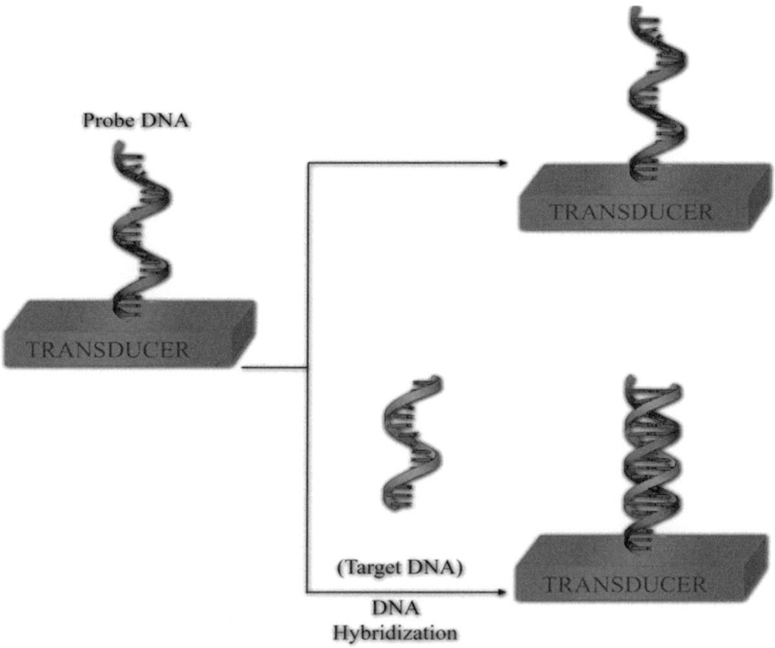

Fig. 8.4 Scheme of an electrochemical hybridization biosensor

et al. 2003; Ansari et al. 2009). In their work, they combined the nanoparticles with a conducting polymer (redox polymer), PVF^+. This polymer shows a simple and good electrochemistry because of its ferrocene/ferrocenium groups in its structure. PVF^+-modified electrodes have been used for the same purpose previously (Kuralay et al. 2008, 2009) and in the presence of SnO_2 nanoparticles more sensitive results were obtained. Scanning electron microscopy (SEM) was used to differentiate the modifications on the pencil graphite electrodes (PGEs), as well as electrochemical experiments (Fig. 8.4). Electrochemical behaviors of the PGEs were investigated by differential pulse voltammetry (DPV) and electrochemical impedance spectroscopy (EIS). The change in the guanine oxidation signals was used as the indicator of DNA hybridization. Different modifications in the probe DNA and probe DNA concentration were examined in order to obtain optimum working conditions for improving sensitivity and selectivity. After optimization studies, DNA hybridization was performed in the case of complementary hepatitis B virus (HBV), mismatch (MM), and noncomplementary (NC) sequences. The SNP—polymer-modified PGE showed high selectivity and specificity to its complementary DNA in the concentration range of 20–140 μg mL^{-1} with a detection limit of 1.82 μg mL^{-1}.

Zinc oxide (ZnO) nanoparticles (ZNPs) enriched with PVF^+ hybrids were used for electrochemical nucleic acid hybridization related to HBV by Yumak et al. (2011) using PGEs as the electrode materials. ZNPs have different applications due to their wide band gap and large excitation energy (Na et al. 2008; Sun et al. 2009). ZNPs (approximately 30 nm) were synthesized by the hydrothermal method and characterized by X-ray diffraction (XRD), Braun–Emmet–Teller (BET) N_2 adsorption analysis and transmission electron microscopy (TEM). SEM was used to identify different modifications on the PGEs. Electrochemical experiments included DPV and EIS techniques. The change in the guanine signals was evaluated and used as the indicator of DNA hybridization. Various modifications in DNA oligonucleotide types and probe concentrations were examined in order to optimize the electrochemical signals. After the optimization studies, the sequence-selective DNA hybridization was investigated for the cases of a complementary amino-acid-linked probe (target), NC sequences, or target and MM mixture in the ratio of 1:1. The detection limit was calculated as 11.7 μg mL^{-1}.

Besides this work there have been many attempts in this attractive topic. For example, Fang et al. investigated label-free electrochemical detection method for DNA–peptide nucleic acid (PNA) hybridization using ferrocene-functionalized polythiophene transducer and ssPNA probes on a nanogold-modified electrode (Fang et al. 2008). DNA hybridization using gold nanoparticles based on assembly of alternating DNA and poly(dimethyldiallylammonium chloride) multilayer films by layer-by-layer electrostatic adsorption has been studied by Chang et al. (2008). DNA hybridization detection was performed by Zhang et al. using ZNPs, multi-walled carbon nanotubes (MWCNTs), and chitosan hybrids with methylene blue (MB) indicator (Zhang et al. 2008a, b). Electrochemical detection of DNA hybridization based on carbon nanotubes, nano-zirconium dioxide (ZrO_2), and chitosan-modified electrodes was studied by Yang et al. (2007) using glassy carbon electrodes (GCEs) with DPV. The detection limit ($S/N=3$) was found to be 75 pM. Sun et al. (2010) performed DNA hybridization using nano-V_2O_5, MWCNTs, and chitosan nanocomposite materials modified *N*-hexylpyridinium hexafluorophosphate carbon ionic liquid (CILE) as binder with graphite powder. The electrochemical indicator MB was used to monitor the hybridization event with DPV. The detection limit ($S/N=3$) was found to be 1.76 pM (Sun et al. 2010). Cationic poly-L-lysine (pLys) and Au–CNT hybrid was used as a DNA hybridization biosensor for detection of the phosphinothricin acetyltransferase (PAT) gene with MB indicator (Du et al. 2009). An electrochemical DNA biosensor based on silver nanoparticles/poly(*trans-3*-(3-pyridyl) acrylic acid) (PPAA)/MWCNTs-COOH-modified GCEs has been prepared by Zhang et al. (2009). The DNA hybridization was monitored using intercalator adriamycin by DPV with a detection limit of 3.2 pM ($S/N=3$). Polypyrrole-poly(3,4-ethylenedioxythiophene)-Ag (PPy-PEDOT-Ag) nanocomposite films for label-free electrochemical DNA sensing were prepared by Radhakrishan et al. (2013). The detection limit was found to be 5.4 fM.

This sub-section summarizes the importance of polymer–nanoparticle hybrids, mainly in electrochemical DNA hybridization biosensors. A general introduction to the topic has been given, then the applications of these biosensors were presented. The applications have shown that electrochemical DNA detection provided sensitive, selective, reliable, low-cost methods when combined with polymer–nanoparticle technology. These works will definitely be useful for future works in different areas including medicine, pharmacy, forensic applications, environmental monitoring, bioterrorism and food applications.

8.4 Conclusions

The integration of bio-recognition and signal transduction elements has been an important area of biosensor research for several decades. However, recent advances in the fabrication of nanomaterials have greatly improved their performance. Metallic and metal oxide nanoparticles (MNPs), carbon–metal hybrid carrier molecules and carbon nanotube (CNT)– or graphene–metallic nanoparticle materials have all greatly enhanced the sensitivity, linear sensing range, and limit of detection of electrochemical biosensors. These nanomaterials display a high degree of catalytic activity, conductivity, and biocompatibility that act in a synergetic manner to improve biosensor performance. For example, vast improvements in electrical conductivity and catalytic performance have been shown by reducing material size from the bulk to the nanoscale. Furthermore, the use of these nanomaterials creates a unique microenvironment that is well suited for biological stability and biological–inorganic interaction, while the use of covalent (e.g., thiol binding, cross-linking) and noncovalent biofunctionalization schemes with biorecognition agents (peptides, proteins, and nucleic acids) transforms these nanomaterials into highly sensitive probes/electrodes capable of both in vivo and ex vivo biosensing. Such nanostructured biosensors have demonstrated utility in a wide range of fields and applications including those associated with health care, environmental monitoring, security surveillance, food safety, and biodefense. Thus, we envisage that the combination of hybrid metallic nanoparticles with CNTs, graphene, and organic conjugation will continue to improve and transform the field of biosensing for years to come.

References

Ajayan PM (1999) Nanotubes from carbon. Chem Rev 99:1787–1799

Alwarappan S, Erdem A, Liu C, Li CZ (2009) Probing the electrochemical properties of graphene nanosheets for biosensing applications. J Phys Chem C 113:8853–8857

Ansari AA, Kausik A, Pratima RS, Malhotra BD (2009) Electrochemical cholesterol sensor based on tin oxide-chitosan nanobiocomposite film. Electroanalysis 21:965–972

Antiochia R, Vinci G, Lo G (2013) Rapid and direct determination of fructose in food: a new osmium-polymer mediated biosensor. Food Chem 140(4):742–747

Antonietti M, Wenz E, Bronstein L, Seregina M (1995) Synthesis and characterization of noble metal colloids in block copolymer micelles. Adv Mater 7(12):1000–1005

Arora P, Sindhu A, Dilbaghi N, Chaudhury A (2011) Biosensors as innovative tools for the detection of food borne pathogens. Biosens Bioelectron 28:1–12

Batra B, Pundir CS (2013) An amperometric glutamate biosensor based on immobilization of glutamate oxidase onto carboxylated multiwalled carbon nanotubes/gold nanoparticles/chitosan composite film modified Au electrode. Biosens Bioelectron 47:496–501

Batra B, Lata S, Sharma M, Pundir CS (2013) An acrylamide biosensor based on immobilization of hemoglobin onto multiwalled carbon nanotube/copper nanoparticles/polyaniline hybrid film. Anal Biochem 433(2):210–217

Campuzano S, Wang J (2011) Nanobioelectroanalysis based on carbon/inorganic hybrid nanoarchitectures. Electroanalysis 23:1289–1300

Cao Q, Zhao H, Yang YM, He YJ, Ding N, Wang J, Wu ZJ, Xiang KX, Wang GW (2011) Electrochemical immunosesor for casein based on gold nanoparticles and poly(L-arginine)/

multi-walled carbon nanotubes composite film functionalized interface. Biosens Bioelectron 26(8):3469–3474
Cao S, Zhang L, Chai Y, Yuan R (2013) Electrochemistry of cholesterol biosensor based on a novel Pt-Pd bimetallic nanoparticle decorated graphene catalyst. Talanta 109:167–172
Caruso F (2001) Nanoengineering of particle surfaces. Adv Mater 13:11–22
Cesarino I, Moraes FC, Lanza MRV, Machado SAS (2012) Electrochemical detection of carbamate pesticides in fruit and vegetables with a biosensor based on acetylcholinesterase immobilised on a composite of polyaniline-carbon nanotubes. Food Chem 135(3):873–879
Cevik E, Senel M, Baykal A, Abasiyanik MF (2012) A novel amperometric phenol biosensor based on immobilized HRP on poly(glycidylmethacrylate)-grafted iron oxide nanoparticles for the determination of phenol derivatives. Sens Actuat B 173:396–405
Chang Z, Chen M, Fan H, Zhao K, Zhuang S, He P, Fang Y (2008) Multilayer membranes via layer-by-layer deposition of PDDA and DNA with Au nanoparticles tags for DNA biosensing. Electrochim Acta 53:2939–2945
Chang JB, Mo S, Zhang Y, Cui SM, Zhou GH, Wu XG, Yang CH, Chen JH (2013) Ultrasonic-assisted self-assembly of monolayer graphene oxide for rapid detection of *Escherichia coli* bacteria. Nanoscale 5(9):3620–3626
Chen YS, Huang JH (2010) Arrayed CNT-Ni nanocomposites grown directly on Si substrate for amperometric detection of ethanol. Biosens Bioelectron 26(1):207–212
Chen X, Cheng GJ, Dong SJ (2001) Amperometric tyrosinase biosensor based on a sol–gel-derived titanium oxide-copolymer composite matrix for detection of phenolic compounds. Analyst 126(10):1728–1732
Claussen JC, Franklin AD, Ul Haque A, Porterfield DM, Fisher TS (2009) Electrochemical biosensor of nanocube-augmented carbon nanotube networks. ACS Nano 3(1):37–44
Claussen JC, Kim SS, Haque AU, Artiles MS, Porterfield DM, Fisher TS (2010) Electrochemical glucose biosensor of platinum nanospheres connected by carbon nanotubes. J Diabetes Sci Technol 4(2):312–319
Claussen JC, Wickner MM, Fisher TS, Porterfield DM (2011a) Transforming the fabrication and biofunctionalization of gold nanoelectrode arrays into versatile electrochemical glucose biosensors. ACS Appl Mater Interfaces 3(5):1765–1770
Claussen JC, Hengenius JB, Wickner MM, Fisher TS, Umulis DM, Porterfield DM (2011b) Effects of carbon nanotube-tethered nanosphere density on amperometric biosensing: simulation and experiment. J Phys Chem C 115:20896–20904
Claussen JC, Artiles MS, McLamore ES, Mohanty S, Shi J, Rickus JL, Fisher TS, Porterfield DM (2011c) Electrochemical glutamate biosensing with nanocube and nanosphere augmented single-walled carbon nanotube networks: a comparative study. J Mater Chem 21(30): 11224–11231
Claussen JC, Kumar A, Jaroch DB, Khawaja MH, Hibbard AB, Porterfield DM, Fisher TS (2012) Nanostructuring platinum nanoparticles on multilayered graphene petal nanosheets for electrochemical biosensing. Adv Func Mater 22(16):3399–3405
Claussen JC, Algar WR, Hildebrandt N, Susumu K, Ancona MG, Medintz IL (2013) Biophotonic logic devices based on quantum dots and temporally-staggered Förster energy transfer relays. Nanoscale 5:12156–12170
Claussen JC, Hildebrandt N, Susumu K, Ancona MG, Medintz IL (2014) Complex logic functions implemented with quantum dot bionanophotonic circuits. ACS Appl Mater Interfaces 6(6): 3771–3778
Crespilho FN, Ghica ME, Florescu M, Nart FC, Oliveira ON, Brett CMA (2006) A strategy for enzyme immobilization on layer-by-layer dendrimer-gold nanoparticle electrocatalytic membrane incorporating redox mediator. Electrochem Commun 8(10):1665–1670. doi:10.1016/j.elecom.2006.07.032
Cui Y, Zhang B, Liu B, Chen H, Chen G, Tang D (2011) Sensitive detection of hydrogen peroxide in foodstuff using an organic–inorganic hybrid multilayer-functionalized graphene biosensing platform. Microchim Acta 174:137–144

Dai LM (2007) Electrochemical sensors based on architectural diversity of the pi-conjugated structure: recent advancements from conducting polymers and carbon nanotubes. Aust J Chem 60:472–483

Daniele MA, Shaughnessy ML, Roeder R, Childress A, Bandera YP, Foulger S (2013) Magnetic nanoclusters exhibiting protein-activated near-infrared fluorescence. Acs Nano 7(1):203–213. doi:10.1021/Nn3037368

Daniel MC, Astruc D (2004) Gold nanoparticles: assembly, supramolecular chemistry, quantum-size-related properties, and applications toward biology, catalysis, and nanotechnology. Chem Rev 104(1):293–346

Ding SN, Shan D, Xue HG, Cosnier S (2010) A promising biosensing-platform based on bismuth oxide polycrystalline-modified electrode: characterization and its application in development of amperometric glucose sensor. Bioelectrochemistry 79(2):218–222

Dreaden EC, Alkilany AM, Huang X, Murphy CJ, El-Sayed MA (2012) The golden age: gold nanoparticles for biomedicine. Chem Soc Rev 41(7):2740–2779

Du M, Yang T, Zhang Y, Jiao K (2009) Sensitively electrochemical sensing for sequence-specific detection of *Phosphinothricin acetyltransferase* gene: layer-by-layer films of poly-L-lysine and Au-carbon nanotube hybrid. Electroanalysis 21:2521–2526

Eder D (2010) Carbon nanotube-inorganic hybrids. Chem Rev 110:1348–1385

Eissa S, L'Hocine L, Siaj M, Zourob M (2013) A graphene-based label-free voltammetric immunosensor for sensitive detection of the egg allergen ovalbumin. Analyst 138:4378–4384

Eremia SV, Vasilescu I, Radoi A, Litescu SC, Radu GL (2013) Disposable biosensor based on platinum nanoparticles-reduced graphene oxide-laccase biocomposite for the determination of total polyphenolic content. Talanta 110:164–170

Fang B, Jiao S, Li M, Qu Y, Jiang X (2008) Label-free electrochemical detection of DNA using ferrocene-containing cationic polythiophene and PNA probes on nanogold modified electrodes. Biosens Bioelectron 23:1175–1179

Ferey G (2008) Hybrid porous solids: past, present, future. Chem Soc Rev 37(1):191–214

Franklin AD, Janes DB, Claussen JC, Fisher TS, Sands TD (2008) Independently addressable fields of porous anodic alumina embedded in SiO on Si. Appl Phys Lett 92:013122

Gan N, Zhou J, Xiong P, Futao H, Cao Y, Li T, Jiang Q (2013a) An ultrasensitive electrochemiluminescent immunoassay for aflatoxin M1 in milk, based on extraction by magnetic graphene and detection by antibody-labeled CdTe quantum dots-carbon nanotubes nanocomposite. Toxins 5:865–883

Gan T, Sun J, Lin Z, Li Y (2013b) Highly sensitive determination of orange II based on the dual amplified electrochemical signal of graphene and mesoporous TiO2. Anal Methods 5:2964–2970

Gan T, Sun J, Wu Q, Jing Q, Yu S (2013c) Graphene decorated with nickel nanoparticles as a sensitive substrate for simultaneous determination of sunset yellow and tartrazine in food samples. Electroanalysis 25:1505–1512

Garcia-Aljaro C, Cella LN, Shirale DJ, Park M, Munoz FJ, Yates MV, Mulchandani A (2010) Carbon nanotubes-based chemiresistive biosensors for detection of microorganisms. Biosens Bioelectron 26(4):1437–1441

Ge SG, Liu WY, Ge L, Yan M, Yan JX, Huang JD, Yu JH (2013) In situ assembly of porous Au-paper electrode and functionalization of magnetic silica nanoparticles with HRP via click chemistry for Microcystin-LR immunoassay. Biosens Bioelectron 49:111–117

Gooding JJ (2002) Electrochemical DNA hybridization biosensors. Electroanalysis 14: 1149–1156

Gooding JJ (2005) Nanostructuring electrodes with carbon nanotubes: a review on electrochemistry and applications for sensing. Electrochim Acta 50:3049–3060

Guo Y, Li J, Dong S (2011) Hemin functionalized graphene nanosheets-based dual biosensor platforms for hydrogen peroxide and glucose. Sens Actuat B Chem 160:295–300

Han J, Zhuo Y, Chai Y et al (2012) Simultaneous electrochemical detection of multiple tumor markers based on dual catalysis amplification of multi-functionalized onion-like mesoporous graphene sheets. Anal Chim Acta 746:70–76

He H, Zhang Y, Gao C, Wu JY (2009) 'Clicked' magnetic nanohybrids with a soft polymer interlayer. Chem Commun 13:1655–1657

Hong W, Bai H, Xu Y, Yao Z, Gu Z, Shi G (2010) Preparation of gold nanoparticle/graphene composites with controlled weight contents and their application in biosensors. J Phys Chem C 114:1822–1826

Horak D, Babic M, Mackova H, Benes MJ (2007) Preparation and properties of magnetic nano- and microsized particles for biological and environmental separations. J Sep Sci 30(11): 1751–1772

Hu X, Dou W, Fu L, Zhao G (2013) A disposable immunosensor for *Enterobacter sakazakii* based on an electrochemically reduced graphene oxide-modified electrode. Anal Biochem 434:218–220

Huang ST, Shi Y, Li NB, Luo HQ (2012) Sensitive turn-on fluorescent detection of tartrazine based on fluorescence resonance energy transfer. Chem Commun 48:747–749

Huang KJ, Wang L, Li J, Yu M, Liu YM (2013) Electrochemical sensing of catechol using a glassy carbon electrode modified with a composite made from silver nanoparticles, polydopamine, and graphene. Microchim Acta 180(9–10):751–757

Jia LP, Wang HS (2013) Preparation and application of a highly sensitive nonenzymatic ethanol sensor based on nickel nanoparticles/Nafion/graphene composite film. Sens Actuat B Chem 177:1035–1042

Jung JH, Cheon DS, Liu F, Lee KB, Seo TS (2010) A graphene oxide based immuno-biosensor for pathogen detection. Angew Chem Int Ed 49(33):5708–5711

Katz E, Willner I (2004) Biomolecule-functionalized carbon nanotubes: applications in nanobioelectronics. Chemphyschem 5:1085–1104

Kievit FM, Zhang M (2011) Surface engineering of iron oxide nanoparticies for targeted cancer therapy. Acc Chem Res 44(10):853–862

Kim JH, Cho JH, Cha GS, Lee CW, Kim HB, Paek SH (2000) Conductimetric membrane strip immunosensor with polyaniline-bound gold colloids as signal generator. Biosens Bioelectron 14(12):907–915

Kim HJ, Park SH, Park HJ (2010) Synthesis of a new electrically conducting nanosized Ag-polyaniline-silica complex using gamma-radiolysis and its biosensing application. Radiat Phys Chem 79(8):894–899

Kim B, Lim D, Jin HJ, Lee HY, Namgung S, Ko Y, Park SB, Hong S (2012a) Family-selective detection of antibiotics using antibody-functionalized carbon nanotube sensors. Sens Actuat B 166:193–199

Kim JE, Shin JY, Cho MH (2012b) Magnetic nanoparticles: an update of application for drug delivery and possible toxic effects. Arch Toxicol 86(5):685–700

Ko JW, Woo JM, Jinhong A, Cheon JH, Lim JH, Kim SH, Chun H, Kim E, Park YJ (2011) Multi-order dynamic range DNA sensor using a gold decorated SWCNT random network. ACS Nano 5(6):4365–4372

Kong Y, Ren X, Huo Z, Wang G, Tao Y, Yao C (2013) Electrochemical detection of pyrosine with electrochemically reduced graphene oxide modified glassy carbon electrode. Eur Food Res Technol 236:955–961

Kozitsina AN, Shalygina ZV, Dedeneva SS, Rusinov GL, Tolshchina SG, Verbitskiy EV, Brainina KZ (2009) Catalytic systems based on the organic nickel(II) complexes in chronoamperometric determination of urea and creatinine. Russian Chem Bull 58:1119–1125

Kuralay F, Erdem A, Abacı S, Özyörük H, Yıldız A (2008) Electrochemical biosensing of DNA immobilized poly(vinylferrocenium) modified electrode. Electroanalysis 20:2563–2570

Kuralay F, Erdem A, Abacı S, Özyörük H, Yıldız A (2009) Poly(vinylferrocenium) coated disposable pencil graphite electrode for DNA hybridization. Electrochem Commun 11:1242–1246

Labroo P, Cui Y (2013) Flexible graphene bio-nanosensor for lactate. Biosens Bioelectron 41:852–856

Lai G, Wu J, Ju H et al (2011) Streptavidin-functionalized silver-nanoparticle-enriched carbon nanotube tag for ultrasensitive multiplexed detection of tumor markers. Adv Funct Mater 21: 2938–2943

Lata S, Batra B, Kumar P, Pundir CS (2013) Construction of an amperometric D-amino acid biosensor based on D-amino acid oxidase/carboxylated multiwalled carbon nanotube/copper nanoparticles/polyalinine modified gold electrode. Anal Biochem 437(1):1–9

Lee N, Hyeon T (2012) Designed synthesis of uniformly sized iron oxide nanoparticles for efficient magnetic resonance imaging contrast agents. Chem Soc Rev 41(7):2575–2589

Lee YJ, Lyn YK, Choi HN, Lee WY (2007) Amperometric tyrosinase biosensor based on carbon nanotube-titania-Nafion composite film. Electroanalysis 19(10):1048–1054

Lei J, Ju H (2012) Signal amplification using functional nanomaterials for biosensing. Chem Soc Rev 41:2122–2134

Li Y, Wei Y, Shi G, Xian Y, Jin L (2011) Facile synthesis of leaf-like CuO nanoparticles and their application on glucose biosensor. Electroanalysis 23(2):497–502

Li Y, Cheng P, Gong JH, Fang LC, Deng J, Liang WB, Zheng JS (2012) Amperometric immunosensor for the detection of *Escherichia coli* O157:H7 in food specimens. Anal Biochem 421(1):227–233

Li L, Lu H, Deng L (2013) A sensitive NADH and ethanol biosensor based on graphene-Au nanorods nanocomposites. Talanta 113:1–6

Lidong W, Deng D, Jin J, Lu X, Chen J (2012) Nanographene-based tyrosinase biosensor for rapid detection of bisphenol A. Biosens Bioelectron 35(1):193–199

Lim JH, Park J, Ahn JH, Jin HJ, Hong S, Park TH (2013) A peptide receptor-based bioelectronic nose for the real-time determination of seafood quality. Biosens Bioelectron 39(1):244–249

Lin X, Ni Y, Kokot S (2013) Glassy carbon electrodes modified with gold nanoparticles for the simultaneous determination of three food antioxidants. Anal Chim Acta 765:54–62

Liu G, Lin Y (2007) Nanomaterial labels in electrochemical immunosensors and immunoassays. Talanta 74:308–317

Liu LH, Dietsch H, Schurtenberger P, Yan MD (2009) Photoinitiated coupling of unmodified monosaccharides to iron oxide nanoparticles for sensing proteins and bacteria. Bioconjug Chem 20(7):1349–1355

Liu HY, Malhotra BD, Peczuh MW, Rusling JF (2010) Electrochemical immunosensors for antibodies to peanut allergen Ara h2 using gold nanoparticle-peptide films. Anal Chem 82(13): 5865–5871

Liu F, Choi KS, Park TJ, Lee SY, Seo TS (2011a) Graphene-based electrochemical biosensor for pathogenic virus detection. Biochip J 5(2):123–128

Liu F, Laurent S, Fattahi H, Elst LV, Muller RN (2011b) Superparamagnetic nanosystems based on iron oxide nanoparticles for biomedical imaging. Nanomedicine 6(3):519–528

Liu DB, Chen WW, Wei JH, Li XB, Wang Z, Jiang XY (2012) A highly sensitive, dual-readout assay based on gold nanoparticles for organophosphorus and carbamate pesticides. Anal Chem 84(9):4185–4191

Lu J, Drzal LT, Worden RM, Lee I (2007) Simple fabrication of a highly sensitive glucose biosensor using enzymes immobilized in exfoliated graphite nanoplatelets Nafion membrane. Chem Mater 19:6240–6246

Lu J, Do I, Drzal LT, Worden RM, Lee I (2008) Nanometal-decorated exfoliated graphite nanoplatelet based glucose biosensors with high sensitivity and fast response. ACS Nano 2:1825–1832

Lu CH, Yang HH, Zhu CL, Chen X, Chen GN (2009) A graphene platform for sensing biomolecules. Angew Chem Int Ed 121:4879–4881

Ma X, Chao M, Wang Z (2013) Voltammetric determination of sudan I in food samples at graphene modified glassy carbon electrode based on the enhancement effect of sodium dodecyl sulfate. Food Chem 35(1):739–744

Manesh KM, Halámek J, Pitab M, Zhoub J, Tamb TK, Santhosha P, Chuanga M-C, Windmiller JR, Abidina D, Katz E, Wang J (2011) Enzyme logic gates for the digital analysis of physiological level upon injury. Biosens Bioelectron 24(12):3569–3574

McCreery RL (2008) Advanced carbon electrode materials for molecular electrochemistry. Chem Rev 108:2646–2687

McLamore ES, Shi J, Jaroch D, Claussen JC, Uchida A, Jiang Y, Zhang W, Donkin SS, Banks MK, Buhman KK, Teegarden D, Rickus JL, Porterfield DM (2011) A self referencing platinum

nanoparticle decorated enzyme-based microbiosensor for real time measurement of physiological glucose transport. Biosens Bioelectron 26(5):2237–2245
Merkoçi A (2010) Nanoparticles-based strategies for DNA, protein and cell sensors. Biosens Bioelectron 26:1164–1177
Mirkin CA, Letsinger RL, Mucic RC, Storhoff JJ (1996) A DNA-based method for rationally assembling nanoparticles into macroscopic materials. Nature 382(6592):607–609
Mohanty N, Berry V (2008) Graphene-based single-bacterium resolution biodevice and DNA transistor: interfacing graphene derivatives with nanoscale and microscale biocomponents. Nano Lett 8:4469–4476
Monosik R, Stredansky M, Greif G, Sturdik E (2012a) A rapid method for determination of L-lactic acid in real samples by amperometric biosensor utilizing nanocomposite. Food Control 23(1):238–244
Monosik R, Ukropcova D, Stredansky M, Sturdik E (2012b) Multienzymatic amperometric biosensor based on gold and nanocomposite planar electrodes for glycerol determination in wine. Anal Biochem 421(1):256–261
Monosik R, Stred'ansky M, Sturdik E (2013) A biosensor utilizing L-glutamate dehydrogenase and diaphorase immobilized on nanocomposite electrode for determination of L-glutamate in food samples. Food Anal Methods 6(2):521–527
Muti M, Kuralay F, Erdem A, Abaci S, Yumak T, Sinağ A (2010) Tin oxide nanoparticles-polymer modified single-use sensors for electrochemical monitoring of label-free DNA hybridization. Talanta 82:1680–1686
Muraviev DN, Pividori MI, Soto JLM, Alegret S (2006) Extractant assisted synthesis of polymer stabilized platinum and palladium metal nanoparticles for sensor applications. Solvent Extr Ion Exc 24(5):731–745. doi:10.1080/07366290600851588
Na S-I, Kim S-S, Hong W-K, Park J-W, Jo J, Nah Y-C, Lee KD-Y (2008) Fabrication of TiO_2 nanotubes by using electrodeposited ZnO nanorod template and their application to hybrid solar cells. Electrochim Acta 53:2560–2566
Nandini S, Nalini S, Manjunatha R, Shanmugam S, Melo JS, Suresh GS (2013) Electrochemical biosensor for the selective determination of hydrogen peroxide based on the co-deposition of palladium, horseradish peroxidase on functionalized-graphene modified graphite electrode as composite. J Electroanal Chem 689:233–242
Netto CGCM, Toma HE, Andrade LH (2013) Superparamagnetic nanoparticles as versatile carriers and supporting materials for enzymes. J Mol Catal B 85–86:71–92
Njagi J, Andreescu S (2007) Stable enzyme biosensors based on chemically synthesized Au-polypyrrole nanocomposites. Biosens Bioelectron 23(2):168–175
Ohno Y, Maehashi K, Yamashiro U, Matsumoto K (2009) Electrolyte-gated graphene field-effect transistors for detecting pH protein adsorption. Nano Lett 9:3318–3322
Oliveira TMBF, Barroso MF, Morais S, Araujo M, Freire C, Lima-Neto P, Correia AN, Oliveira MBPP, Delerue-Matos C (2013a) Laccase-prussian blue film-graphene doped carbon paste modified electrode for carbamate pesticides quantification. Biosens Bioelectron 47:292–299
Oliveira TMBF, Barroso MF, Morais S, de Lima-Neto P, Correia AN, Oliveira MBPP, Delerue-Matos C (2013b) Biosensor based on multi-walled carbon nanotubes paste electrode modified with laccase for pirimicarb pesticide quantification. Talanta 106:137–143
Omidfar K, Zarei H, Gholizadeh F, Larijani B (2012) A high-sensitivity electrochemical immunosensor based on mobile crystalline material-41-polyvinyl alcohol nanocomposite and colloidal gold nanoparticles. Anal Biochem 421(2):649–656
Palaniappan A, Goh WH, Fam DWH, Rajaseger G, Chan CEZ, Hanson BJ, Moochhala SM, Mhaisalkar SG, Liedberg B (2013) Label-free electronic detection of bio-toxins using aligned carbon nanotubes. Biosens Bioelectron 43:143–147
Palecek E (2002) Past, present and future of nucleic acids electrochemistry. Talanta 56:809–819
Pandey CM, Singh R, Sumana G, Pandey MK, Malhotra BD (2011) Electrochemical genosensor based on modified octadecanethiol self-assembled monolayer for Escherichia coli detection. Sens Actuat B 151(2):333–340

Pang X, He D, Luo S, Cai Q (2009) An amperometric glucose biosensor fabricated with Pt nanoparticle-decorated carbon nanotubes/TiO_2 nanotube arrays composite. Sens Actuat B 137(1):134–138

Park J, Lim JH, Jin HJ, Namgung S, Lee SH, Park TH, Hong S (2012) A bioelectronic sensor based on canine olfactory nanovesicle-carbon nanotube hybrid structures fast assessment of food quality. Analyst 137(14):3249–3254

Pei X, Zhang B, Tang J et al (2013) Sandwich-type immunosensors and immunoassays exploiting nanostructure labels: a review. Anal Chim Acta 758:1–18

Peng X, Chen J, Misewich JA et al (2009) Carbon nanotube–nanocrystal heterostructures. Chem Soc Rev 38:1076–1098

Peng HP, Liang RP, Zhang L, Qiu JD (2013) Facile preparation of novel core-shell enzyme-Au-polydopamine-Fe_3O_4 magnetic bionanoparticles for glucosesensor. Biosens Bioelectron 42:293–299

Pingarrón JM, Yáñez-Sedeño P, González-Cortés A (2008) Gold nanoparticle-based electrochemical biosensors. Electrochim Acta 53:5848–5866

Prakash S, Chakrabarty T, Singh AK, Shahi VK (2013) Polymer thin films embedded with metal nanoparticles for electrochemical biosensors applications. Biosens Bioelectron 41:43–53

Pyun J (2012) Self-assembly and colloidal polymerization of polymer-nanoparticle hybrids into mesoscopic chains. Angew Chem Int Ed 51:12408–12409

Qi B, Ye LF, Stone R, Dennis C, Crawford TM, Mefford OT (2013) Influence of ligand-precursor molar ratio on the size evolution of modifiable iron oxide nanoparticles. J Phys Chem C 117(10):5429–5435

Radhakrishnan S, Sumathi C, Umar A, Kim SJ, Wilson J, Dharuman V (2013) Polypyrrole-poly(3,4-ethylenedioxythiophene)-Ag (PPy-PEDOT-Ag) nanocomposite films for label-free electrochemical DNA sensing. Biosens Bioelectron 47:133–140

Ramesh GV, Porel S, Radhakrishnan TP (2009) Polymer thin films embedded with in situ grown metal nanoparticles. Chem Soc Rev 38(9):2646–2656

Rao CRK (2012) Polyelectrolyte-aided synthesis of gold and platinum nanoparticles: implications in electrocatalysis and sensing. J Appl Polym Sci 124(6):4765–4771

Rossi LM, Quach AD, Rosenzweig Z (2004) Glucose oxidase-magnetite nanoparticle bioconjugate for glucose sensing. Anal Bioanal Chem 380(4):606–613. doi:10.1007/s00216-004-2770-3

Ruan C, Li T, Niu Q, Lu M, Lou J, Gao W, Sun W (2012) Electrochemical myoglobin biosensor based on graphene-ionic liquid-chitosan bionanocomposites: direct electrochemistry and electrocatalysis. Electrochim Acta 64:184–189

Sandhu P, Yang J, Xu CQ (2010) In-Fiber michelson interferometer with polymeric/nanoparticle thin-film overlay as a platform for biosensing. Ieee J Sel Top Quant 16(3):685–690. doi:10.1109/Jstqe.2009.2029874

Saha S, Arya SK, Singh SP, Sreenivas K, Malhotra BD, Gupta V (2009) Nanoporous cerium oxide thin film for glucose biosensor. Biosens Bioelectron 24(7):2040–2045

Shaidarova LG, Budnikov GK (2008) Chemically modified electrodes based on noble metals, polymer films, or their composites in organic voltammetry. J Anal Chem 63(10):922–942

Shan CS, Yang HF, Song JF, Han DX, Ivaska A, Niu L (2009) Direct electrochemistry of glucose oxidase and biosensing for glucose based on graphene. Anal Chem 81:2378–2382

Shan C, Yang H, Han D et al (2010) Graphene/AuNPs/chitosan nanocomposites film for glucose biosensing. Biosens Bioelectron 25:1070–1074

Shang NG, Papakonstantinou P, McMullan M, Chu M, Stamboulis A, Potenza A, Dhesi SS, Marchetto H (2008) Catalyst-free efficient growth, orientation and biosensing properties of multilayer graphene nanoflake films with sharp edge planes. Adv Funct Mater 18:3506–3514

Shi X, Gu W, Li B, Chen N, Zhao K, Yuezhong X (2013) Enzymatic biosensors based on the use of metal oxide nanoparticles. Microchim Acta 181:1–22

Si P, Dong X-C, Chen P, Kim D-H (2013) A hierarchically structured composite of Mn3O4/3D graphene foam for flexible nonenzymatic biosensors. J Mater Chem B 1:110–115

Singh A, Sinsinbar G, Choudhary M, Kumar V, Pasricha R, Verman HN, Singh SP, Arora K (2013a) Graphene oxide-chitosan nanocomposite based electrochemical DNA biosensor for detection of typhoid. Sens Actuat B 185:675–684
Singh C, Srivastava S, Ali MA, Gupta TK, Sumana G, Srivastava A, Mathur RB, Malhotra BD (2013b) Carboxylated multiwalled carbon nanotubes based biosensor for aflatoxin detection. Sens Actuat B 185:258–264
Spatz JP, Roescher A, Moller M (1996) Gold nanoparticles in micellar poly(styrene)-b-poly(ethylene oxide) film-ize and interparticle distance control in monoparticulate films. Adv Mater 8(4):337–340
Sperling RA, Parak WJ (2010) Surface modification, functionalization and bioconjugation of colloidal inorganic nanoparticles. Phil Trans R Soc A 368:1333–1383
Srivastava S, Kumar V, Ali MD, Solanki PR, Srivastava A, Sumana G, Saxena PS, Joshi AG, Malhotra BD (2013) Electrophoretically deposited reduced graphene oxide platform for food toxin detection. Nanoscale 5:3043–3051
Storhoff JJ, Mirkin CA (1999) Programmed materials synthesis with DNA. Chem Rev 99:1849–1862
Sun W, Zhai Z, Wang D, Liu S, Jiao K (2009) Electrochemistry of hemoglobin entrapped in a nafion/nano-ZnO film on carbon ionic liquid electrode. Colloids Surf B 24:295–300
Sun W, Qin P, Gao H, Li G, Jiao K (2010) Electrochemical DNA biosensor based on chitosan/nano-V_2O_5/MWCNTs composite film modified carbon ionic liquid electrode and its application to the LAMP product of *Yersinia enterocolitica* gene sequence. Biosens Bioelectron 25:1264–1270
Sun W, Li L, Lei B, Li T, Ju X, Wang X, Li G, Sun Z (2013) Fabrication of graphene-platinum nanocomposite for the direct electrochemistry and electrocatalysis of myoglobin. Mater Sci Eng 33(4):1907–1913
Taguchi M, Ptitsyn A, McLamore ES, Claussen JC (2014) Nanomaterial-mediated biosensors for monitoring glucose. J Diabetes Sci Technol 48(2):403–411
Tang J, Tang D, Su B et al (2011) Enzyme-free electrochemical immunoassay with catalytic reduction of *p*-nitrophenol and recycling of *p*-aminophenol using gold nanoparticles-coated carbon nanotubes as nanocatalysts. Biosens Bioelectron 26:3219–3226
Tang J, Hou L, Tang D, Zhou J, Zhouping W, Li J, Chen G (2012) Magneto-controlled electrochemical immunoassay of brevetoxin B in seafood based on guanine-functionalized graphene nanoribbons. Biosens Bioelectron 38:86–93
Temur E, Zengin A, Boyaci IH, Dudak FC, Torul H, Tamer U (2012) Attomole sensitivity of Staphylococcal Enterotoxin B detection using an aptamer-modified surface-enhanced Raman scattering probe. Anal Chem 84(24):10600–10606
Ugelstad J, El-Aasser MS, Vanderhoff JWJ (1973) Emulsion polymerization: initiation of polymerization in monomer droplets. J Polym Sci 11(8):503–513
Veiseh O, Gunn JW, Zhang M (2010) Design and fabrication of magnetic nanoparticles for targeted drug delivery and imaging. Adv Drug Deliv Rev 62(3):284–304
Villalonga R, Villalonga ML, Diez P, Pingarron JM (2011) Decorating carbon nanotubes with polyethylene glycol-coated magnetic nanoparticles for implementing highly sensitive enzyme biosensors. J Mater Chem 21(34):12858–12864
Villalonga R, Díez P, Eguílaz M, Martínez P, Pingarrón JM (2012) Supramolecular immobilization of xanthine oxidase on electropolymerized matrix of functionalized hybrid gold nanoparticles/single-walled carbon nanotubes for the preparation of electrochemical biosensors. ACS Appl Mater Interfaces 4(8):4312–4319
Wan Y, Lin ZF, Zhang D, Wang Y, Hou BR (2011) Impedimetric immunosensor doped with reduced graphene sheets fabricated by controllable electrodeposition for the non-labelled detection of bacteria. Biosens Bioelectron 26(5):1959–1964
Wang J (2000) From DNA biosensors to gene chips. Nucl Acid Res 28:3011–3016
Wang J (2002) Electrochemical detection for microscale analytical systems: a review. Talanta 56:223–231
Wang J (2003) Nanoparticle-based electrochemical DNA detection. Anal Chim Acta 500:247–257

Wang J (2007) Nanoparticles for electrochemical bioassays. In: Mirkin CA, Niemeyer CM (eds) Nanobiotechnology II: more concepts and applications. Wiley-VCH, Weinheim, pp 125–133

Wang T, Ma Z, Xu F, Jiang Z (2003) The one-step preparation and electrochemical characteristics of tin oxide nanocrystalline materials. Electrochem Commun 5:599–602

Wang Y, Li YM, Tang LH, Lu J, Li JH (2009) Application of graphene-modified electrode for selective detection of dopamine. Electrochem Commun 11:889–892

Wang X, Li J, Weidong Q, Chen G (2011) Fabrication of graphene/poly(methyl methacrylate) composite electrode for capillary electrophoretic determination of bioactive constituents in Herba Geranii. J Chromatog A 1218:5542–5548

Wang L, Ye YJ, Zhu HZ, Song YH, He SJ, Xu FG, Hou HQ (2012a) Controllable growth of Prussian blue nanostructures on carboxylic group-functionalized carbon nanofibers and its application for glucose biosensing. Nanotechnology 23(45):455502

Wang Q, Wang Y, Liu S, Wang L, Gao F, Gao F, Sun W (2012b) Voltammetric detection of bisphenol a by a chitosan-graphene composite modified carbon ionic liquid electrode. Thin Solid Films 520:4459–4464

Wang X, Dong J, Ming H, Ai S (2013) Sensing of glycoprotein via a biomimetic sensor based on molecularly imprinted polymers and graphene-Au nanoparticles. Analyst 138(4):1219–1225

Wei Q, Zhao Y, Du B, Wu D, Li H, Yang M (2012) Ultrasensitive detection of kanamycin in animal derived foods by label-free electrochemical immunosensor. Food Chem 134:1601–1606

Wen J, Zhou S, Yuan Y (2013) Graphene oxide as nanogold carrier for ultrasensitive electrochemical immunoassay of *Shewanella oneidensis* with silver enhancement strategy. Biosens Bioelectron 52:44–49

Willner I, Willner B (2002) Functional nanoparticle architectures for sensoric, optoelectronic, and bioelectronics applications. Pure Appl Chem 74:1773–1783

Wu B, Kuang Y, Zhang X et al (2011) Noble metal nanoparticles/carbon nanotubes nanohybrids: synthesis and applications. Nano Today 6:75–90

Xie Y, Li Y, Niu L, Wang H, Qian H, Yao W (2012) A novel surface-enhanced Raman scattering sensor to detect prohibited colorants in food by graphene/silver nanocomposite. Talanta 100:32–37

Xu P, Zeng GM, Huang DL, Feng CL, Hu S, Zhao MH, Lai C, Wei Z, Huang C, Xie GX, Liu ZF (2012) Use of iron oxide nanomaterials in wastewater treatment: a review. Sci Total Environ 424:1–10

Yáñez-Sedeño P, Riu J, Pingarrón JM et al (2010) Electrochemical sensing based on carbon nanotubes. Trends Anal Chem 29:939–953

Yang Y, Wang Z, Yang M, Li J, Zheng F, Shen G, Yu R (2007) Electrochemical detection of deoxyribonucleic acid hyridzation based on carbon-nanotubes/nano zirconium dioxide/chitosan modified electrodes. Anal Chim Acta 584:268–274

Yang LQ, Ren XL, Tang FQ, Zhang L (2009) A practical glucose biosensor based on Fe_3O_4 nanoparticles and chitosan/Nafion composite film. Biosens Bioelectron 25(4):889–895

Yang MH, Sun S, Bruck HA, Kostov Y, Rasooly A (2010a) Electrical percolation-based biosensor for real-time direct detection of staphylococcal enterotoxin B. Biosens Bioelectron 25(12):2573–2578

Yang W, Ratinac KR, Ringer SP, Thordarson P, Gooding JJ, Braet F (2010b) Carbon nanomaterials in biosensors: should you use nanotubes or graphene? Angew Chem Int Ed 49:2114–2138

Yang M, Javadi A, Gong S (2011) Sensitive electrochemical immunosensor for the detection of cancer biomarker using quantum dot functionalized graphene sheets as labels. Sens Actuators B 155:357–360

Yang M, Peng ZH, Ning Y, Chen YZ, Zhou Q, Deng L (2013a) Highly specific and cost efficient detection of Salmonella Paratyphi A combining aptamers with single-walled carbon nanotubes. Sensors 13(5):6865–6881

Yang X-H, Kong W-J, Yang M-H, Zhao M, Ouyang Z (2013b) Application of aptamer identification technology in rapid analysis of mycotoxins. Chin J Anal Chem 41(2):297–306

Ye X, Du Y, Lu D, Wang C (2013) Fabricaiton of b-cyclodextrin-coated poly(diallyldimethylammonium chloride)-functionalized graphene composite film modified glassy carbon-rotating disk electrode and its application for simultaneous electrochemical determination colorants of sunset yellow and tartrazine. Anal Chim Acta 779:22–34

Yumak T, Kuralay F, Erdem M, Sinag A, Erdem A, Abaci S (2011) Preparation and characterization of zinc oxide nanoparticles and their sensor applications for electrochemical monitoring of nucleic acid hybridization. Colloids Surf B 86:397–403

Zelada-Guillen GA, Bhosale SV, Riu J, Rius FX (2010) Real-time potentiometric detection of bacteria in complex samples. Anal Chem 82(22):9254–9260

Zeng Q, Cheng J-S, Liu X-F et al (2011) Palladium nanoparticle/chitosan-grafted graphene nanocomposites for construction of a glucose biosensor. Biosens Bioelectron 26:3456–3463

Zhang M, Smith A, Gorski W (2004) Carbon nanotube – chitosan system for electrochemical sensing based on dehydrogenase enzymes. Anal Chem 76:5045–5050

Zhang LH, Zhai YM, Gao N, Wen D, Dong SJ (2008a) Sensing H_2O_2 with layer-by-layer assembled Fe_3O_4-PDDA nanocomposite film. Electrochem Commun 10(10):1524–1526

Zhang W, Yang T, Huang DM, Jıao K (2008b) Eectrochemical sensing of DNA immobilization and hybridization based on carbon nanotubes/nanozinc oxide/chitosan composite film. Chin Chem Lett 19:589–591

Zhang Y, Zhang K, Ma H (2009) Electrochemical DNA biosensor based on silver nanoparticles and poly(3-(3-pyridyl) acrylic acid)/carbon nanotubes modified electrode. Anal Biochem 387:13–19

Zhang B, Cui Y, Chen H, Liu B, Chen G, Tang D (2011) A new electrochemical biosensor for determination of hydrogen peroxide in food based on well-dispersive gold nanoparticles on graphene oxide. Electroanalysis 23(8):1821–1829

Zhang WJ, Shi XH, Huang J, Zhang YX, Wu ZR, Xian YZ (2012) Bacitracin-conjugated superparamagnetic iron oxide nanoparticles: synthesis, characterization and antibacterial activity. Chemphyschem 13(14):3388–3396

Zhang J, Sun Y, Xu B, Zhang H, Gao Y, Zhang H, Song D (2013) A novel surface plasmon resonance biosensor based on graphene oxide decorated with gold nanorod-antibody conjugates for determination of transferrin. Biosens Bioelectron 45:230–236

Zhang FH, Cho SS, Yang SH, Seo SS, Cha GS, Nam H (2006) Gold nanoparticle-based mediatorless biosensor prepared on microporous electrode. Electroanal 18(3):217–222. doi:10.1002/elan.200503401

Zhao K, Zhuang S, Chang Z, Songm H, Dai L, He P, Fanga Y (2007) Amperometric glucose biosensor based on platinum nanoparticles combined aligned carbon nanotubes electrode. Electroanalysis 19(10):1069–1074

Zhao Y, Wei Q, Xu C, Li H, Wu D, Cai Y, Mao K, Cui Z, Du B (2011) Label-free electrochemical immunosensor for sensitive detection of kanamycin. Sens Actuat B 155:618–625

Zhong Z, Wu W, Wang D et al (2010) Nanogold-enwrapped graphene nanocomposites as trace labels for sensitivity enhancement of electrochemical immunosensors in clinical immunoassays: carcinoembryonic antigen as a model. Biosens Bioelectron 25:2379–2383

Zhou Y, Wang SX, Xie YY, Guan W, Ding BJ, Yang ZM, Jiang XY (2008) 1,3-Dipolar cycloaddition as a general route for functionalization of Fe(3)O(4) nanoparticles. Nanotechnology 19(17):175601

Zhou S, Wei D, Shi H, Feng X, Xue K, Zhang F, Song W (2013) Sodium dodecyl benzene sulfonate functionalized graphene for confined electrochemical growth of metal/oxide nanocomposites for sensing application. Talanta 107:349–355

Zhuo Y, Yuan PX, Yuan R, Chai YQ, Hong CL (2009) Bienzyme functionalized three-layer composite magnetic nanoparticles for electrochemical immunosensors. Biomaterials 30(12):2284–2290

Ziyatdinova GK, Nizamova AM, Aytuganova II, Budnikov HC (2013) Voltammetric evaluation of the antioxidant capacity of tea on electrodes modified with multi-walled carbon nanotubes. J Anal Chem 68(2):132–139

Zou C, Fu YC, Xie QJ, Yao SZ (2010) High-performance glucose amperometric biosensor based on magnetic polymeric bionanocomposites. Biosens Bioelectron 25(6):1277–1282

Chapter 9
Novel Nanobiosensing Using a Focused Laser Beam

Hiroyuki Yoshikawa

Abstract Lasers are very useful in nanobiosensing. The laser beam is spatially coherent and can be tightly focused by lenses. Interactions of photons with matter in the laser focus enable micro- and nanoscale analyses of a small objective in a specific area. In addition, focusing a laser beam induces photochemical reactions and optical forces in a sub-micron region of molecular solutions and nanoparticle dispersions. These effects of a focused laser beam are utilized for developing rapid, sensitive, compact, and low-cost biosensors, which are required for applications in point-of-care diagnostics, food safety controls, and environmental monitoring. In the first part of this chapter, a biosensing technique is introduced which works by simply focusing a single laser beam and detecting its reflection intensity. The polymer nanostructure deposited in a laser focus due to self-catalytic oxidative photopolymerization converts the enzyme reactions of horseradish peroxidase into a back-reflected intensity of the focused laser beam. A reliable optical quantification of glucose can be performed in a short time on a small sample volume. Another biosensing technique based on optical trapping of Ag nanoparticles is effective for sensitive biomolecular detection in solution. A focused laser beam immobilizes Ag nanoparticles with analyte molecules at a local spot on a polymer substrate. Surface-enhanced Raman scattering of analyte molecules can be measured by irradiation of a visible-light excitation laser beam. Adenine molecules are detected quantitatively in a concentration range from 0.01 to 1 μM.

Keywords Glucose • Optical trap • Photopolymerization • Silver nanoparticles • Surface-enhanced Raman scattering (SERS)

H. Yoshikawa (✉)
Department of Applied Physics, Osaka University, Osaka, Japan
e-mail: yosikawa@ap.eng.osaka-u.ac.jp

M.C. Vestergaard et al. (eds.), *Nanobiosensors and Nanobioanalyses*,
DOI 10.1007/978-4-431-55190-4_9

9.1 Introduction

Laser technologies have made great contributions to various scientific fields, including physics, chemistry, biology, and medical science. Confocal laser scanning microscopy reveals the three-dimensional structures of cells and tissues. Flow cytometry is used for cell sorting by detecting the light scattering of cells crossing the laser beam path. Matrix-assisted laser deposition/ionization (MALDI) technique is used in the mass spectroscopy of protein and peptides. Optical and photonic biosensing methods benefit from laser technologies. For example, sensitive biosensors based on surface plasmon resonance (SPR), microring resonator, and fluorescence are realized by using laser light sources (Anker et al. 2008; Ramachandran et al. 2008; Pickup et al. 2005). Small and cheap laser systems like diode laser (LD) and DPSS laser are available nowadays. In addition, the energy of a laser beam can be localized in a submicrometer spot easily just by focusing with a lens. This is advantageous for increasing the laser power (photon density) and reducing the sample volume needed, which is important for the development of highly sensitive and compact biosensors. We show that a focused visible-light laser beam induces characteristic nanostructure formation due to the oxidative polymerization of o-phenylenediamine molecules at its focus (Yoshikawa et al. 2012). This phenomenon is applied to the detection of horseradish peroxidase enzyme reactions, which has conventionally been detected by optical absorption due to chromogenic reactions.

A focused laser beam can exert physical force to trap micro- and nanoparticles in the laser focus. This is called optical tweezers or optical trapping. Nanoparticles whose diameter is much less than the laser wavelength can be trapped as a group in the laser focus and their aggregates formed (Hosokawa et al. 2005; Yoshikawa et al. 2004; Tanaka et al. 2009). We utilize the aggregation of Ag nanoparticles in an optical trap to analyze molecules adsorbed on the aggregates based on surface-enhanced Raman scattering (SERS) spectroscopy. Raman spectroscopy is one of the dominant methods of molecular identification, because the Raman scattering spectrum gives structural and conformational information of molecules without any pretreatment. Thus, this technique is applicable to label-free biomolecular sensing and analysis in a small region of microfluidic chip devices.

9.2 Single Beam Optical Biosensors with Enzyme Reactions

Most sensitive optical biosensors require spectroscopy to quantitatively analyze optical signals such as optical absorption, fluorescence, and Raman scattering. The size and cost of optical biosensors can be reduced to only a limited extent owing to the fact that the detection of optical signals necessarily requires optical components or instruments for spectroscopy such as a spectrometer, multichannel detector, or optical filters. On the other hand, detection of reflected light is quite simple. A part of the light incident on the interface between media with different refractive indices reflects back in the incident medium. Normal incident light reflects back in the

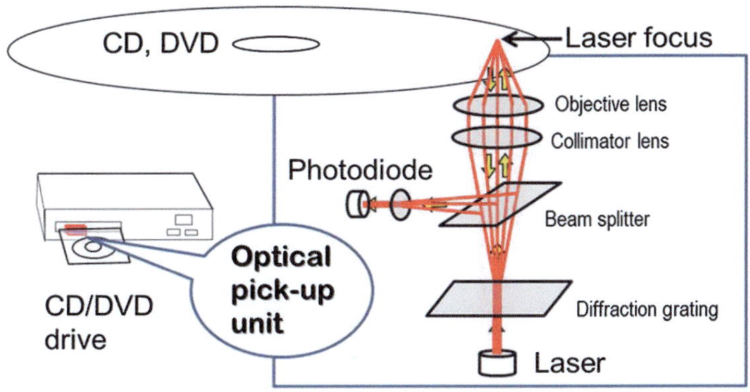

Fig. 9.1 Schematic illustrate of an optical pickup unit used in CD/DVD recorders and players

opposite direction to the incident angle, and so goes back to the light source. Detection of this "back-reflection" is used to read pit marks recorded on optical discs like CDs and DVDs. We focus on the fact that the optical pickup unit of optical storage drives is compact and cheap, because it just detects the back-reflection intensity of a laser beam focused on an optical disk (Fig. 9.1), so biosensing in a manner similar to the optical pickup unit is ideal for point-of-care testing devices.

Horseradish peroxidase (HRP) catalyzes a reaction in which hydrogen peroxide oxidizes organic and inorganic compounds (Veitch 2004). Chromogenic substrates such as tetramethylbenzidine (TMB), Amplex Red, and *o*-phenylenediamine are converted into colored products by HRP catalytic reactions, yielding a characteristic color change that is detectable by spectroscopic methods. Such HRP reactions are used in reagents for enzyme assays and immunoassays. For example, hydrogen peroxide produced by a glucose oxidase reaction can be quantified by a HRP chromogenic reaction. An antibody conjugated to HRP is used to detect a small amount of a specific protein in immunoassays like Western blotting and ELISA. The HRP-labeled antibody or antigen is used to detect the protein of interest, because it produces a detectable color change in the presence of a substrate (Voller et al. 1978). HRP is widely used for biosensing applications because it is small, stable, and inexpensive.

We propose a method of detecting HRP enzyme reactions based on the measurement of the back-reflection intensity of the focused laser beam by using the self-catalytic oxidative photopolymerization (SCOPP) of *o*-phenylenediamine (*o*-PD). This discussion is based on our recent work (Yoshikawa et al. 2012). The mechanism of SCOPP of *o*-PD is schematically explained in Fig. 9.2. *o*-PD molecules form dimers (diaminophenazine) due to oxidative polymerization. When a visible-light laser beam is focused in the *o*-PD solution including a small amount of dimers, *o*-PD molecules are oxidatively polymerized by radical oxygen species produced by photoabsorption of dimers. The polymerized products (dimer, oligomer, polymer) also absorb the laser beam, so that the oxidative polymerization proceeds in a self-catalytic way in the laser focus. As a result, a nanostructure of polymer aggregates is formed

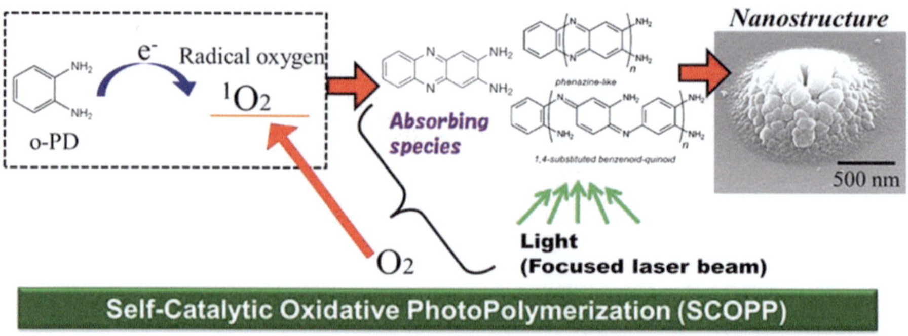

Fig. 9.2 Schematic diagram of self-catalytic oxidative photopolymerization of *o*-phenylenediamine (*o*-PD)

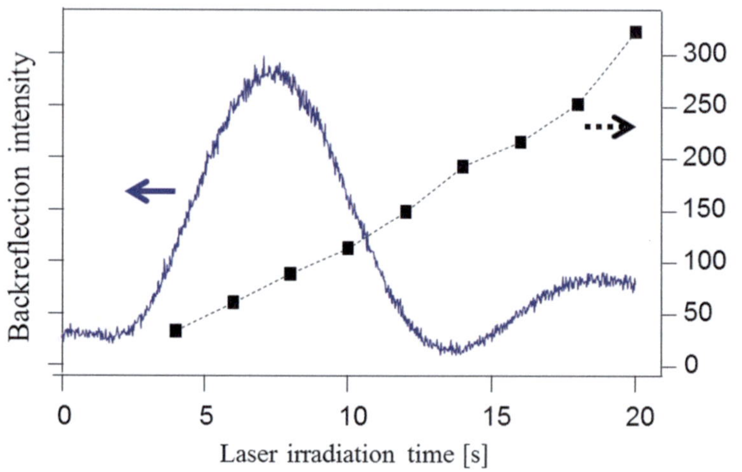

Fig. 9.3 Back-reflected intensity of the focused laser beam and height of the nanostructure plotted as a function of the laser irradiation time

in the laser focus. The size and the formation speed of the nanostructure are strongly dependent on the original concentration of dimers. Therefore, the dimer concentration included in the original solution can be detected quantitatively by measuring the change in the nanostructure size.

Generally, such nanostructures are evaluated by atomic force microscopy (AFM) or scanning electron microscopy (SEM). However, it is very hard to find the nanostructure formed in the laser focus on a glass substrate, like finding a sesame seed in a tennis court. Fortunately, we do not need to do such laborious work with expensive equipment. The back-reflection of the focused laser beam brings information on the nanostructure size. Figure 9.3 shows the temporal change in the back-reflection intensity of the focused laser beam. The *o*-PD solution including a small amount of dimers is placed on the glass substrate and the laser is focused at the glass–solution interface. The intensity shows increases and decreases, i.e. clear oscillations.

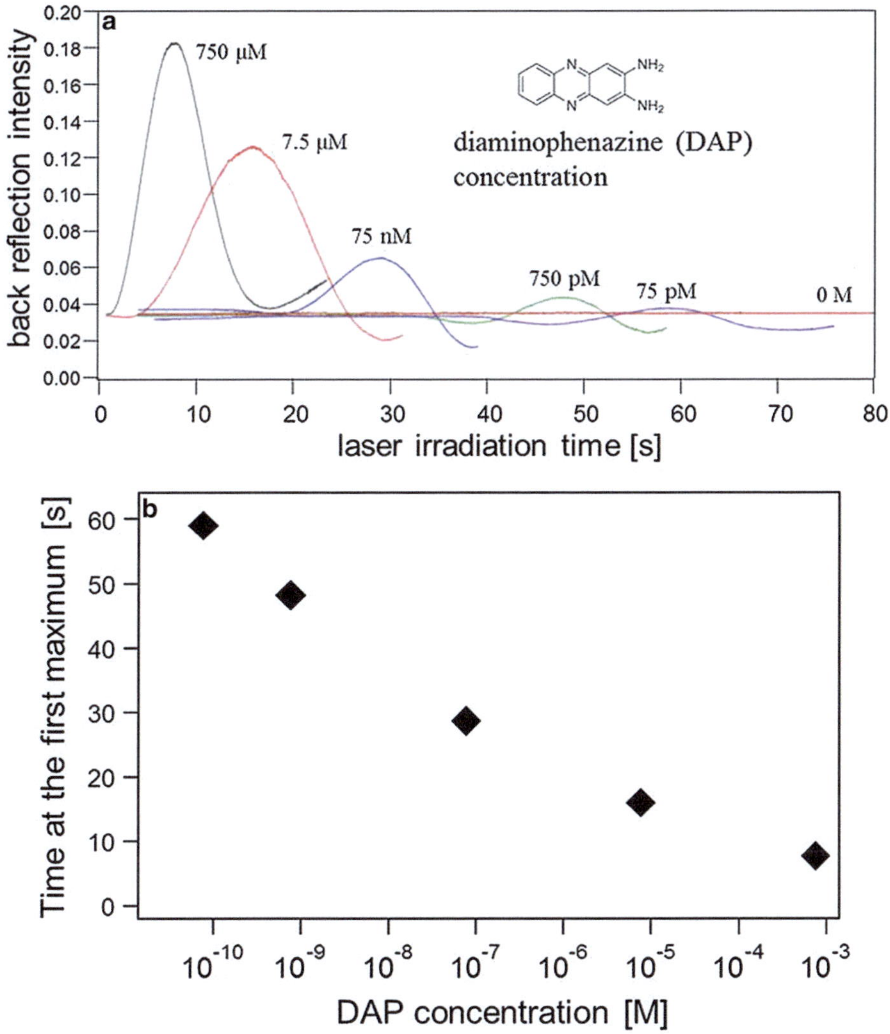

Fig. 9.4 (**a**) Temporal variations in the back-reflected light at different dimer (diaminophenazyne) concentrations. (**b**) Peak times of temporal variation curves vs. dimer concentrations

To investigate the reason for this oscillation, aggregates deposited with different laser irradiation times were measured by AFM. It was found that the height of the aggregate increases monotonicly with the laser irradiation time, whereas the back-reflection intensity fluctuates. The back-reflection intensity reaches a maximum when the height of the aggregate is ~80 nm and a minimum at a height of ~180 nm. The relationship between the back-reflection intensity and the height of the aggregate suggests a mechanism based on the interference.

Figure 9.4 shows the temporal change in the back-reflection intensity at different dimer concentrations. A higher concentration of dimer gives a higher rhythm and peak of oscillation, demonstrating that the growth speed of the aggregate increases with the dimer concentration. It should be noted that the nanoscale growth of the aggregate can be measured in real time by such a simple approach based on the optical interference. This laser-induced reaction realizes single-beam biosensing of the HRP reaction as follows.

Glucose oxidase (GOD) is a popular enzyme which oxidizes β-D-glucose to D-glucono-Δ-lactone. Because hydrogen peroxide is produced as a result of the glucose oxidation, the glucose concentration can be quantified by detecting the concentration of hydrogen peroxide. The oxidative polymerization of *o*-PD catalyzed by HRP produces dimers, whose concentration is proportional to that of hydrogen peroxidase. Therefore the glucose concentration can be quantified by the amount of produced dimer, which is reflected in the temporal change in the back-reflected intensity of a focused laser beam. An *o*-PD solution (1 mM) including a small amount of dimers was prepared to have an absorbance of ~0.08 at 450 nm. 20 μL of glucose solution (0.001–1 mM) was added to the same amount of the enzyme solution, which was prepared by dissolving GOD and HRP in the citrate buffer. After incubation for 1 min, 20 μL of the above *o*-PD solution was mixed into it. 20 μL of the mixed solution was placed on a glass plate, and the laser beam was focused at the solution–glass interface. As shown in Fig. 9.5, the temporal variations in the back-reflected laser intensity obviously depended on the glucose concentration. The peak time of each curve was plotted as a function of the glucose concentration in Fig. 9.5b, demonstrating that the glucose concentrations are quantitatively determined in the range between 1 mM and 100 nM from the peak times (Yoshikawa et al. 2012). In the conventional approach, the light absorption produced by *o*-PD dimers is used to determine the substrate concentration. The sensitivity of our method is quite high compared with that of other glucose biosensors in previous reports. In addition, the total measurement is carried out in a short time compared with other methods, which require longer incubation time, background calibration, and baseline measurement.

9.3 SERS Detection of Biomolecules with Optical Trapping of Ag Nanoparticles

Spectroscopy combined with optical trapping is a powerful technique for investigating micro- and nanomaterials dispersed in solution. Particles trapped at the focal spot of a laser beam can be analyzed by various spectroscopic methods. In particular, Raman scattering spectroscopy provides molecular information on optically trapped materials (Lankers et al. 1994; Houlne et al. 2002; Xie et al. 2005; Tsuboi et al. 2005; Creely et al. 2005). The exceedingly weak Raman signal can be greatly enhanced on noble metallic nanostructures. This phenomenon is called surface-enhanced Raman scattering (SERS) (Fleischm et al. 1974; Kneipp et al. 1996; Nie and Emery 1997; Michaels et al. 1999). Recently, SERS spectroscopy with the optical trapping technique has attracted considerable attention (McNay et al. 2004; Svedberg and Kall 2006; Itoh et al. 2006; Yoshikawa et al. 2007).

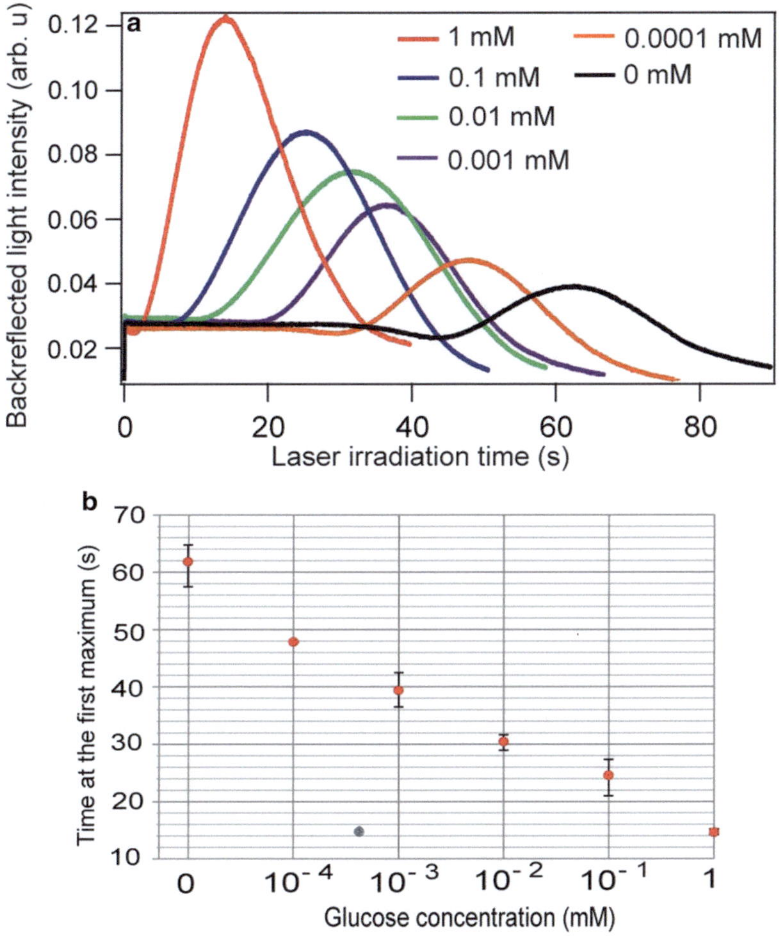

Fig. 9.5 Demonstration of enzymatic glucose sensing. (**a**) Temporal variations in the back-reflected light at different glucose concentrations. (**b**) Peak times of temporal variation curves vs. glucose concentrations

The potential depth U of the optical trap is proportional to the polarizability α:

$$U = -\frac{1}{2}\alpha E^2 \tag{9.1}$$

where E is the electric field of a laser beam. In the case of a colloidal system, the local colloidal concentration c_L in the optical trap is described as:

$$c_L = c \cdot \exp\left(\frac{|U|}{kT}\right) = c \cdot \exp\left(\frac{\alpha E^2}{2kT}\right) \tag{9.2}$$

where c is the concentration before laser irradiation (or apart from the focal spot), k is the Bolzmann constant, and T is temperature. Colloidal particles gathered in the laser focus have many chances to make aggregates due to the high concentration. It is known that the electromagnetic field of the irradiated light is strongly enhanced at nanometer-scale gaps formed in the aggregate of silver nanoparticles when the light wavelength matches the resonance wavelength of surface plasmon. Such a localized site where enhancement of the electromagnetic field occurs is called a "hot spot". SERS is attributed to the Raman scattering of molecules adsorbed at hot spots. Therefore, we can expect that aggregates of Ag nanoparticles are formed and SERS is induced by focusing a laser beam in the silver colloid.

Both the enhancement factor of SERS and the optical trap force depend on the size, shape, and aggregate structure of the Ag nanoparticles. Because the trap force (gradient force) is proportional to the polarizability of a particle, a large particle experiences a strong trap (Bowman and Padgett 2013). However, large Ag particles and their aggregates strongly scatter the light, producing an optical radiation pressure toward the propagation direction of the laser beam. Thus there are suitable sizes and shapes of Ag nanoparticles for optical trapping. Synthesis methods for preparing Ag nanoparticles with controlled size and shape have been developed by some researchers (Jana et al. 2001; Potara et al. 2011; Wang et al. 2013).

Polymer or surfactant molecules used in some preparation methods adsorb on the surface of Ag nanoparticles and prevent the adsorption of analyte molecules. Because this situation causes the interference signal due to SERS of surfactant molecules, the surfactant-free synthesis of nanoparticles is an important issue in SERS application (Bao et al. 2013). We prepared Ag nanoparticles suitable for SERS measurement and optical trapping by an original synthesis method without surfactant. Figure 9.6 shows SEM images and absorption spectra of Ag colloids prepared by conventional chemical reduction and our original methods. A typical Ag colloid whose average diameter is ~20 nm has an absorption peak at ~390 nm in water (Fig. 9.6b). This peak is attributed to the localized surface plasmon resonance (LSPR). On the other hand, our Ag colloid has two distinct peaks (Fig. 9.6d), indicating the anisotropic shape of the particles. Ag particles with triangular shape are included in the colloidal solution as shown in a SEM image (Fig. 9.6c).

We evaluate SERS measurement of molecules adsorbed on Ag nanoparticle aggregates by using optical trapping. The procedure is illustrated schematically in Fig. 9.7. Ag colloid is mixed with the target molecule, adenine. Adenine has been used in much research on SERS, because SERS-based analysis is expected to have potential in label-free sequencing of DNA/RNA (Feng et al. 2009; Papadopoulou and Bell 2010). The mixed solution is added to a sample plate with a small chamber fabricated by an acrylic plate, a cover glass, and silicon rubber spacers. The sample plate is set on an inverted microscope and an infrared laser beam (wavelength: 1,064 nm) is focused in the colloidal solution via an objective lens (100×, N.A. 1.3). The optically trapped Ag nanoparticles are contacted on an acrylic top plate by approaching the focal spot to the plate with a stage control (Fig. 9.7a, b). Figure 9.8a, b shows optical and emission images of an acrylic plate surface after the immobilization of Ag nanoparticles at five positions. Clear SERS spectra of adenine from immobilized

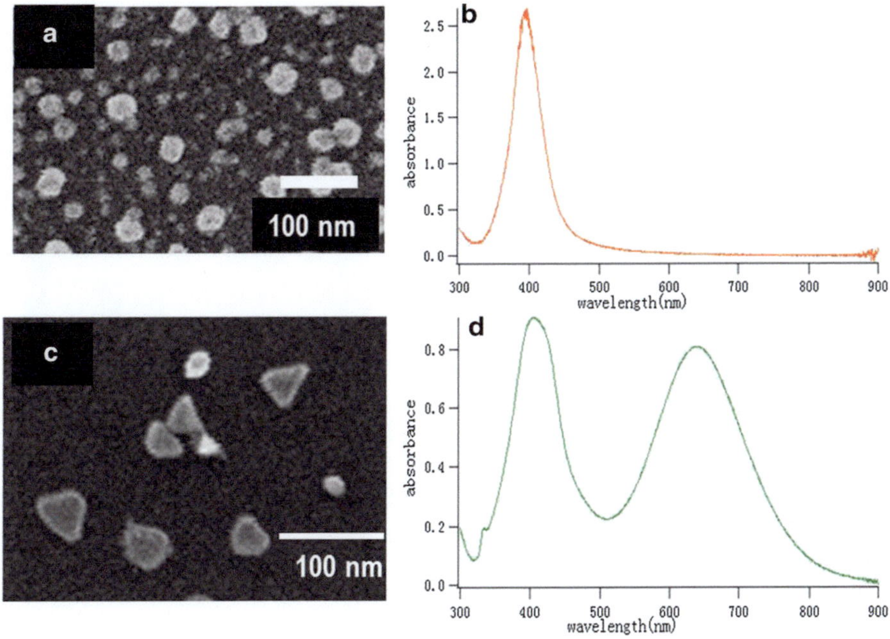

Fig. 9.6 (**a**, **c**) SEM images and (**b**, **d**) absorption spectra of silver colloidal particles prepared by (**a**, **b**) citrate and tannic acid reduction and (**c**, **d**) our original method

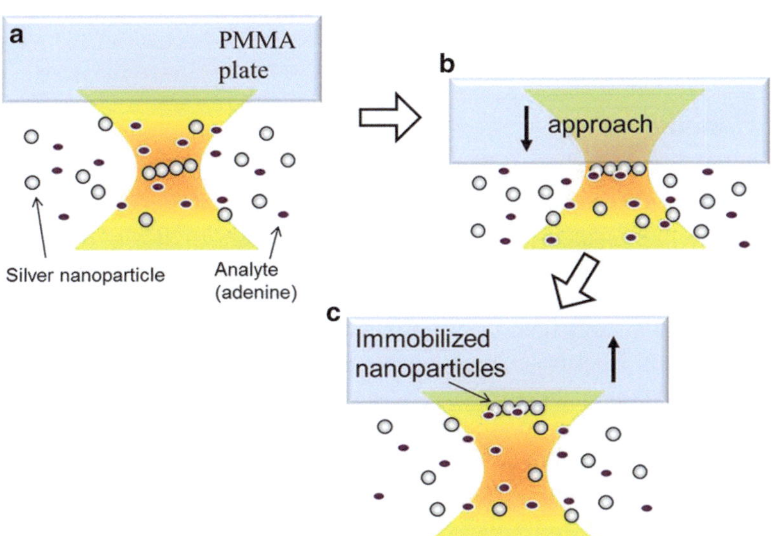

Fig. 9.7 Schematic procedure of the immobilization of SERS-active aggregates of Ag nanoparticles with analyte molecules

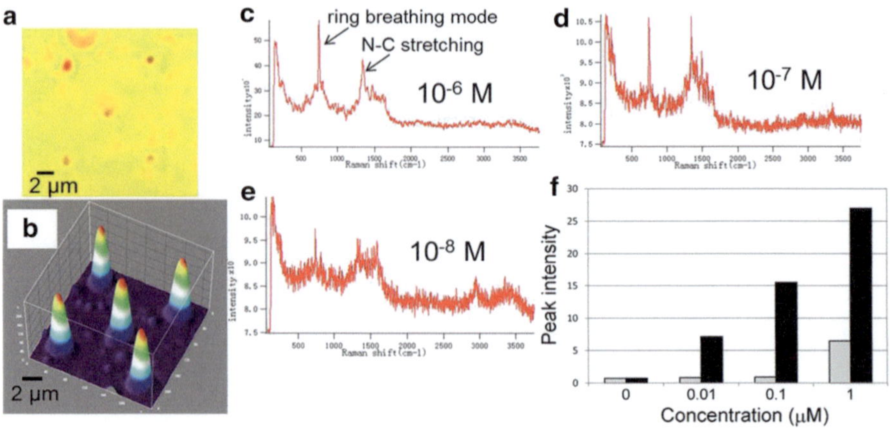

Fig. 9.8 (**a**, **b**) Bright-field and emission images of immobilized aggregates of Ag nanoparticles. (**c**–**e**) SERS spectra of Ag aggregates immobilized with different concentrations of adenine. (**f**) Concentration dependence of the peak intensity of ring breathing mode

spots were obtained as shown in Fig. 9.8c–e. We plotted the peak intensity of the ring breathing mode (735 cm^{-1}) at each adenine concentration in Fig. 9.8f. Ag colloid prepared by the original synthesis method gives a good result (high sensitivity and reproducibility) as compared to the normal method. Clear concentration dependence of the SERS intensity suggests that this method is applicable to the quantitative analysis of various molecules. The sample volume needed can be less than a nanoliter, making this method suitable for biomolecular analysis in the microflow chip. Further study to seek applicable target molecules is necessary to enhance generality.

9.4 Conclusion

In this chapter, novel nanobiosensing methods using focused laser beams are introduced. We found a characteristic modulation of the back-reflected laser intensity induced by focusing a laser beam on an *o*-PD solution, and applied this phenomenon to enzymatic glucose detection. A reliable optical quantification of glucose can be performed in a short time (~2 min including incubation) with a small sample volume (<20 μL). Because enzyme reactions using HRP are widely used for biosensing applications, particularly in enzyme-linked immunosorbent assays (ELISAs), this technique should also be applicable to various biomolecules. In addition, the sensing system can be constructed by using simple components without requiring any complex optics and expensive spectroscopic detectors. Laser focusing and back-reflection detection are simple techniques that are even performed by the read/write heads, called optical (or laser) pickups, of optical storage drives. We also introduced another sensing method using a focused laser beam to quantify and identify biomolecules with high sensitivity based on surface-enhanced

Raman spectroscopy. A focused laser beam produces a sub-micrometer sensing spot of Ag nanoparticles and immobilizes analyte molecules on it due to optical force. It is common sense in biosensing research that a sensing spot to convert biomolecular interactions into detectable signals is fabricated and then the analyte or bioreceptor is immobilized on it. The SERS analysis with optical trapping provides an innovative way of fabricating the sensing spot and immobilizing molecules at the same time. We hope that various biosensing devices based on this new design and concept are realized by technologies using focused laser beams.

References

Anker JN, Hall WP, Lyandres O, Shah NC, Zhao J, Van Duyne RP (2008) Biosensing with plasmonic nanosensors. Nat Mater 7(6):442–453. doi:10.1038/Nmat2162

Bao ZY, Dai JY, Lei DY, Wu YC (2013) Maximizing surface-enhanced Raman scattering sensitivity of surfactant-free Ag-Fe_3O_4 nanocomposites through optimization of silver nanoparticle density and magnetic self-assembly. J Appl Phys 114(12):7. doi:10.1063/1.4823732

Bowman RW, Padgett MJ (2013) Optical trapping and binding. Rep Prog Phys 76(2):026401. doi:10.1088/0034-4885/76/2/026401

Creely CM, Singh GP, Petrov D (2005) Dual wavelength optical tweezers for confocal Raman spectroscopy. Opt Commun 245(1–6):465–470. doi:10.1016/j.optcom.2004.10.011

Feng F, Zhi G, Jia HS, Cheng L, Tian YT, Li XJ (2009) SERS detection of low-concentration adenine by a patterned silver structure immersion plated on a silicon nanoporous pillar array. Nanotechnology 20(29):295501. doi:10.1088/0957-4484/20/29/295501

Fleischm M, Hendra PJ, Mcquilla AJ (1974) Raman-spectra of pyridine adsorbed at a silver electrode. Chem Phys Lett 26(2):163–166. doi:10.1016/0009-2614(74)85388-1

Hosokawa C, Yoshikawa H, Masuhara H (2005) Cluster formation of nanoparticles in an optical trap studied by fluorescence correlation spectroscopy. Phys Rev E 72(2):021408. doi:10.1103/PhysRevE.72.021408

Houlne MP, Sjostrom CM, Uibel RH, Kleimeyer JA, Harris JM (2002) Confocal Raman microscopy for monitoring chemical reactions on single optically trapped, solid-phase support particles. Anal Chem 74(17):4311–4319. doi:10.1021/Ac020325t

Itoh T, Ozaki Y, Yoshikawa H, Ihama T, Masuhara H (2006) Hyper-Rayleigh scattering and hyper-Raman scattering of dye-adsorbed silver nanoparticles induced by a focused continuous-wave near-infrared laser. Appl Phys Lett 88(8):084102. doi:10.1063/1.2172733

Jana NR, Gearheart L, Murphy CJ (2001) Seed-mediated growth approach for shape-controlled synthesis of spheroidal and rod-like gold nanoparticles using a surfactant template. Adv Mater 13(18):1389–1393. doi:10.1002/1521-4095(200109)13:18<1389::Aid-Adma1389>3.0.Co;2-F

Kneipp K, Wang Y, Kneipp H, Itzkan I, Dasari RR, Feld MS (1996) Population pumping of excited vibrational states by spontaneous surface-enhanced Raman scattering. Phys Rev Lett 76(14):2444–2447. doi:10.1103/PhysRevLett.76.2444

Lankers M, Popp J, Kiefer W (1994) Raman and fluorescence-spectra of single optically trapped microdroplets in emulsions. Appl Spectrosc 48(9):1166–1168. doi:10.1366/0003702944029569

McNay G, Docherty FT, Graham D, Smith WE, Jordan P, Padgett M, Leach J, Sinclair G, Monaghan PB, Cooper JM (2004) Visual observations of SERRS from single silver-coated silica microparticles within optical tweezers. Angew Chem Int Ed 43(19):2512–2514. doi:10.1002/anie.200352999

Michaels AM, Nirmal M, Brus LE (1999) Surface enhanced Raman spectroscopy of individual rhodamine 6G molecules on large Ag nanocrystals. J Am Chem Soc 121(43):9932–9939. doi:10.1021/Ja992128q

Nie SM, Emery SR (1997) Probing single molecules and single nanoparticles by surface-enhanced Raman scattering. Science 275(5303):1102–1106. doi:10.1126/science.275.5303.1102

Papadopoulou E, Bell SE (2010) Surface enhanced Raman evidence for Ag+ complexes of adenine, deoxyadenosine and 5′-dAMP formed in silver colloids. Analyst 135(12):3034–3037. doi:10.1039/c0an00612b

Pickup JC, Hussain F, Evans ND, Rolinski OJ, Birch DJ (2005) Fluorescence-based glucose sensors. Biosens Bioelectron 20(12):2555–2565. doi:10.1016/j.bios.2004.10.002

Potara M, Gabudean AM, Astilean S (2011) Solution-phase, dual LSPR-SERS plasmonic sensors of high sensitivity and stability based on chitosan-coated anisotropic silver nanoparticles. J Mater Chem 21(11):3625–3633. doi:10.1039/C0jm03329d

Ramachandran A, Wang S, Clarke J, Ja SJ, Goad D, Wald L, Flood EM, Knobbe E, Hryniewicz JV, Chu ST, Gill D, Chen W, King O, Little BE (2008) A universal biosensing platform based on optical micro-ring resonators. Biosens Bioelectron 23(7):939–944. doi:10.1016/j.bios.2007.09.007

Svedberg F, Kall M (2006) On the importance of optical forces in surface-enhanced Raman scattering (SERS). Faraday Discuss 132:35–44. doi:10.1039/B509301p

Tanaka Y, Yoshikawa H, Itoh T, Ishikawa M (2009) Laser-induced self-assembly of silver nanoparticles via plasmonic interactions. Opt Express 17(21):18760–18767

Tsuboi Y, Nishino M, Sasaki T, Kitamura N (2005) Poly(N-isopropylacrylamide) microparticles produced by radiation pressure of a focused laser beam: a structural analysis by confocal Raman microspectroscopy combined with a laser-trapping technique. J Phys Chem B 109(15):7033–7039. doi:10.1021/Jp044894b

Veitch NC (2004) Horseradish peroxidase: a modern view of a classic enzyme. Phytochemistry 65(3):249–259. doi:10.1016/j.phytochem.2003.10.022

Voller A, Bartlett A, Bidwell DE (1978) Enzyme immunoassays with special reference to ELISA techniques. J Clin Pathol 31(6):507–520

Wang Y, Wan D, Xie S, Xia X, Huang CZ, Xia Y (2013) Synthesis of silver octahedra with controlled sizes and optical properties via seed-mediated growth. ACS Nano 7(5):4586–4594. doi:10.1021/Nn401363e

Xie C, Mace J, Dinno MA, Li YQ, Tang W, Newton RJ, Gemperline PJ (2005) Identification of single bacterial cells in aqueous solution using conflocal laser tweezers Raman spectroscopy. Anal Chem 77(14):4390–4397. doi:10.1021/Ac0504971

Yoshikawa H, Matsui T, Masuhara H (2004) Reversible assembly of gold nanoparticles confined in an optical microcage. Phys Rev E 70(6):061406. doi:10.1103/PhysRevE.70.061406

Yoshikawa H, Adachi T, Sazaki G, Matsui T, Nakajima K, Masuhara H (2007) Surface-enhanced hyper-Raman spectroscopy using optical trapping of silver nanoparticles for molecular detection in solution. J Opt A 9(8):S164–S171. doi:10.1088/1464-4258/9/8/s08

Yoshikawa H, Imura S, Tamiya E (2012) Single-beam optical biosensing based on enzyme-linked laser nanopolymerization of o-phenylenediamine. Anal Chem 84(22):9811–9817. doi:10.1021/ac301951w

Chapter 10
Semiconductor Quantum Dots and Energy Transfer for Optical Sensing and Bioanalysis: Principles

Miao Wu and W. Russ Algar

Abstract Semiconductor quantum dots (QDs) are very promising materials for optical sensing and bioanalysis. This chapter serves as a primer for Chap. 11, which focuses on bioanalysis, by describing QD materials and their unique and highly advantageous properties with respect to photoluminescence spectroscopy, imaging, and energy transfer. These properties include, among others, broad and strong light absorption with spectrally narrow and tunable luminescence—a combination ideal for optical multiplexing. The interfacial chemistry and bioconjugation of QDs are then described, as these topics are critical considerations for designing sensors and bioanalysis methods. Energy transfer mechanisms including Förster resonance energy transfer (FRET), bioluminescence and chemiluminescence resonance energy transfer (BRET and CRET), nanosurface energy transfer (NSET), and charge transfer are reviewed with discussion of the advantages and disadvantages of QDs in the role of donor or acceptor. The importance and utility of these concepts are then illustrated in the context of specific examples of optical sensing and bioanalysis in the next chapter.

Keywords Assays • Biosensing • Charge transfer • Quantum dot • Resonance energy transfer

10.1 Introduction

As this book attests, nanomaterials have had a substantial impact on the field of biosensing and bioanalysis. Of the many different nanomaterials that have been investigated for such applications, brightly luminescent colloidal semiconductor

---

M. Wu • W.R. Algar (✉)
Department of Chemistry, University of British Columbia,
2036 Main Mall, Vancouver, BC V6T 1Z1, Canada
e-mail: algar@chem.ubc.ca

M.C. Vestergaard et al. (eds.), *Nanobiosensors and Nanobioanalyses*,
DOI 10.1007/978-4-431-55190-4_10

nanocrystals, called quantum dots (QDs), are among the most promising. The intention of this chapter and Chap. 11 is to provide an introduction to the properties of QDs that make them advantageous for biosensing and bioanalysis, and, in particular, illustrate how QDs can be combined with different mechanisms of energy transfer for a wide variety of bioanalytical applications. There has been widespread interest in such approaches and numerous developments have been reported in the scientific literature. Here, we first discuss the unique optical properties of QDs, the fundamental chemistry needed to render QDs biofunctional, and the essential concepts for dipole–dipole energy transfer and charge transfer with QDs.

10.2 Quantum Dots

10.2.1 Materials and Optical Properties

QDs are colloidal semiconductor nanocrystals that are typically between 2 and 10 nm in diameter (Chan et al. 2002; Medintz et al. 2005; Michalet et al. 2005). They comprise hundreds to thousands to tens of thousands of atoms arranged in a periodic lattice analogous to the corresponding bulk crystalline material (see Fig. 10.1). The size and material of this core nanocrystal are responsible for the optical properties of the QD, which arise through quantum confinement (Alivisatos 1996). When bulk semiconductors absorb light, the photon energy is used to generate an exciton via promotion of an electron across the band gap, from the valence band to the conduction band. The same process occurs in QDs with the distinction that, as the particle size becomes comparable to the Bohr exciton radius for the material, the band gap energy becomes dependent on the size of the core nanocrystal (Nirmal and Brus 1999). As the nanocrystal size decreases, the band gap energy increases. Changes in the band gap energy result in changes in wavelengths of light absorbed and emitted by the QD (see Fig. 10.2).

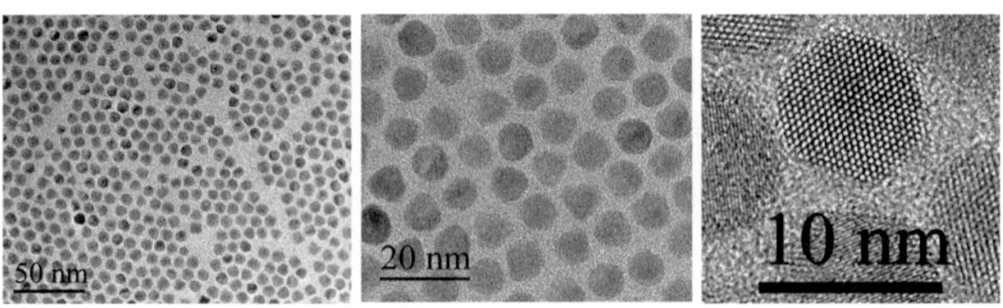

Fig. 10.1 High-resolution TEM images of CdSe/ZnS QDs. Images courtesy of Dr. Eunkeu Oh and Dr. Vaibhav Jain, U.S. Naval Research Laboratory. Copyright 2013 Eunkeu Oh and Vaibhav Jain

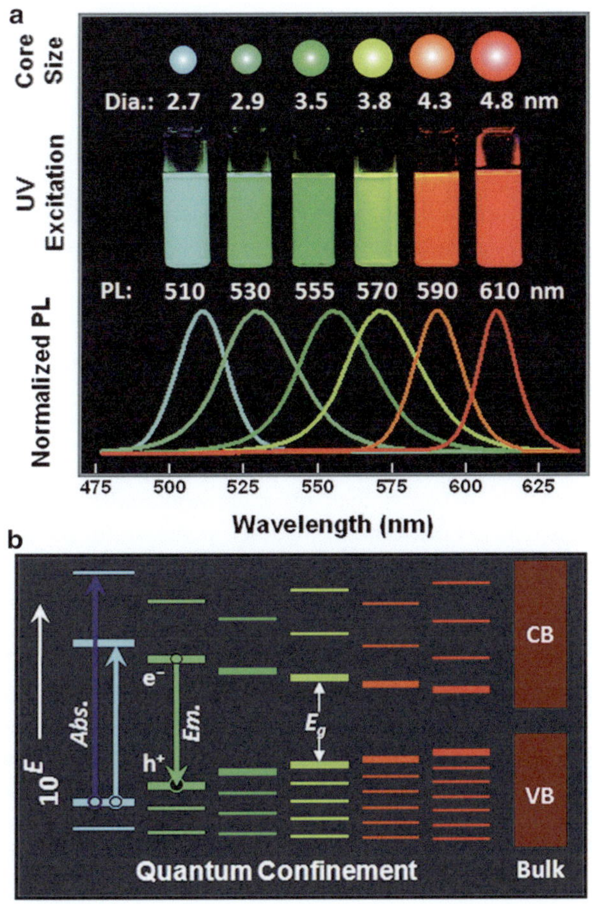

Fig. 10.2 (**a**) Correlation between core size and emission color (shown as both a photograph under UV light and PL emission spectra) for CdSe QDs. (**b**) Qualitative energy level diagram for CdSe QDs of different size, illustrating the effect of quantum confinement on transitions for absorption (*Abs.*) and emission (*Em.*) from recombination of the electron (e^-) and hole (h^+) across the band gap (E_g). Reprinted with permission from Algar et al. (2011b). Copyright 2011 American Chemical Society

Many highly advantageous optical properties arise from the nanoscale size of QDs (Algar et al. 2011b). For example, many QD materials exhibit bright visible and near-infrared photoluminescence (PL) that results from radiative recombination of the exciton across the quantum confined bandgap. This bright PL is a consequence of strong, broad absorption spectra and quantum yields that are comparable to common fluorescent dyes. QDs frequently have molar absorption coefficients in the range of 10^4–10^7 M^{-1} cm^{-1} (cf. 10^4–10^5 for organic dyes) with the lowest values at the absorption edge (i.e., the band gap energy) and increasing at shorter wavelengths of light. The two-photon absorption cross-sections of QDs (10^3–10^4 GM; 1 GM = 10^{-50} cm^4 s $photon^{-1}$) also tend to be much larger than those of organic dyes (10^1–10^2 GM). QD PL is spectrally narrow, with full-width-at-half-maxima typically in the range of 25–35 nm for relatively monodisperse samples, and its spectral peak position can be tuned through control of nanocrystal size and composition (see Fig. 10.3), shape, and structure. This chapter is mainly concerned with tuning

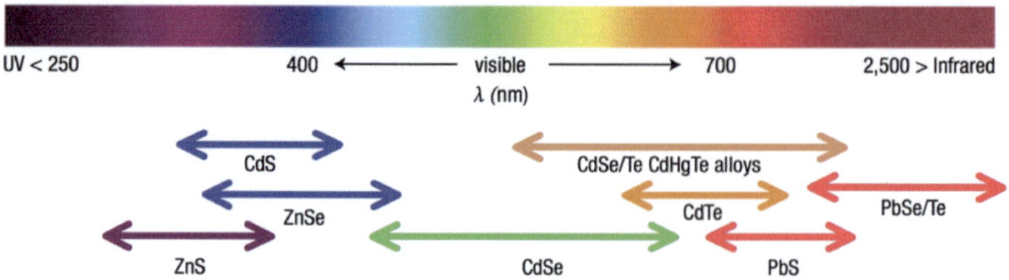

Fig. 10.3 Approximate size-tunable emission ranges for various QD materials. Reprinted with permission from Macmillan Publishers Ltd: Nature Materials (Medintz et al. 2005). Copyright 2005

PL through the size and composition of the QD nanocrystal, as these approaches are most widely used (and commercially available) for optical sensing and bioanalysis. Other interesting optical properties of QDs include their longer excited state lifetimes (10^1–10^2 ns) than most conventional organic fluorophores (<10 ns), their resistance to photobleaching, their potential for photobrightening and photodarkening, and, at the single particle level, their propensity for PL intermittency (i.e., blinking) (Lee and Osborne 2009; Algar et al. 2011b).

The most widely utilized QD material for bioanalysis and bioimaging applications is CdSe/ZnS with a core/shell structure (see Fig. 10.4) (Petryayeva et al. 2013). This material provides narrow PL across the visible spectrum as the core nanocrystal size varies between approximately 2 and 6 nm (Medintz et al. 2005). Methods for synthesizing high-quality, crystalline, and monodisperse CdSe cores are well known, and further growth of a thin (typically ≤1 nm) ZnS shell around the CdSe core is able to protect and enhance the optical properties of the latter. ZnS is, in turn, the most widely utilized shell material because of its adequate structural compatibility with common core materials and its relatively large band gap energy (Smith and Nie 2010). The structural compatibility helps passivate surface trap states that decrease the quantum yield of the core nanocrytal and the larger band gap energy helps confine a photogenerated exciton to the core nanocrystal, minimizing leakage of the exciton wavefunction into the surrounding environment (Smith and Nie 2010). In addition to CdSe, other common core materials include semiconductors such as CdS, CdTe, PbS, PbSe, InAs, InP, ZnO, and Si, among several others (Michalet et al. 2005; Algar et al. 2011b; Medintz et al. 2005). Alloyed core materials such as $CdSe_xS_{1-x}$ and $CdSe_xTe_{1-x}$ are also utilized (Swafford et al. 2006), as are QD materials such as ZnS or ZnSe doped with manganese (albeit that these QDs emit via dopant phosphorescence) (Norris et al. 2001). The core nanocrystal composition determines the spectral range over which PL can be tuned through changes in nanocrystal size, and the PL from alloyed QDs can also be tuned through changes in composition.

Shell materials are less diverse than core materials, with CdS and ZnSe being the most common alternatives to ZnS. In addition to core/shell structures where the shell material has a larger band gap energy than the core material (a type-I configuration), core/shell structures with different band alignments can be synthesized with

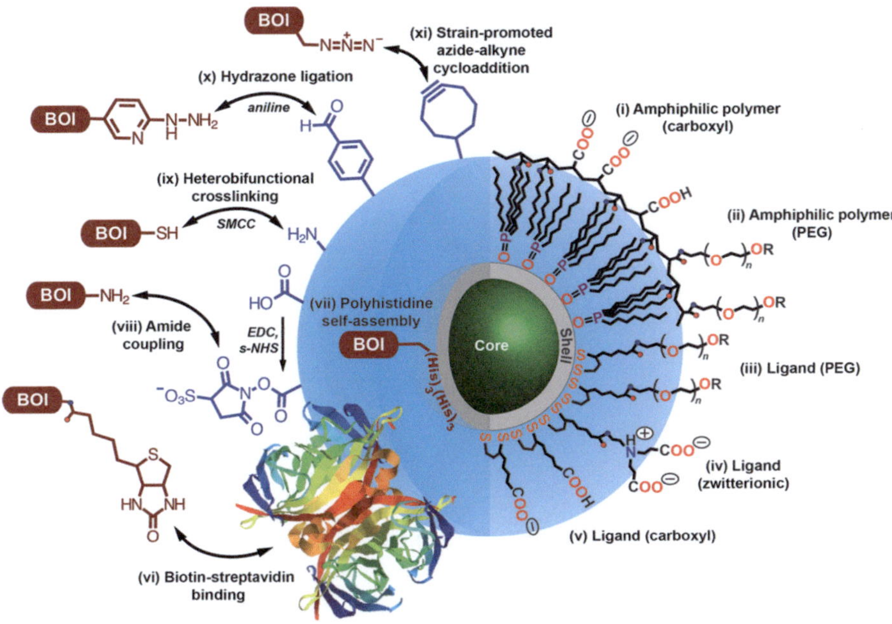

Fig. 10.4 Illustration of the essential elements of interfacial chemistry for making QD biofunctional. A core/shell QD (e.g., CdSe/ZnS) is either encapsulated in an amphiphilic polymer (*i–ii*) or modified with various hydrophilic, bifunctional ligands (*iii–v*). Biomolecules of interest (BOI) are then conjugated to the QD using affinity interactions (*vi–vii*) or covalent coupling (*viii–xi*). Representative examples of chemistries are shown, but many others have been utilized. Not drawn to scale. Reprinted with permission from Petryayeva et al. (2013). Copyright 2013 Society for Applied Spectroscopy

different effects on QD PL (Petryayeva et al. 2013). However, the photophysical properties of such varieties of QDs (non-type-I) tend to better suited to photovoltaic applications than bioanalysis. Core/shell/shell QDs and core/gradient shell QDs (a smooth, rather than abrupt transition between core and shell material) can be synthesized with potentially superior optical properties to core/shell QDs (Talapin et al. 2004; Chen et al. 2008), but are currently less utilized in bioanalysis applications.

The physical and optical properties of QDs offer several potential advantages for optical sensing and bioanalysis. The excellent photostability of high-quality QDs permits measurements over extended periods of time, and their high brightness can provide a sensitivity advantage, which can be further augmented by a reduction in background from the large *effective* Stokes shift (i.e., the difference between the excitation wavelength and emission wavelength, see Fig. 10.5). In addition, the ability to continuously tune PL emission over a wide range using a common material and synthetic procedure (see below) allows optimization with, for example, available bandpass filters. Synergistically, the broad absorption of QDs permits efficient excitation with a wide variety of sources and wavelengths. QDs also provide several technical advantages for multiplexed bioanalysis: multiple colors of QD can be

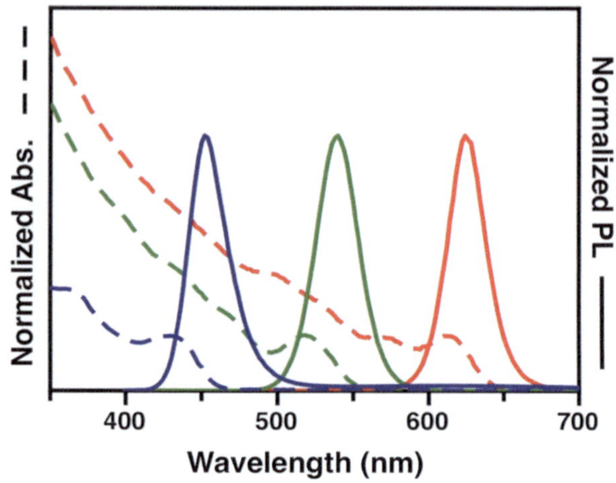

Fig. 10.5 Normalized absorption (*dashed lines*) and emission spectra (*solid lines*) for various compositions of CdSeS/ZnS QD, each approximately the same size. The spectra for CdSe/ZnS QDs of different sizes are qualitatively similar

excited using a common excitation wavelength, and the narrow, symmetric, and tunable emission profiles help maximize resolution and, when necessary, facilitate deconvolution of more signals in a given spectral range than is typically possible with the broader, red-tailed emission of fluorescent dyes. Another potential advantage of QDs that should not be overlooked is their high surface area-to-volume ratio. QDs are small enough to enter cells and tissues while providing an interface that can be chemically tailored and biofunctionalized. Whereas the modification of fluorescent dyes is frequently restricted to the addition of a reactive group for conjugation to a biomolecule of interest, QDs can be modified with diverse functional coatings and can be conjugated with multiple copies of a particular biomolecule or even multiple copies of different biomolecules.

10.2.2 Synthesis

Colloidal QDs, synthesized from the bottom up, were first reported in 1983 by Brus and coworkers (Rossetti et al. 1983). These QDs were prepared through an aqueous method but were of very low quality. Although QDs synthesized through aqueous routes have been occasionally utilized for bioanalysis applications, the quality of these QD remains relatively poor with low quantum yields and broad size distributions leading to broad PL spectral widths. A possible exception is CdTe QDs synthesized in aqueous media, which can have good brightness, but are not as monodisperse as QDs synthesized through other methods (Rogach et al. 2007). The breakthrough in the synthesis of high-quality, monodisperse QDs with bright PL came with the development of methods that relied on the injection of organic metallic precursors into hot coordinating solvents (Steigerwald et al. 1988; Murray et al. 1993), and the ability to synthesize core/shell QDs using these methods (Hines and Guyot-Sionnest 1996; Dabbousi et al. 1997). Other major advances were the modification of these methods to use more benign and less volatile precursors than the original reports (Li et al. 2003; Peng and Peng 2001; Qu et al. 2001), and the elucidation of the important role of impurities in technical grade solvents to achieve greater control and reproducibility in synthetic reactions. Consequently, non-coordinating solvents such as octadecene can now be used for the synthesis of high-quality QDs (Yu and Peng 2002; Bullen and Mulvaney 2004). The details of the above synthetic methods are beyond the scope of this chapter; however, interested readers can find more information in several published reviews (Mattoussi et al. 2012; Samokhvalov et al. 2013; Sowers et al. 2013). For the purposes of this chapter, the key point is that high-quality, monodisperse, and bright QDs are prepared in hydrophobic media and are stabilized with ligands such as alkyl amines, fatty acids, and alkyl phosphonic acids, phosphines, or phosphine oxides. As-synthesized QDs capped thusly are dispersible in organic solvents such as decane, toluene, and chloroform, but aggregate in aqueous or other polar media, and are therefore not suitable for bioanalysis applications without additional functionalization.

10.2.3 Interfacial Chemistry

Given that high-quality semiconductor QDs are synthesized in organic solvents and are not compatible with aqueous solution, the unique properties of QDs did not attract widespread interest for biological applications until 1998 when Chan and Nie and Bruchez et al. reported methods for rendering high-quality core/shell QDs water-soluble and biofunctional through interfacial chemistry (Chan and Nie 1998; Bruchez et al. 1998). In the context of developing energy transfer-based probes for bioanalysis, the ideal criteria for such QD coatings include: (1) high affinity for the QD; (2) the ability to provide long-term colloidal stability under biologically relevant conditions; (3) capacity for bioconjugation; (4) minimal thickness; and (5) resistance to the non-specific adsorption of biomolecules (Algar et al. 2011b). Other ideal criteria may be important in other applications with QDs, but are generally similar to the criteria above.

The two most established methods for coating QDs are encapsulation with amphiphilic polymers and ligand exchange with bifunctional molecules (see Fig. 10.4) (Petryayeva et al. 2013; Zhang and Clapp 2011; Resch-Genger et al. 2008). The first example of ligand exchange was the replacement of hydrophobic phosphine (oxide) ligands on as-synthesized CdSe/ZnS QDs with mercaptoacetic acid (MAA) (Chan and Nie 1998). This process was driven by mass action and the high affinity of the thiol group for the inorganic surface of the QDs. The carboxylate group afforded stable colloidal dispersion in aqueous solution and provided a functional group for bioconjugation with protein. Unfortunately, the stability of QDs coated with thioalkyl acids is limited to basic pH and low ionic strength. These ligands may also gradually desorb from the surface of the QD, although this effect can be largely mitigated with bidentate thiol ligands such as dihydrolipoic acid (DHLA) (Mattoussi et al. 2000). The limitations of electrostatic stabilization can be overcome by the use of poly(ethylene glycol) (PEG) or zwitterionic derivatives, which provide colloidal stability over a much broader range of ionic strength and pH, and better resist the non-specific adsorption of biomolecules (Zhang and Clapp 2011; Algar et al. 2011b). A significant advantage of ligand coatings is their compact size, which permits QDs and fluorescent dyes or other optically active materials to be tethered in close proximity, increasing the efficiency of energy transfer (see below). A drawback of this approach is that QDs tend to have a diminished quantum yield after ligand exchange and transfer to aqueous media. In contrast, encapsulation of as-synthesized QDs within an amphiphilic polymer coating better retains the quantum yield from the initial synthesis, albeit at the expense of a much larger size (Smith et al. 2006). Amphiphilic polymers are characterized by a mixture of pendant alkyl chains and pendant hydrophilic groups such as carboxylic acids and PEG chains. The alkyl chains interdigitate with the hydrophobic ligands on the surface of the QD from synthesis (Mattoussi et al. 2012; Algar et al. 2011b). Amphiphilic polymer coatings are somewhat less frequently used than ligand coatings for energy transfer-based bioanalyses with QDs, but are the underlying layer in commercially available streptavidin-coated QDs. Hybrid coatings that comprise a polymer backbone with pendant hydrophilic groups and pendant groups that coordinate to the inorganic interface of the QDs have also been developed (Liu et al. 2010; Yildiz et al. 2010). These coatings are more compact than amphiphilic polymers.

In addition to aqueous compatibility, bioconjugate chemistry is needed to render QDs biofunctional through the attachment of antibodies, enzymes, other proteins, peptides, oligonucleotides, aptamers, and many other biomolecular probes (Algar et al. 2011a). The most common bioconjugate methods can generally be categorized as either covalent coupling or self-assembly (see Fig. 10.4). Covalent coupling involves the formation of new chemical bonds between the coating on the QD and the biomolecule(s) of interest. Such reactions generally target amine, carboxyl, and thiol groups on the target biomolecule and the coating of the QD via carbodiimide activation, succinimidyl esters, maleimides, and homo- or hetero-bifunctional cross-linkers based on these reactive groups. Other reactions, such as chemoselective ligations (Blanco-Canosa et al. 2010), azide-alkyne "click" chemistry (Bernardin et al. 2010), tetrazine-based cycloadditions (Han et al. 2010), and other bioorthogonal reactions are also gaining popularity (Algar et al. 2011a).

Self-assembly methods rely on non-covalent interactions such as the formation of dative bonds or biological complexes. Examples of the former include the ability of thiol- and polyhistidine-terminated linkers to coordinate to the inorganic surface of QDs (Blanco-Canosa et al. 2013). For this interaction to occur, the linker must be able to penetrate the coating of the QD. In practice, peptides and oligonucleotides with these modifications can bind to QDs with a wide array of ligand coatings; however, proteins, while able to bind to DHLA-coated QDs, may not be able to bind to QDs coated with bulkier PEG-appended DHLA ligands without a sufficiently long, engineered linker (Boeneman et al. 2013). Polymer coatings do not generally support self-assembly to the QD surface, but can support the self-assembly of polyhistidine-appended biomolecules to pendant Ni^{2+}-nitrilotriacetic acid (Ni-NTA) groups.

Carboxylated polymer coatings can directly coordinate Ni^{2+} for the same purpose (Yao et al. 2007). Bulky ligands can also be terminated with Ni-NTA to permit self-assembly (Susumu et al. 2010), although mixed modes of binding to the QD may exist (Dennis et al. 2010). Perhaps the most common bioconjugation strategy is the spontaneous binding of biotinylated biomolecules with commercial streptavidin-coated QDs. Biotin–streptavidin is one of the strongest known non-covalent interactions (Green 1990) and biotinylated biomolecules are very widely available.

Important considerations for bioconjugate chemistry include the degree of control over the number of conjugated biomolecules per QD, the orientation of those biomolecules, the retention of their biological activity, the stability of the linkage, the mildness of the reaction conditions, the kinetics and yield of the coupling reaction, the propensity for competing reactions (e.g., hydrolysis) and undesirable cross-linking, and compatibility with other bioconjugate chemistries (Algar et al. 2011a). In general, QDs are a polyvalent interface and the final ensemble of conjugated QDs will generally exhibit a Poisson distribution in the number of biomolecules per QD. The bioconjugate chemistry selected for the preparation of QD probes can have a significant impact on the analytical figures ultimately achieved.

10.3 Energy Transfer Mechanisms

10.3.1 Förster Resonance Energy Transfer

Förster (or fluorescence) resonance energy transfer (FRET) is a non-radiative, through-space energy transfer mechanism that is mediated by dipole–dipole interactions between an excited state donor fluorophore and a ground state acceptor chromophore. FRET is one of the most versatile fluorescence techniques available, and is particularly prominent in the study of biochemical and biophysical systems. Several excellent books and reviews have been written on the topic of FRET (Valeur 2001; Roy et al. 2008; Cheung 1991; Jares-Erijman and Jovin 2003; Sahoo 2011; Van der Meer 2013; Lakowicz 2006). Here, we provide only a basic introduction that is suitable for understanding the utility of FRET in biosensing and bioanalyses with QDs.

For a donor–acceptor pair separated by a distance, r, the rate of FRET between a donor fluorophore and an acceptor chromophore, k_{FRET}, is given by Eq. (10.1), where $\tau_{\mathrm{D}} = (k_{\mathrm{r}} + k_{\mathrm{nr}})^{-1}$ is the inverse decay rate of the donor, k_{r} is the radiative decay rate, k_{nr} is the non-radiative decay rate, R_0 is the Förster distance (see below), Φ_{D} is the quantum yield of the donor, κ^2 is the orientation factor, $J(\lambda)$ is the spectral overlap integral, N_A is Avogadro's number, and n is the refractive index of the medium between the donor and the acceptor.

$$k_{\mathrm{FRET}} = \frac{1}{\tau_{\mathrm{D}}}\left(\frac{R_0}{r}\right)^6 = \frac{1}{\tau_{\mathrm{D}}}\frac{9(\ln 10)\Phi_{\mathrm{D}}\kappa^2 J(\lambda)}{128\pi^5 N_A n^4 r^6} \tag{10.1}$$

The spectral overlap integral is defined by Eq. (10.2), where $I_D(\lambda)$ is the wavelength-dependent emission intensity of the donor, $\varepsilon_A(\lambda)$ is the wavelength-dependent absorption coefficient of the acceptor, and λ is the wavelength.

$$J(\lambda) = \frac{\int I_D(\lambda)\varepsilon_A(\lambda)\lambda^4 d\lambda}{\int I_D(\lambda) d\lambda} \tag{10.2}$$

The efficiency of energy transfer is given by Eq. (10.3), where it is seen that R_0 is the distance between donor and acceptor that corresponds to $k_{FRET} = \tau_D^{-1}$. By definition, the FRET efficiency is 50 % when $r = R_0$.

$$E = \frac{k_{FRET}}{k_r + k_{nr} + k_{FRET}} = \frac{R_0^6}{r^6 + R_0^6} \tag{10.3}$$

Since the rate of FRET decreases as the inverse-sixth power of the distance between the donor and the acceptor, energy transfer is only appreciable between approximately $r = 0.5R_0$ and $r = 1.5R_0$. Förster distances for typical donor–acceptor pairs are in the range of ca. 3–6 nm, making FRET an excellent tool for studying processes on the size scale of biomolecules and many nanoparticles. This scaling, combined with the fast, sensitive, and non-invasive nature of fluorescence measurements, makes FRET ideal for monitoring dynamic biological processes at the ensemble and single-molecule levels (Sahoo 2011).

Experimentally, FRET efficiency E is most commonly measured from quenching of the donor emission intensity or lifetime, Eq. (10.4), where I is an emission intensity, τ is an emission lifetime, D is a subscript denoting a donor-only reference state, and DA is a subscript denoting a donor quantity measured in the presence of acceptor.

$$E = 1 - \frac{I_{DA}}{I_D} = 1 - \frac{\tau_{DA}}{\tau_D} \tag{10.4}$$

Alternatively, FRET efficiency can be measured from the relative amounts of quenched donor emission, I_{DA}, and FRET-induced acceptor emission, I_{AD}, Eq. (10.5), where I_A is acceptor emission from direct optical excitation, and Φ is a quantum yield. In many studies, the ratio I_{AD}/I_{DA} is used as a semi-quantitative measure of the amount of FRET without explicit calculation of the FRET efficiency.

$$E = \frac{(I_{AD} - I_A)}{(\Phi_A / \Phi_D) I_{DA} + (I_{AD} - I_A)} \tag{10.5}$$

QDs are highly advantageous as both donors and acceptors in FRET pairs for many reasons (Algar et al. 2010, 2013; Medintz and Mattoussi 2009). As donors, the broad absorption spectrum of QDs affords a wide range of possible excitation wavelengths, permitting straightforward selection of a wavelength that minimizes direct excitation of the acceptor while still efficiently exciting the QD donor and providing a large *effective* Stokes shift. The photostability of QDs is also advantageous for monitoring FRET over time. Furthermore, the narrow, tunable PL emission from QDs permits maximization of the spectral overlap integral, $J(\lambda)$, without introducing problematic emission crosstalk. The minimal crosstalk makes ratiometric detection, which is relatively insensitive to changes in excitation intensity or variation in concentration, easier to implement and more sensitive. Finally, since QDs have an interface (i.e., surface area) that can be chemically or physically modified, it is possible to associate multiple acceptors per QD to increase the FRET efficiency according to Eq. (10.6), where N is the number of equivalent acceptors per QD. Since QDs are, to a first approximation, centrosymmetric, configurations with multiple, equivalent acceptors are easily accessed in experiments.

$$E = \frac{NR_0^6}{r^6 + NR_0^6} \tag{10.6}$$

As acceptors, QDs are advantageous because of their broad absorption spectra and large molar absorption coefficients, which afford equally large spectral overlap integrals and commensurately large Förster distances, with potentially large spectral separation between the donor emission and QD emission. The caveat is that, because the efficient direct excitation of QDs is unavoidable, and because QDs tend to have a longer excited state lifetime than most organic fluorescent dyes, the ground state of QDs is poorly accessible. Consequently, and despite their a priori advantages, QDs are poor acceptors in practice for organic fluorophores. Nonetheless, QDs are excellent acceptors for luminescent lanthanide complexes with comparatively long excited state lifetimes (microseconds to milliseconds) (Algar et al. 2013). In practice, FRET is measured in a time-gated mode where a time delay is introduced between flash excitation and acquisition of emission signals to permit decay of the initially excited QDs to their ground states. The added advantage of such a system is time-gated rejection of background scattering or autofluorescence that occurs on the nanosecond timescale (Charbonnière and Hildebrandt 2008). QDs are also good acceptors when an excited state donor is generated chemically rather than through optical excitation (see Sect. 10.3.2).

10.3.2 Chemiluminescence and Bioluminescence Resonance Energy Transfer

Bioluminescence resonance energy transfer (BRET) and chemiluminescence resonance energy transfer (CRET) are both non-radiative, Förster-type resonance energy transfer processes that have growing importance in bioanalysis. The distinction between BRET/CRET and FRET is that the excited state donor is not generated through optical excitation. In BRET, an excited state donor is the product of an enzyme-catalyzed biochemical reaction; in CRET, the excited state donor is the product of a chemical reaction. Otherwise, the formal mechanism of energy transfer is analogous to FRET. CRET and BRET are advantageous in that they offer a near-zero background, free of interference from scattered excitation light. The trade-off is the need to introduce additional reactants to the system under study.

The most common chemiluminescent systems for CRET are those based on the oxidation of luminol to produce emission at ca. 430 nm; for example, oxidation with H_2O_2 catalyzed by horseradish peroxidase (HRP) or transition metals. In the case of BRET, *Renilla* luciferase (Rluc), apoaequorin, and firefly luciferase (Fluc) from *Hotaria parvula* have been widely used as BRET donors (Xia and Rao 2009; Rowe et al. 2009). Rluc functions as a catalyst for the oxidation of coelenterazine to excited state coelenteramide with bioluminescent emission at 480 nm (Xia and Rao 2009). Apoaequorin reacts with coelenterazine and O_2, in the presence of Ca^{2+}, with emission at 460 nm, and Fluc catalyzes the oxidization of D-luciferin with ATP and O_2, in the presence of Mg^{2+}, with emission at ca. 560 nm (Roda et al. 2009). Analogous to FRET experiments with QD donors, luciferase enzymes, HRP, or luminol can be conjugated to QDs to effectively serve as donors, with an enhancement of QD-sensitization rates as the number of donors per QD increases.

10.3.3 Nanosurface Energy Transfer

Nanosurface energy transfer (NSET) is a non-radiative mechanism of energy transfer that can occur between a fluorescent molecule and a nanoscale metal surface, most commonly that of a gold nanoparticle (Au NP). The interaction is between the oscillating dipole of the fluorophore and an induced image dipole associated with the Au NP. The quantitative model for NSET has been developed and refined by Strouse and coworkers, building from the seminal contributions of others made toward understanding energy transfer to bulk metal surfaces (Jennings et al. 2006a, b; Singh and Strouse 2010; Yun et al. 2005; Breshike et al. 2013). Depending on the size of the Au NP and its extinction coefficient at wavelengths of donor emission, the effective range of NSET can extend to 15–20 nm and even up to 40 nm versus <10 nm for conventional FRET (Yun et al. 2005; Breshike et al. 2013). This longer range arises from the different dimensionality of NSET, which is a dipole–surface interaction, whereas FRET is a dipole–dipole interaction. The efficiency of NSET, E_{NSET}, scales with distance according to Eqs. (10.7) and (10.8), where d_0 is conceptually analogous to the Förster distance, and d is the separation distance between the donor and the nanosurface. The value of d_0 is calculated from Eqs. (10.8) and (10.9), where $\alpha = [(9/2)^{1/4}]/4\pi$ is the orientation of the donor with respect to the metal plasmon vector, λ is the peak emission wavelength for the donor, Φ is the quantum yield of the donor, A is the absorptivity of the Au NP, n_{m} is the refractive index of the medium, n_{r} is the refractive index of the metal, ε_1 is the dielectric constant of the medium, and ε_2 is the complex dielectric function of the Au NP. The complex dielectric function, ε_2, includes the bulk, Drude, and interband contributions to the full complex dielectric constant of the Au NP. Details of the calculation of complex dielectric function are beyond our scope here, but can be found elsewhere (Breshike et al. 2013). For the absorptivity, Eq. (10.9), ε_λ is the extinction coefficient of the Au NP at the peak emission wavelength of the donor, r is the radius of the NP (in units of cm), N_A is Avogadro's number, V is the volume of the NP (in units of cm^3), and δ_{skin} is the skin depth of the NP.

$$E_{\mathrm{NSET}} = \frac{d_0^4}{d_0^4 + d^4} \tag{10.7}$$

$$d_0 = \frac{\alpha\lambda}{n_\mathrm{m}}(A\phi)^{1/4}\left[\frac{n_\mathrm{r}}{2n_\mathrm{m}}\left(1+\frac{\varepsilon_1^2}{|\varepsilon_2|^2}\right)\right]^{1/4} \tag{10.8}$$

$$A = \left[10^3 \ln(10)\right]\frac{\varepsilon_\lambda 2r\left[\frac{2r}{\delta_{\mathrm{skin}}}\right]}{N_A V} \tag{10.9}$$

Studies have suggested that QDs engage in NSET with Au NPs <3 nm in size. For example, Pons et al. measured the quenching of QDs by proximal Au NPs at distances between 5 and 20 nm by adjusting the length of a rigid polypeptide linker (Pons et al. 2007). The quenching efficiency versus distance curve had better agreement with a NSET model than a FRET model. Quenching of QD PL by Au NPs was observed at distances beyond ~15 nm whereas quenching by an organic dye, Cy3, was only observed out to distances of ~6.5 nm. Using a double-stranded DNA (dsDNA) spacer, Li et al. investigated the effect of the Au NP diameter and distance on QD PL quenching efficiency (Li et al. 2011). For 3-nm NPs, the data set, albeit limited, was consistent with NSET, with an apparent d^{-4} dependence. In contrast, for 80-nm NPs, the data set was consistent with FRET, with an apparent d^{-6} dependence. The scaling for 15-nm NPs was more ambiguous, although Han et al. found that quenching of QD PL by 12-nm Au NPs was consistent with NSET (Han et al. 2012). The model developed by Strouse and coworkers has not yet been validated for large Au NPs (>10 nm diameter).

Regardless of the precise nature of the dipolar coupling and energy transfer, it is clear that Au NPs are highly efficient dark quenchers of QD PL and can provide non-trivial quenching at distances beyond those possible with organic dyes. In the context of bioanalysis, this can be advantageous if large biomolecular probes (e.g., long oligonucleotides, antibodies) are necessary but hinder efficient FRET. This challenge is especially relevant with QDs since the donor–acceptor distance is measured from the center of the QD, and the separation distance is determined by the radius of the QD, the thickness of its coating, and the length of any intervening linker and biomolecular probe between the QD and the dye.

10.3.4 Charge Transfer

QD luminescence originates from the radiative recombination of an exciton across the quantum-confined bandgap after excitation. Photoinduced electron transfer (PET) is a process that can compete with radiative recombination and quench QD PL. Proximal redox-active molecules can transfer an electron into the quantum-confined valence band state (i.e., hole transfer from the QD) or accept an electron from the quantum-confined conduction band state when their highest-occupied molecular orbital (HOMO) or lowest-unoccupied molecular orbital (LUMO), respectively, are intermediate in energy to the quantum-confined states of the QD (Algar et al. 2010). Alternatively, ground state electron or hole transfer can also quench QD PL through enhancement of non-radiative Auger recombination pathways (Medintz et al. 2008). The fast rate of Auger recombination in charged QDs is attributed to their small size and confinement of carriers, which increases Coulomb interactions. Charge transfer (CT) processes are distance-dependent and require close association between the QD and redox-active moiety. However, little is known about the details of these processes in bioanalytical contexts (Petryayeva et al. 2013). There has been much physical study of CT processes with QDs in abiotic environments with characterization of the process using Marcus theory. Here, the QDs tend to be core-only structures and are studied in organic solvent with non-polar ligands and large numbers of adsorbed redox-active moieties. In contrast, bio-analyses based on CT quenching of QD PL utilize aqueous core/shell structures with hydrophilic ligands and biomolecular linkers between the QD and the redox-active moiety. These systems, which are relevant to bioanalysis, are much more complex, more diverse, and exhibit wider variation and sometimes inconsistent results between studies (Petryayeva et al. 2013). Variations appear to arise, in part, from the surface properties of the QD and their putative importance in the CT process. Redox active molecules that have been successfully utilized for CT quenching in bioanalysis formats have included Ru^{2+}-phenanthroline (Sandros et al. 2006; Medintz et al. 2008), ferrocene (Opperwall et al. 2012), bipyridinium dyes (Yildiz et al. 2006), and quinones (Medintz et al. 2010), although many other redox-active complexes and molecules have been found to engage in CT with QDs in non-bioanalytical contexts (Petryayeva et al. 2013).

10.4 Summary

The brightness, spectrally broad light absorption, and tunable, spectrally narrow PL offered by QDs is a unique combination of properties that is further enhanced by the ability to chemically and biologically tailor their interface. As a consequence of these features, QDs are highly advantageous components for assembling energy transfer configurations based on FRET, BRET, CRET, and NSET. QDs provide new levels of control over energy transfer efficiencies, optimization of signal-to-background ratios, and minimization of crosstalk. In addition to dipolar mechanisms of energy transfer, another mechanism, albeit less understood, is CT. Chap. 11 will describe in detail how all of these energy transfer mechanisms can be exploited for biosensing and bioanalysis.

References

Algar WR, Tavares AJ, Krull UJ (2010) Beyond labels: a review of the application of quantum dots as integrated components of assays, bioprobes, and biosensors utilizing optical transduction. Anal Chim Acta 673:1–25

Algar WR, Prasuhn DE, Stewart MH, Jennings TL, Blanco-Canosa JB, Dawson PE, Medintz IL (2011a) The controlled display of biomolecules on nanoparticles: a challenge suited to bioorthogonal chemistry. Bioconjug Chem 22:825–858

Algar WR, Susumu K, Delehanty JB, Medintz IL (2011b) Semiconductor quantum dots in bioanalysis: crossing the valley of death. Anal Chem 83:8826–8837

Algar WR, Kim H, Medintz IL, Hildebrandt N (2013) Emerging non-traditional Förster resonance energy transfer configurations with semiconductor quantum dots; investigations and applications. Coord Chem Rev 263–264:65–85

Alivisatos AP (1996) Perspectives on the physical chemistry of semiconductor nanocrystals. J Phys Chem 100:13226–13239

Bernardin A, Cazet A, Guyon L, Delannoy P, Vinet F, Bonnaffe D, Texier I (2010) Copper-free click chemistry for highly luminescent quantum dot conjugates: application to in vivo metabolic imaging. Bioconjug Chem 21:583–588

Blanco-Canosa JB, Medintz IL, Farrell D, Mattoussi H, Dawson PE (2010) Rapid covalent ligation of fluorescent peptides to water solubilized quantum dots. J Am Chem Soc 132:10027–10033

Blanco-Canosa JB, Wu M, Susumu K, Petryayeva E, Jennings TL, Dawson PE, Algar WR, Medintz IL (2013) Recent progress in the bioconjugation of quantum dots. Coord Chem Rev. doi:10.1016/j.ccr.2013.1008.1030

Boeneman K, Deschamps JR, Delehanty JB, Susumu K, Stewart MH, Glaven RH, Anderson GP, Goldman ER, Huston AL, Medintz IL (2013) Optimizing protein coordination to quantum dots with designer peptidyl linkers. Bioconjug Chem 24:269–281

Breshike CJ, Riskowski RA, Strouse GF (2013) Leaving Förster resonance energy transfer behind: nanometal surface energy transfer predicts the size-enhanced energy coupling between a metal nanoparticle and an emitting dipole. J Phys Chem C 117:23942–23949

Bruchez M, Moronne M, Gin P, Weiss S, Alivisatos AP (1998) Semiconductor nanocrystals as fluorescent biological labels. Science 281:2013–2016

Bullen CR, Mulvaney P (2004) Nucleation and growth kinetics of CdSe nanocrystals in octadecene. Nano Lett 4:2303–2307

Chan WCW, Nie SM (1998) Quantum dot bioconjugates for ultrasensitive nonisotopic detection. Science 281:2016–2018

Chan WCW, Maxwell DJ, Gao X, Bailey RE, Han M, Nie S (2002) Luminescent quantum dots for multiplexed biological detection and imaging. Curr Opin Biotechnol 13:40–46

Charbonnière LJ, Hildebrandt N (2008) Lanthanide complexes and quantum dots: a bright wedding for resonance energy transfer. Eur J Inorg Chem 2008:3241–3251

Chen Y, Vela J, Htoon H, Casson JL, Werder DJ, Bussian DA, Klimov VI, Hollingsworth JA (2008) "Giant" multishell CdSe nanocrystal quantum dots with suppressed blinking. J Am Chem Soc 130:5026–5027

Cheung HC (1991) Resonance energy transfer. In: Lakowicz JR (ed) Topics in fluorescence spectroscopy, vol 2. Plenum, New York, pp 127–176

Dabbousi BO, Viejo-Rodriguez J, Mikulec FV, Heine JR, Mattoussi H, Ober R, Jensen KF, Bawendi MG (1997) (CdSe)ZnS core-shell quantum dots: synthesis and characterization of a size series of highly luminescent nanocrystallites. J Phys Chem B 101:9463–9475

Dennis AM, Sotto DC, Mei BC, Medintz IL, Mattoussi H, Bao G (2010) Surface ligand effects on metal-affinity coordination to quantum dots: implications for nanoprobe self-assembly. Bioconjug Chem 21:1160–1170

Green NM (1990) Avidin and streptavidin. Method Enzymol 184:51–67

Han HS, Devaraj NK, Lee J, Hilderbrand SA, Weissleder R, Bawendi MG (2010) Development of a bioorthogonal and highly efficient conjugation method for quantum dots using tetrazine-norbornene cycloaddition. J Am Chem Soc 132:7838–7839

Han H, Valle V, Maye MM (2012) Probing resonance energy transfer and inner filter effects in quantum dot–large metal nanoparticle clusters using a DNA-mediated quench and release mechanism. J Phys Chem C 116:22996–23003

Hines MA, Guyot-Sionnest P (1996) Synthesis and characterization of strongly luminescing ZnS-Capped CdSe nanocrystals. J Phys Chem 100:468–471

Jares-Erijman EA, Jovin TM (2003) FRET imaging. Nat Biotechnol 21:1387–1395

Jennings TL, Schlatterer JC, Singh MP, Greenbaum NL, Strouse GF (2006a) NSET molecular beacon analysis of hammerhead RNA substrate binding and catalysis. Nano Lett 6:1318–1324

Jennings TL, Singh MP, Strouse GF (2006b) Fluorescent lifetime quenching near d = 1.5 nm gold nanoparticles: probing NSET validity. J Am Chem Soc 128:5462–5467

Lakowicz JR (2006) Principles of fluorescence spectroscopy, 3rd edn. Springer, New York

Lee SF, Osborne MA (2009) Brightening, blinking, bluing and bleaching in the life of a quantum dot: friend or foe? Chemphyschem 10:2174–2191

Li JJ, Wang YA, Gu W, Keay JC, Mishima TD, Johnson MB, Peng X (2003) Large-scale synthesis of nearly monodisperse CdSe/CdS core/shell nanocrystals using air-stable reagents via successive ion layer adsorption and reaction. J Am Chem Soc 125:12567–12575

Li M, Cushing SK, Wang Q, Shi X, Hornak LA, Hong Z, Wu N (2011) Size-dependent energy transfer between CdSe/ZnS quantum dots and gold nanoparticle. J Phys Chem Lett 2:2125–2129

Liu W, Greytak AB, Lee J, Wong CR, Park J, Marshall LF, Jiang W, Curtin PN, Ting AY, Nocera DG, Fukumura D, Jain RK, Bawendi MG (2010) Compact biocompatible quantum dots via RAFT-mediated synthesis of imidazole-based random copolymer ligand. J Am Chem Soc 132:472–483

Mattoussi H, Mauro JM, Goldman ER, Anderson GP, Sundar VC, Mikulec FV, Bawendi MG (2000) Self-assembly of CdSe-ZnS quantum dot bioconjugates using an engineered recombinant protein. J Am Chem Soc 122:12142–12150

Mattoussi H, Palui G, Na HB (2012) Luminescent quantum dots as platforms for probing in vitro and in vivo biological processes. Adv Drug Deliv Rev 64:138–166

Medintz IL, Mattoussi H (2009) Quatum dot-based resonance energy transfer and its growing application in biology. Phys Chem Chem Phys 11:17–45

Medintz IL, Uyeda HT, Goldman ER, Mattoussi H (2005) Quantum dot bioconjugates for imaging, labelling and sensing. Nat Mater 4:435–445

Medintz IL, Pons T, Trammell SA, Grimes AF, English DS, Blanco-Canosa JB, Dawson PE, Mattoussi H (2008) Interactions between redox complexes and semiconductor quantum dots coupled via a peptide bridge. J Am Chem Soc 130:16745–16756

Medintz IL, Stewart MH, Trammell SA, Susumu K, Delehanty JB, Mei BC, Melinger JS, Blanco-Canosa JB, Dawson PE, Mattoussi H (2010) Quantum-dot/dopamine bioconjugates function as redox coupled assemblies for in vitro and intracellular pH sensing. Nat Mater 9:676–684

Michalet X, Pinaud FF, Bentolila LA, Tsay JM, Doose S, Li JJ, Sundaresan G, Wu AM, Gambhir SS, Weiss S (2005) Quantum dots for live cells, in vivo imaging, and diagnostics. Science 307:538–544

Murray CB, Norris DJ, Bawendi MG (1993) Synthesis and characterization of nearly monodisperse CdE (E = S, Se, Te) semiconductor nanocrystallites. J Am Chem Soc 115:8706–8715

Nirmal M, Brus L (1999) Luminescence photophysics in semiconductor nanocrystals. Acc Chem Res 32:407–414

Norris DJ, Yao N, Charnock FT, Kennedy TA (2001) High-quality manganese-doped ZnSe nanocrystals. Nano Lett 1:3–7

Opperwall SR, Divakaran A, Porter EG, Christians JA, DenHartigh AJ, Benson DE (2012) WIde dynamic range sensing with single quantum dot biosensors. ACS Nano 6:8078–8086

Peng ZA, Peng XG (2001) Formation of high-quality CdTe, CdSe, and CdS nanocrystals using CdO as precursor. J Am Chem Soc 123:183–184

Petryayeva E, Algar WR, Medintz IL (2013) Quantum dots in bioanalysis: a review of applications across various platforms for fluorescence spectroscopy and imaging. Appl Spectrosc 67:215–252

Pons T, Medintz IL, Sapsford KE, Higashiya S, Grimes AF, English DS, Mattoussi H (2007) On the quenching of semiconductor quantum dot photoluminescence by proximal gold nanoparticles. Nano Lett 7:3157–3164

Qu L, Peng ZA, Peng X (2001) Alternative routes toward high quality CdSe nanocrystals. Nano Lett 1:333–337

Resch-Genger U, Grabolle M, Cavaliere-Jaricot S, Nitschke R, Nann T (2008) Quantum dots versus organic dyes as fluorescent labels. Nat Methods 5:763–775

Roda A, Guardigli M, Michelini E, Mirasoli M (2009) Bioluminescence in analytical chemistry and in vivo imaging. Trends Anal Chem 28:307–322

Rogach AL, Franzl T, Klar TA, Feldmann J, Gaponik N, Lesnyak V, Shavel A, Eychmüller A, Rakovich YP, Donegan JF (2007) Aqueous synthesis of thiol-capped CdTe nanocrystals: state-of-the-art. J Phys Chem C 111:14628–14637

Rossetti R, Nakahara S, Brus LE (1983) Quantum size effects in the redox potentials, resonance raman-spectra, and electronic-spectra of CdS crystallites in aqueous-solution. J Chem Phys 79:1086–1088

Rowe L, Dikici E, Daunert S (2009) Engineering bioluminescent proteins: expanding their analytical potential. Anal Chem 81:8662–8668

Roy R, Hohng S, Ha T (2008) A practical guide to single-molecule FRET. Nat Methods 5:507–516

Sahoo H (2011) Forster resonance energy transfer – a spectroscopic nanoruler: principle and applications. J Photochem Photobiol C 12:20–30

Samokhvalov P, Artemyev M, Nabiev I (2013) Basic principles and current trends in colloidal synthesis of highly luminescent semiconductor nanocrystals. Chem Eur J 19:1534–1546

Sandros MG, Shete V, Benson DE (2006) Selective, reversible, reagentless maltose biosensing with core-shell semiconducting nanoparticles. Analyst 131:229–235

Singh MP, Strouse GF (2010) Involvement of the LSPR spectral overlap for energy transfer between a dye and Au nanoparticle. J Am Chem Soc 132:9383–9391

Smith AM, Nie S (2010) Semiconductor nanocrystals: structure, properties, and band gap engineering. Acc Chem Res 43:190–200

Smith AM, Duan H, Rhyner MN, Ruan G, Nie S (2006) A systematic examination of surface coatings on the optical and chemical properties of semiconductor quantum dots. Phys Chem Chem Phys 8:3895–3903

Sowers KL, Swartz B, Krauss TD (2013) Chemical mechanisms of semiconductor nanocrystal synthesis. Chem Mater 25:1351–1362

Steigerwald ML, Alivisatos AP, Gibson JM, Harris TD, Kortan R, Muller AJ, Thayer AM, Duncan TM, Douglass DC, Brus LE (1988) Surface derivatization and isolation of semiconductor cluster molecules. J Am Chem Soc 110:3046–3050

Susumu K, Medintz IL, Delehanty JB, Boeneman K, Mattoussi H (2010) Modification of poly(ethylene glycol)-capped quantum dots with nickel nitrilotriacetic acid and self-assembly with histidine-tagged proteins. J Phys Chem C 114:13526–13531

Swafford LA, Weigand LA, Bowers MJ, McBride JR, Rapaport JL, Watt TL, Dixit SK, Feldman LC, Rosenthal SJ (2006) Homogeneously alloyed CdSxSe1-x nanocrystals: synthesis, characterization, and composition/size- dependent band gap. J Am Chem Soc 128:12299–12306

Talapin DV, Mekis I, Götzinger S, Kornowski A, Benson O, Weller H (2004) CdSe/CdS/ZnS and CdSe/ZnSe/ZnS core-shell-shell nanocrystals. J Phys Chem B 108:18826–18831

Valeur B (2001) Molecular fluorescence: principles and applications. Wiley-VCH, Toronto

Van der Meer BW (2013) Förster theory. In: Hildebrandt N, Medintz IL (eds) FRET - Förster resonance energy transfer, vol 1. Wiley-VCH, Weinheim, pp 23–62

Xia ZY, Rao JH (2009) Biosensing and imaging based on bioluminescence resonance energy transfer. Curr Opin Biotechnol 20:37–44

Yao H, Zhang Y, Xiao F, Xia Z, Rao J (2007) Quantum dot/bioluminescence resonance energy transfer based highly sensitive detection of proteases. Angew Chem Int Ed 46:4346–4349

Yildiz I, Tomasulo M, Raymo FM (2006) A mechanism to signal receptor-substrate interactions with luminescent quantum dots. Proc Natl Acad Sci U S A 103:11457–11460

Yildiz I, Deniz E, McCaughan B, Cruickshank SF, Callan JF, Raymo FM (2010) Hydrophilic CdSe-ZnS core-shell quantum dots with reactive functional groups on their surface. Langmuir 26:11503–11511

Yu WW, Peng XG (2002) Formation of high-quality CdS and other II-VI semiconductor nanocrystals in noncoordinating solvents: tunable reactivity of monomers. Angew Chem Int Ed 41: 2368–2371

Yun CS, Javier A, Jennings T, Fisher M, Hira S, Peterson S, Hopkins B, Reich NO, Strouse GF (2005) Nanometal surface energy transfer in optical rulers, breaking the FRET barrier. J Am Chem Soc 127:3115–3119

Zhang YJ, Clapp A (2011) Overview of stabilizing ligands for biocompatible quantum dot nanocrystals. Sensors 11:11036–11055

Chapter 11
Semiconductor Quantum Dots and Energy Transfer for Optical Sensing and Bioanalysis: Applications

Miao Wu and W. Russ Algar

Abstract Semiconductor quantum dots (QDs) are very promising materials for optical sensing and bioanalysis. This chapter builds on Chap. 10, which reviewed the optical properties of QDs and their benefits for energy transfer, by illustrating the utility of QDs and energy transfer for optical sensing and bioanalysis. Representative examples of different in vitro assays and cellular probes from the literature are described. Energy transfer mechanisms including Förster resonance energy transfer (FRET), bioluminescence and chemiluminescence resonance energy transfer (BRET and CRET), nanosurface energy transfer (NSET), and charge transfer can be used for optical signal generation in homogeneous assays, single-particle assays, and heterogeneous assays targeting bioanalytes as diverse as nucleic acids, proteins, small molecules, ions, and the activity of enzymes such as proteases, kinases, and nucleases. The importance and versatility of QDs in optical sensing and bioanalysis has been growing steadily since their introduction and will continue to grow in the near future as QD-based assays are optimized and applied to new problems, and new capabilities are developed.

Keywords Assays • Biosensing • Charge transfer • Quantum dot • Resonance energy transfer

11.1 Introduction

As this book attests, nanomaterials have had a substantial impact on the field of biosensing and bioanalysis. Of the many different nanomaterials that have been investigated for such applications, brightly luminescent colloidal semiconductor nanocrystals, called quantum dots (QDs), are among the most promising. The intention of Chap. 10 and this chapter is to provide an introduction to the properties of

---

M. Wu • W.R. Algar (✉)
Department of Chemistry, University of British Columbia,
2036 Main Mall, Vancouver, BC V6T 1Z1, Canada
e-mail: algar@chem.ubc.ca

M.C. Vestergaard et al. (eds.), *Nanobiosensors and Nanobioanalyses*,
DOI 10.1007/978-4-431-55190-4_11

QDs that make them advantageous for biosensing and bioanalysis, and, in particular, illustrate how QDs can be combined with different mechanisms of energy transfer for a wide variety of bioanalytical applications. There has been widespread interest in such approaches and numerous developments have been reported in the scientific literature. Here, we illustrate, using representative examples from the literature, how the convergence of the concepts in Chap. 10 can be used to create a variety of probes for a diverse set of target analytes, across multiple assay formats. In each case, the modulation of QD photoluminescence (PL) by energy transfer is coupled to biorecognition processes that alter the energy transfer efficiency. QDs are an ideal platform material for this purpose because of their nanoscale size, surface area that can be chemically and biologically functionalized, and their bright luminescence and capacity for sensitive, multiplexed detection (see Chap. 10). Regrettably, there are many interesting and useful advances that could not be included in this chapter. The interested reader is encouraged to explore these and related topics in many published review articles (Kim and Kim 2012; Algar et al. 2010, 2013; Medintz and Mattoussi 2009; Clapp et al. 2006; Charbonnière and Hildebrandt 2008; Grigsby et al. 2012).

11.2 FRET-Based Bioanalyses with Quantum Dots

This section describes several representative examples of bioanalyses that utilize QDs and Förster resonance energy transfer (FRET) for detection. These examples do not comprise a comprehensive review of the field and, indeed, there are many other examples published in the literature. Throughout the sections below, QDs are denoted by, for example, QD525 or QD605, representing QDs with peak PL at 525 and 605 nm, respectively. Peak PL values are rounded to the nearest 5 nm. If the QD material is not explicitly mentioned, then it should be assumed to be CdSe/ZnS.

11.2.1 Homogeneous Assays

11.2.1.1 Hybridization Assays

The innate selectivity of hybridization between two complementary strands of nucleic acid to form a double helix provides a means of detecting genes and other sequences of interest. In hybridization assays, a probe sequence complementary to the target sequence to be assayed is introduced and conditions such as temperature, ionic strength, and concentration of denaturant are optimized to provide maximum selectivity for hybridization. These oligonucleotide probe sequences can be chemically modified at one of their termini to facilitate conjugation to QDs. To date, the most common methods for conjugating oligonucleotides to QDs have been terminal modifications with either a thiol, dithiol, or polyhistidine tag for self-assembly to the ZnS shell of QDs (Gill et al. 2005; Medintz et al. 2007), or biotinylation for binding to streptavidin-coated QDs (Zhang et al. 2005). In FRET-based assays, hybridization between probe and target generates the necessary proximity between QDs and a suitable donor or acceptor for energy transfer. Several different hybridization assay formats have been demonstrated with QDs and FRET; examples of each are described below and are listed in Table 11.1.

The simplest format for a QD-FRET hybridization assay is for the DNA target to be directly labeled with an acceptor dye that can form a FRET pair with the QD. The drawback, however, is that target DNA must be pre-labeled with the dye, restricting the applicability of this format. An early example of this format was reported by Algar et al. and was also one of the first demonstrations of a multiplexed QD-FRET assay (Algar and Krull 2007). Mercaptoacetic acid (MAA)-coated CdSe/ZnS QD525 and QD605 were each conjugated to different amine-terminated oligonucleotide probes through carbodiimide coupling. The target sequence paired with the QD525 was labeled with Cy3 and the target sequence paired with QD605 was labeled with Alexa Fluor 647 (A647). The amount of FRET-sensitized Cy3 or A647 emission provided the analytical signal. The QD525-Cy3 FRET pair could detect down to 40 nM of its target (4 % of its effective probe concentration) and the QD605-A647 FRET pair could detect down to 12 nM of its target (10 % of its effective probe concentration). Homogeneous, ensemble multiplexing with these two FRET pairs using a single excitation wavelength was demonstrated. The study also demonstrated that the non-specific adsorption of DNA could be a challenge due to interactions between the DNA and the MAA-coated QD. To address this issue, ethidium bromide, an intercalating dye that undergoes a quantum yield enhancement in the presence dsDNA, was used as an acceptor and improved the signal-to-background ratio by more than fivefold while also avoiding direct labeling of the target (Algar and Krull 2007). Subsequent work by other researchers found the adsorption could be ameliorated with PEGylated surface ligands on the QD (Zhou et al. 2008) or coating with DHLA.

Another strategy that avoids both direct labeling of targets and the use of intercalating dyes (which are often less favourable acceptors than conventional dyes) is a competitive hybridization assay. For example, Vannoy et al. prepared conjugates of streptavidin-coated QD605 and biotinylated probes, pre-hybridizing those probes with an A647-labeled reporter sequence prior to assaying the sample (Vannoy et al. 2013). Target sequences in the sample competed with the reporter sequence, eventually displacing the reporter from the QD and turning off the FRET signal (see Fig. 11.1). Displacement was made energetically favourable by choosing the sequence of the reporter to be partially mismatched with the probe, permitting detection of nanomolar concentrations of target, although equilibrium displacement was not reached for several hours.

Molecular beacons are another strategy that has been adopted for FRET-based detection of unlabeled targets with QDs. First developed by Tyagi and Kramer in 1996 (Tyagi and Kramer 1996), molecular beacons comprise a "stem-loop" or "hairpin" DNA probe labeled at opposite termini with a donor and an acceptor,

Table 11.1 Energy transfer-based hybridization assays with QDs

Analyte	Format	Mechanism	Donor	Acceptor	LOD	Reference
Oligos (19/24-mer) (M)	HMG	FRET	CdSe/ZnS QD525	Cy3	40 nM	Algar and Krull (2007)
				Ethidium bromide	165 nM	
			CdSe/ZnS QD605	Alexa Fluor 647	12 nM	
Oligo (30-mer)	HMG	FRET	CdSe/ZnS QD530	Alexa Fluor 594	–	Zhou et al. (2008)
				Ethidium bromide	1 nM	
Oligo (98-mer)	HMG	FRET	CdSe/ZnS QD605	Alexa Fluor 647	<12 nM	Vannoy et al. (2013)
Oligo	HMG	FRET	CdSe/ZnS QD490	DABCYL	–	Kim et al. (2004)
Oligo (21-mer)	HMG	FRET	CdSe/ZnS QD510	TAMRA	–	Medintz et al. (2007)
			CdSe/ZnS QD590	Cy5		
Oligos (21-mer) (M)	HMG	FRET	CdSe/ZnS QD620	BHQ-2	1 pM	Freeman et al. (2011a)
			CdSe/ZnS QD540	BHQ-1		
Oligo (30-mer)	HMG-SP	FRET	CdSe/ZnS QD605	Cy5	5 fM	Zhang et al. (2005)
PCR product						
Oligos (25-mer) (M)	HMG-SP	FRET	CdSe/ZnS QD605	Alexa Fluor 647	–	Zhang and Hu (2010)
Oligos (19/39-mer) (M)	HTG	FRET	CdSe/ZnS QD530	Cy3	1 nM	Algar and Krull (2009, 2010a)
			CdSe/ZnS QD620	Alexa Fluor 647		
Oligos (19/39-mer) (M)	HTG	FRET	CdSe/ZnS QD530	Cy3	1 nM	Algar and Krull (2010a, c)
				Rhodamine Red		

Oligo (19/24-mer)	HTG	FRET	CdSeS/ZnS QD525	Cy3	4 nM	Petryayeva et al. (2013)
			CdSeS/ZnS QD625	Alexa Fluor 647		
Oligos (19/24-mer) (M)	HTG	FRET	CdSe/ZnS QD525	Cy3	5 fmol	Noor et al. (2013c)
			CdSe/ZnS QD605	Alexa Fluor 647		
Oligo (19-mer)	HTG	FRET	CdSeS/ZnS QD525	Cy3	150 nM	Noor et al. (2013b)
Oligos (18-mer) (M)	HMG	CRET	Luminol	CdSe/ZnS QD490	<10 nM	Freeman et al. (2011b)
				CdSe/ZnS QD560		
				CdSe/ZnS QD620		
mRNA	HMG	FRET	CdSe/ZnS QD580	Cy5		Bakalova et al. (2005)
Oligo (24-mer)	HMG	FRET	CdTe QD540	Rhodamine B	0.08 nM	Xu et al. (2011)
miRNA	HMG-SP	FRET	CdSe/ZnS QD605	Cy5	0.1 aM	Zhang and Zhang (2012)
Oligo (22-mer)	HMG	BRET	Rluc	CdTe/ZnS QD710	20 nM	Cissell et al. (2008)
Oligo (M)	HMG	FRET	Tb cryptate	CdSe/ZnS QD625	–	Algar et al. (2012d)
			CdSe/ZnS QD625	Alexa Fluor 647		
Oligo (22-mer)	HMG	RET	CdTe QD590	GO	12 nM	Dong et al. (2010)
Oligos (20-mer) (M)	HMG	RET	CdSe/ZnS QD510	MWNT	0.2 pM	Cui et al. (2008)
			CdSe/ZnS QD555			
			CdSe/ZnS QD600			

HMG homogeneous, *HTG* heterogeneous, *M* multiplexed, *SP* single particle

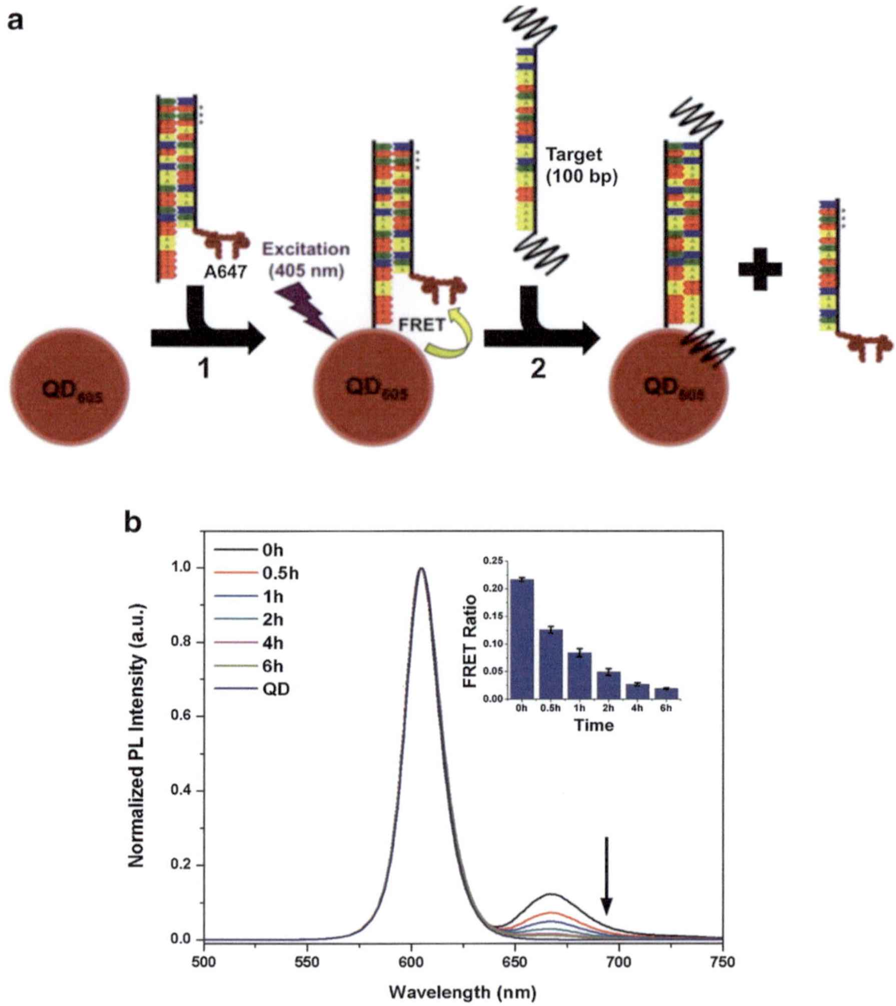

Fig. 11.1 (**a**) Displacement hybridization assay strategy using QDs and FRET. (**b**) PL spectra showing a decrease in A647 PL emission over time after the addition of target DNA. The *inset* shows the corresponding change in the A647/QD PL ratio. Adapted with permission from Vannoy et al. (2013). Copyright 2013 Elsevier

respectively. In the absence of target, hybridization between the terminal stem regions provides the proximity for FRET. The loop region is complementary to the target and, due to the greater thermodynamic favourability, target hybridization opens the hairpin, separating the donor and acceptor to disrupt FRET. Kim and coworkers reported one of the first QD-based MBs, conjugating hairpin probes to QDs coated with MAA through carbodiimide coupling (Kim et al. 2004). CdSe/ZnS QD490 were paired with DABCYL, a dark quencher, and compared to a more

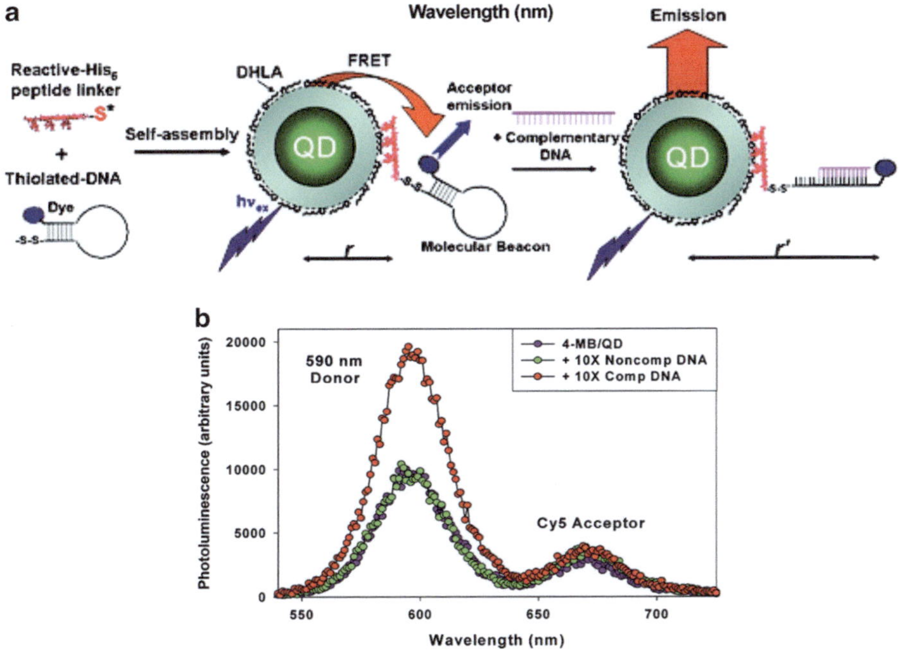

Fig. 11.2 (**a**) Design of a QD-based molecular beacon. (**b**) Representative data showing the response of a QD-based molecular beacon with CdSe/ZnS QD590 and Cy5-labeled hairpin probe to complementary and non-complementary DNA. Reprinted with permission from Medintz et al. (2007). Copyright 2007 American Chemical Society

conventional carboxyfluorescein (FAM)-DABCYL dye–dye MB. Discrimination of a single base-pair mismatch with the QD-MB was possible and the QD-MB had greater photostability than the FAM-DABCYL MB. Medintz et al. reported a self-assembled QD-MB where the hairpin probe was ligated with a polyhistidine tag at one terminus and labeled with a fluorescent dye at the opposite terminus (Medintz et al. 2007). Here, CdSe/ZnS QD510 and QD590 were paired with TAMRA and Cy5 acceptors, respectively (see Fig. 11.2).

An enzyme-amplified hybridization assay strategy with QDs and FRET was developed by Freeman and coworkers (Freeman et al. 2011a). CdSe/ZnS QDs coated with glutathione were self-assembled with oligonucleotide probes modified at one terminus with a thiol linker and modified at the other terminus with Black Hole Quencher 2 (BHQ-2). In the absence of target, BHQ-2 quenched the QD PL. Probe–target hybridization provided a double-stranded DNA helix that exonuclease III (*Exo*III) could digest. *Exo*III selectively cleaves from the 3′ terminus of one strand in a DNA duplex. In this case, the sequences were designed such that the probe was digested and the target, which had a 3′ single-stranded overhang, remained intact and could be recycled between different probe molecules (~6–10 per QD), leading to signal amplification with multiple turnovers by *Exo*III (see Fig. 11.3).

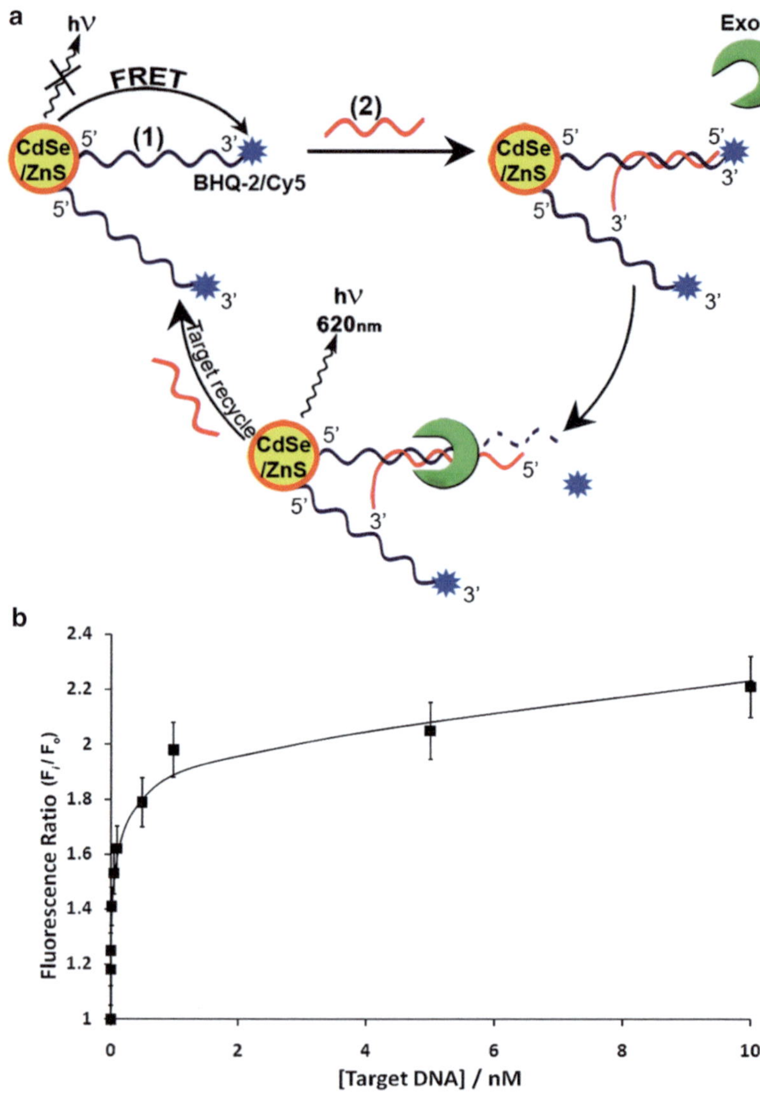

Fig. 11.3 (**a**) Scheme showing the *Exo*III-amplified detection of target DNA (2) using QD-probe (1) conjugates and FRET. (**b**) Plot showing the recovery of QD PL as a function of the concentration of target DNA. Adapted with permission from Freeman et al. (2011a). Copyright 2011 American Chemical Society

As little as 1 pM of target could be detected within 2 h using QD-probe conjugates (10 nM) and 20 units of *Exo*III. Multiplexed detection of two different target sequences was possible by pairing QD620 and QD540 with BHQ-2 and a second quencher, BHQ-1, respectively.

11.2.1.2 Aptamers for Protein and Small Molecule Detection

Aptamers are a class of oligonucleotides that bind target biomolecules, organic molecules, and inorganic molecules with high affinity and specificity (Hermann and Patel 2000). A combinatorial process called the *systematic evolution of ligands by exponential enrichment* (SELEX) is used to produce and isolate aptamer sequences (Wilson and Szostak 1999). Aptamers are a potential alternative to antibodies in that they can provide similar molecular recognition without the need for expression in cell lines or animals, are amenable to chemical modification, and are generally more robust. A detailed description of the advantages and disadvantages (e.g., lower affinity) of aptamers relative to antibodies can be found elsewhere (Keefe et al. 2010); however, the benefits of aptamers are such that they have been adopted for a wide range of analytical applications (Tombelli et al. 2005). Several applications with aptamers, QDs, and FRET are listed in Table 11.2, and some are described in more detail below.

Levy and coworkers developed one of the first QD-based bioassays incorporating an aptamer probe (Levy et al. 2005). Biotinylated thrombin-binding aptamer (TBA) was hybridized with a short complementary oligonucleotide that was labeled with Eclipse, a commercially available dark quencher dye, and then conjugated to streptavidin-coated CdSe/ZnS QD525 to provide the proximity for FRET. Upon the introduction of thrombin (a protease important in blood clotting and cancer), the TBA bound the thrombin and underwent a conformational change that displaced the Eclipse-labeled oligonucleotide. The corresponding loss of FRET resulted in an increase in the QD PL intensity, which provided an analytical signal proportional to the amount of thrombin introduced (see Fig. 11.4). A similar strategy with the displacement of a dye-labeled, complementary oligonucleotide has been used for the detection of ATP with ATP-binding aptamer (Chen et al. 2008).

Another strategy, which has been used for the detection of thrombin (Chi et al. 2011) and, more recently, Mucin 1 (Muc1, a protein biomarker useful for the early detection of cancer), is a QD-based aptamer beacon (Shin et al. 2012). For thrombin detection, TBA-probes were designed to fold into a hairpin structure, thereby providing a short segment of dsDNA for the intercalation of a fluorescent dye, BOBO-3 (Chi et al. 2011). The quantum yield of BOBO-3 was enhanced when intercalated in dsDNA, and its fluorescence was sensitized by FRET from the proximal CdSe/ZnS QD565. Sensitized emission from the BOBO-3 was used as the analytical signal and was inversely proportional to the amount of thrombin. The QD-FRET strategy provided significantly enhanced photostability (>80 min) compared to a BOBO-3-only aptamer beacon, which quickly decayed to 50 % of its initial intensity (within ~20 min).

A third strategy used for aptamer-based sensing with QDs is target-stabilized association of aptamer subunits. Freeman et al. conjugated one subunit of cocaine-binding aptamer to CdSe/ZnS QD570 and labeled the other subunit with Atto 590 fluorescent dye as an acceptor for the QD (Freeman et al. 2009b). The subunits were designed with limited complementarity at their 3′ and 5′ termini, and could only associate with the added stabilization from cocaine binding. The association provided

Table 11.2 Energy transfer-based assays with QDs and aptamer probes

Analyte	Format	Mechanism	Donor	Acceptor	LOD	Reference
Adenosine (M)	HMG	NSET	CdSe/ZnS QD525	13 nM AuNPs	50 μM	Liu et al. (2007)
Cocaine (M)			CdSe/ZnS QD585		120 μM	
Adenosine triphosphate (ATP)	HMG	FRET	CdSe/ZnS QD605	Cy5	24 μM	Chen et al. (2008)
Adenosine triphosphate (ATP)	HMG	CRET	Luminol	CdSe/ZnS QD620	100 nM	Freeman et al. (2011b)
Cocaine	HMG	FRET	CdSe/ZnS QD570	Atto590	1 μM	Freeman et al. (2009b)
Cocaine	HMG-SP	FRET	CdSe/ZnS QD605	Cy5	0.5 μM	Zhang and Johnson (2009)
Mucin-1	HMG	FRET	CdSe/ZnS QD565	BOBO-3	50 nM	Shin et al. (2012)
Platelet-derived growth factor	HMG	FRET	CdSe/ZnS QD600	BHQ-2	0.4 nM	Kim et al. (2009a)
Prostate-specific membrane antigen (PSMA)	HMG (cell)	FRET	CdSe/ZnS QD490	Doxorubicin	–	Bagalkot et al. (2007)
Thrombin	HMG	FRET	CdSe/ZnS QD525	Eclipse	–	Levy et al. (2005)
Thrombin	HMG	FRET	CdSe/ZnS QD565	BOBO-3	1 nM	Chi et al. (2011)
Thrombin	HMG	CT	Ru-phen	CdSe/ZnS QD545	–	Swain et al. (2008)

HMG homogeneous, *M* multiplexed, *SP* single particle

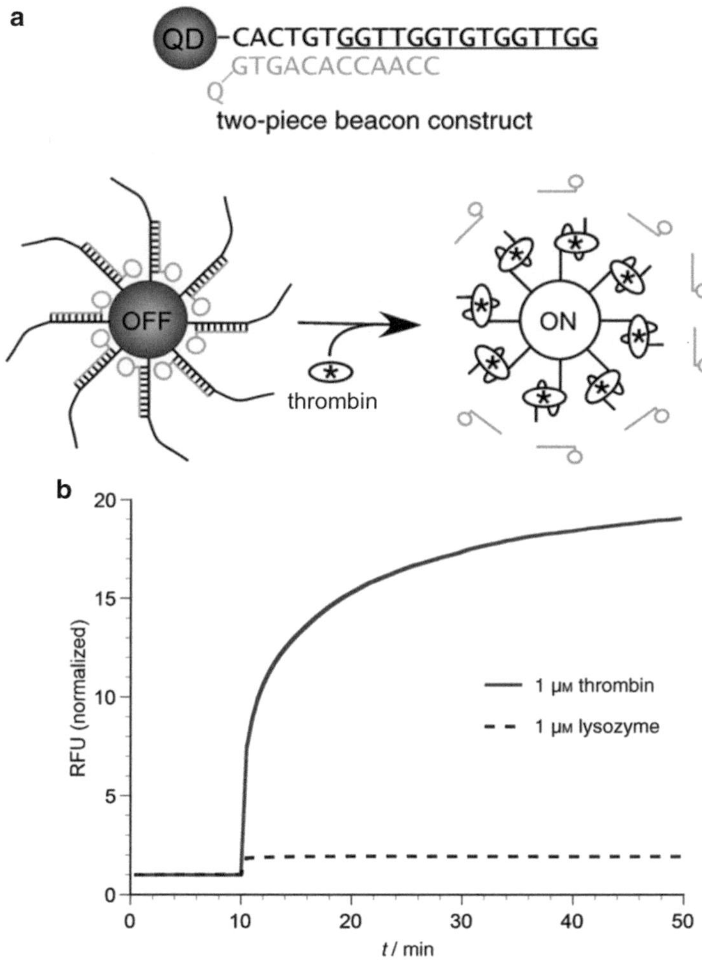

Fig. 11.4 (**a**) Design of a QD-FRET aptamer beacon for the detection of thrombin. (**b**) Chronofluorimetric response of the QD-FRET aptamer beacon to thrombin target and lysozyme (negative control). Adapted with permission from Levy et al. (2005). Copyright 2005 Wiley-VCH Verlag GmbH & Co. KGaA, Weinheim

the proximity for FRET and the degree of quenching of the QD PL was used as the analytical signal. The response was linear over approximately three orders of magnitude and the LOD of the assay was 1 μM.

11.2.1.3 Immunoassays

The specific binding between an antibody and its target antigen or hapten is also used in the design of FRET-based probes and assays with QDs. However, the number of reports in the literature of this type is significantly less than would be expected from the ubiquity of antibodies in assays and in vitro diagnostics, including those without FRET (e.g., enzyme-linked immunosorbent assays) and with FRET (e.g., time-resolved FRET assays with lanthanides). The advantage of antibodies is that they can be selected for a wide range of proteins, toxins, pathogens, and haptens (e.g., drugs, food additives, persistent organic pollutants, explosives, mycotoxins, metabolites). Table 11.3 lists several examples of immunoassays based on QDs and FRET.

Goldman et al. developed one of the first immunoassays with QDs and FRET, employing a competitive format to detect aqueous 2,4,6-trinitrotoluene (TNT) (Goldman et al. 2005). An anti-TNT single-chain Fv antibody fragment (scFv) was engineered with a polyhistidine tag and self-assembled to CdSe/ZnS QD530 as a capture probe. The QD-scFv conjugate was mixed with Black Hole Quencher 10 (BHQ-10)-labeled trinitrobenzene, a TNT analog, which was bound by the scFv, resulting in quenching of QD PL via FRET. Upon the addition of sample, any TNT present competed with and displaced the analog, thereby disrupting FRET. The recovery of QD PL was proportional to the concentration of TNT. This general immunoassay strategy has also been extended to the detection of aqueous *Aspergillus amstelodami* spores using QDs and FRET. Kattke et al. conjugated aminopegylated CdSe/ZnS QD625 with anti-*Aspergillus* antibodies and *A. fumigatus* spores were labeled with Black Hole Quencher 3 (BHQ-3) to serve as a competitive reporter (Kattke et al. 2011). The latter species of spore was selected empirically based on its much lower binding affinity with anti-*Aspergillus* antibodies. *A. amstelodami* spores present in samples displaced BHQ-3-labeled *A. fumigatus* spores from the QD-anti-*Aspergillus* antibody conjugates, yielding a proportional recovery of QD PL with loss of FRET between the QD and BHQ-3 (see Fig. 11.5). The LOD was 10^3 *A. amstelodami* spores/mL. One of the advantages of the QDs in this assay was their brightness, which helped overcome the large autofluoresence background associated with some mold species.

An alternative strategy for competitive immunoassays is to label the QD with antigen or hapten and to label the detection antibody with a FRET acceptor dye. In one such example, the analyte was a pesticide, 2,4-dichlorophenoxyacetic acid (2,4-D) (Long et al. 2012a). Bovine serum albumin (BSA) was labeled with 2,4-D (BSA-2,4-D), then conjugated to CdSe/ZnS QD605. The QD–BSA-2,4-D conjugates were mixed with samples and Cy5.5-labeled anti-2,4-D monoclonal antibodies. The QD–BSA-2,4-D conjugates competed with any 2,4-D in the sample to bind with the anti-2,4-D (see Fig. 11.6). The Cy5.5 was a good FRET acceptor for the QD605 donors and the system was interrogated at a wavelength that selectively excited the QD605. Increasing amounts of 2,4-D in the sample resulted in a decrease in FRET-sensitized Cy5.5 emission. A microfluidic platform was used for detection and provided an LOD of 0.5 nM.

Another immunoassay format with QDs and FRET is a sandwich assay. Here, two antibodies targeting two different epitopes of the same analyte are used: one antibody is conjugated to QDs and serves as a capture probe; the second antibody is labeled with an acceptor dye or another moiety suitable for FRET with QDs. With the addition of analyte, the two antibodies form a sandwich structure that brings the

Table 11.3 Energy transfer-based immunoassays with QDs

Analyte	Format	Mechanism	Donor	Acceptor	LOD	Reference
2,4-Dichlorophenoxyacetic acid	HMG	FRET	CdSe/ZnS QD605	Cy5.5	0.5 nM	Long et al. (2012a)
2,4,6-Trinitrotoluene	HMG	FRET	CdSe/ZnS QD530	BHQ-10	20 ng/mL	Goldman et al. (2005)
A. amstelodami	HMG	FRET	CdSe/ZnS QD625	BHQ-3	10^3 spores/mL	Kattke et al. (2011)
α-Fetoprotein	HMG	RET	CdTe QD510	GO	0.15 ng/mL	Liu et al. (2010)
α-Fetoprotein	HMG	FRET	Tb complex	CdSe/ZnS QD605	0.4 ng/mL	Chen et al. (2012b)
Breast cancer associated antigen 1	HMG	RET	CdSe/ZnS QD600	MWNT	0.4 nM	Cui et al. (2008)
Cortisol	HMG	FRET	CdSe/ZnS QD605	Cy5	–	Nikiforov and Beechem (2006)
Estrogen receptor-β	HMG	FRET	CdSe/ZnS QD565	Alexa Fluor 568	0.05 nM	Wei et al. (2006)
				Alexa Fluor 633		
Human cardiac troponin I	HMG	FRET	CdSe/ZnS QD545	Alexa Fluor 546	55 nM	Stringer et al. (2008)
Human enterovirus 71 (M)	HMG	RET	CdSe/ZnS QD525	GO	0.42 ng/mL	Chen et al. (2012a)
Coxsackievirus B3 (M)			CdSe/ZnS QD605		0.39 ng/mL	
IgG	HMG	NSET	Mn:CdTe QD655/720	Au nanorod	50 nM	Liang et al. (2009)
Prostate-specific antigen (PSA)	HMG	FRET	Tb cryptate	CdSe/ZnS QD605	0.05 nM	Wegner et al. (2013)
				CdSe/ZnS QD650		

HMG homogeneous, *M* multiplexed

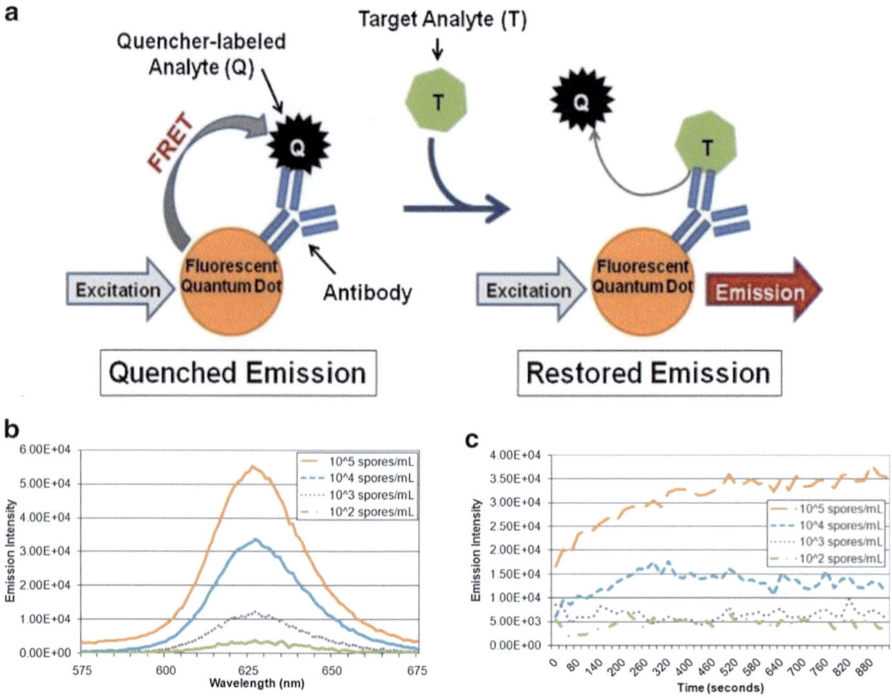

Fig. 11.5 (**a**) Competitive immunoassay strategy for the detection of mold (*A. amstelodami*) spores using QDs and FRET. Changes in QD PL intensity as a function of (**b**) spore concentration and (**c**) time. Adapted under the Creative Commons Attribution license from Kattke et al. (2011). Copyright 2011 The Authors

QDs and dyes sufficiently close to one another for FRET. Wei developed such a sandwich immunoassay for the detection of an antigen, estrogen receptor β (ER-β), which is a putative biomarker and therapeutic target for cancer (Wei et al. 2006). An anti-ER-β monoclonal antibody labeled with CdSe/ZnS QD565 and an anti-ER-β polyclonal antibody labeled with either Alexa Fluor 568 or Alexa Fluor 633 formed the FRET pair. The dye/QD PL ratio provided the analytical signal and permitted detection of ER-β at concentrations as low as 0.05 nM. One of the main challenges with the sandwich format is the long donor–acceptor distances imposed by the large size of full antibodies, often resulting in low-efficiency FRET. To address this challenge, Wegner et al. demonstrated a QD-FRET sandwich immunoassay for prostate-specific antigen (PSA) using luminescent terbium complexes (Tb) as FRET donors and CdSe/ZnS QD605 and QD650 as FRET acceptors (see Fig. 11.7) (Wegner et al. 2013). The millisecond excited state lifetime of the Tb permitted use of the QDs as acceptors (see above) with Förster distances of ~10.7 nm (QD650) and ~7.7 nm (QD605), and effectively extended the range of FRET to 15–20 nm. Furthermore, Fab fragments targeting PSA were used instead of full IgG molecules, thereby decreasing the average donor–acceptor distance. One monoclonal Fab

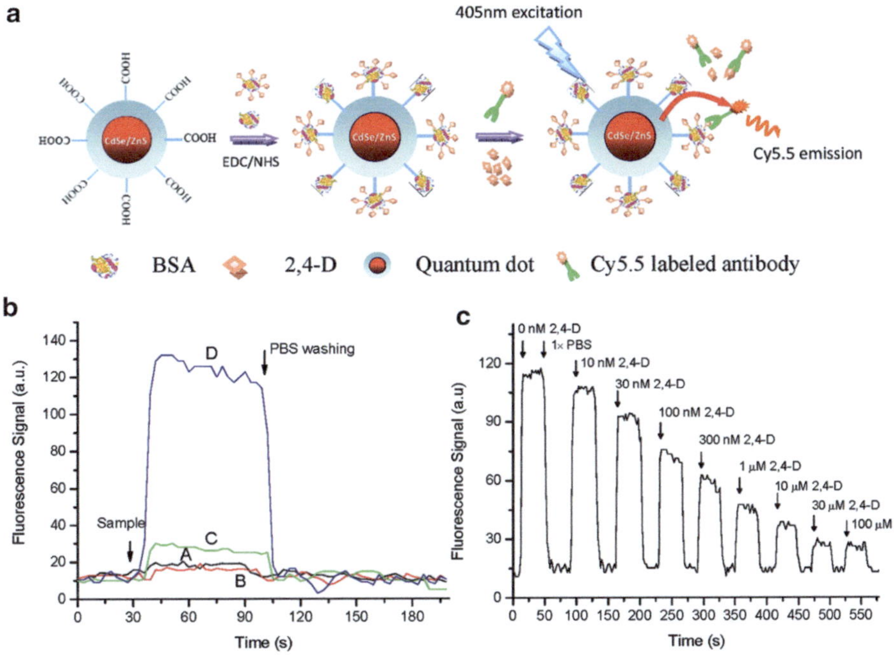

Fig. 11.6 (**a**) Design of a competitive immunoassay strategy for the detection of 2,4-D using a microfluidic platform. (**b**) Representative data showing the selectivity of the device after injection of different reagents: (*trace A*) QD–BSA-2,4-D conjugate (i.e., immunoprobe); (*trace B*) immunoprobe and Cy5.5-labeled rabbit-anti-mouse antibody; (*traces C and D*) immunoprobe and Cy5.5-labeled anti-2,4-D antibody (2 and 10 nM). (**c**) Representative data for a competitive immunoassay with different sample concentrations of 2,4-D. Adapted with permission from Long et al. (2012a). Copyright 2012 American Chemical Society

fragment was conjugated to the QDs (~16 fragments per QD) and a second monoclonal Fab fragment was labeled with Tb, resulting in an estimated 62 Tb in close proximity to each QD. This configuration provided the sensitivity to detect between 0.05 and 15 nM of PSA in spiked serum samples (50 μL) by using the ratio of QD and Tb PL intensity measured between 0.1 and 0.9 ms after pulsed excitation.

11.2.1.4 Proteolysis Assays

Proteases play critical roles in numerous biochemical and signaling processes (López-Otín and Overall 2002). Consequently, these enzymes are important targets for disease diagnostics and therapeutics. FRET has been widely used as a method for the detection of protease activity with dye-labeled peptides. For example, Whitney et al. reported a FRET-based method for detecting thrombin activity in vivo using hairpin peptide probes with a Cy7-labeled polyanionic domain and a Cy5-labeled polycationic domain (Whitney et al. 2013). Thrombin activity cleaved the

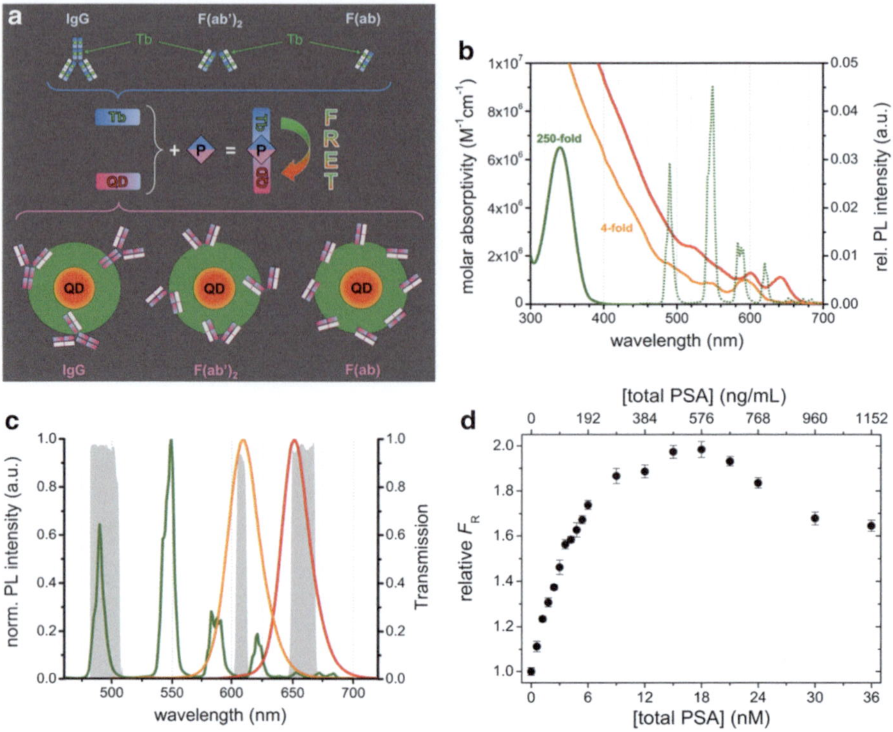

Fig. 11.7 (**a**) Principle of a sandwich immunoassay based on FRET between Tb-cryptate donors and QD acceptors using antibodies or antibody fragments. (**b**) Absorption spectra for two colors of QD and the absorption and emission spectra for the Tb-cryptate. (**c**) Emission spectra for two colors of QD and the Tb-cryptate. The *shaded regions* represent the transmission ranges of optical bandpass filters used to define detection channels. (**d**) Calibration curve plotting time-gated QD/Tb PL intensity ratios versus antigen (prostate-specific antigen) concentration. Adapted with permission from Wegner et al. (2013). Copyright 2013 American Chemical Society

peptide at a recognition site between the two labeled domains, disengaging FRET and restoring Cy5 PL. A similar format has been used with QDs, where a peptide bridges the QD and a suitable acceptor/donor for FRET (see Fig. 11.8). These systems are typically multivalent and benefit from the optical properties of the QD. It has been suggested that the peptide substrate should include a linker that has affinity to the surface or coating of QDs (e.g., a polyhistidine tag or thiol group, both of which have affinity for the ZnS shell of a QD, or biotin, which has affinity for streptavidin-coated QDs) or is suitable for covalent coupling; a spacer following the linker to promote access to the recognition sites (e.g., a helix-forming motif such as a polyproline or an alanine/α-aminoisobutyric acid-rich sequence); a recognition site for cleavage of the peptide by the target protease; and a terminal modification with a dye or quencher (Medintz et al. 2006a; Shi et al. 2006). Chang et al., Medintz et al., and Shi et al. reported the first examples of the above designs (Chang et al. 2005; Medintz et al. 2006a; Shi et al. 2006). In each case, enzyme-induced changes

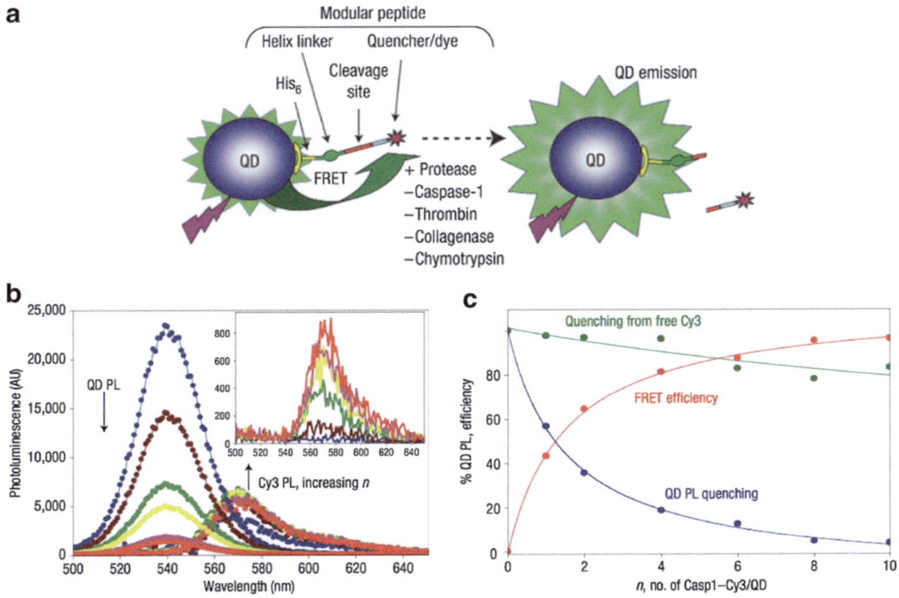

Fig. 11.8 (**a**) Design of a QD-FRET assay for sensing proteolytic activity. (**b**) Emission spectra showing quenching of QD PL and sensitization of Cy3 PL via FRET as the number of Cy3-labeled peptide substrates per QD increases. (**c**) Corresponding plot of FRET efficiency versus the number of peptides. With proteolysis, the number of peptides per QD decreases, resulting in a loss of FRET. Adapted with permission from Macmillan Publishers Ltd: Nature Materials (Medintz et al. 2006a). Copyright 2006

in QD or dye PL were related to proteolytic activity. For example, in the report by Chang et al., collagenase activity was detected through loss of QD PL quenching associated with energy transfer from CdSe/CdS QD620 to 1.4-nm AuNPs that were distal labels on peptide substrates covalently coupled to the QDs (Chang et al. 2005). Shi et al. solubilized QD545 in aqueous solution using short, thiol-terminated and Rhodamine Red-X dye-labeled peptides as ligands (Shi et al. 2006). These QD–peptide conjugates were used for qualitative detection of the activity of matrix metalloproteinases (MMPs) secreted into the culture medium of HTB 126 breast cancer cells. The in vitro assay showed a clear difference in extracellular proteolytic activity between cancerous cells and non-cancerous cells. In the work by Medintz et al., DHLA-coated QDs were conjugated with a peptide substrate for one of four different proteases (Medintz et al. 2006a). The peptides were labeled with either Cy3 or QXL, a dark quencher, and paired with QD510, QD520, or QD540 (see Fig. 11.8). The QD–peptide FRET probes were useful for measuring initial proteolytic rates and estimating apparent Michaelis–Menten parameters, and model protease inhibitor screening assays. Algar et al. later developed a rigorous method for measuring proteolysis kinetics associated with QD–peptide FRET probes in real time, including application of an integrated Michaelis–Menten kinetic model (Algar et al. 2012b). These preliminary results suggested that the classical

Michaelis–Menten model was not directly applicable with QD–peptide conjugates and that the interactions between the protease and the surface of the QD played an important role in the rate of proteolytic digestion of the peptides. A "hopping" mechanism was proposed where protease molecules would, on average, transiently associate with the QD conjugates and successively and rapidly digest all peptide substrates conjugated to the QD before dissociating and diffusing to a fresh QD–peptide conjugate. Overall, some of the proteases targeted for QD-FRET sensing have included caspase-1, caspase-3, chymotrypsin, collagenase, matrix metalloproteinase-7, thrombin, trypsin, and others (Algar et al. 2010).

Algar et al. have developed multiplexed QD probes for sensing proteolytic activity using so-called "concentric FRET relays." Multiple and sequential energy transfer pathways exist within these configurations, which are created by assembling different donors and/or acceptors around each QD. One such configuration co-assembled a luminescent Tb complex and A647 around CdSe/ZnS QD625. The QD served as a donor for A647 and an acceptor for the Tb (Algar et al. 2012c). This FRET relay was interrogated in two modes due to the direct excitation of the QDs: (1) a prompt mode (~0 μs delay after flash excitation), where the QD was excited and transferred energy to the A647; and (2) a time-gated mode (>50 μs after flash excitation), where the QD had relaxed to its ground state, was re-excited by energy transfer from the Tb, and re-transferred that energy to the A647 via FRET. The time-gated QD/Tb and prompt A647/QD PL ratios provided analytical signals proportional to the number of Tb and A647 per QD (see Fig. 11.9). In another configuration, Alexa Fluor 555 (A555) and A647 were co-assembled around a CdSe/ZnS QD525 (Algar et al. 2012a). The QD and A555 formed a good FRET pair, the QD and A647 formed a moderate FRET pair, and the A555 and A647 formed a good FRET pair. The combined A555/QD and A647/QD PL ratios provided a unique set of signals for each unique combination of different numbers of A555 and A647 per QD. Multiplexed detection of proteolytic activity was possible in both configurations by using different peptide bridges between the QDs and different donors/acceptors, with each peptide having a sequence recognized and cleaved by its target protease. Proof-of-concept was demonstrated using trypsin and chymotrypsin as proteases, including assays where the activation of pro-chymotrypsin to chymotrypsin by trypsin was quantitatively tracked in real time.

Examples of enzyme assays with QDs and FRET, including those for the activity of proteases (above), kinases, and nucleases (see below), are summarized in Table 11.4.

11.2.1.5 Kinase Assays

Protein kinase enzymes, like proteases, play critical roles in regulating cellular processes and signal transduction; however, rather than hydrolyzing residues in proteins, kinases modulate their activity through phosphorylation (Hunter 1995). Freeman et al. reported a method for assaying kinase activity with QDs and FRET that involved conjugating a peptide with a serine residue to a CdSe/ZnS QD560

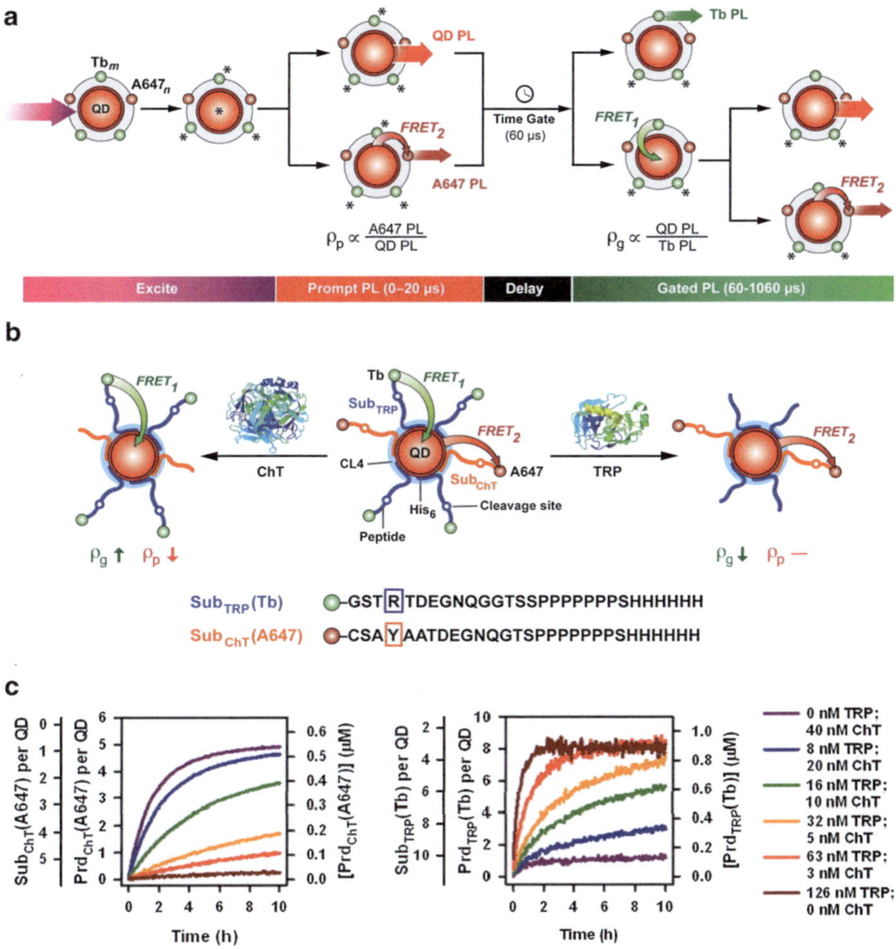

Fig. 11.9 (**a**) Principle of a time-gated, concentric FRET relay using a QD intermediary between a Tb-cryptate donor and an Alexa Fluor 647 (A647) acceptor. Prompt and time-gated measurement domains provide two detection channels with two analytical signals: the prompt A647/QD PL ratio and the time-gated QD/Tb PL ratio. (**b**) Design of a multiplexed probe for trypsin (*TRP*) activity and chymotrypsin (*ChT*) activity. (**c**) Representative progress curves showing multiplexed detection of TRP and ChT activity. Adapted with permission from Algar et al. (2012c). Copyright 2012 American Chemical Society

using carbodiimide chemistry, followed by mixing with ATP that was labeled with ATTO-590 dye at its terminal phosphate group (Freeman et al. 2010). Casein kinase activity phosphorylated the serine residue, concomitantly transferring the ATTO-590 to the peptide, providing the proximity for FRET (see Fig. 11.10). These authors also reported a similar configuration where unlabeled ATP was used in combination with an ATTO-590 labeled anti-phosphoserine antibody. Both methods could detect

Table 11.4 Energy transfer-based assays of enzyme activity

Analyte	Format	Mechanism	Donor	Acceptor	LOD	Reference
Botulinum neurotoxin serotype A light chain protease	HMG	FRET	CdSe/ZnS QD550	Cy3	0.35 nM	Sapsford et al. (2011)
Casein kinase	HMG	FRET	CdSe/ZnS QD560	Atto-590	0.1 units	Freeman et al. (2010)
Caspase-1	HMG	FRET	CdSe/ZnS QD510	Cy3	–	Medintz et al. (2006a)
Chymotrypsin			CdSe/ZnS QD520	QXL		
Collagenase			CdSe/ZnS QD540	QXL		
Thrombin			CdSe/ZnS QD520	QXL		
Caspase-3	HTG	NSET	CdSe/ZnS QD525	1.4-nm AuNP	20 ng/mL	Kim et al. (2008)
MMP-7			CdSe/ZnS QD605		10 ng/mL	
Thrombin			CdSe/ZnS QD655		1 U/mL	
Caspase-3	HMG	FRET	CdSe/ZnS QD550	mCherry	20 pM	Boeneman et al. (2009)
Chymotrypsin (M)	HMG	FRET	Tb cryptate	CdSe/ZnS QD625	<8 nM	Algar et al. (2012c)
Trypsin (M)			CdSe/ZnS QD625	Alexa Fluor 647		
Chymotrypsin (M)	HMG	FRET	CdSe/ZnS QD525	Alexa Fluor 555	<1 nM	Algar et al. (2012a)
Trypsin (M)				Alexa Fluor 647		
Chymotrypsin (M)	HTG	FRET	CdSeS/ZnS QD525	Alexa Fluor 555	1 nM	Petryayeva and Algar (2013)
Enterokinase (M)						
Trypsin (M)						

Chymotrypsin	HMG	CT	Ru-phen	CdSe/ZnS QD590	–	Medintz et al. (2008)
Thrombin				CdSe/ZnS QD540		
Collagenase	HMG	FRET	CdSe/ZnS QD545	Rhodamine Red-X	–	Shi et al. (2006)
Collagenase	HMG	NSET	CdSe/CdS QD620	1.4-nm AuNP	–	Chang et al. (2005)
DNase I	HMG	FRET	CdSe/ZnS QD580	Texas Red	–	Gill et al. (2005)
HER2 (M)	HMG	FRET	CdSe/ZnS QD655	Alexa Fluor 660	7.5 nM	Lowe et al. (2012)
uPA		NSET	CdSe/ZnS QD525	1.4-nm AuNP	50 ng/mL	
Human myosin Va	HTG-SP	FRET	CdSe/ZnS QD520	Cy3	–	Sugawa et al. (2010)
MMP-2	HMG	BRET	Rluc	CdSe/ZnS QD655	2 ng/mL	Yao et al. (2007)
MMP-2	HMG	BRET	Rluc	CdSe/ZnS QD655	5 ng/mL	Xia et al. (2008)
MMP-7 (M)				CdSe/ZnS QD655	1 ng/mL	
uPA (M)				CdTe/ZnS QD705	500 ng/mL	
MNase	HMG	FRET	CdSe/ZnS QD550	6-Carboxy-X-rhodamine	0.06 units	Huang et al. (2008)
Renin	HMG-SP	FRET	CdSe/ZnS QD605	Cy5	25 pM	Long et al. (2012b)

HMG homogeneous, *HTG* heterogeneous, *M* multiplexed, *SP* single particle

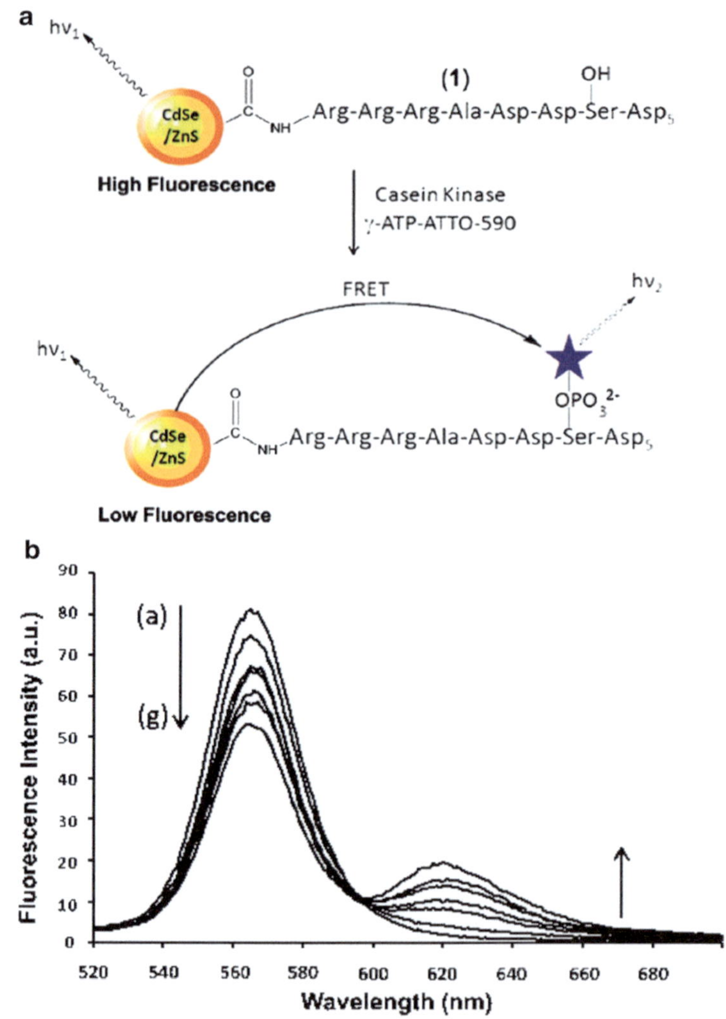

Fig. 11.10 (**a**) Design of a QD-FRET assay for kinase activity. (**b**) Time-dependent spectral changes (7-min intervals) after adding 1 unit of casein kinase (*b–g*). The original PL spectrum is (*a*). Adapted with permission from Freeman et al. (2010). Copyright 2010 American Chemical Society

down to ~0.1 units of casein kinase. Lowe and coworkers developed a homogeneous, multiplexed assay capable of detecting both protease and kinase activity (Lowe et al. 2012). The key to this assay was orthogonality in bioconjugation: a biotinylated peptide substrate for urokinase-type plasminogen activator (uPA), a protease, was bound to streptavidin-coated QD525, whereas a polyhistidine-appended peptide substrate for human epidermal growth factor receptor 2 (HER2),

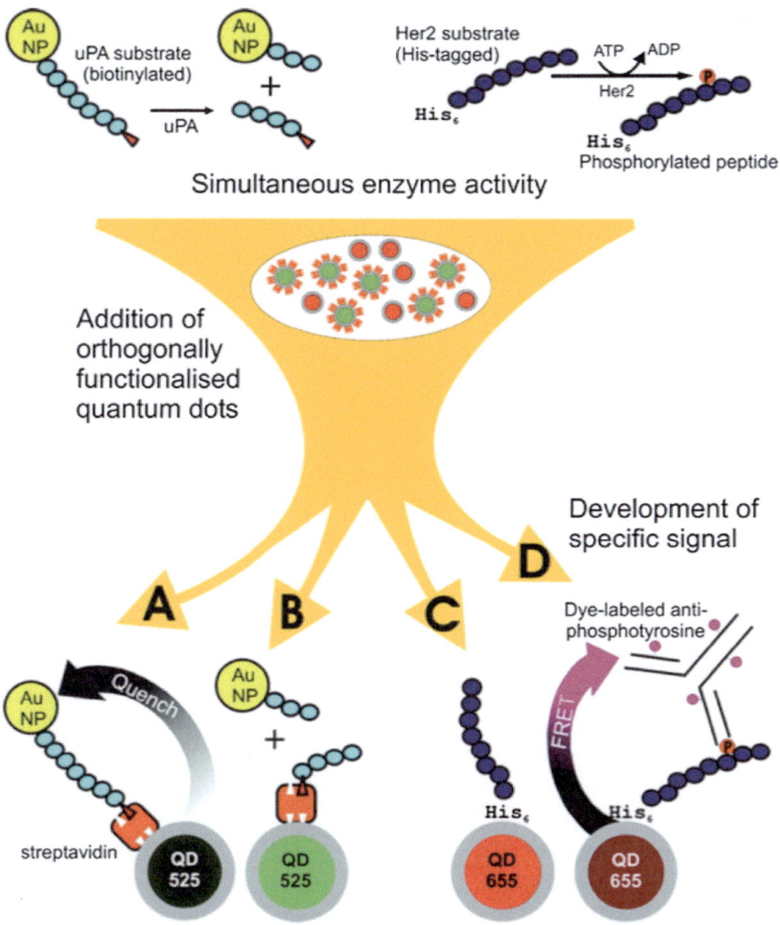

Fig. 11.11 Design of a multiplexed assay for protease activity and kinase activity using QDs and FRET/NSET: (**a**) no protease activity; (**b**) protease activity; (**c**) no kinase activity; (**d**) kinase activity. Adapted with permission from Lowe et al. (2012). Copyright 2012 American Chemical Society

a tyrosine kinase was assembled to mercaptopropionic acid-coated QD655 (see Fig. 11.11). The uPA substrate peptide was labeled with a 1.4-nm AuNP to quench QD PL via NSET (see below). After mixing substrate with uPA-containing samples, then adding QD525, the observed QD525 PL intensity was proportional to the uPA activity. In contrast, the HER2 substrate was indirectly labeled after phosphorylation though binding with Alexa Fluor 660 (A660)-labeled anti-phosphotyrosine antibodies. The ratio of FRET-sensitized A660/QD655 PL ratio was proportional to the HER2 activity. The limits of detection were 50 ng/mL for uPA and 7.5 nM for HER2 kinase.

11.2.1.6 Nuclease Assays

Assays for nuclease activity with QDs and FRET are designed analogously to assays for proteolytic activity. A QD and a suitable FRET acceptor are held in close proximity by a nucleic acid bridge, which is cleaved by nuclease activity, disrupting FRET. One of the first such assays was reported by Gill et al., who self-assembled thiol-terminated DNA to CdSe/ZnS QD580 (Gill et al. 2005). This DNA was hybridized with its complement, which was labeled at its proximal terminus with Texas Red as a FRET acceptor. With the addition of DNase I, FRET decreased from digestion of the duplex DNA and loss of the FRET acceptor. Huang et al. reported a similar strategy for sensing micrococcal nuclease (MNase) activity (Huang et al. 2008). Streptavidin-coated CdSe/ZnS QD550 was assembled with a DNA probe that was biotinylated at one terminus and labeled with 6-carboxy-X-rhodamine (ROX) at the opposite terminus. Hydrolysis of the DNA by MNase resulted in the loss of FRET between the QD and ROX. As little as 0.06 units of MNase activity could be detected and the QD-DNA probe was used to monitor MNase activity in the culture media of *S. aureus*.

11.2.1.7 Carbohydrates

The first homogeneous assay using QDs and FRET was reported by Medintz et al. and targeted maltose (Medintz et al. 2003). QDs were self-assembled with polyhistidine-terminated maltose binding protein (MBP). Dark quencher (QSY9)-labeled β-cyclodextrin (β-CD) was bound by the MBP and quenched CdSe/ZnS QD560 PL through FRET. The addition of maltose displaced the QSY9-β-CD with concomitant recovery of QD PL (see Fig. 11.12). An alternative configuration was also developed using a QD530-to-Cy3-to-Cy3.5 two-step FRET relay, where the MBP was labeled with Cy3 to provide a relay site for energy transfer to bound β-CD-Cy3.5 (Medintz et al. 2003). This configuration increased energy transfer from the QD to >90 % efficiency versus 65 % with an approximately equivalent number of QSY9-β-CD. The enhanced efficiency was a product of both the shorter distance for initial energy transfer and the stronger net spectral overlap for the energy transfer steps between the QD, Cy3, and Cy3.5. Both configurations had micromolar dissociation constants for binding maltose.

In addition to receptor proteins, another common molecular recognition approach for the detection of sugars is the use of boronic acids, which reversibly bind with the *cis*-diols found in sugars (and other molecules). To this end, Freeman et al. coated QD570 with boronic acid ligands and labeled galactose or dopamine with ATTO-590 dye (Freeman et al. 2009a). Binding between the boronic acid-functionalized QD and the ATTO-590-labeled galactose/dopamine resulted in efficient FRET. Glucose, galactose, or dopamine in a sample could be measured in a competitive binding assay format with increasing loss of FRET with increasing sugar or dopamine concentrations in the sample (see Fig. 11.13). The LODs for glucose, galactose, and dopamine were 1, 50, and 100 μM, respectively. Table 11.5 summarizes several examples of assays for carbohydrates that rely on QDs and FRET.

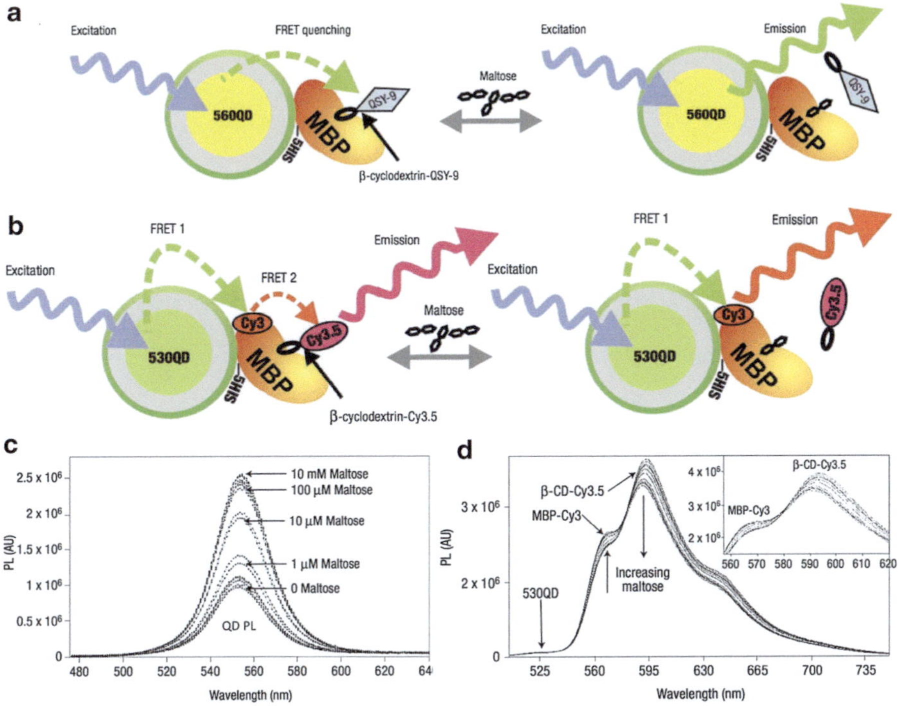

Fig. 11.12 Schemes for competitive binding assays of maltose using (**a**) dark quencher-labeled β-cyclodextrin with maltose binding protein (MBP) conjugated to CdSe/ZnS QD560, and (**b**) a FRET relay with Cy3-labeled MBP conjugated to CdSe/ZnS QD530 with Cy3.5-labeled β-cyclodextrin. (**c**) Representative data for the configuration in (**a**). (**d**) Representative data for the configuration in (**b**). Adapted with permission from Macmillan Publishers Ltd: Nature Materials (Medintz et al. 2003). Copyright 2003

11.2.1.8 Ions and Dissolved Gases

Probes for several different ions have been developed using QD-FRET for transduction and several examples are listed in Table 11.6. The key cog in most of these sensing mechanisms is an ion-sensitive dye. In some instances, the ion-sensitive dye is a poor acceptor at low ion activity and a good acceptor at high ion activity, or vice versa, whereas in other instances, the ion activity does not affect the acceptor-capability of the dye but rather its emission quantum yield. Both formats offer ratiometric detection of ion activity coupled with the advantages of QDs.

QD-FRET probes for pH have been developed and are, for example, promising for imaging tumour microenvironments. One mechanism of pH sensing is to attach a color-changing, pH-sensitive dye to a QD donor, where the pH-dependent color change brings the dye in and out of resonance (i.e., spectral overlap) with the QD donor. For example, Snee et al. directly conjugated a squaraine dye (fluorescence emission peak at 650 nm) to an amphiphilic polymer coating on CdSe/ZnS QD615

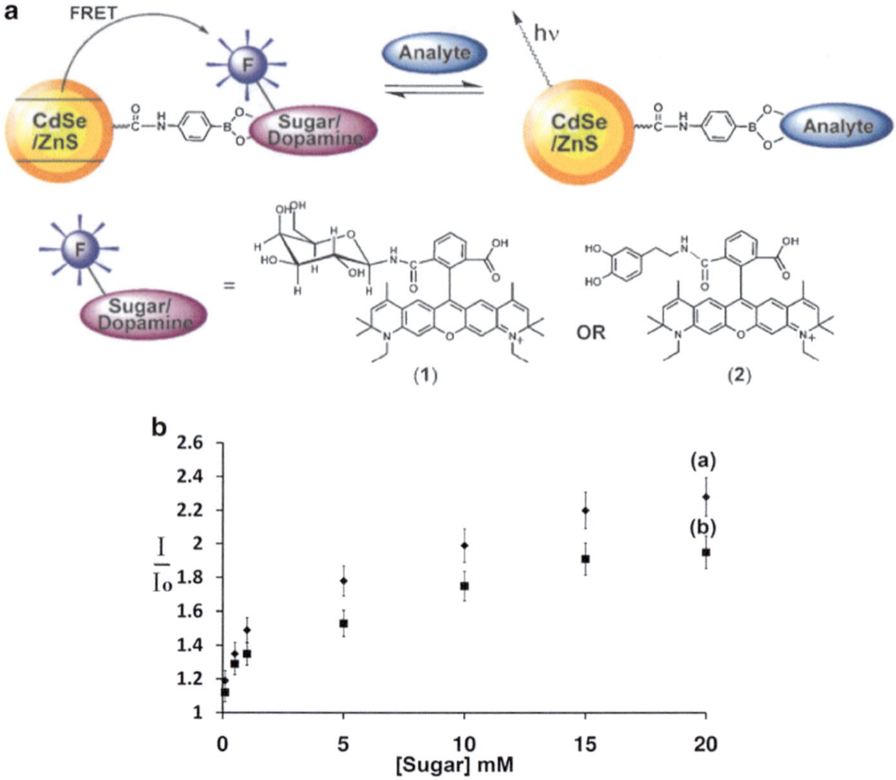

Fig. 11.13 (**a**) Design of a QD-FRET assay for sugars and dopamine using boronic acid ligands. (**b**) Relative changes in QD PL intensity in response to increasing concentrations of (*a*) galactose and (*b*) glucose. Adapted with permission from (Freeman et al. 2009a). Copyright 2009 Royal Society of Chemistry

to create a ratiometric pH sensor (Snee et al. 2006). With increases in pH, the squaraine dye exhibited a small hypsochromic shift in its absorption spectrum, a decrease in its absorption coefficient, and a decrease in its fluorescence quantum yield, whereas the QD PL intensity was largely unaffected by the changes in pH. Consequently, the dye/QD PL ratio decreased as the solution pH became more basic in the range of pH 6–10. Subsequently, SNARF-5 F dye was used as an acceptor for CdSe/CdZnS QD525 and permitted high-resolution sensing of pH within the physiologically relevant range of 6.0–8.0 with both one-photon and two-photon excitation (see Fig. 11.14) (Somers et al. 2012). The SNARF-5 F dye was conjugated to DHLA ligands appended with a first-generation poly(amido amine) (PAMAM) dendrimer to permit conjugation of a large number of dyes per QD. In a similar example of pH sensing, Dennis et al. assembled mOrange fluorescent proteins to QD520 (Dennis et al. 2012). The absorption spectrum of the mOrange was pH-sensitive and had good spectral overlap with the QD donor at basic pH and

Table 11.5 Energy transfer-based assays of carbohydrates with QDs

Analyte	Format	Mechanism	Donor	Acceptor	LOD	Reference
Galactose	HMG	FRET	CdSe/ZnS QD570	ATTO-590	50 μM	Freeman et al. (2009a)
Glucose					1 μM	
Dopamine					100 μM	
Glucose	HMG	CRET	Luminol	CdSe/ZnS QD620	5 mM	Niazov et al. (2011)
Glucose	HMG	NSET	CdSe/ZnS QD525	AuNP	50 nM	Tang et al. (2008)
Glycosylated protein	HMG	NSET	CdSe/ZnS QD605	3-nm AuNP	–	Oh et al. (2006)
Maltose	HMG	FRET	CdSe/ZnS QD560	QSY-9	–	Medintz et al. (2003)
Maltose	HTG	FRET	CdSe/ZnS QD530	Cy3	–	Medintz et al. (2006b)
Maltose	HMG	CT	Ru-phen	CdSe/ZnS QD565		Sandros et al. (2006)
Maltose	HMG-SP	CT	Ferrocene	CdSe/ZnS QD560	100 pM	Opperwall et al. (2012)
Sialoglycans	HMG	NSET	CdSe/ZnS QD610	AuNP	~1 μM	Kikkeri et al. (2013)

HMG homogeneous, *HTG* heterogeneous, *SP* single particle

Table 11.6 FRET-based assays of ions and dissolved gases

Analyte	Format	Mechanism	Donor	Acceptor	LOD	Reference
Ca^{2+}	HMG	FRET	CdSe/ZnS QD580	CaRuby	2 μM	Prasuhn et al. (2010)
Hg^{2+}	HMG	FRET	CdTe QD550	Rhodamine dye	52 ppb	Liu et al. (2012)
Hg^{2+}	HMG	NSET	CdSe/ZnS QD575	3-nm AuNP	1.2 ppb	Li et al. (2011)
Pb^{2+}	HMG	CT	Ru-phen	CdSe/ZnS QD545	100 pM	Shete and Benson (2009)
				InGaP/ZnS QD660		
Pb^{2+} (M)	HMG	FRET	CdSe/ZnS QD530	Iowa Black FQ	0.2 nM	Wu et al. (2010)
Cu^{2+} (M)			CdSe/ZnS QD625	Iowa Black RQ	0.5 nM	
pH	HMG	FRET	CdSe/ZnS QD615	Squaraine dye	–	Snee et al. (2006)
pH	HMG	FRET	CdSe/ZnS QD525	SNARF-5 F	–	Somers et al. (2012)
pH	HMG	FRET	CdSe/ZnS QD520	mOrange	–	Dennis et al. (2012)
pH	HMG	CT	CdSe/ZnS QD550	Dopamine	–	Medintz et al. (2010)
pO_2	HMG	FRET	CdZnSe/CdZnS QD550	Os(II)-polypyridyl complex	–	McLaurin et al. (2009)
Zn^{2+}	HMG	FRET	CdSe/ZnS QD540	Zincon	0.6 μM	Ruedas-Rama and Hall (2009)
			CdSe/ZnS QD620			
Zn^{2+}	HMG	CT	Cyclam	CdSe/ZnS QD620	1 μM	Ruedas-Rama and Hall (2008)
			Cyclen			

HMG homogeneous, *M* multiplexed

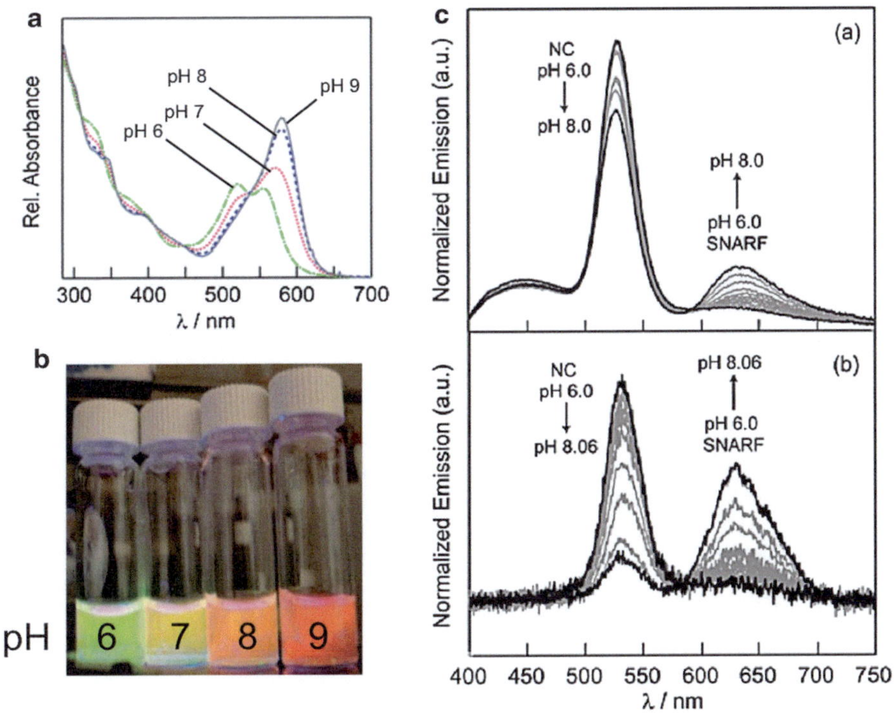

Fig. 11.14 (**a**) Absorption spectra of QD–SNARF dye conjugates at pH 6, 7, 8, and 9. The dye absorption band shifts out of resonance with the QD PL when the pH decreases. (**b**) Photograph of the net PL of the conjugates in solutions buffered at the indicated pH. (**c**) One-photon (*top*) and two-photon (*bottom*) PL emission spectra as a function of physiologically relevant changes in pH. Adapted with permission from Somers et al. (2012). Copyright 2012 Royal Society of Chemistry

poor spectral overlap with the QD donor at acidic pH. The mOrange/QD PL ratio increased 12-fold between pH 6 and 8 and 20-fold between pH 5 and 10. The QD-conjugate had better photostability than mOrange alone and, when expressed with a C-terminal polyarginine sequence to facilitate endosomal uptake by HeLa cells, enabled observation of the acidification of endocytic vesicles as they progressed over time to early endosomes and then to late endosomes.

Prasuhn et al. developed a QD-FRET probe for calcium (Ca^{2+}) by conjugating a Ca^{2+}-sensitive dye, Calcium Ruby (CaRuby), to the C-terminus of a peptide using azide-alkyne click chemistry (Prasuhn et al. 2010). The peptide had a polyhistidine sequence at its N-terminus for self-assembly to CdSe/ZnS QD580, which served as a FRET donor. The CaRuby/QD PL ratio provided an analytical signal that increased with increases in the concentration of Ca^{2+} between approximately 2 and 100 μM. Ruedas-Rama et al. developed a probe for zinc (Zn^{2+}) using QD540 or QD620 electrostatically assembled with zincon (Ruedas-Rama and Hall 2009). Upon the addition of Zn^{2+}, formation of a colored Zn^{2+}–zincon complex (from a substantial bathochromic shift of its absorption spectrum) resulted in quenching of QD PL and detection of as little as 0.6 μM of Zn^{2+} with good selectivity against several other divalent ions and biologically relevant cations. Liu et al. created a probe for mercury (Hg^{2+}) by embedding CdTe QD550 in silica nanoparticles that were then surface-modified with a spirolactam rhodamine derivative as an ion recognition element (Liu et al. 2012). The rhodamine dye undergoes ring opening in the presence of Hg^{2+} ions, altering its optical properties to become a good acceptor for the QD. The LOD was 52 ppb of Hg^{2+} ions. The silica nanoparticle was necessary for "turn-on" sensing because Hg^{2+} and some other metals (e.g., Cu^{2+}) can quench QD PL directly without any built-in selectivity.

McLaurin et al. developed a pO_2 sensor by coupling a phosphorescent osmium(II) polypyridyl complex (OsPP) to $Cd_xZn_{1-x}Se/Cd_yZn_{1-y}S$ QD550 (McLaurin et al. 2009). The QD functioned to sensitize the OsPP via FRET; however, transduction was via the quenching of the OsPP phosphorescence by dissolved oxygen and not changes in FRET efficiency. The principal advantage of the QD antenna was that it provided a 60-fold larger two-photon absorption cross-section than the OsPP.

11.2.2 Single-Particle Measurements

QD-based probes and sensors that function at the single-molecule level can resolve dynamic processes that occur asynchronously in the ensemble or exist as a distribution of similar but nonetheless distinct populations. In both cases, measurements are not possible in the ensemble because of bulk averaging. Single-particle measurements are made in one of two formats: in homogeneous solutions under dilute conditions, or as a heterogeneous system with particles sparsely immobilized on a solid support. Although the heterogeneous format is common for biophysical studies and dynamic measurements, the homogeneous format is preferred for assays, avoids surface effects, and has higher throughput.

Pons et al. were among the first to physically study solution-phase, single-QD FRET (Pons et al. 2006). MBP with a polyhistidine tag was labeled with Rhodamine Red and self-assembled to CdSe/ZnS QD540, and the conjugates were measured in the ensemble and at the single-particle level using burst analysis. One important outcome of this study was confirmation that the MBP self-assembled with the QDs according to a Poisson distribution. That is, mixing N equivalents of MBP with QDs resulted in conjugates with … $N-2, N-1, N, N+1, N+2,$ … MBP per QD, according to Eq. (11.1), where N is the average stoichiometry and n is the valence of an individual conjugate.

$$p(N,n) = N^n \exp(-N) / n! \tag{11.1}$$

A second important result was that FRET-based sensing was comparable between ensemble and single-particle measurements. For this experiment, CdSe/ZnS QD520 were conjugated with MBP that was labeled at an allosteric site with Cy3. Upon binding maltose, changes in the conformation of the MBP induced changes in the quantum yield of the Cy3 acceptor, resulting in a change in the Cy3/QD PL ratio. From this PL data, approximately the same binding constants for MBP and maltose could be derived from single-molecule and ensemble measurements.

Zhang et al. were the first to develop an assay in a single-molecule QD-FRET format (Zhang et al. 2005). The basis of the assay was sandwich hybridization between a biotinylated probe oligonucleotide, the target DNA sequence, and a Cy5-labeled reporter oligonucleotide. The probe and reporter were complementary to adjacent regions along the target sequence. The sandwich hybrids were mixed with streptavidin-coated CdSe/ZnS QD605 and the resulting QD–DNA complexes were passed through a microcapillary that was integrated with a two-colour single-molecule fluorescence microscope with appropriate filters for the QD605 and Cy5 emission, with coincident QD and Cy5 emission bursts signaling the presence of probe–target hybrids (see Fig. 11.15). The advantage of interrogating each QD-Cy5 FRET-pair individually was exceptional sensitivity and rejection of signal from excess reporter. A target concentration as small as 5 fM could be detected, and this value was approximately 100-fold better than a conventional dye-based molecular beacon. The QD-FRET strategy was combined with an oligonucleotide ligation assay and successfully applied to the detection of Kras point mutations in clinical samples taken from ovarian cancer patients. This group has also adopted this format for the quantitative detection of methylation-specific PCR products (Bailey et al. 2009). Zhang et al. also demonstrated multiplexed sensing in a single-particle format by combining FRET with direct excitation of fluorescence (Zhang and Hu 2010). The HIV-1 and HIV-2 genes were selected as targets and were recognized by two different biotinylated capture probes. Two corresponding reporter oligonucleotides were labeled with Alexa 488 and Alexa 647, respectively, and the resulting sandwich hybrids were assembled on streptavidin-coated QD605. The samples were again passed through a microcapillary with three-color burst coincidence detection of Alexa 488, QD605, and Alexa 647 emission. Zhang and Johnson also developed a single QD probe to monitor the binding between HIV-1 Rev responsive element (RRE) RNA and Rev peptide (Zhang and Johnson 2007). This interaction is important for HIV replication, and the QD probe is potentially useful for inhibitor screening. The format was largely the same as for DNA detection, with the biotinylated RRE binding to streptavidin-coated QD605 and interacting with Cy5-labeled Rev peptide. Zhang and various coworkers have made several other contributions to single-particle assays with QDs and FRET (Zhang and Johnson 2006, 2009; Zhang and Zhang 2012).

Turning to a heterogeneous single-particle format, Sugawa et al. monitored individual ATP hydrolysis reactions by recombinant human myosin Va (Sugawa et al. 2010). Peptide-coated QD520s were conjugated with myosin Va using a heterobifunctional cross-linker and adsorbed on a glass substrate. Cy3-labeled ATP was then introduced and the substrate imaged using total internal reflection fluorescence (TIRF). Although TIRF largely limits excitation to the interface of the glass substrate, high background can result from high concentrations of freely diffusing dye in bulk solution, limiting the concentration of Cy3-ATP that can be utilized and

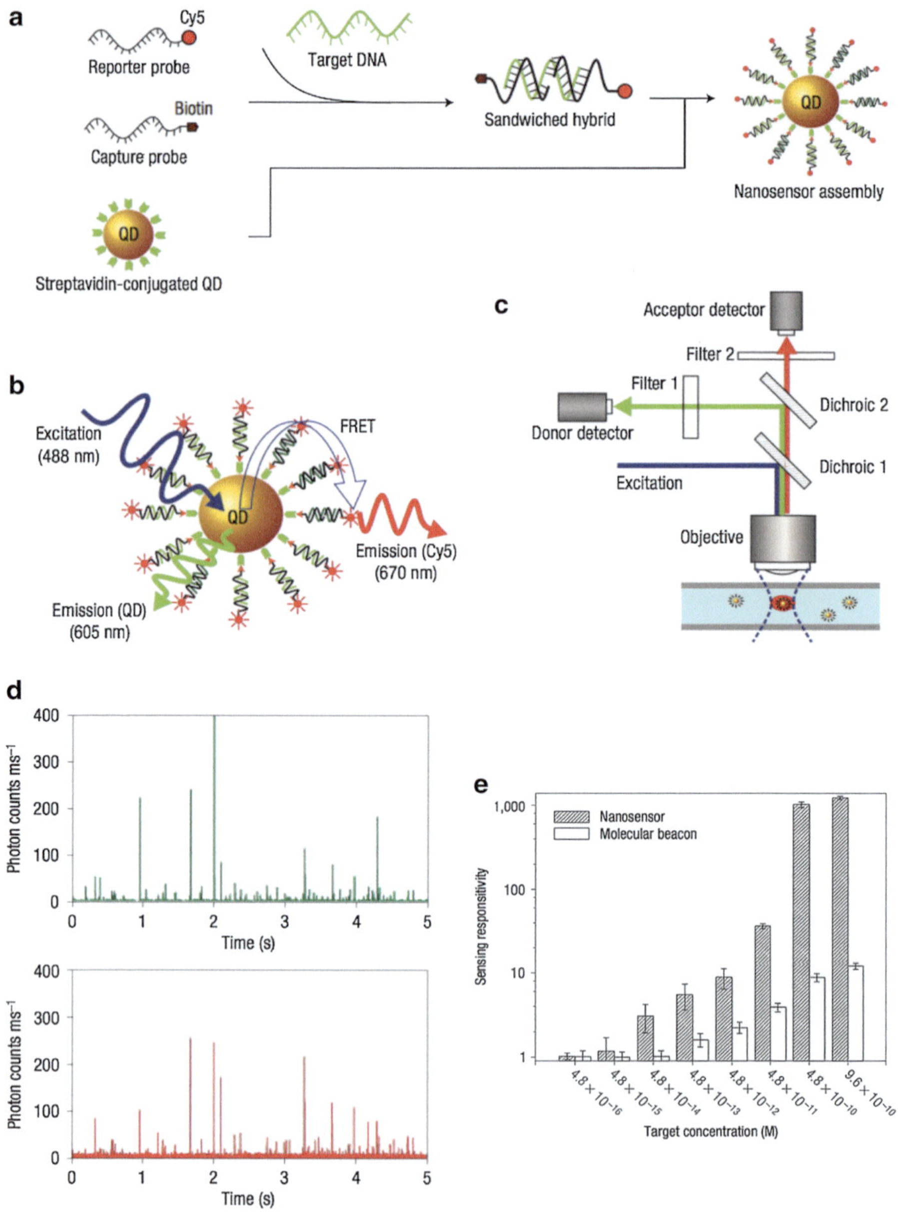

Fig. 11.15 (**a**) Design of a sandwich DNA hybridization assay using assembly with QDs. (**b**) FRET induced by DNA hybridization. (**c**) Schematic of the instrument used for single-particle detection. (**d**) Burst coincidence analysis in the presence of DNA target. The *top panel* is QD emission; the *bottom panel* is Cy5 emission. (**e**) Comparison between the QD-FRET DNA nanosensor and a conventional molecular beacon. Adapted with permission from Macmillan Publishers Ltd: Nature Materials (Zhang et al. 2005). Copyright 2005

preventing access to high substrate concentrations (>100 nM). The application of FRET with QDs solved this problem as FRET is restricted to <10 nm and, in this case, 405 nm excitation did not directly excite the Cy3, thus avoiding high fluorescence background. The photostability of the QDs also permitted measurement of enzyme–substrate binding over several minutes. The anticorrelation of the QD and Cy3 PL trajectories permitted derivation of off-rates that reflected both substrate turnover and dissociation from the enzyme without catalysis.

11.2.3 Heterogeneous Assays

The majority of QD-FRET assays developed to date have been in a homogeneous format. There are many reasons for this apparent preference: homogeneous assays avoid the need to develop chemistry for the immobilization of QDs in a manner suitable for FRET-based transduction; homogenous assays can have benefits in terms of speed and simplicity; and solution-phase probes are more amenable to intracellular or in vivo applications. Nonetheless, heterogeneous assays also have their advantages: immobilized QDs are not prone to aggregation under non-ideal conditions, bioconjugate reactions can be done with simple washing steps to remove excess reagents (cf. chromatography), washing steps can improve signal-to-background ratios in the analysis of complex biological samples, reusability is possible, and near-field optical interrogation techniques (e.g., total internal reflection) can be utilized, among other case-by-case advantages.

Sapsford, Medintz, and coworkers were able to adapt QD–MBP conjugates to heterogeneous assays for maltose (Medintz et al. 2006b; Sapsford et al. 2004). In one configuration, a rod-like β-strand peptide was biotinylated at its C-terminus for attachment to a NeutrAvidin-modified surface and had a polyhistidine sequence at its N-terminus for binding and immobilizing QDs. Polyhistidine-MBP labeled at an allosteric site with Cy3 could be assembled on the immobilized CdSe/ZnS QD530 to create a reagentless, solid-phase assay for maltose, although the heterogeneous format had lower sensitivity than the corresponding homogeneous format (Medintz et al. 2005).

Algar et al. developed a solid-phase nucleic acid hybridization assay using QDs and FRET (Algar and Krull 2009). MPA-coated CdSe/ZnS QDs were immobilized on the exterior of fused silica optical fibers using multidentate surface ligand exchange with a thin film of silane-based thiols. The QDs were coated with a layer of NeutrAvidin, conjugated with biotinylated oligonucleotide probes, and the surface blocked with BSA. FRET could be generated by probe–target hybridization using DNA targets directly labeled with an acceptor dye or through sandwich hybridization assays with labeled reporter oligonucleotides and unlabeled target. Multiplexed sensing was possible by co-immobilizing QD530 and QD620 and two different probe oligonucleotides, both in direct and sandwich assays, with Cy3 and Alexa 647 acceptor dyes (see Fig. 11.16). Alternatively, QD530 could be combined with two spectrally distinct acceptors, Cy3 and Rhodamine Red-X (Algar and Krull 2010c). In both cases, the acceptor/QD PL ratios provided a quantitative analytical

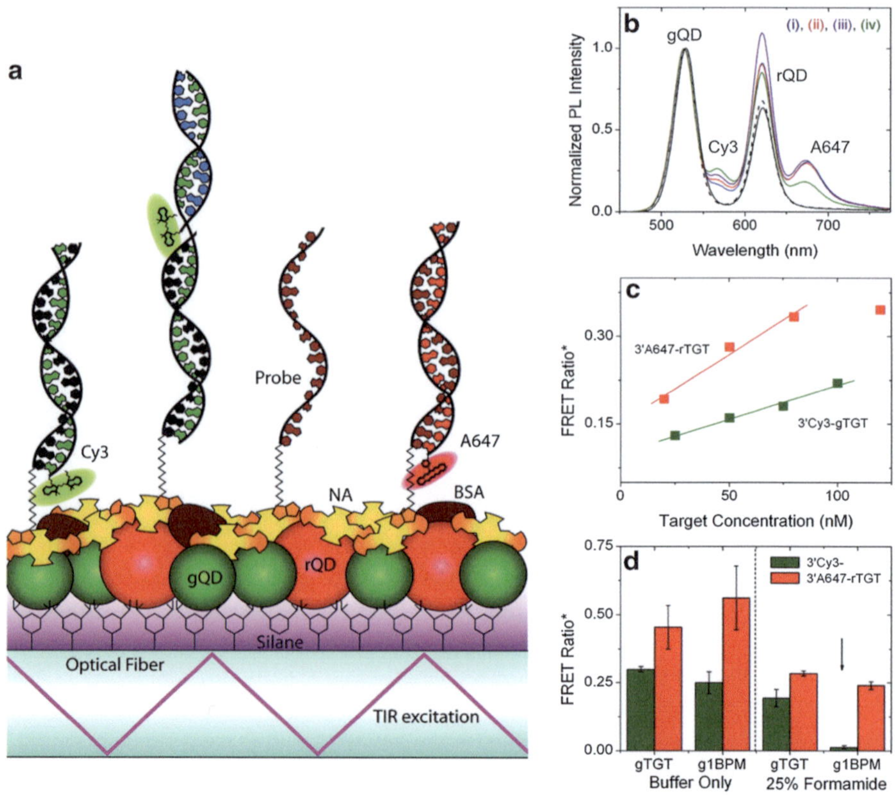

Fig. 11.16 (**a**) Design of a solid-phase, multiplexed DNA hybridization assay with two colors of immobilized CdSe/ZnS QD (gQD = QD530; rQD = QD620). (**b**) PL emission spectra showing the response to four samples with varying amounts of two target DNA sequences. (**c**) Calibration data showing changes in the Cy3/gQD or A647/rQD PL ratio as a function of target concentration. (**d**) Discrimination between fully complementary target (TGT) and a single base-pair mismatch (1BPM) through control of stringency. Adapted with permission from Algar and Krull (2009). Copyright 2009 American Chemical Society

signal with LODs as low as 1 nM and the discrimination of single nucleotide polymorphisms with contrast ratios of 31:1 between fully complementary and mismatched target. Three-plex and four-plex detection were possible by combining various donor–acceptor pairs with direct excitation of Pacific Blue, a blue-emitting fluorescent dye that was efficiently excited at the same wavelength as the QDs (Algar and Krull 2010a, b). Algar et al. further explored how to optimize the solid-phase multiplexing sensing system to obtain a lower LOD, better selectivity, regeneration, and signal enhancement through a reduction of the active sensing area and through the use of doubly labeled reporter probes for sandwich assays (Algar and Krull 2011). This work, and some of the work described below, has been reviewed in detail elsewhere (Noor et al. 2013a).

Petryayeva et al. (2013) extended the above strategy to hybridization assays in microtiter plates. Two colors of QD were immobilized in polystyrene microwells

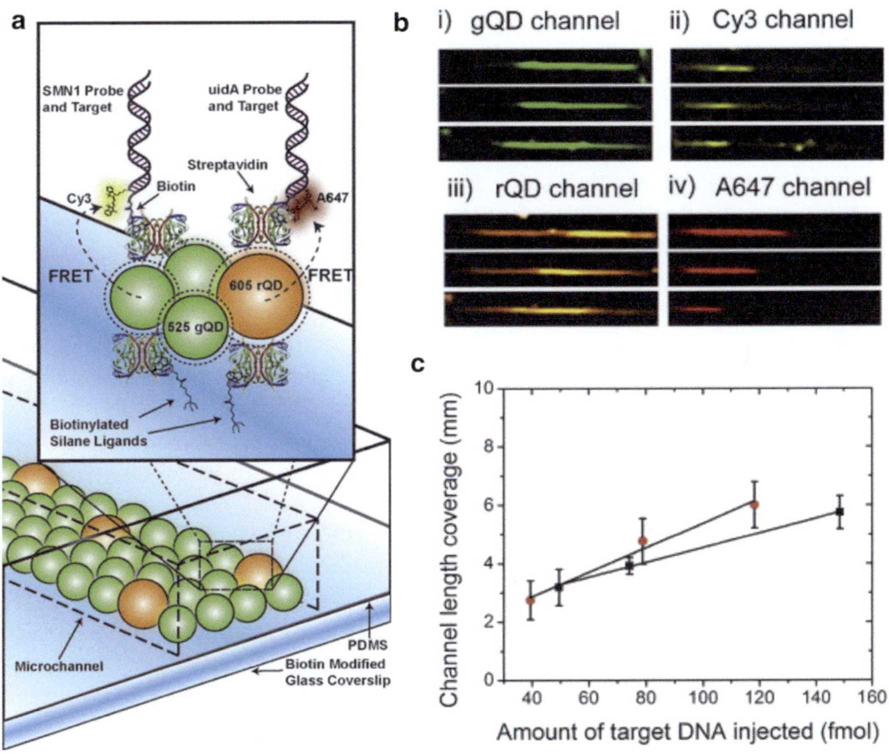

Fig. 11.17 (**a**) Design of a solid-phase, multiplexed DNA hybridization assay within a microfludic channel. (**b**) Images of donor and acceptor PL along the channel length. The length of the zone of acceptor emission (Cy3, A647) provides quantitative data. (**c**) Calibration data plotting the length of FRET-sensitized acceptor PL within the channel versus the amount of target injected into the channel. Adapted with permission from Noor et al. (2013c). Copyright 2013 Elsevier

functionalized with multidentate imidazole surface ligands. Here, a green CdSeS/ZnS QD525–Cy3 pair was used as an internal standard channel and a CdSeS/ZnS QD625–A647 pair served as the detection channel. An LOD of 4 nM was possible in combination with the high-throughput format of the microtiter plate. Similarly, Noor and coworkers (Noor et al. 2013c) adapted the solid-phase QD-FRET nucleic acid hybridization assay format to a microfluidic chip format. Streptavidin-coated CdSe/ZnS QD525 and QD605 were immobilized on a biotinylated glass surface inside a glass-polydimethylsiloxane (PDMS) microfluidic channel. The QDs were modified with two oligonucleotide probe sequences complementary to two target sequences labeled with either Cy3 or Alexa 647. The QDs were delivered to the channel through electro-osmotic flow and immobilized through the biotin–streptavidin interaction. Biotinylated oligonucleotide probe was delivered through electrophoretic injection and immobilized on the QDs. The target was also introduced by electrophoretic injection. In contrast to previous studies where the dye/QD PL ratio was used for quantitative detection, the length of a zone of FRET-sensitized emission along the channel was used for quantitative measurements (see Fig. 11.17) since it was proportional to the amount of target, which was rapidly pulled from solution by hybridization with available probes within the confines of the microfluidic channel. It was possible to detect femtomole quantities of target in this format. The integration of microfluidics and the solid-phase assay format is advantageous due to its relatively low cost, small sample volume, and the ability to detect smaller absolute quantities of material. Multiplexed sensing was realized in this design with an 8:1 contrast ratio for discrimination between perfectly matched and three-base-pair mismatched sequences.

Independently, Noor et al. and Petryayeva et al. developed QD-FRET assays on paper substrates (Noor et al. 2013b; Petryayeva and Algar 2013). Paper has a number of advantages as a solid support as it is able to transport fluid through capillary wicking, it is compatible with many biomolecules, and its fibres can be modified with functional groups amenable to bioconjugation (Noor et al. 2013b). In both cases, cellulose chromatography paper was oxidized and derivatized with surface ligands to immobilize QDs. Noor et al. used imidazole ligands to immobilize GSH-coated CdSeS/ZnS QD525 that had been pre-conjugated with oligonucleotide probes (Noor et al. 2013b). Hybridization assays with Cy3-labeled target permitted detection within minutes with a dynamic range from 300 fmol to 5 pmol (150 nM). The paper platform helped minimize interference from non-complementary DNA with a contrast ratio of 19:1. In contrast, Petryayeva et al. modified paper with thiol ligands to immobilize CdSeS/ZnS QD525 (Petryayeva and Algar 2013). The immobilized QDs were further modified with peptide probes for detecting proteolytic activity. Spots of QDs with peptide probes for different proteases of interest were arrayed on paper substrates for multiplexed detection of up to three target proteases, as well as detection of a two-step enzyme activation cascade. Importantly, the dye/QD PL ratio, which served as the analytical signal, could be detected using a low-cost digital camera or smartphone camera with violet excitation light from a low-power light-emitting diode (LED) (see Fig. 11.18). Further work by this group demonstrated the utility of the paper-based QD-FRET format for protease inhibition assays, and demonstrated that the paper environment enhanced FRET rates by three-fold and dye/QD PL ratios by seven-fold relative to bulk solution (Kim et al. 2014). Overall, the paper substrate combined with QD-FRET is a potentially promising platform for point-of-care tests and diagnostics.

11.3 CRET and BRET-Based Bioanalysis with Quantum Dots

CRET and BRET are analogous to FRET except that the excited state donor is not created through optical excitation but rather through a chemical or biochemical reaction. That is, the donor is chemiluminescent or bioluminescent. QDs are an ideal BRET or CRET acceptor owing to their broad absorption spectra, which generates large spectral overlap. Importantly, the absence of optical excitation in CRET

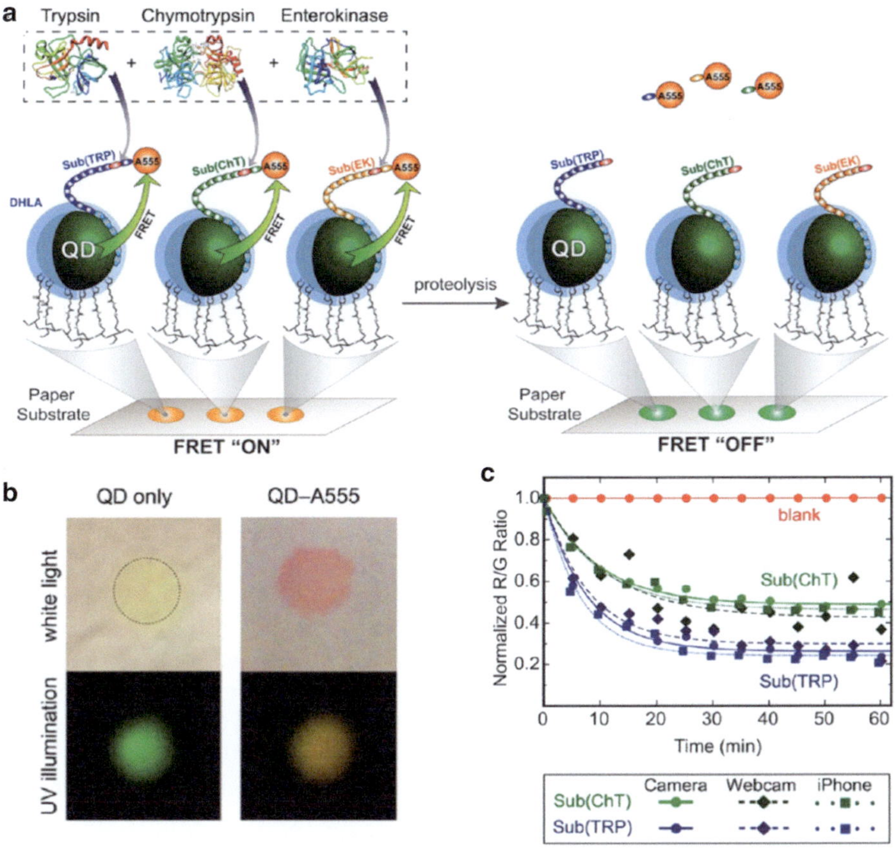

Fig. 11.18 (**a**) Design of QD-FRET, array-based assays for proteolytic enzymes on paper substrates. (**b**) Photographs of immobilized QDs and QD-peptide conjugates (QD–A555) under white light and UV illumination. (**c**) Progress curves, measured as a red/green (A555/QD) emission ratio, for the digestion of peptide substrates by trypsin (*TRP*) and chymotrypsin (*ChT*), measured using a low-cost digital camera, a consumer webcam, and a smartphone camera (iPhone). These measurements were collected through imaging with illumination from a low-cost, low-power violet LED. Adapted with permission from Petryayeva and Algar (2013). Copyright 2013 American Chemical Society

and BRET avoids direct excitation of the QD acceptors. CRET with QD acceptors was first investigated by Huang et al. utilizing a luminol–H_2O_2 system as a CRET donor (Huang et al. 2006). Horseradish peroxidase (HRP) was conjugated to CdTe QDs coated with MPA using carbodiimide coupling. CRET could be observed with several colors of CdTe QD (QD560, QD590, QD620, QD660). Detection of the formation of a rudimentary immunocomplex using QD-CRET was demonstrated using antigen-coated QDs and an HRP-labeled antibody. Examples of CRET-based assays are also listed in Tables 11.1, 11.2, 11.3, 11.4, 11.5, and 11.6.

As an alternative to HRP, the catalytic reactions of some DNAzymes generate chemiluminescence (Kosman and Juskowiak 2011). Willner's group has developed several assays based on a HRP-mimicking DNAzyme. This DNAzyme comprises a Hemin/G-quadruplex unit and catalyzes the chemiluminescent reaction between luminol and H_2O_2. For example, Niazov et al. developed a DNAzyme-based CRET probe for sensing glucose (Niazov et al. 2011). Glucose oxidase (GOx) was conjugated to GSH-coated CdSe/ZnS QD620 using a homobifunctional cross-linker, and the GOx was further conjugated with the DNAzyme using a heterobifunctional cross-linker. In the presence of glucose, the GOx produces H_2O_2, which is consumed in a DNAzyme-catalyzed reaction with luminol, generating chemiluminescence in close proximity to the QD with concomitant CRET. The concentration of glucose controls the amount of H_2O_2 produced by GOx and thus the intensity of CRET-sensitized QD luminescence. This design does not require an external excitation source, provides near-zero background signal, and has a LOD of 5 mM glucose. The design can also be modified for other oxidation reactions or biocatalytic processes that can directly or indirectly produce H_2O_2. Additional work done by Willner's group has shown that catalysis of the luminol–H_2O_2 reaction by the HRP-mimicking hemin/G-quadruplex DNAzyme to produce CRET-sensitized QD emission is useful for the detection of aptamer substrates such as thrombin and ATP, as well as multiple DNA sequences (Freeman et al. 2011b; Liu et al. 2011). For the latter, the DNAzyme sequence was hidden within stem–loop oligonucleotides that opened in the presence of complementary target DNA to enable self-assembly of the hemin/G-quadruplex DNAzyme. The analytical signal was evolution of CRET-sensitized QD emission. Three different colors of CdSe/ZnS QD (QD490, QD560, QD620) were functionalized with stem–loop probes complementary to different target sequences for multiplexed detection (see Fig. 11.19). For the detection of ATP, the DNA sequences were designed so that ATP induced self-assembly of two subunits of a hemin/G-quadruplex to realize DNAzyme function and trigger CRET to proximal acceptor QDs. The assembly of the two subunits was driven by the affinity of the aptamer sequence for ATP (see Fig. 11.19).

Rao's group has pioneered much of the work with QD-BRET systems and assays. These researchers have used QDs as acceptors in BRET for in vivo imaging, sensing protease activity, and the detection of protein–protein interactions. For in vivo imaging, CdSe/ZnS QD655 and CdTe/ZnS QD705 and QD800 were conjugated with a *Renilla* luciferase (RLuc) mutant, Luc8, through carbodiimide coupling (So et al. 2006). The BRET-sensitized, long-wavelength luminescence was within the biological tissue window and offered a signal-to-noise ratio of $>10^3$ for imaging 5 pmol of conjugate. The broad absorption spectra and tunable emission of QDs permitted pairing of multiple colors of QD with Luc8 for multicolor imaging. This group also developed a QD-BRET format for sensing proteolytic activity using either polyhistidine self-assembly (Yao et al. 2007) or intein-mediated conjugation to attach Luc8 to QDs for efficient BRET (Xia et al. 2008). In the former study, the luciferase was expressed with a polyhistidine-terminated peptide linker that doubled as a substrate for matrix metalloproteinase-2 (MMP-2) (Yao et al. 2007). MMP-2 overexpression is strongly correlated with cancer and the detection of its

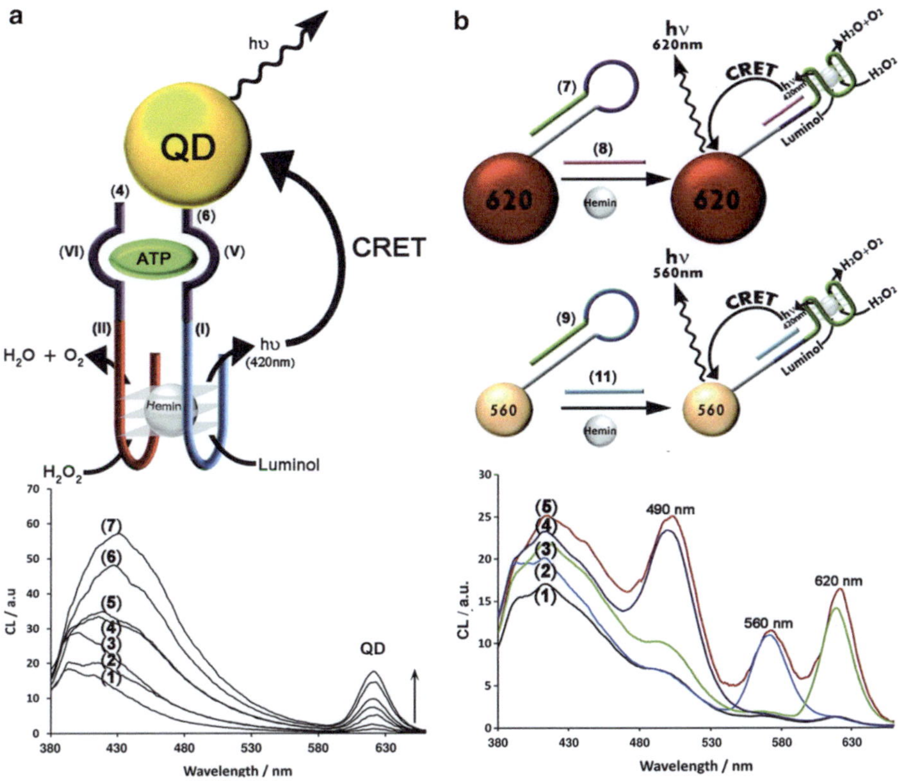

Fig. 11.19 (**a**) Scheme and representative data for CRET-based detection of ATP using QD–aptamer conjugates and a catalytic hemin/G-quadruplex DNAzyme to generate chemiluminescence (CL). The *lower panel* shows changes in emission spectra as the ATP concentration is varied between 0.1 and 100 μM. (**b**) Scheme and representative data for the multiplexed detection of target DNA sequences. QD620, QD560, and QD490 (not shown) are conjugated with different hairpin probes for three-plex detection. The *lower panel* shows emission spectra obtained from samples with (*1*) absence of target, (*2*) target for QD560, (*3*) target for QD620, (*4*) target for QD 490, (*5*) target for all three QDs. Adapted with permission from Freeman et al. (2011b). Copyright 2011 American Chemical Society

activity is of clinical value. In the presence of MMP-2, hydrolysis of the substrate broke the proximity and changed the BRET ratio as a signal of MMP-2 activity and amount (see Fig. 11.20). The LOD of this assay was 2 ng mL^{-1} with a 24-h incubation period and showed good selectivity. A similar assay for proteolytic activity was developed using site-specific intein-mediated chemical ligation for conjugation to QDs (Xia et al. 2008). Signaling was analogous to the previous probe for MMP-2, but here MMP-7 detection was possible down to 5 ng mL^{-1}. The BRET ratio was log-linear with respect to MMP-7 concentration. Multiplexed detection of MMP-2 and uPA activity was also possible using QD655 and QD705 conjugates with Luc8 mutants expressed with a suitable peptide linker. LODs of 1 and 500 ng mL^{-1}, respectively, were achieved.

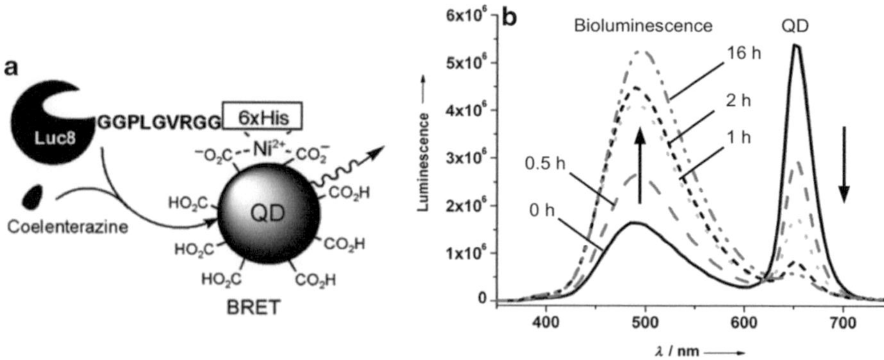

Fig. 11.20 (**a**) Design of a QD-BRET probe for assaying MMP-2 proteolytic activity. (**b**) Changes in emission spectra after addition of QDs to Luc8 fusion proteins after incubation with MMP-2 for 0, 0.5, 1, 2, and 16 h. Adapted with permission from Yao et al. (2007). Copyright 2007 Wiley-VCH Verlag GmbH & Co. KGaA, Weinheim

11.4 Bioanalyses with Energy Transfer Between Quantum Dots and Other Nanomaterials

A variety of energy transfer-based assays involving QDs and other nanomaterials such as gold nanoparticles (AuNPs), carbon nanotubes (CNTs), and graphene oxide (GO) have been reported in the literature. Of these materials, AuNPs are the most widely utilized. An example of the quenching of QDs with AuNPs was noted previously in Sect. 11.2.1.4 for assaying collagenase activity. Several other examples are listed in Tables 11.1, 11.2, 11.3, 11.4, 11.5, and 11.6.

Oh et al. developed one of the first assays based on the quenching of QD PL by AuNPs. QD605 were modified with dextran and 3-nm AuNPs were modified with concanavalin A (ConA), which has high affinity for mannosyl and glucosyl groups (Oh et al. 2006). In the absence of glycosylated protein, the ConA and dextran could bind one another without competition, thus bringing the AuNPs and QDs in close proximity with PL quenching of the latter. Using this assay format, it was possible to measure the degree of chemical glycosylation of BSA (see Fig. 11.21) and differentiate between hyperglycosylated and hypoglycosylated recombinant glucose oxidase from yeast strains. This assay was converted to a heterogeneous format by coupling ConA-modified QD525 or QD605 onto an NHS-activated hydrogel-coated glass slide with binding to dextran-modified AuNPs (Kim et al. 2009b). This group also developed a heterogeneous protease assay with QD525, QD605, and QD655 immobilized on an activated glass slide (Kim et al. 2008). Streptavidin-coated QDs were immobilized in a multicolor array format and conjugated with peptides that were biotinylated at one terminus, labeled at the opposite terminus with 1.4-nm AuNP, and contained an amino-acid sequence recognized and cleaved by either MMP-7, caspase-3, or thrombin. QD PL was quenched by >80 % in the absence of proteolytic activity and the recovery of QD PL could be used to detect as little as 10 ng/mL MMP-7, 20 ng/mL caspase-3, and 1 U/mL of thrombin.

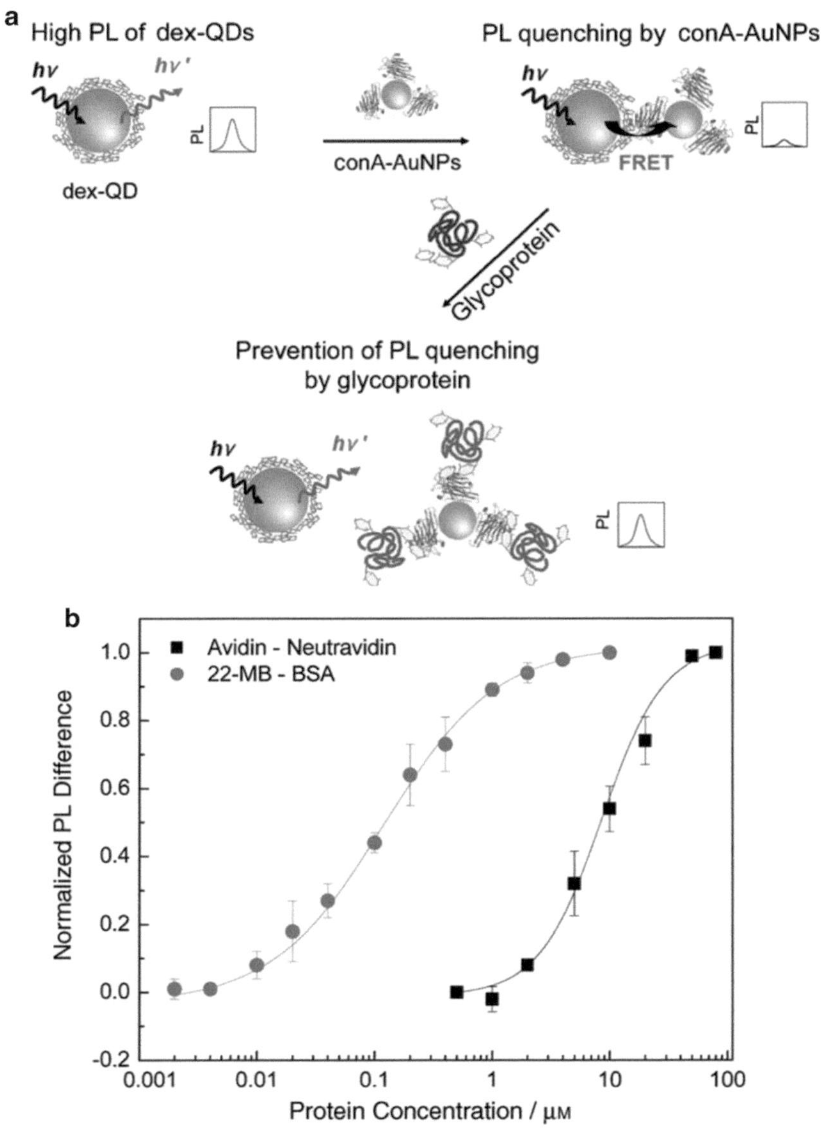

Fig. 11.21 (**a**) Competitive binding assay for measurement of protein glycosylation using QDs and NSET with AuNPs. (**b**) Changes in normalized PL for glycosylated (avidin, 22-MB) versus non-glycosylated proteins (NeutrAvidin, BSA). The normalized PL was calculated as, for example, $(PL_{22\text{-}MB} - PL_{BSA})/(PL_{22\text{-}MB} - PL_{BSA})_{max}$. Adapted with permission from Oh et al. (2006). Copyright 2006 WILEY-VCH Verlag GmbH & Co. KGaA, Weinheim

Li et al. developed a sensitive assay for the detection of mercury (Hg^{2+}) using CdSe/ZnS QD575 modified with a thiol-terminated oligonucleotide probe and a reporter oligonucleotide labeled with a 3-nm AuNP (Li et al. 2011). The probe and reporter were both 10-mers and were complementary except for a central three-base

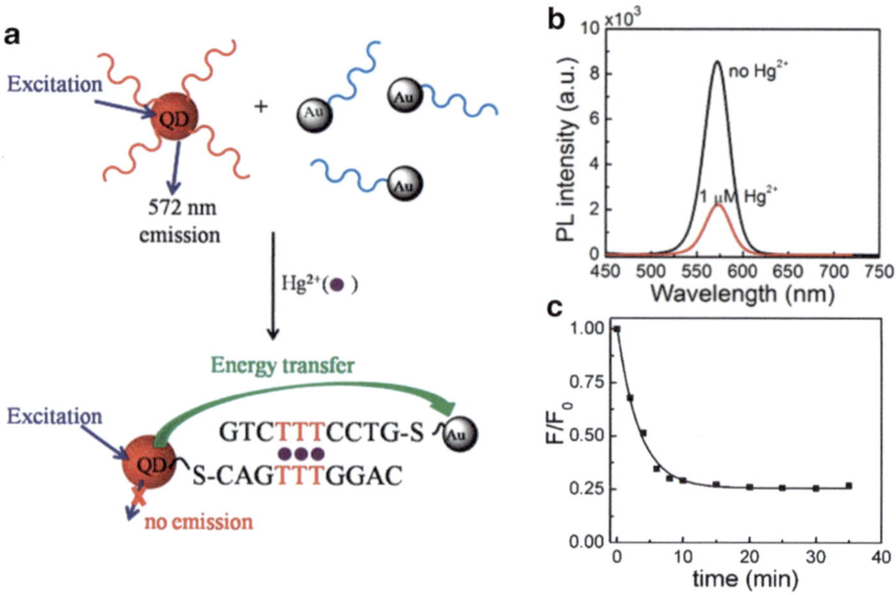

Fig. 11.22 (**a**) QD-NSET method for assaying Hg^{2+} (aq) based on its interaction with thymine residues. (**b**) Quenching of QD PL via NSET in the presence of Hg^{2+} (aq). (**c**) Time-dependent quenching of QD PL after the addition of Hg^{2+} to the probes. Adapted with permission from Li et al. (2011). Copyright 2011 American Chemical Society

repeat of a thymine–thymine mismatch. Only in the presence of Hg^{2+} could these sequences form double-stranded DNA stabilized by thymine–Hg^{2+}–thymine complexes, thus bringing the AuNP and QD in close proximity with quenching of QD PL via a putative NSET mechanism (see Fig. 11.22). As little as 1.2 ppb of Hg^{2+} in river water could be detected using this assay. Kikkeri et al. developed a competitive binding assay for quantifying specific sialoglycans on proteins and in serum (Kikkeri et al. 2013). CdSe/ZnS QD610 were modified with sialic acid binding proteins and AuNPs were modified with sialic acids. Sialoglycans in a sample compete for binding to the sialic acid binding proteins and thus decrease the number QD–AuNP conjugates formed and quenched by NSET, leading to a recovery of QD PL proportional to the concentration of sialoglycan. Detection limits were of the order of micromolar concentrations. Different binding proteins were able to detect different linkages and forms of sialic acids.

Similar to AuNPs, carbon nanomaterials such as GO and CNTs have the ability to quench QD PL through dipolar energy transfer mechanisms, although the precise mechanism may be surface energy transfer (SET) rather than classical FRET. This phenomenon has been used to develop several assays with QDs. For example, Dong et al. modified CdTe QD590 with oligonucleotide hairpin probes (Dong et al. 2010). These conjugates adsorbed to GO and their PL was quenched. Introduction of complementary target and hybridization to form double-stranded DNA resulted in the recovery of QD PL due to a weaker tendency for adsorption and a greater distance between the QD and the GO. These authors also adapted this assay format to the detection of thrombin using thrombin-binding aptamer. In another study, Liu et al. reported a sandwich assay for α-fetoprotein (AFP, a hepatocellular carcinoma biomarker) with capture antibodies on GO and reporter antibodies conjugated to CdTe QD510 (Liu et al. 2010). In the presence of AFP, and with 2 % BSA to mitigate non-specific adsorption, formation of a sandwich immunocomplex brought the QD and GO in close proximity, resulting in quenching of QD PL. The LOD was 0.15 ng/mL of AFP. Multi-walled CNTs have also been used to develop a sandwich immunoassay for breast cancer-associated antigen 1 (BRCAA1) and for multiplexed sandwich hybridization assays. BRCAA1 could be detected at levels as low as 0.4 nM (Cui et al. 2008). The three-plex hybridization assay used CdSe/ZnS QD510, QD555, and QD600 reporters for the detection of target DNA sequences in the range of 0.2–200 pM.

11.5 Charge Transfer-Based Bioanalyses with Quantum Dots

In bioanalysis applications, CT quenching of QD PL is used analogously to FRET. Biorecognition events can alter the distance between a QD and redox-active moiety, resulting in changes in the efficiency of CT quenching. Potential advantages of CT over FRET include greater distance and environmental sensitivity, no requirement for spectral overlap (i.e., a potentially universal quencher, see Fig. 11.23), and the potential to completely eliminate reliance on organic dyes (e.g., through the use of inorganic complexes as quenchers) (Medintz et al. 2009; Aryal and Benson 2006). However, as noted earlier, the precise CT mechanism(s) with aqueous QD bioconjugates is not well established and bioanalysis with QDs and CT quenching is still very much a nascent area of research compared to bioanalysis with QDs and FRET. Some examples of this technique are listed in Tables 11.1, 11.2, 11.3, 11.4, 11.5, and 11.6.

Benson's group has developed CT quenching-based sensing methods with QDs for several different target analytes, including maltose (Sandros et al. 2005, 2006), lead (Shete and Benson 2009), thrombin (Swain et al. 2008), and fatty acids (Aryal and Benson 2006). In their initial work, maltose was targeted using CdSe and CdSe/ZnS QD565–MBP conjugates, where the MBP was site-specifically labeled with a ruthenium(II) phenanthroline complex, [(tetraamine)(5-maleimido-phenanthroline) ruthenium]-$(PF_6)_2$ (Ru-phen), as an electron donor. Upon photoexcitation, the Ru-phen was able to transfer an electron to the QD and quench its PL. With the introduction and subsequent binding of maltose, MBP underwent a conformational change, altering the position of the Ru-phen relative to the QD and decreasing the CT quenching efficiency (see Fig. 11.24). A 22 % increase in QD PL was observed with the addition of maltose and reversible sensing was possible. This format was later adapted to single-particle measurements using ferrocene-labeled MBP

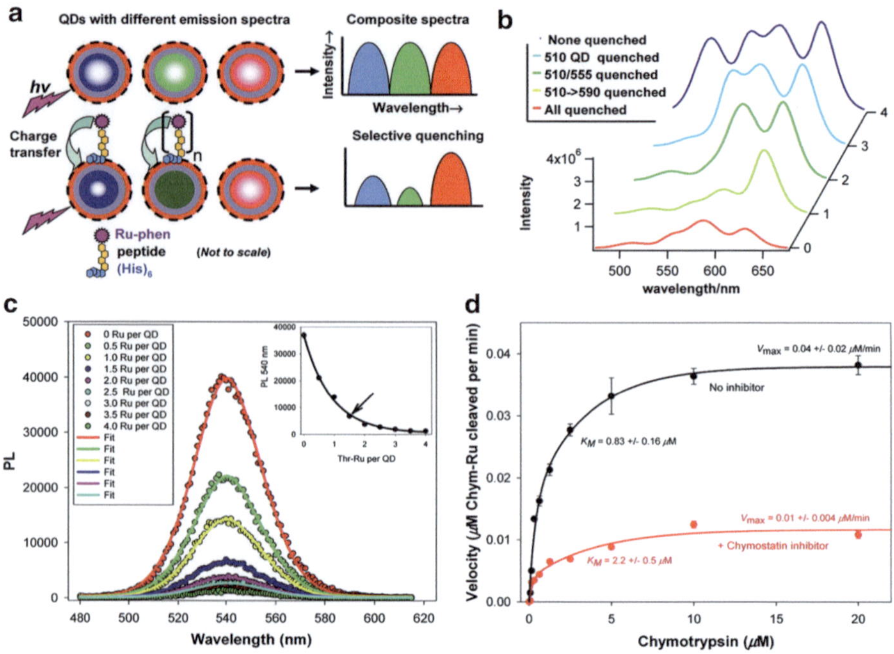

Fig. 11.23 (**a**) Scheme showing the application of Ru-phen-labeled peptides as a universal CT quencher of QD PL. (**b**) Representative spectra show selective quenching of different colors of QD by pre-assembly with Ru-phen-labeled peptide before mixing. Adapted with permission from Medintz et al. (2009). Copyright 2009 American Chemical Society. (**c**) Spectra showing quenching of QD PL by assembly of an increasing amount of Ru-phen-labeled peptide per CdSe/ZnS QD540. (**d**) Assays of chymotrypsin activity and inhibition using recovery of QD PL after proteolytic relief of CT quenching. The activity was measured with and without an inhibitor and analyzed using a classic Michaelis–Menten model. Adapted with permission from Medintz et al. (2008). Copyright 2008 American Chemical Society

immobilized on a glass cover slip, with detection of between 100 pM and 10 μM of maltose (Opperwall et al. 2012). Re-engineering of phosphate-binding protein to bind Pb^{2+} and labeling of the protein with Ru-phen permitted detection of Pb^{2+} based on similar conformational changes with analyte binding (Shete and Benson 2009). A probe for thrombin was developed based on the conformational change of a Ru-phen-labeled aptamer upon binding to thrombin. Thiol-terminated double-stranded DNA with the TBA sequence was self-assembled to QDs and the opposite terminus was labeled with Ru-phen. Binding of thrombin induced a conformational change with concomitant dehybridization of the distal portion of the dsDNA, changing the separation between the Ru-phen and the QDs to alter the CT quenching efficiency, although the direction of this change depended on the number of dsDNA per QD (Swain et al. 2008). While the above examples relied on putative changes in the separation between QDs and Ru-phen to provide an analytical signal, changes in the solvation of Ru-phen have also been reported alter CT quenching efficiency—something that has not been demonstrated with FRET. Aryal et al. suggested that

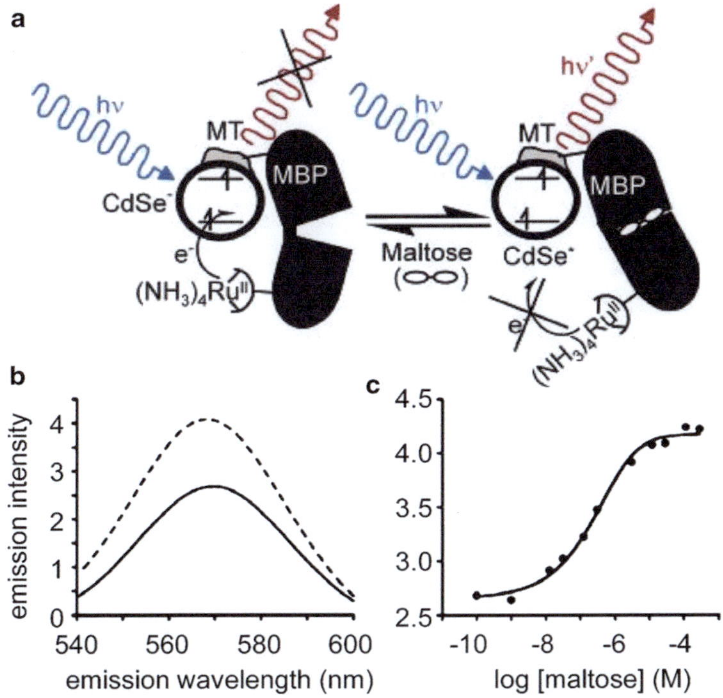

Fig. 11.24 (**a**) Scheme for the detection of maltose through conformational changes in MBP labeled with Ru-phen. The conformational change upon binding reduces the rate of CT to the QD. (**b**) PL emission spectra for the QD–MBP-Ru-phen conjugate without maltose (*solid line*) and with 1 mM maltose (*dashed line*). (**c**) QD PL intensity (a.u.) versus concentration of maltose. The association constant derived from the binding isotherm is 1.9×10^6 M^{-1}. Adapted with permission from Sandros et al. (2005). Copyright 2005 American Chemical Society

changes in solvation were responsible for 32–58 % change in QD PL when palmitate was bound to intestinal fatty acid binding protein labeled with Ru-phen at its binding pocket (Aryal and Benson 2006). This configuration could be used for sensing palmitate.

Medintz et al. have reported CT-based detection of protease activity using Ru-phen-labeled peptides (Medintz et al. 2008). QD PL is quenched from transfer of an electron from the Ru-phen to the QD. Peptide substrates for thrombin and chymotrypsin were appended with a polyhistidine sequence to self-assemble onto different colors of QDs. Ru-phen was conjugated to the peptides at the opposite terminus and an increasing number of labeled peptides per QD progressively quenched QD PL with efficiencies >70 %, where the degree of quenching per Ru-phen depended on the QD to which it was conjugated. With the addition of thrombin and chymotrypsin, these peptides were cleaved and Ru-phen was released from the QD with loss of CT quenching and recovery of QD PL. Protease inhibition assays were also possible in this format (see Fig. 11.23). The possibility of multiplexed sensing using CT-quenching was also evaluated (Medintz et al. 2009).

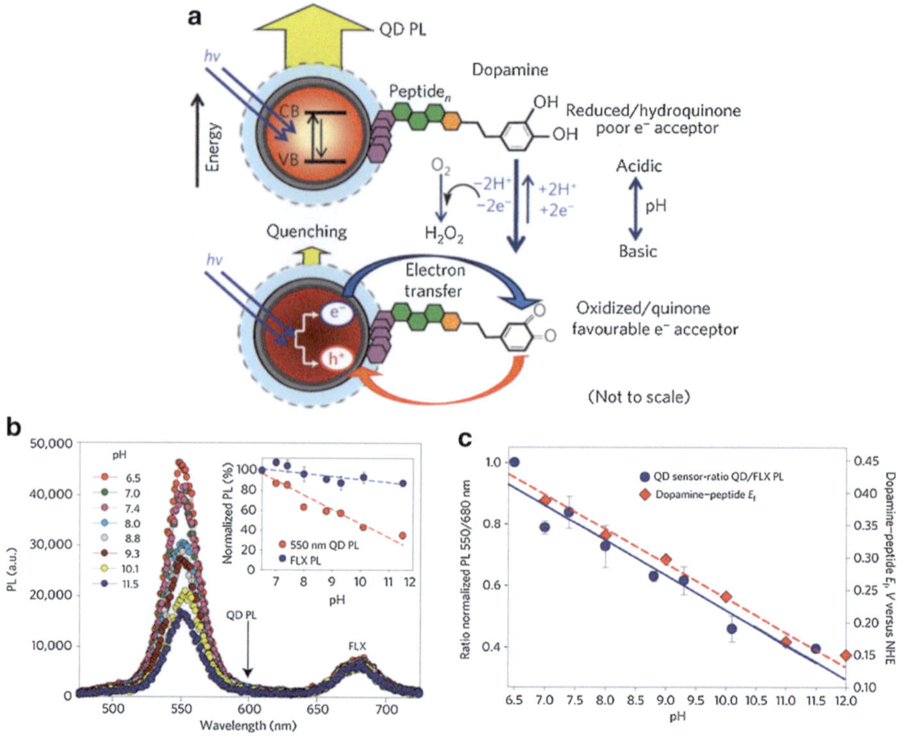

Fig. 11.25 (**a**) Design of a CT-based pH sensor using QDs conjugated with dopamine-modified peptides. (**b**) Change in QD PL as a function of pH in a mixture with pH-insensitive FLX fluorescent nanospheres. (**c**) Plot of QD/FLX PL ratio versus pH. The formal redox potential is also plotted as a function of pH. Adapted with permission from Macmillan Publishers Ltd: Nature Materials (Medintz et al. 2010). Copyright 2010

Multiple colors of QDs emitting across the visible spectrum were quenched simultaneously, but with different efficiencies, by Ru-phen. The quenching of up to eight different colors of QD PL could be resolved using a Gaussian deconvolution algorithm.

Medintz et al. developed an in vitro and *intracellular* pH sensor based on CT between QDs and the oxidized form of dopamine, which functioned as an electron acceptor (Medintz et al. 2010). A peptide labeled with dopamine at its N-terminus and appended with a polyhistidine tag at its C-terminus was self-assembled onto the surface of QDs. The relative amount of oxidized dopamine (quinone) and reduced dopamine (hydroquinone) per QD was strongly pH-dependent. Since only the quinone form, which dominated at basic pH, was a good electron acceptor, the ensemble efficiency of electron transfer quenching varied in proportion to the pH (see Fig. 11.25). At acidic pH, dopamine existed mostly as a hydroquinone, and there was very little CT quenching. The QD–dopamine conjugates exhibited a linear change in PL intensity between pH 6.5 and 11.5 when referenced against pH-sensitive fluorescent microspheres (FLX) as an internal standard. This design was also tested intracellularly where nystatin, an antifungal drug, was used to permeablize COS-1 cells and permit changes in the cytosolic pH by exchanging the extracellular buffer. The QD–dopamine conjugates and FLX internal standard were micro-injected into the cells beforehand and pH changes were detected through the change in the relative QD PL intensity as a readout signal for CT. This format provided real-time tracking of changes in cystosolic pH.

Tomasulo et al. developed a QD-CT-based pH indicator (Tomasulo et al. 2006). [1,3]Oxazine ligands incorporating an electron-rich indole group and a 4-nitrophenylazophenoxy chromophore were able to quench QD PL via either CT or energy transfer, with the former being the dominant mechanism. Under basic conditions, the [1,3]oxazine ring was opened to generate a 4-nitrophenylazophenolate chromophore that (1) had better spectral overlap with the QD PL, improving energy transfer efficiency, and (2) had a more negative oxidation potential, improving charge transfer efficiency. The result was an overall 85 % decrease in the QD PL quantum yield. In contrast, acidic conditions increased the QD PL quantum yield by 33 %. In another example, Ruedas-Rama et al. conjugated QDs with different azamacrocycles (cyclam, cyclen, and 1,4,7-triaza-cyclononane) that could quench QD PL by >80 % through electron transfer (Ruedas-Rama and Hall 2008). When zinc (Zn^{2+}) was introduced and bound by the azamacrocycles, the QD PL was recovered because the electron transfer pathway was blocked by coordination to the Zn^{2+} ion. The LOD was approximately 1–2 μM Zn^{2+} with a dynamic range of almost two orders of magnitude and good selectivity against other physiologically relevant cations.

11.6 Conclusions and Outlook

This chapter has described and illustrated several different bioanalysis formats that combine the unique and advantageous optical properties of QDs with the utility of energy transfer. The examples discussed have included homogeneous assays, heterogeneous assays, and single-molecule assays, both in vitro and in cellular milieu, targeting analytes as diverse as small molecules and drugs, ions, carbohydrates, antigens, enzymes, and nucleic acids. The most widely used energy transfer mechanism has been FRET, although applications of BRET, CRET, NSET, and CT are growing. In each case, the modulation of QD PL by energy transfer is coupled to biorecognition processes that alter the energy transfer efficiency. QD energy transfer is an established field but is nonetheless still rapidly growing. Emerging areas of research and application include the development of bioanalysis probes that are composites of QDs and other nanomaterials with unique properties; the integration of QDs into devices suitable for point-of-care diagnostics and supporting personalized medicine; more sophisticated application of QD probes to the analysis and elucidation of biochemical and cellular processes; and, most challengingly, developing QD probes suitable for in vivo analyses and, ultimately, theranostics, the combination of therapy (e.g., drug delivery) and diagnostics (e.g., biosensing). QDs continue to offer many opportunities for future advances in biosensing and bioanalysis.

References

Algar WR, Krull UJ (2007) Towards multi-colour strategies for the detection of oligonucleotide hybridization using quantum dots as energy donors in fluorescence resonance energy transfer (FRET). Anal Chim Acta 581:193–201

Algar WR, Krull UJ (2009) Toward a multiplexed solid-phase nucleic acid hybridization assay using quantum dots as donors in fluorescence resonance energy transfer. Anal Chem 81: 4113–4120

Algar WR, Krull UJ (2010a) Developing mixed films of immobilized oligonucleotides and quantum dots for the multiplexed detection of nucleic acid hybridization using a combination of fluorescence resonance energy transfer and direct excitation of fluorescence. Langmuir 26:6041–6047

Algar WR, Krull UJ (2010b) FRET-based solid-phase three-color and four-color hybridization assays using mixed films of quantum dots and oligonucleotides. Mater Res Soc Symp Proc 1241:45–55

Algar WR, Krull UJ (2010c) Multiplexed interfacial transduction of nucleic acid hybridization using a single color of immobilized quantum dot donor and two acceptors in fluorescence resonance energy transfer. Anal Chem 82:400–405

Algar WR, Krull UJ (2011) Interfacial chemistry and the design of solid-phase nucleic acid hybridization assays using immobilized quantum dots as donors in fluorescence resonance energy transfer. Sensors 11:6214–6236

Algar WR, Tavares AJ, Krull UJ (2010) Beyond labels: a review of the application of quantum dots as integrated components of assays, bioprobes, and biosensors utilizing optical transduction. Anal Chim Acta 673:1–25

Algar WR, Ancona MG, Malanoski AP, Susumu K, Medintz IL (2012a) Assembly of a concentric Förster resonance energy transfer relay on a quantum dot scaffold: characterization and application to multiplexed protease sensing. ACS Nano 6:11044–11058

Algar WR, Malanoski A, Deschamps JR, Blanco-Canosa JB, Susumu K, Stewart MH, Johnson BJ, Dawson PE, Medintz IL (2012b) Proteolytic activity at quantum dot-conjugates: kinetic analysis reveals enhanced enzyme activity and localized interfacial "hopping". Nano Lett 12:3793–3802

Algar WR, Malanoski AP, Susumu K, Stewart MH, Hildebrandt N, Medintz IL (2012c) Multiplexed tracking of protease activity using a single color of quantum dot vector and a time-gated Förster resonance energy transfer relay. Anal Chem 84:10136–10146

Algar WR, Wegner D, Huston AL, Blanco-Canosa JB, Stewart MH, Armstrong A, Dawson PE, Hildebrandt N, Medintz IL (2012d) Quantum dots as simultaneous acceptors and donors in time-gated Forster resonance energy transfer relays: characterization and biosensing. J Am Chem Soc 134:1876–1891

Algar WR, Kim H, Medintz IL, Hildebrandt N (2013) Emerging non-traditional Förster resonance energy transfer configurations with semiconductor quantum dots; Investigations and applications. Coord Chem Rev 263–264:65–85

Aryal BP, Benson DE (2006) Electron donor solvent effects provide biosensing with quantum dots. J Am Chem Soc 128:15986–15987

Bagalkot V, Zhang L, Levy-Nissenbaum E, Jon S, Kantoff PW, Langer R, Farokhzad OC (2007) Quantum dot–aptamer conjugates for synchronous cancer imaging, therapy, and sensing of drug delivery based on bi-fluorescence resonance energy transfer. Nano Lett 7:3065–3070

Bailey VJ, Easwaran H, Zhang Y, Griffiths E, Belinsky SA, Herman JG, Baylin SB, Carraway HE, Wang TH (2009) MS-qFRET: a quantum dot-based method for analysis of DNA methylation. Genome Res 19:1455–1461

Bakalova R, Zhelev Z, Ohba H, Baba Y (2005) Quantum dot-conjugated hybridization probes for preliminary screening of siRNA sequences. J Am Chem Soc 127:11328–11335

Boeneman K, Mei BC, Dennis AM, Bao G, Deschamps JR, Mattoussi H, Medintz IL (2009) Sensing caspase 3 activity with quantum dot-fluorescent protein assemblies. J Am Chem Soc 131:3828–3829

Chang E, Miller JS, Sun J, Yu WW, Colvin VL, Drezek R, West JL (2005) Protease-activated quantum dot probes. Biochem Biophys Res Commun 334:1317–1321

Charbonnière LJ, Hildebrandt N (2008) Lanthanide complexes and quantum dots: a bright wedding for resonance energy transfer. Eur J Inorg Chem 2008:3241–3251

Chen Z, Li G, Zhang L, Jiang JF, Li Z, Peng ZH, Deng L (2008) A new method for the detection of ATP using a quantum-dot-tagged aptamer. Anal Bioanal Chem 392:1185–1188

Chen L, Zhang X, Zhou G, Xiang X, Ji X, Zheng Z, He Z, Wang H (2012a) Simultaneous determination of human enterovirus 71 and coxsackievirus B3 by dual-color quantum dots and homogeneous immunoassay. Anal Chem 84:3200–3207

Chen MJ, Wu YS, Lin GF, Hou JY, Li M, Liu TC (2012b) Quantum-dot-based homogeneous time-resolved fluoroimmunoassay of alpha-fetoprotein. Anal Chim Acta 741:100–105

Chi CW, Lao YH, Li YS, Chen LC (2011) A quantum dot-aptamer beacon using a DNA intercalating dye as the FRET reporter: Application to label-free thrombin detection. Biosens Bioelectron 26:3346–3352

Cissell KA, Campbell SA, Deo SK (2008) Rapid, single-step nucleic acid detection. Anal Bioanal Chem 391:2577–2581

Clapp AR, Medintz IL, Mattoussi H (2006) Förster resonance energy transfer investigation using quantum-dot fluorophores. Chem Phys Chem 7:47–57

Cui D, Pan B, Zhang H, Gao F, Wu R, Wang J, He R, Asahi T (2008) Self-assembly of quantum dots and carbon nanotubes for ultrasensitive DNA and antigen detection. Anal Chem 80:7996–8001

Dennis AM, Rhee WJ, Sotto D, Dublin SN, Bao G (2012) Quantum dot-fluorescent protein FRET probes for sensing intracellular pH. ACS Nano 6:2917–2924

Dong HF, Gao WC, Yan F, Ji HX, Ju HX (2010) Fluorescence resonance energy transfer between quantum dots and graphene oxide for sensing biomolecules. Anal Chem 82:5511–5517

Freeman R, Bahshi L, Finder T, Gill R, Willner I (2009a) Competitive analysis of saccharides or dopamine by boronic acid-functionalized CdSe-ZnS quantum dots. Chem Commun 764–766

Freeman R, Li Y, Tel-Vered R, Sharon E, Elbaz J, Willner I (2009b) Self-assembly of supramolecular aptamer structures for optical or electrochemical sensing. Analyst 134:653–656

Freeman R, Finder T, Gill R, Willner I (2010) Probing protein kinase (CK2) and alkaline phosphatase with CdSe/ZnS quantum dots. Nano Lett 10:2192–2196

Freeman R, Liu X, Willner I (2011a) Amplified multiplexed analysis of DNA by the exonuclease III-catalyzed regeneration of the target DNA in the presence of functionalized semiconductor quantum dots. Nano Lett 11:4456–4461

Freeman R, Liu XQ, Willner I (2011b) Chemiluminescent and chemiluminescence resonance energy transfer (CRET) detection of DNA, metal ions, and aptamer-substrate complexes using hemin/G-quadruplexes and CdSe/ZnS quantum dots. J Am Chem Soc 133:11597–11604

Gill R, Willner I, Shweky I, Banin U (2005) Fluorescence resonance energy transfer in CdSe/ZnS-DNA conjugates: probing hybridization and DNA cleavage. J Phys Chem B 109: 23715–23719

Goldman ER, Medintz IL, Whitley JL, Hayhurst A, Clapp AR, Uyeda HT, Deschamps JR, Lassman ME, Mattoussi H (2005) A hybrid quantum dot-antibody fragment fluorescence resonance energy transfer-based TNT sensor. J Am Chem Soc 127:6744–6751

Grigsby CL, Ho YP, Leong KW (2012) Understanding nonviral nucleic acid delivery with quantum dot-FRET nanosensors. Nanomedicine 7:565–577

Hermann T, Patel DJ (2000) Biochemistry-adaptive recognition by nucleic acid aptamers. Science 287:820–825

Huang X, Li L, Qian H, Ding C, Ren J (2006) A resonance energy transfer between chemiluminescent donors and luminescent quantum dots as acceptors. Angew Chem Int Ed 45:5140–5143

Huang S, Xiao Q, He ZK, Liu Y, Tinnefeld P, Su XR (2008) A high sensitive and specific QDs FRET bioprobe for MNase. Chem Commun 5990–5992

Hunter T (1995) Protein kinases and phosphatases: the Yin and Yang of protein phosphorylation and signaling. Cell 80:225–236

Kattke MD, Gao EJ, Sapsford KE, Stephenson LD, Kumar A (2011) FRET-based quantum dot immunoassay for rapid and sensitive detection of *Aspergillus amstelodami*. Sensors 11:6396–6410

Keefe AD, Pai S, Ellington AD (2010) Aptamers as therapeutics. Nat Drug Discovery Rev 9:537–550

Kikkeri R, Padler-Karavani V, Diaz S, Verhagen A, Yu H, Cao HZ, Langereis MA, Groot RJD, Chen X, Varkit A (2013) Quantum dot nanometal surface energy transfer based biosensing of sialic acid compositions and linkages in biological samples. Anal Chem 85:3864–3870

Kim GB, Kim YP (2012) Analysis of protease activity using quantum dots and resonance energy transfer. Theranostics 2:127–138

Kim JH, Morikis D, Ozkan M (2004) Adaptation of inorganic quantum dots for stable molecular beacons. Sens Actuators B 102:315–319

Kim YP, Oh YH, Oh E, Ko S, Han MK, Kim HS (2008) Energy transfer-based multiplexed assay of proteases by using gold nanoparticle and quantum dot conjugates on a surface. Anal Chem 80:4634–4641

Kim GI, Kim KW, Oh MK, Sung YM (2009a) The detection of platelet derived growth factor using decoupling of quencher-oligonucleotide from aptamer/quantum dot bioconjugates. Nanotechnol 20:175503

Kim YP, Park S, Oh E, Oh YH, Kim HS (2009b) On-chip detection of protein glycosylation based on energy transfer between nanoparticles. Biosens Bioelectron 24:1189–1194

Kim H, Petryayeva E, Algar WR (2014) Enhancement of quantum dot Förster resonance energy transfer with paper matrices and application to proteolytic assays. IEEE J Sel Top Quant Electron 20:7300211

Kosman J, Juskowiak B (2011) Peroxidase-mimicking DNAzymes for biosensing applications: a review. Anal Chim Acta 707:7–17

Levy M, Cater SF, Ellington AD (2005) Quantum-dot aptamer beacons for the detection of proteins. Chem Bio Chem 6:2163–2166

Li M, Wang QY, Shi XD, Hornak LA, Wu NQ (2011) Detection of mercury(II) by quantum dot/DNA/gold nanoparticle ensemble based nanosensor via nanometal surface energy transfer. Anal Chem 83:7061–7065

Liang GX, Pan HC, Li Y, Jiang LP, Zhang JR, Zhu JJ (2009) Near infrared sensing based on fluorescence resonance energy transfer between Mn:CdTe quantum dots and Au nanorods. Biosens Bioelectron 24:3693–3697

Liu J, Lee JH, Lu Y (2007) Quantum dot encoding of aptamer-linked nanostructures for one-pot simultaneous detection of multiple analytes. Anal Chem 79:4120–4125

Liu M, Zhao H, Quan X, Chen S, Fan X (2010) Distance-independent quenching of quantum dots by nanoscale-graphene in self-assembled sandwich immunoassay. Chem Commun 46: 7909–7911

Liu XQ, Freeman R, Golub E, Willner I (2011) Chemiluminescence and chemiluminescence resonance energy transfer (CRET) aptamer sensors using catalytic hemin/G-quadruplexes. ACS Nano 5:7648–7655

Liu BY, Zeng F, Wu GF, Wu SZ (2012) Nanoparticles as scaffolds for FRET-based ratiometric detection of mercury ions in water with QDs as donors. Analyst 137:3717–3724

Long F, Gu CM, Gu AZ, Shi HC (2012a) Quantum dot/carrier-protein/haptens conjugate as a detection nanobioprobe for FRET-based immunoassay of small analytes with all-fiber microfluidic biosensing platform. Anal Chem 84:3646–3653

Long Y, Zhang LF, Zhang Y, Zhang CY (2012b) Single quantum dot based nanosensor for renin assay. Anal Chem 84:8846–8852

López-Otín C, Overall CM (2002) Protease degradomics: a new challenge for proteomics. Nat Rev Mol Cell Biol 3:509–519

Lowe SB, Dick JAG, Cohen BE, Stevens MM (2012) Multiplex sensing of protease and kitnase enzyme activity via orthogonal coupling of quantum dot peptide conjugates. ACS Nano 6:851–857

McLaurin EJ, Greytak AB, Bawendi MG, Nocera DG (2009) Two-photon absorbing nanocrystal sensors for ratiometric detection of oxygen. J Am Chem Soc 131:12994–13001

Medintz IL, Mattoussi H (2009) Quatum dot-based resonance energy transfer and its growing application in biology. Phys Chem Chem Phys 11

Medintz IL, Clapp AR, Mattoussi H, Goldman ER, Fisher B, Mauro JM (2003) Self-assembled nanoscale biosensors based on quantum dot FRET donors. Nat Mater 2:630–638

Medintz IL, Clapp AR, Melinger JS, Deschamps JR, Mattoussi H (2005) A reagentless biosensing assembly based on quantum dot-donor Forster resonance energy transfer. Adv Mater 17: 2450–2455

Medintz IL, Clapp AR, Brunel FM, Tiefenbrunn T, Uyeda HT, Chang EL, Deschamps JR, Dawson PE, Mattoussi H (2006a) Proteolytic activity monitored by fluorescence resonance energy transfer through quantum-dot-peptide conjugates. Nat Mater 5:581–589

Medintz IL, Sapsford KE, Clapp AR, Pons T, Higashiya S, Welch JT, Mattoussi H (2006b) Designer variable repeat length polypeptides as scaffolds for surface immobilization of quantum dots. J Phys Chem B 110:10683–10690

Medintz IL, Berti L, Pons T, Grimes AF, English DS, Alessandrini A, Facci P, Mattoussi H (2007) A reactive peptidic linker for self-assembling hybrid quantum dot-DNA bioconjugates. Nano Lett 7:1741–1748

Medintz IL, Pons T, Trammell SA, Grimes AF, English DS, Blanco-Canosa JB, Dawson PE, Mattoussi H (2008) Interactions between redox complexes and semiconductor quantum dots coupled via a peptide bridge. J Am Chem Soc 130:16745–16756

Medintz IL, Farrell D, Susumu K, Trammell SA, Deschamps JR, Brunel FM, Dawson PE, Mattoussi H (2009) Multiplex charge-transfer interactions between quantum dots and peptide-bridged ruthenium complexes. Anal Chem 81:4831–4839

Medintz IL, Stewart MH, Trammell SA, Susumu K, Delehanty JB, Mei BC, Melinger JS, Blanco-Canosa JB, Dawson PE, Mattoussi H (2010) Quantum-dot/dopamine bioconjugates function as redox coupled assemblies for in vitro and intracellular pH sensing. Nat Mater 9:676–684

Niazov A, Freeman R, Girsh J, Willner I (2011) Following glucose oxidase activity by chemiluminescence and chemiluminescence resonance energy transfer. Sensors 11:10388–10397

Nikiforov TT, Beechem JM (2006) Development of homogeneous binding assays based on fluorescence resonance energy transfer between quantum dots and Alexa Fluor fluorophores. Anal Biochem 357:68–76

Noor MO, Petryayeva E, Tavares AJ, Uddayasankar U, Algar WR, Krull UJ (2013a) Building from the "Ground" up: developing interfacial chemistry for solid-phase nucleic acid hybridization assays based on quantum dots and fluorescence resonance energy transfer. Coord Chem Rev 263–264:25–52

Noor MO, Shahmuradyan A, Krull UJ (2013b) Paper-based solid-phase nucleic acid hybridization assay using immobilized quantum dots as donors in fluorescence resonance energy transfer. Anal Chem 85:1860–1867

Noor MO, Tavares AJ, Krull UJ (2013c) On-chip multiplexed solid-phase nucleic acid hybridization assay using spatial profiles of immobilized quantum dots and fluorescence resonance energy transfer. Anal Chim Acta 788

Oh E, Lee D, Kim YP, Cha SY, Oh DB, Kang HA, Kim J, Kim HS (2006) Nanoparticle-based energy transfer for rapid and simple detection of protein glycosylation. Angew Chem Int Ed 45:7959–7963

Opperwall SR, Divakaran A, Porter EG, Christians JA, DenHartigh AJ, Benson DE (2012) WIde dynamic range sensing with single quantum dot biosensors. ACS Nano 6:8078–8086
Petryayeva E, Algar WR (2013) Proteolytic assays on quantum-dot-modified paper substrates using simple optical readout platforms. Anal Chem 85:8817–8825
Petryayeva E, Algar WR, Krull UJ (2013) Adapting fluorescence resonance energy transfer with quantum dot donors for solid-phase hybridization assays in microtiter plate format. Langmuir 29:977–987
Pons T, Medintz IL, Wang X, English DS, Mattoussi H (2006) Solution-phase single quantum dot fluorescence resonance energy transfer. J Am Chem Soc 128:15324–15331
Prasuhn DE, Feltz A, Blanco-Canosa JB, Susumu K, Stewart MH, Mei BC, Yakovlev AV, Loukov C, Mallet JM, Oheim M, Dawson PE, Medintz IL (2010) Quantum dot peptide biosensors for monitoring caspase 3 proteolysis and calcium ions. ACS Nano 4:5487–5497
Ruedas-Rama MJ, Hall EA (2008) Azamacrocycle activated quantum dot for zinc ion detection. Anal Chem 80:8260–8268
Ruedas-Rama MJ, Hall EAH (2009) Multiplexed energy transfer mechanisms in a dual-function quantum dot for zinc and manganese. Analyst 134:159–169
Sandros MG, Gao D, Benson DE (2005) A modular nanoparticle-based system for reagentless small molecule biosensing. J Am Chem Soc 127:12198–12199
Sandros MG, Shete V, Benson DE (2006) Selective, reversible, reagentless maltose biosensing with core-shell semiconducting nanoparticles. Analyst 131:229–235
Sapsford KE, Medintz IL, Golden JP, Deschamps JR, Uyeda HT, Mattoussi H (2004) Surface-immobilized self-assembled protein-based quantum dot nanoassemblies. Langmuir 20:7720–7728
Sapsford KE, Granek J, Deschamps JR, Boeneman K, Blanco-Canosa JB, Dawson PE, Susumu K, Stewart MH, Medintz IL (2011) Monitoring botulinum neurotoxin A activity with peptide-functionalized quantum dot resonance energy transfer sensors. ACS Nano 5:2687–2699
Shete VS, Benson DE (2009) Protein design provides lead(II) ion biosensors for imaging molecular fluxes around red blood cells. Biochemistry 48:462–470
Shi LF, De Paoli V, Rosenzweig N, Rosenzweig Z (2006) Synthesis and application of quantum dots FRET-based protease sensors. J Am Chem Soc 128:10378–10379
Shin S, Nam HY, Lee EJ, Jung W, Hah SS (2012) Molecular beacon-based quantitation of epithelial tumor marker mucin 1. Bioorg Med Chem Lett 22:6081–6084
Snee PT, Somers RC, Nair G, Zimmer JP, Bawendi MG, Nocera DG (2006) A ratiometric CdSe/ZnS nanocrystal pH sensor. J Am Chem Soc 128:13320–13321
So MK, Xu C, Loening AM, Gambhir SS, Rao J (2006) Self-illuminating quantum dot conjugates for in vivo imaging. Nat Biotechnol 24:339–343
Somers RC, Lanning RM, Snee PT, Greytak AB, Jain RK, Bawendi MG, Nocera DG (2012) A nanocrystal-based ratiometric pH sensor for natural pH ranges. Chem Sci 3:2980–2985
Stringer RC, Hoehn D, Grant SA (2008) Quantum dot-based biosensor for detection of human cardiac troponin I using a liquid-core waveguide. IEEE Sens J 8:295–300
Sugawa M, Nishikawa S, Iwane AH, Biju V, Yanagida T (2010) Single-molecule FRET imaging for enzymatic reactions at high ligand concentrations. Small 6:346–350
Swain MD, Octain J, Benson DE (2008) Unimolecular, soluble semiconductor nanoparticle-based biosensors for thrombin using charge/electron transfer. Bioconjug Chem 19:2520–2526
Tang B, Lao L, Xu K, Zhuo L, Ge J, Li Q, Yu L (2008) A new nanobiosensor for glucose with high sensitivity and selectivity in serum based on fluorescence resonance energy transfer (FRET) between CdTe quantum dots and Au nanoparticles. Chem Eur J 14:3637–3644
Tomasulo M, Yildiz I, Kaanumalle SL, Raymo FM (2006) pH-sensitive ligand for luminescent quantum dots. Langmuir 22:10284–10290
Tombelli S, Minunni M, Mascini M (2005) Analytical applications of aptamers. Biosens Bioelectron 2005:2424–2434
Tyagi S, Kramer FR (1996) Molecular beacons: probes that fluoresce upon hybridization. Nat Biotechnol 14:303–308

Vannoy CH, Chong L, Le C, Krull UJ (2013) A competitive displacement assay with quantum dots as fluorescence resonance energy transfer donors. Anal Chim Acta 759:92–99

Wegner KD, Jin Z, Lindén S, Jennings TL, Hildebrandt N (2013) Quantum-dot-based förster resonance energy transfer immunoassay for sensitive clinical diagnostics of low-volume serum samples. ACS Nano 7:7411–7419

Wei Q, Lee M, Yu X, Lee EK, Seong GH, Choo J, Cho YW (2006) Development of an open sandwich fluoroimmunoassay based on fluorescence resonance energy transfer. Anal Biochem 358:31–37

Whitney M, Svariar EN, Friedman B, Levin RA, Crisp JL, Glasgow HL, Lefkowitz R, Adams SR, Steinbach P, Nashi N, Nguyen QT, Tsien RY (2013) Ratiometric activatable cell-penetrating peptides provide rapid in vivo readout of thrombin activation. Angew Chem Int Ed 52:325–330

Wilson DS, Szostak JW (1999) In vitro selection of functional nucleic acids. Annu Rev Biochem 68:611–647

Wu CS, Oo MKK, Fan X (2010) Highly sensitive multiplexed heavy metal detection using quantum-dot-labeled DNAzymes. ACS Nano 4:5897–5904

Xia Z, Xiang Y, So MK, Koh AL, Sinclair R, Rao J (2008) Multiplex detection of protease activity with quantum dot nanosensors prepared by intein-mediated specific bioconjugation. Anal Chem 80:8649–8655

Xu L, Zhu Y, Ma W, Kuang H, Liu L, Wang L, Xu C (2011) Sensitive and specific DNA detection based on nicking endonuclease-assisted fluorescence resonance energy transfer amplification. J Phys Chem C 115:16315–16321

Yao H, Zhang Y, Xiao F, Xia Z, Rao J (2007) Quantum dot/bioluminescence resonance energy transfer based highly sensitive detection of proteases. Angew Chem Int Ed 46:4346–4349

Zhang CY, Hu J (2010) Single quantum dot-based nanosensor for multiple DNA detection. Anal Chem 82:1921–1927

Zhang CY, Johnson LW (2006) Quantum dot-based fluorescence resonance energy transfer with improved FRET efficiency in capillary flows. Anal Chem 78:5532–5537

Zhang CY, Johnson LW (2007) Quantifying RNA-eptide interaction by single-quantum dot-based nanosensor: an approach fair drug screening. Anal Chem 79:7775–7781

Zhang CY, Johnson LW (2009) Single quantum-dot-based aptameric nanosensor for cocaine. Anal Chem 81:3051–3055

Zhang Y, Zhang CY (2012) Sensitive detection of microRNA with isothermal amplification and a single-quantum-dot-based nanosensor. Anal Chem 84:224–231

Zhang CY, Yeh HC, Kuroki MT, Wang TH (2005) Single-quantum-dot-based DNA nanosensor. Nat Mater 4:826–831

Zhou D, Ying L, Hong X, Hall EA, Abell C, Klenerman D (2008) A compact functional quantum dot-DNA conjugate: preparation, hybridization, and specific label-free DNA detection. Langmuir 24:1659–1664

Chapter 12
Nanoparticle-Based Detection of Protein Phosphorylation

Sanela Martic and Heinz-Bernhard Kraatz

Abstract Protein kinases are critical catalytic proteins in humans and regulate a huge number of biological functions and cellular transformations. Understanding the role and mechanism of protein kinases is of major interest. Moreover, hyperactive protein kinases have been identified in certain diseases, which has triggered the search for protein kinase inhibitors. Since protein kinases are critical and "druggable" targets, alternative methodologies for monitoring protein kinase activity and inhibitor screening are needed. The relatively old techniques for kinome analysis, immunoassay and radiography, have been replaced by new optical and electrochemical methods, some of which will be discussed here. In this chapter, the focus is on the development of nanoparticle-based strategies for efficient monitoring of protein kinase activity and inhibition.

Keywords ATP • Nanoparticles • Protein kinase

12.1 Introduction

Protein phosphorylations by protein kinases drive the cellular signaling pathways and are involved in all stages of cell life: division, proliferation, etc. (Hunter 2000; Manning et al. 2002). Protein kinases catalytically transfer a phosphate group from adenosine triphosphate to a substrate, typically a protein with Ser, Thr, or Tyr groups. The phosphorylation of the substrate may trigger a change in its electrostatic profile, conformation, or activity. For example, the phosphorylation of STAT3

S. Martic (✉)
Department of Chemistry, Oakland University, 2200 North Squirrel Road, Rochester, MI 48309, USA
e-mail: martic@oakland.edu

H.-B. Kraatz
Department of Chemistry, University of Toronto, 80 St. George Street, Toronto, Canada M5S 3H6

Department of Physical and Environmental Sciences, University of Toronto Scarborough, 1265 Military Trail, Toronto, ON M1C 1A, Canada
e-mail: bernie.kraatz@utoronto.ca

M.C. Vestergaard et al. (eds.), *Nanobiosensors and Nanobioanalyses*,
DOI 10.1007/978-4-431-55190-4_12

by sarcoma-related kinase induces the formation of the dimer from two pSTAT3 molecules (Bromberg et al. 1999; Cimica et al. 2012). Once phosphorylated, pSTAT3 dimerizes and becomes activated. The aforementioned activation causes the signaling cascade effect downstream in the kinase pathway. In another example, the phosphorylation of tau protein substrate, which is related to neurobiological disorders, by glycogen synthase kinase causes a change in its conformation and promotes its self-association into toxic neurofibrillary tangles and filaments (Rankin et al. 2007). The phosphorylation reaction is relatively simple and involves three distinct partners, all of which are required for the post-translation to take place: protein kinase, substrate, and co-substrate as depicted in Fig. 12.1.

The dephosphorylation is carried out by protein phosphatases and the balance between kinases/phosphatases is biologically important and critical for normal function.

The most commonly used analytical method for monitoring protein kinase activity is the radiolabeled assay based on ^{32}P-ATP (Houseman et al. 2002). Recent advances have been made towards the screening of protein kinases, and involve the use of specific antiphospho-antibodies (Umezava 2005), fluorescently labeled peptide substrates or kinase (Schultz et al. 2005) thiol (Allen et al. 2007), or ferrocene conjugates (Martic et al. 2013). Alternative label-free methods for the detection of protein kinase activity and inhibitor screening are of interest because they eliminate the need for chemical derivatization of protein kinase, substrate, or co-substrate. In this chapter, the label-free approaches are discussed as they pertain to the use of nanoparticles (NPs) for kinome studies. This chapter will systematically cover the use of NPs for kinome analysis, but will not include the use of NPs for phosphopeptide enrichment for mass spectrometry analysis (Grimsrud et al. 2010).

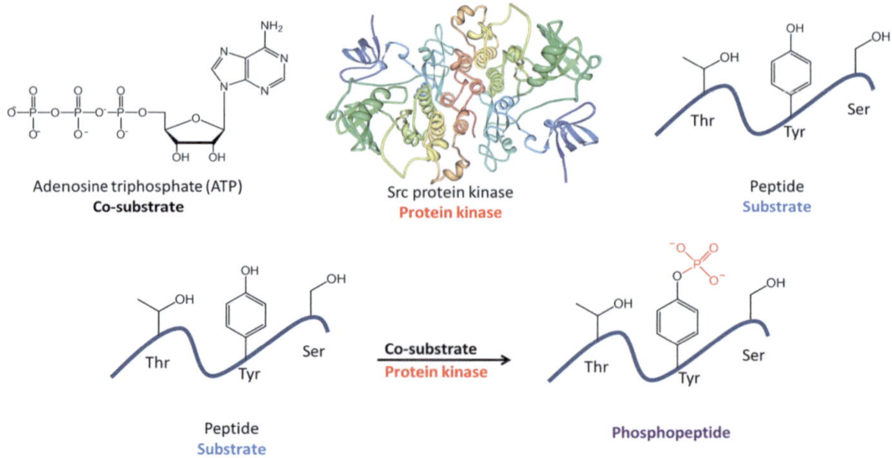

Fig. 12.1 Illustration of the co-substrate, protein kinase and substrate required for the kinase-catalyzed phosphorylation reaction. The overall reaction depicts phosphorylation of Tyr residue by a given kinase (reprinted with permission)

12.2 Biochemical Role of Protein Kinase Phosphorylations

More than 500 protein kinases regulate metabolic and cellular pathways by phosphorylation of a substrate and represent almost 2 % of total proteins in our bodies. The growth factors which are involved in the signal transduction pathways are highly dependent on their receptors, which are composed of extracellular, transmembrane, and cytoplasmic tyrosine kinase domains. The growth receptor binding to the kinase domain activates that kinase, a process which is strictly regulated in normal cells. However, dysregulation and overexpression or overactivation of receptor tyrosine kinases has been linked to wide range of cancers. The overexpression or upregulation of certain kinases has been linked to diseases, such as cancer, inflammation, and diabetes (Manning et al. 2002). Other tyrosine kinases may not be receptors, and they relay intracellular signals. The complexity of the kinase activity and relationships which drive normal cellular function is shown in Fig. 12.2.

12.3 Protein Kinase Inhibition

Since protein kinase activity has been linked to certain diseases, it is not surprising that the identification of potential kinase inhibitors is driving kinome therapeutics (Davies et al. 2000; Takeuchi and Ito 2011). For example, cancer targeting of the receptor tyrosine kinase is achieved via targeting those kinases which are major regulators of cancer cell survival. Hence, in breast cancer, over-expressed HER2 occurs in ~25 % of patients and is a viable target for drug development. In addition, epidermal growth factor receptor (EGFR) is widely up-regulated in solid tumors and is another viable target. The drug targeting approach towards inhibition of the protein kinases is divided into (1) small molecule inhibitors and (2) antibody therapies. The latter approach is specific to receptor tyrosine kinases, because therapeutic antibody binds the specific kinase and prevents the growth hormone from activating it (Grant 2009). The immunotherapies based on trastuzumab, cetuximab and panitumumab work for HER2 and EGFR protein kinases. Small-molecule drug targeting has been successful for other protein kinases like Abl and Src, and includes imatinib and nilotinib, among others (Fig. 12.3).

Currently, a number of ATP-competitive inhibitors are commercially available: imatinib, nilotinib, and ruxolitinib (Medves and Demoulin 2012). The drug target screening for protein kinases is complicated due to the non-specific inhibitor binding domain which is an ATP-binding site. The generic nature of the ATP-binding site makes the search for specific inhibitors targeting protein kinase challenging. The current criticism for ATP-competitive inhibitors is that they may not be able to compete with high intracellular ATP concentrations and are not kinase-selective. These inhibitors have affinity for kinases in the 10–300 micromolar range while intracellular concentrations of ATP are in the millimolar range (Grant 2009). To improve kinase selectivity, new inhibitors are being developed which bind the ATP site but also extend beyond the ATP pocket to include other points of

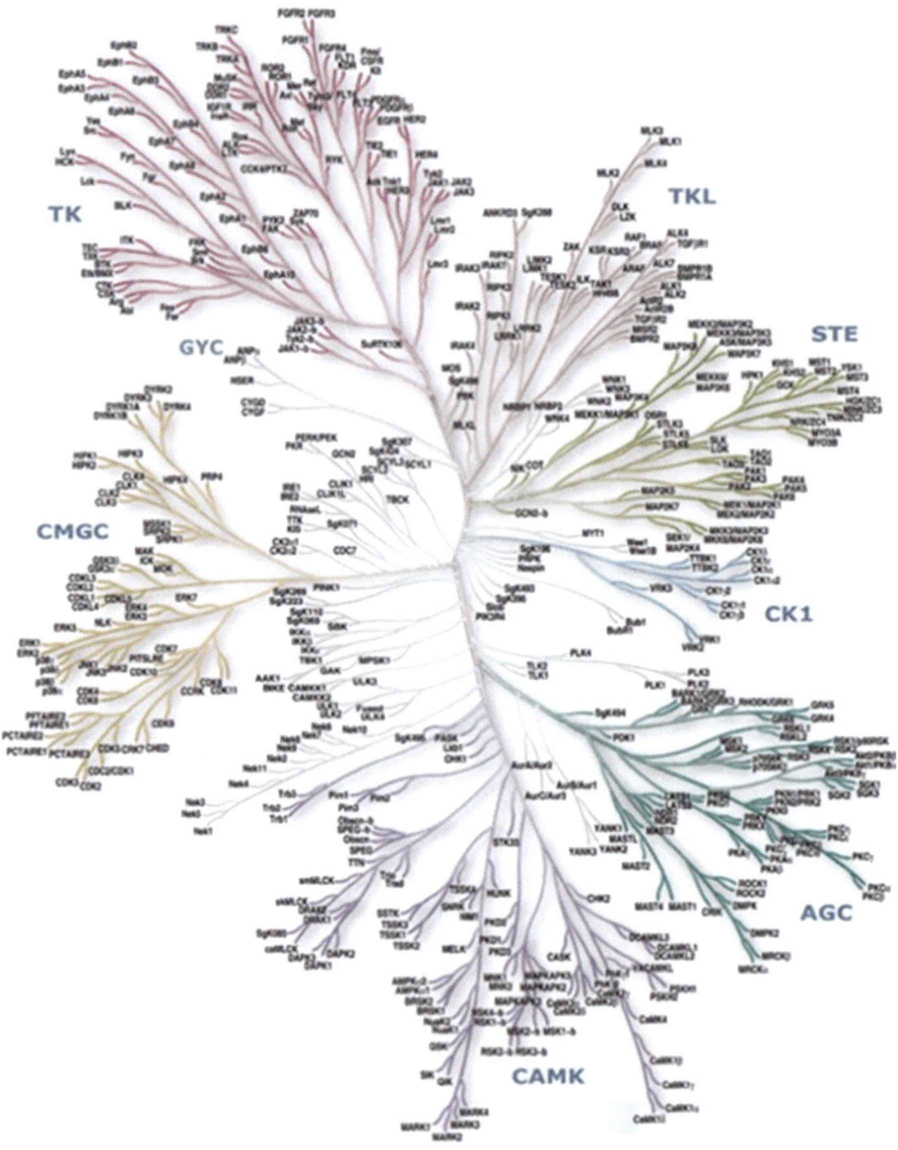

Fig. 12.2 Human protein kinase signaling pathways (reprinted with permission from SignalChem 2013)

interactions which are specific to the amino-acid sequences of a given kinase. Additional efforts toward druggable kinases include the organometallic analogs of the kinase inhibitors, like staurosporin and its derivatives, as shown in Fig. 12.3 (Pagano et al. 2009). Development of new drug targets fuels the need for alternative inhibitor screening methodologies, some of which will be described below.

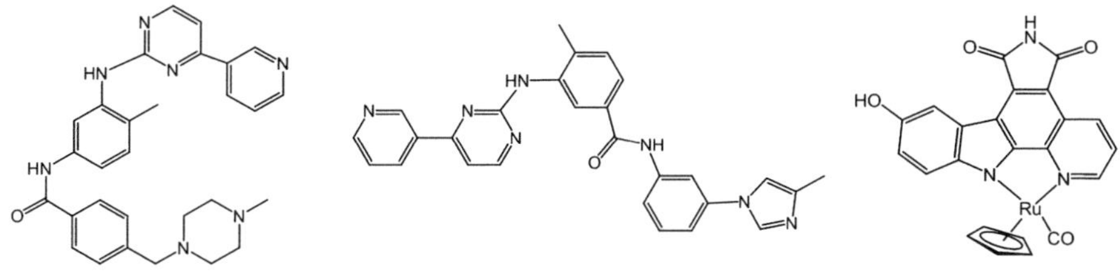

Fig. 12.3 Structures of the small molecule protein kinase inhibitors

12.4 Nanoparticle-Based Platforms for Kinome Analysis and Inhibitor Screening

In recent decades, nanomaterials have been exploited for materials, biomedical, and analytical applications due to their excellent attributes such as ease of preparation, biocompatibility, stability, optical and electrochemical properties, etc. The NP-based sensors fall into two categories: (1) unmodified NPs and (2) modified NPs. The bioassays based on NPs are carried out in three ways: (1) spectroscopy, (2) microscopy, and (3) electrochemistry. In this section inorganic and organic NPs will be discussed as they pertain to kinome analysis through a variety of methods based on unmodified and modified NPs.

12.4.1 Gold NPs

Gold nanoparticles (AuNPs) have been widely used in bioassays of protein kinases in electrochemical or optical platforms. The electrochemical approach may use redox labeling or the oxidation/reduction of gold directly for detection of phosphorylation. Optical methods for phosphorylation studies have employed the surface plasmon resonance of AuNPs or AuNPs as surface supports for the reagents in the reaction. Examples of these strategies will be discussed below.

12.4.1.1 Electrochemical Detection with AuNPs

Label-free electrochemical detection of kinase activity was demonstrated using an immobilized peptide substrate on gold electrode and zirconium cation-mediated signaling (Xu et al. 2009). Upon phosphorylation in the presence of ATP and protein kinase A (PKA), the phosphates bound the zirconium ions via multi-coordination between the metal ion and phosphate groups. This zirconium-rich surface was subsequently exposed to the DNA functionalized-AuNPs containing the 5′-phosphate ends which cap the surface zirconium ions. The presence of DNA-AuNPs on the surface increased the redox charge of the $[Ru(NH_3)_6]^{2+/3+}$ solution probe which was detected by chronocoulometry. The lowest detectable kinase activity was at 5 U mL^{-1}, and the IC_{50} of the H-89 inhibitor was at 33 nM.

Due to specific binding between sulfur and gold, AuNPs were used for binding to thiophosphorylated peptides on carbon electrodes. The electrochemical biosensor was based on adenosine 5′-[γ-thio]triphosphate (S-ATP) for kinase-catalyzed phosphorylation in the presence of PKC kinase (Kerman and Kraatz 2007). The target biotin-labeled peptides were immobilized on the streptavidin-coated carbon electrode, and subsequently exposed to S-ATP and PKC. The thiophosphate groups were attached to the surface which was subsequently exposed to AuNPs. The square-wave voltammetry was used to monitor oxidation of Au^0 to $[AuCl_4]^-$, and subsequent reduction to Au^0. This method allowed for a detection limit for PKC of 10 U mL^{-1} with a linear range up to 50 U mL^{-1}.

12.4.1.2 Optical Detection with AuNPs

Typically, optical response is monitored by surface plasmon resonance wherein dispersed AuNPs appear red and aggregated AuNPs appear purple (or blue) in solution. The aggregation of AuNPs leads to a red shift of interparticle surface plasmon absorbance. The latest development in AuNPs regarding kinase analysis was introduced by a chemoluminescence biosensor using S-ATP as a co-substrate (Xu et al. 2010). ECL combines electrochemical and luminescence methods and measures the light emission process in a redox reaction. The peptide substrate, kemptide, was immobilized via a HS-group to the gold electrode prior to kinase activity. In the event of the successful thiophosphorylation of keptide, in the presence of PKA and S-ATP, AuNPs bind the phosphopeptide surface and mediate the ECL of indicators such as luminol. Using the ECL biosensor, the optimal S-ATP and PKA concentrations were determined to be 30 μM and 60 U mL^{-1}. The inhibition of PKA activity was also determined with ellagic acid at an IC_{50} of 5 μM. The obvious shortcoming of this approach is the required use of S-ATP, which is a non-natural co-substrate and precludes cellular studies.

In an array format, biotinylated AuNPs were used with immobilized peptides for the detection of biotin-phosphorylation of peptide substrates (Sun et al. 2007). For Resonance Light Scattering (RLS) detection, silver ions were used for signal enhancement and the kinase activity was measured. This methodology, however, requires the use of biotin-labeled ATP and introduces another level of complexity. A next-generation RLS biosensor for kinases was developed based on this method but it utilized ATP rather than biotin-ATP. The immobilized peptide substrate was combined with biotinylated antiphospo-antibody, dye-modified avidin, and biotin-AuNPs for detection of peptide phosphorylation by PKA and lymphocyte-specific protein tyrosine kinase (LCK) in a format depicted in Fig. 12.4 (Li et al. 2010).

Briefly, the immobilized phosphorylated kemptide or LCK peptide substrate was exposed to biotinylated antibody, specific for phosphoserine or phosphotyrosine sites. Subsequently, avidin-fluorescein was introduced to bind immobilized biotin-peptide,

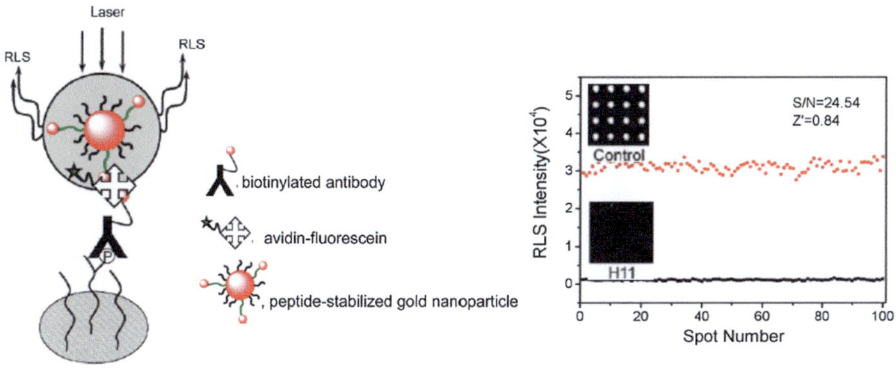

Fig. 12.4 Schematic representation of one spot on the microarray showing stepwise buildup of biosensors based on AuNPs and RLS detection; plot of RLS intensities for phosphorylation reactions in the absence (*red symbols*) and presence (*black symbols*) of staurosporin-based inhibitor (H11). Images of microarrays show control reactions free of inhibitors and H11 reactions with inhibitor. (Reprinted with permission from Li et al. (2010). Copyright (2014) American Chemical Society)

on which fluorescence detection may be carried out. For dual output, the RLS detection was achieved by using the avidin-biotin interactions with peptide-stabilized AuNPs containing biotin. Due to the relatively low light-scattering ability of AuNPs, the final step in the biosensor design was a silver enhancement step. The excellent feature of the assay is its high-throughput nature and capability for monitoring hundreds of enzymatic reactions per kinase. However, a higher loading of protein kinases is required (100 units). In addition, over 80 kinase inhibitors were screened using this assay and were mostly in agreement with the literature data, when available.

Using fluorescence imaging, the peptide substrate-coated AuNPs on glass were phosphorylated in the usual way and exposed to antiphospho-antibodies and fluorescently labeled secondary antibodies for imaging (Kim et al. 2008). The introduction of AuNP layers produced higher surface area, higher surface density, and improved detection and sensitivity over conventional self-assembled monolayers on surfaces. The lowest detectable protein kinase concentration was 0.05 ng mL^{-1} with a linear detection range up to 10 ng mL^{-1}. No fluorescence quenching from Au was reported in this work.

A solution approach based on an AuNP platform may typically use the aggregation propensity of AuNPs as the detection method in a calorimetric assay. The first calorimetric kinase assay of its kind was reported by Brust et al., who utilized 13-nm AuNPs decorated with peptide substrate for phosphorylation detection, as shown in Fig. 12.5 (Wang et al. 2006).

For the phosphorylation reactions, γ-biotin-ATP was used with PKA or CaMKII kinases in addition to avidin-coated AuNPs. In the event of biotin-phosphorylation of the peptide substrate, avidin-AuNPs bind and produce aggregation resulting in a color change. The UV–Vis absorption band at 536 nm (red colour) prior to phosphorylation was blue-shifted to 520 nm (blue/grey) upon phosphorylation due to AuNP aggregation. The kinase inhibition was detected by UV–Vis and transmission electron microscopy in the presence of H89, KN62, and SB203580 inhibitors.

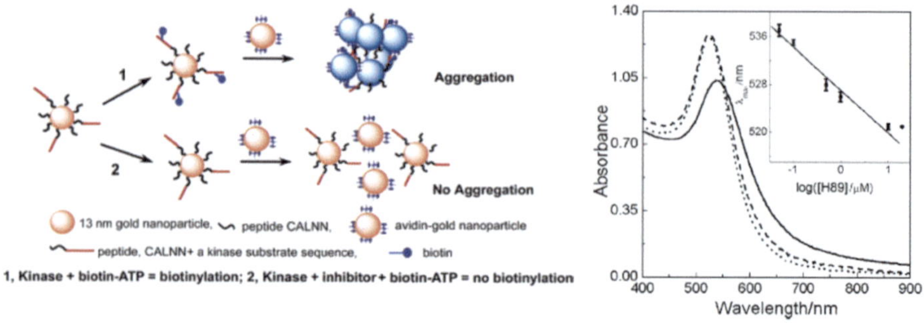

Fig. 12.5 Schematic illustration of kinase-catalyzed aggregation of AuNPs modified with target peptide during biotin-ATP phosphorylations. Absorption spectra of peptide–AuNPs prior to phosphorylation (*dashed line*) and after phosphorylation (*solid line*) with protein kinase A. The *inset* shows the inhibition with protein kinase inhibitor H89. (Reprinted with permission from Wang et al. (2006). Copyright (2014) American Chemical Society)

In a similar aggregation assay, the coagulating ability of cationic peptide substrates with anionic surfaces of AuNPs was compared to the coagulating property of phosphorylated peptide substrates (Oishi et al. 2008). For multiplate detection, peptide was phosphorylated in solution and coagulated AuNPs added, which resulted in a blue-to-red color shift. This dramatic color change was possible only when the peptide substrate concentration was 500 nM, protein kinase was at 50 U mL^{-1}, and specific-size coagulated AuNPs were used (20 or 40 nm diameter). Interestingly, when coagulated AuNPs were 10 or 60 nm, no change in color was observed after phosphorylation.

12.4.2 Silver NPs

Electrochemical detection using silver nanoparticles (AgNPs) was similar to that using AuNPs and includes monitoring the silver oxidation/reduction as a function of phosphorylation. The optical strategy using AgNPs was identical to that using AuNPs. Examples of the electrochemical and optical strategies will be described below.

12.4.2.1 Electrochemical Detection

Electrochemical methodology for detection of phosphorylation based on AgNPs was first demonstrated by Wieckowska et al. (2008). The authors used a peptide–Au-based approach, as shown in Fig. 12.6, to monitor CK-catalyzed phosphorylation of a single Ser residue in the presence of ATP.

In this format, the phosphate sites bind Ag ions and the chemical reduction of bound Ag ions is monitored by square-wave voltammetry. The reduction wave at ~200 mV was attributed to the reduction of Ag^+ to Ag^0 only in the presence of CK

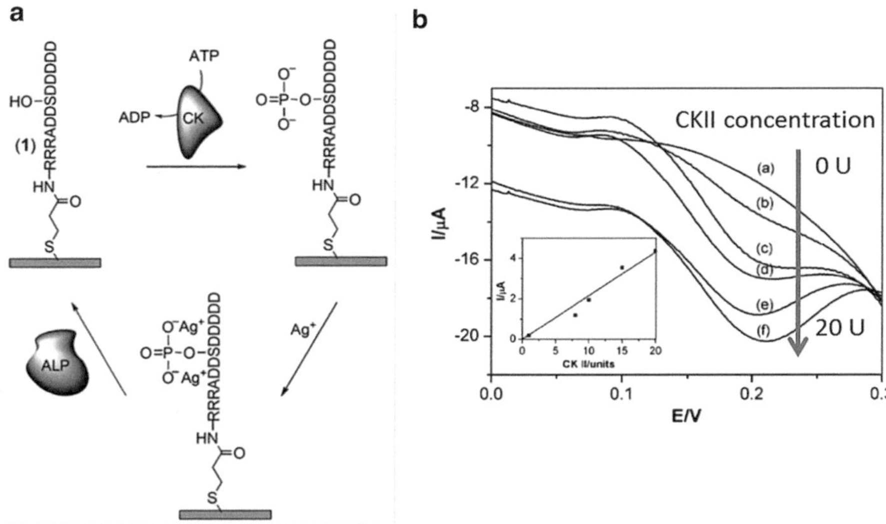

Fig. 12.6 (**a**) Electrochemical kinase assay based on AgNPs. (**b**) Plot of current as a function of potential for phosphorylation reactions at various protein kinase CKII concentrations. *Inset* shows a plot of current as a function of CKII concentrations. (Reproduced in part from Wieckowska et al. (2008) with permission of The Royal Society of Chemistry)

kinase and ATP. With the exclusion of either partner, no reduction signal was observed and the re-oxidation of Ag^0 was possible in the presence of alkaline phosphatase. This methodology was limited to the use of peptide substrates, and has not been applied to protein substrates.

An alternative multi-nanoparticle-based assay was employed in an electrochemical format as illustrated in Fig. 12.7 (Ji et al. 2008). This was a modified Willner approach by combining AgNPs with titanium oxide (TiO_2) particles, which have strong affinity for phosphopeptides, and are used in phosphopeptide enrichment.

The immobilized peptide on Au electrode was phosphorylated and subsequently bound to TiO_2. To elicit an electrochemical signal, the $AgNO_3$ solution was irradiated with 365-nm light and Ag^0 generated bound to TiO_2. Differential pulse voltammetry was used to visualize and quantify the current at ~650 mV stemming from the deposited AgNPs. In this assay, ellagic acid, a model PKA inhibitor, was used and the IC_{50} value determined to be 5.5 μM. The detection limit with this method was 0.2 U mL^{-1} of protein kinase.

12.4.2.2 Optical Detection

While AuNPs have been prevalent in the field of bioassays, AgNPs are also of importance for biological detection. As shown in Fig. 12.8, AgNPs were used in a label-free, sensitive, selective, and simple enzyme colorimetric assay for detection of the enzymatic reactions of phosphorylation and dephosphorylation (Wei et al. 2008).

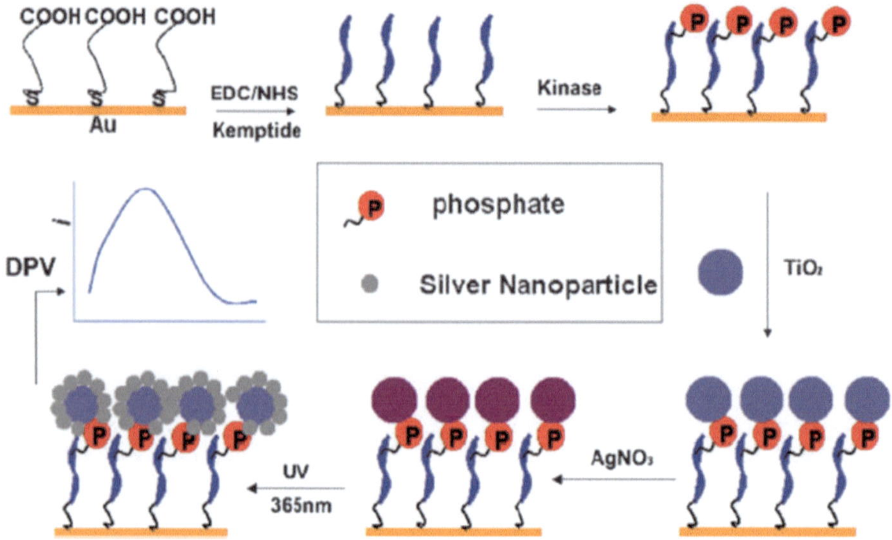

Fig. 12.7 Electrochemical detection of kinase activity based on TiO_2 and AgNPs. (Reproduced in part from Ji et al. (2008) with permission of The Royal Society of Chemistry)

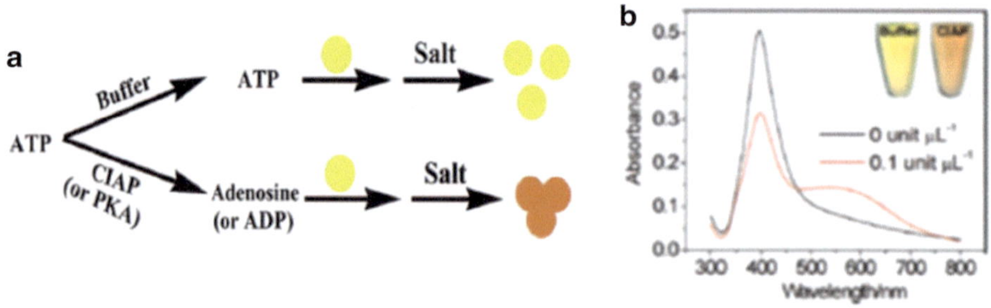

Fig. 12.8 (**a**) Illustration of AgNP methodology for calorimetric detection of kinase activity. (**b**) Absorbance spectra showing AgNPs prior to phosphorylation (0 unit μL^{-1} of kinase, *black line*) and following phosphorylation (0.1 unit μL^{-1} of kinase, *red line*). (Reprinted with permission from Wei et al. (2008). Copyright (2014) American Chemical Society)

PKA and calf intestine alkaline phosphatase (CIAP) were used as model enzymes and the peptide ALRRASLG as a model substrate, in the presence of ATP, as phosphate source. The basic principle behind this assay is the reduced propensity of AgNPs to aggregate in the presence of ATP, but if the ATP levels are depleted then AgNPs aggregation may take place. The depletion of the ATP levels was caused by phosphatase production of adenosine, or kinase production of ADP, both of which were not AgNPs stabilizers. Different colors were associated with the unaggregated and aggregated AgNPs and were monitored by UV–Vis spectroscopy.

The unaggregated AgNP solution in the presence of ATP had an absorption band at 400 nm (Fig. 12.8). Upon phosphorylation by CIAP, the ATP is depleted and adenosine is formed, and the AgNP solution had a second absorption peak at 580 nm, which was indicative of AgNPs aggregation. Notably, the sharp peak at 400 nm was significantly reduced. The absorption ration (A_{550}/A_{397}) was used to compare a variety of proteins—BSA, trypsin, and thrombin—with CIAP, and negligible changes were observed under UV with the control proteins, indicating the specificity of the assay. The phosphorylation by PKA was carried out in a solution of peptide substrate, ATP, PKA, and AgNPs. Notably, before and after phosphorylation, small changes in UV–Vis were observed, as indicated by a shoulder at 520 nm, as seen in Fig. 12.8. This bioassay did not produce two distinct absorption peaks and the ratiometric analysis was challenging. However, the authors were able to monitor the inhibition of CIAP in the presence of sodium orthovanadate with an IC_{50} value of 5 mM, but monitoring PKA inhibition remains a challenge.

12.4.3 CdSe NPs

Quantum dots (QDs) are a class of nanomaterials that has attracted attention due to their exceptional brightness, broad excitation spectra, large effective Stokes' shift, narrow emissions, and tunable emissions. Unlike metallic nanoparticles, QDs are exclusively used in optical assays rather than the electrochemical ones. The use of QDs for phosphorylation detection was originally demonstrated by Willner and coworkers. The CdSe/ZnS QDs conjugated with $Arg_3AlaAspAspSerAsp_5$, a substrate for CK2, showed high fluorescence (Freeman et al. 2010). Phosphorylation in the presence of CK2 and γ-ATP-ATTO-590, fluorescently labeled ATP, resulted in the transfer of phosphor-ATP-590 to peptide-QDs, and quenching of the QDs' fluorescence via energy transfer. As a result the ATTO fluorescence at ~640 nm was intensified, as shown in Fig. 12.9.

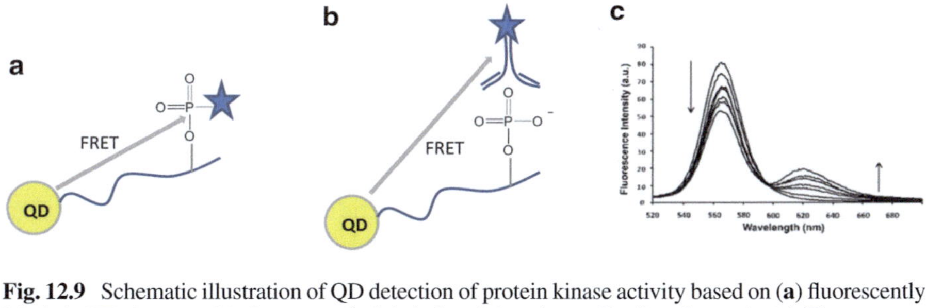

Fig. 12.9 Schematic illustration of QD detection of protein kinase activity based on (**a**) fluorescently labeled ATP (*top*) or (**b**) fluorescently labeled antiphospho-antibodies (*bottom*). (**c**) Fluorescence spectra showing decrease in QD fluorescence and increase in FRET. (Reprinted with permission from Freeman et al. (2010). Copyright (2014) American Chemical Society)

The second platform used by the same authors was the phosphorylation of the substrate peptide-QDs in the presence of CK2 and ATP, followed by immunoreaction with ATTO-590-modified antiphosphoserine antibodies, as shown in Fig. 12.9. Similarly to the ATP-labeled approach, the peptide-QDs' fluorescence at ~570 nm decreased and the FRET energy band at ~620 nm increased in the presence of increasing CK2 concentration (Fig. 12.9). The detection limits of the ATTO-ATP and ATTO-Ab approaches were 0.25 U mL^{-1} and 0.1 U mL^{-1} of CK2, respectively. The above QD approach was extended to include nonreceptor protein tyrosine kinases Abl and Src (Ghadiali et al. 2010). The QDs were conjugated to peptide substrates $EAIYPFAEEH_6$ and $IYGEFKKKH_6$ for Abl and Src, respectively, followed by phosphorylation reaction. The phosphorylation products were detected by AlexaFluor 647-labeled phosphotyrosine antibody (as shown in Fig. 12.9). The antibody dye and QDs underwent energy transfer, which resulted in a measurable fluorescence change. In the absence of phosphorylation, the peptide-QDs were characterized by ~610-nm emission, but following the addition of phosphate this emission band decreased and a new transition appeared at ~650 nm. The appearance of the two peaks allowed for ratiometric detection of protein kinase activity, down to 0.3 nM for Abl and 0.1 nM for Src, and determination of the IC_{50} of staurosporine inhibition at 0.2 μM.

12.4.4 Silica NPs

The hybrid NPs made up of Ru(II)-doped phosphonate-terminated silica were prepared by controlled hydrolysis of tetraethoxysilane and 3-(trihydroxysilyl)-propylmethyl-phosphonate in the presence of Ru(II) metal ions (Chen et al. 2013). For kinome analysis, the affinity of zirconium for phosphate groups and the signal amplification of Ru-SiNPs were combined. The ECL detection principle is based on immobilized phosphopeptides binding to zirconium ions followed by binding to Ru-SiNPs. The ECL of Ru(II) markers encapsulated in the phosphopeptide film was measured to determine protein kinase activity. For the optimal ECL signal, a Zr(IV) concentration of 0.2 mM, a NP loading above 0.1 mg mL^{-1}, and PKA of 10 U mL^{-1} were required. Relatively high Zr(IV) loading is required, which may be undesirable for biochemical studies.

12.4.5 Metal Oxide NPs: Magnetic NPs

Biochemical separation via functionalized magnetic NPs has recently blossomed. An example of magnetic separation for kinome analysis was demonstrated recently using TiO_2-coated Fe_3O_4/SiO_2 magnetic microspheres (Bai et al. 2013). These TiO_2-coated NPs have high binding selectivity for phosphopeptides over nonphosphorylated ones, and may be subsequently separated for post-analysis. In this method the magnetic NPs bind phosphopeptides and are removed from the supernatant.

Next, the phosphopeptide-bound magnetic NPs are eluted and tested by fluorescence and imaged by fluorescence microscopy. An interesting discovery in this study was the nonspecific binding of magnetic NPs to both nonphosphopeptides and phosphopeptides, which was addressed by performing measurements in 100 % organic solvent (such as acetonitrile). While improved selectivity for phosphopeptides was observed in organic solvent, this is a largely undesirable medium for biochemical measurements. The PKA detection limit was 0.001 U μL^{-1}, with a linear detection range up to 0.1 U μL^{-1}. The simultaneous detection of PKA and Akt1 was demonstrated by using FITC-labeled or TAMRA-labeled peptide substrates.

An alternative to TiO_2 is zirconium (Zr(IV)), which binds phosphorylated peptides as well. In a similar design to that described above, Zr-coated magnetic NPs were used for detection of PKA-catalyzed phosphorylation of FITC-labeled peptide substrates (Tan et al. 2013). The isolated phosphopeptides were tested using fluorescence spectroscopy by measuring the FITC fluorescence emission. The detection limit was 0.002 U μL^{-1}, with a linear range up to 0.2 U μL^{-1}.

12.4.6 Organic NPs: Polymeric

Conjugated polymeric nanoparticles (CPNs) based on pentiptycene-poly(*p*-phenylene ethynylene) polymer were conjugated to iminodiacetic acid (Moon et al. 2007). The iminodiacetic acid moiety is well known to bind metal ions, so CPNs were exposed to $GaCl_3$ solution for metal binding followed by complexation with the peptide target. For analysis, the peptide target had to be labeled with rhodamine for this biosensor to work. The CPNs were used for detection of rhodamine-labeled peptides for protein kinase detection by mixing CPNs with rhodamine-labeled kemptide and rhodamine-labeled phosphokemptide and monitoring fluorescence emission. In the absence of the phosphate group, the substrate peptide does not bind CPNs and leads to negligible change in fluorescence and a major emission band at 450 nm. However, when rhodamine-labeled phosphokemptide was used and mixed with CPNs, the FRET between the CPN donor and peptide acceptor decreased the emission at 450 nm and produced a new emission band at 600 nm. Ratiometric detection permitted the measurement of PKA activity levels as low as 0.03 U mL^{-1}.

Responsive organic NPs composed of a hydrophobic peptide substrate (lipopeptide) and a fluorescein-labeled polyanion (pA-Fs) were formed due to electrostatic interactions between the lipopeptides and pA-Fs (Koga et al. 2011). The requirement for the formation of the NPs was the use of lipopeptides with net charge of +2 or +5 to promote electrostatic interactions with polyanionic fluorophore. The organic NP formation led to self-quenching and weak fluorescence overall due to the close proximity of fluorophores to each other. The system of kinase detection was based on phosphorylation triggering the dissociation of organic NPs, as demonstrated in Fig. 12.10, and inducing fluorescence. The exposure to a specific kinase, PKA or PKC, and ATP introduced the divalent negatively charged phosphate group into the organic NPs, leading to reduction of the net cationic charge and NP dissociation.

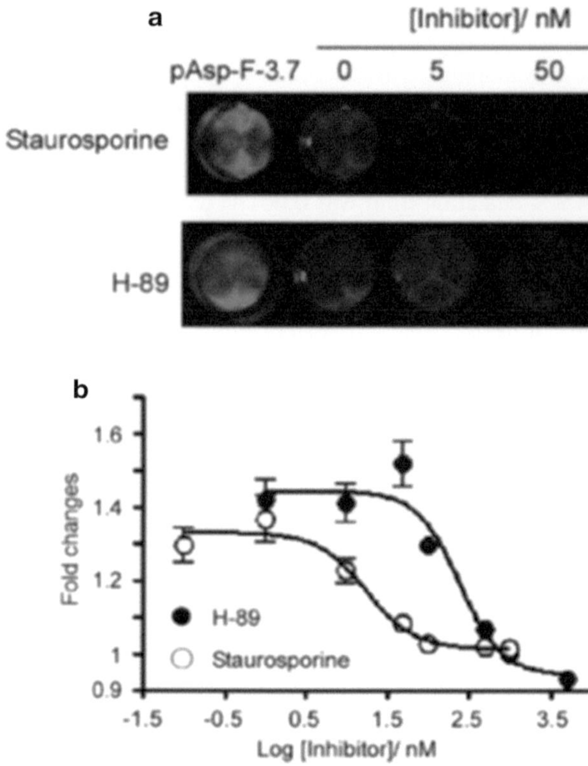

Fig. 12.10 (**a**) Fluorescence optical image of microplate wells obtained for PKA reaction and organic NPs in the presence of increasing inhibitor concentrations: staurosporine or H-89. (**b**) IC_{50} plots for the two inhibitors of PKA activity. (Reprinted with permission from Koga et al. (2011). Copyright (2014) American Chemical Society)

A PKA concentration of 10 U mL^{-1} was required for fluorescence amplification, which is slightly higher than the electrochemical method.

Since the target substrate must be incorporated into NPs, this methodology is limited to using cationic peptides and may not work well for neutral or more anionic substrates.

The second class of polymeric NPs was composed of a peptide substrate, kemptide, conjugated to polyanionic polymer and subsequently mixed with the negatively charged polymer to form NPs (Kim et al. 2007a, b). The negatively charged polymer, poly(aspartic acid), included tetramethyl rhodamine isothiocyanate (TRITC) while the positively charged polymer, poly(ethyleneimine), contained fluorescein isothiocyanate (FITC). In the folded state these NPs exhibited intermolecular FRET, due to the short distances between each FITC and TRITC molecules in NP. However, upon PKA phosphorylation of kemptide, dissociation of NPs takes place due to the introduction of negatively charged phosphate groups. The fluorescence enhancement upon phosphorylation was monitored by fluorescence spectroscopy and fluorescence imaging in a 96-well microtiter plates. This methodology requires covalent linking of the peptide substrate to a positively charged polymer and fluorophore conjugation to positively and negatively charged polymer, and is synthetically demanding. As an alternative, polymeric NPs were developed by using fluorescently labeled cationic polymer decorated with kemptide substrate and combined with poly(aspartic acid) to form polymeric NPs (Kim et al. 2007a, 2007b). In the folded state, these polymeric NPs exhibit low fluorescence due to self-quenching between Cy chromophores. The methodology was based on the principle of charge balance, such that NPs dissolve upon introduction of phosphate groups into kemptide, which in turn triggers near-infrared fluorescence enhancement. Since polymeric NPs are cell-permeable and biocompatible, the cells overexpressing PKA were incubated with polymeric NPs and the near-infrared fluorescence intensities increased within 5 min and increased continuously to 60 min.

12.5 Conclusions

In this chapter the use of NPs for detection of protein kinases and their inhibition was described. The inorganic and organic NPs are flexible functional platforms for kinome analysis. From Table 12.1, it is clear that some optical methods are superior to electrochemical methods in terms of detection limits; however, they require extensive substrate or co-substrate labeling, which are time-consuming, and ample amounts of sample.

Table 12.1 Protein kinase detection limits determined via various optical and electrochemical platforms based on NPs

Method	Kinase activity Detection limit (U mL^{-1})	References
Electrochemical detection		
AgNP/TiO_2	0.2	Ji et al. (2008)
AgNP	1.0	Wieckowska et al. (2008)
AuNP	5.0	Xu et al. (2009)
AuNP/S-ATP	10	Koga et al. (2011)
Optical detection		
AgNP	0.02	
QD/Fl-Antibody	11	Freeman et al. (2010)
QD/Fl-ATP	0.25	Freeman et al. (2010)
AuNP	50	Oishi et al. (2008)
Organic NPs	0.03	Moon et al. (2007)
Magnetic NPs	0.001[a]	Tan et al. (2013)
Silica NPs	10	Chen et al. (2013)

[a]U μL^{-1} units

In contrast, electrochemical assays may be carried out in a label-free environment, further simplifying the analysis. The fast, simple, and sensitive electrochemical methods are promising avenues for detection of protein kinase activity and its inhibition. In the future, new materials based on NPs will be developed to improve selectivity and sensitivity and to extend their use towards in vivo studies of protein kinase activity.

The use of NPs in bioanalysis is not without a challenge: (1) tailored functionalization of NPs, (2) immunogenicity and cytotoxicity of NPs, and (3) sensitivity to low-abundance biomarkers, among others. The application examples described in this chapter were mostly based on in vitro systems; however, it would be of interest to extend the NP methodology to in vivo measurements and the real-time detection of cellular pathways and processes. While NPs are a viable platform for bioanalysis in basic scientific research, their utility in the practical setting has only started to peak, with their use in the point-of-care system for patients and health care providers still largely unestablished.

References

Allen JJ, Li M, Brinkworth CS, Paulson JL et al (2007) A semisynthetic epitope for kinase substrates. Nat Methods 4:511–516

Bai J, Zhao Y, Wang Z et al (2013) Dual-readout fluorescent assay of protein kinase activity by use of TiO_2-coated magnetic microspheres. Anal Chem 85:4813–4821

Bromberg JF, Wrzeszczynska MH, Denga G et al (1999) Stat3 as an oncogene. Cell 98:295–303

Chen Z, He X, Wang Y et al (2013) Ru(II) encapsulated phosphorylate-terminated silica nanoparticles-based electrochemiluminescent strategy for label-free assay of protein kinase activity and inhibition. Biosens Bioelectron 41:519–525

Cimica V, Chen HC, Iyer JK et al (2012) Dynamics of the STAT3 transcription factor: nuclear import dependent on ran and importin-β1. PLoS One 6:e20188

Davies SP, Reddy H, Caivano M et al (2000) Specificity and mechanism of action of sole commonly used kinase inhibitors. Biochem J 351:95–105

Freeman R, Finder T, Gill R et al (2010) Probing protein kinase (CK2) and alkaline phosphatase with CdSe/ZnS quantum dots. Nano Lett 10:2192–2196

Ghadiali JE, Cohen BE, Stevens MM (2010) Protein kinase-actuated resonance energy transfer in quantum dot-peptide conjugates. ACS Nano 4:4915–4919

Grant SK (2009) Therapeutic protein kinase inhibitors. Cell Mol Life Sci 66:1163–1177

Grimsrud PA, Swaney DL, Wenger CD et al (2010) Phosphoproteomics for the masses. ACS Chem Biol 5:105–119

Houseman BT, Huh JH, Kron SJ et al (2002) Peptide chips for the quantitative evaluation of protein kinase activity. Nat Biotechnol 20:270–274

Hunter T (2000) Signaling—2000 and beyond. Cell 100:113–127

Ji J, Yang H, Liu Y et al (2008) TiO_2-assisted silver enhanced biosensor for kinase activity profiling. Chem Commun 1508–1510

Kerman K, Kraatz HB (2007) Electrochemical detection of kinase-catalyzed thiophosphorylation using gold nanoparticles. Chem Commun 5019–5021

Kim JH, Lee S, Kim K et al (2007) Polymeric nanoparticles for protein kinase activity. Chem Commun 1346–1348

Kim JH, Lee S, Park K et al (2007b) Protein-phosphorylation-responsive polymeric nanoparticles for imaging protein kinase activities in single living cells. Angew Chem Int Ed 46:5779–5782

Kim YPY, Oh H, Kim HS (2008) Protein kinase assay on peptide-conjugated gold nanoparticles. Biosens Bioelectron 23:980–986

Koga H, Toita R, Mori T et al (2011) Fluorescent nanoparticles consisting of lipopeptides and fluorescein-modified polyanions for monitoring of protein kinase activity. Bioconjug Chem 22:1526–1534

Li T, Liu D, Wang Z (2010) Screening of kinase inhibitors with a microarray-based fluorescent and resonance light scattering assay. Anal Chem 82:3067–3072

Manning G, Whyte DB, Martinez R et al (2002) The protein kinase complement of the human genome. Science 298:1912–1934

Martic S, Gabriel M, Turowec JP et al (2013) Versatile strategy for biochemical, electrochemical and immunoarray detection of protein phosphorylations. J Am Chem Soc 134:17036–17045

Medves S, Demoulin JB (2012) Tyrosine kinase gene fusion in cancer: translating mechanisms into targeted therapies. J Cell Mol Med 16:237–248

Moon JH, MacLean P, McDaniel W et al (2007) Conjugated polymer nanoparticles for biochemical protein kinase. Chem Commun 4910–4912

Oishi J, Asami Y, Mori T et al (2008) Colorimetric enzymatic activity assay based on noncross-linking aggregation of gold nanoparticles induced by adsorption of substrate peptides. Biocmacromolecules 9:2301–2308

Pagano N, Wong EY, Breiding T et al (2009) From imide to lactam metallo-pyridocarbazoles: distinct scaffolds for the design of selective protein kinase inhibitors. J Org Chem 74:8997–9009

Rankin CA, Sun Q, Gamblin CT (2007) Tau phosphorylation by GSK-3β promotes tangle-like filament morphology. Mol Neurodegener 2:12–26

Schultz MD, Janes KA, Lauffenburger DA et al (2005) A multiplexed homogeneous fluorescence-based assay for protein kinase activity in cell lysate. Nat Methods 2:277–283

SignalChem (2013) http://www.cellsignal.com/reference/kinase/overview.html. Accessed 2 Sept 2013

Sun L, Liu D, Wang Z (2007) Microassay-based kinase inhibition assay by gold nanoparticle probe. Anal Chem 79:773–777

Takeuchi K, Ito F (2011) Target therapy for cancer: anti-cancer drugs targeting growth-factor signaling molecules. Receptor tyrosine kinases and targeted cancer therapeutics. Biol Pharm Bull 34:1774–1780

Tan P, Lei C, Liu X et al (2013) Fluorescent detection of protein kinase based on zirconium ions-immobilized magnetic particles. Anal Chim Acta 780:89–94

Umezava Y (2005) Genetically encoded optical probes for imaging cellular signalling pathways. Biosens Bioelectron 20:2504–2511

Wang Z, Levy R, Fernig DG et al (2006) Kinase-catalyzed modification of gold nanoparticles: a new approach to colorimetric kinase activity screening. J Am Chem Soc 128:2214–2215

Wei H, Chen C, Han B et al (2008) Enzyme colorimetric assay using unmodified silver nanoparticles. Anal Chem 80:7051–7055

Wieckowska A, Li D, Gill R et al (2008) Following protein kinase activity by electrochemical means and contact angle measurements. Chem Commun 2376–2378

Xu X, Nie Z, Chen JY et al (2009) A DNA-based electrochemical strategy for label-free monitoring the activity and inhibition of protein kinase. Chem Commun 6946–6948

Xu S, Li Y, Wang T et al (2010) Highly sensitive electrogenerated chemiluminescence biosensor in profiling protein kinase activity and inhibition using gold nanoparticles as signal transduction probe. Anal Chem 82:9566–9572

Part IV
Applications, Challenges, and Future Outlook

Chapter 13
Carbon Nanotubes: Advances, Integration and Applications to Printable Electrode-Based Biosensors

Vinci Wing Sze Hung and Kagan Kerman

Abstract The development of novel analytical techniques for biosensor application has been an area of intense research efforts in the past decade and, in a field where the demand for rapid and cost-effective analyses is high, the use of screen-printed electrodes as opposed to conventional macroelectrodes can effectively satisfy these demands. With advancements in nanotechnology, the use of nanomaterials in biosensors has also added another level of customizability and sensitivity. In this chapter, we described the use of carbon nanotubes in printable electrode-based biosensors for label-free detection approaches for the detection of environmental contaminants and biologically relevant substrates. A discussion of past developments and recent advancements will highlight the broad range of experimental designs suitable for biosensing applications. With continual innovation in this field, it is expected that screen-printing technology will come to the fore in biosensing technology and progress successfully to overcome fundamental challenges in the biosensor field.

Keywords Biosensor • Carbon nanotube • Electrochemistry • Screen-printed electrodes

13.1 Introduction

The development of biosensors has advanced significantly in the past few decades, with the most successful endeavor being the commercialization of the well-known glucose biosensor, now used by many around the world. In 1956, Clark proposed the first oxygen sensor, with the ability to determine oxygen content in a sample independent of the sample composition. This sensor design integrated an insulated platinum cathode positioned next to a thin polyethylene membrane, a silver wire as the anode, and saturated potassium chloride as the inner electrolyte, all encased

---

V.W.S. Hung • K. Kerman (✉)
Department of Physical and Environmental Sciences, University of Toronto Scarborough, 1265 Military Trail, Toronto, ON M1C 1A4, Canada
e-mail: kagan.kerman@utoronto.ca

M.C. Vestergaard et al. (eds.), *Nanobiosensors and Nanobioanalyses*,
DOI 10.1007/978-4-431-55190-4_13

together in a poly(methyl methacrylate) outer tube (Clark 1956). This type of sensor construction allowed for the determination of oxygen in a sample with neither the cathode nor anode in direct contact with the sample environment, which gave rise to more accurate measurements.

This early innovation in biosensor technology was shortly followed by another report from Clark and Lyons in 1962 where they measured oxygen concentration relative to glucose concentration via the entrapment of glucose oxidase between dialysis membranes inside a Clark oxygen electrode. In this detection scheme, glucose that diffused across the membrane was converted to gluconic acid. Subsequent variations in pH were monitored by a pH-sensitive glass, which gave an indication of the glucose concentration in solution. The determination of glucose concentration can be calibrated for a wide range of variations in the system such as rate of sample flow, temperature and buffering capacity of the sample (Clark and Lyons 1962). The design was further developed by Updike and Hicks with the addition of an enzyme-modified working electrodes for oxygen detection. This design incorporated a sensing mode in which background variations in oxygen level could be accounted for (Updike and Hicks 1967).

Most biosensors consist of three main components. First is a recognition element, which is sensitive to the analyte of choice, which can consist of a variety of biological materials, such as enzymes, antibodies, nucleic acids and receptor molecules. The next component is the transducer, which serves to transform the biochemical signal originating from the recognition element and the target analyte to a quantifiable output. Leading transduction methods include electrochemical, piezoelectrical (Minunni et al. 2003; Tombelli et al. 2009), optical (Wandemur et al. 2014; Khan and Park 2014) and calorimetric (Bhand et al. 2010; Zhang and Tadigadapa 2004). The last component is an amplification module that allows the acquired signal to be visualized on a digital display. A simple schematic of the major components in a biosensor is shown in Fig. 13.1. The analyte being monitored can be customized to the disease that is being studied, and can range from small molecules such as proteins (Kerman et al. 2007, 2008; Kerman and Kraatz.

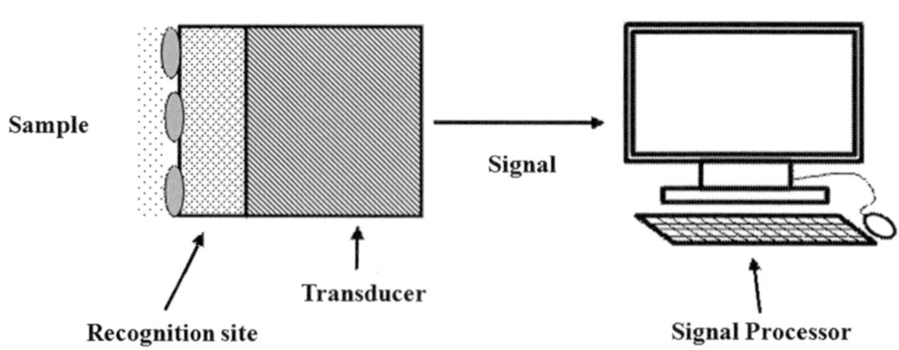

Fig. 13.1 Schematic diagram of a basic biosensor, displaying the biorecognition, interface and transducer components

2009; Vestergaard et al. 2007; Endo et al. 2005) and nucleic acids (Park et al. 2006; Park et al. 2002; Mirkin et al. 1996; Millan and Mikkelsen 1993; Millan et al. 1994; Kerman et al. 2005; Kerman et al. 2004a, b, c; Kerman et al. 2003) to whole cells (Kim et al. 2013; Bohrn et al. 2012). Electrochemical biosensors offer numerous advantages in the field of bio-detection and analytical chemistry. Their ability to combine the sensitivity of analytical techniques with the specificity of bio-recognition elements has allowed for great adaptability in research and modern medicine. With the progression of modern science, the invention of nanomaterials has added another tier of customizability with their inclusion in biosensors. To facilitate this detection platform for larger-scale production in high volumes while keeping manufacturing costs low, methods such as screen-printing have been adapted to the biosensor field.

Screen-printing has been well known since ancient times; it dates back more than a thousand years (Roberts 2006). In recent decades, researchers have effectively utilized this technique to create reproducible electrodes, which satisfied the growing demand for cost-efficient, stable, disposable and portable diagnostic and analytical instruments. Against the ever-increasing demands in clinical and industrial applications, cost-efficiency and portability was crucial. One of the most successful and profitable biosensors to date is the glucose detection meter, which utilizes disposable electrodes to measure glucose level in a small volume of blood taken from the user. Furthermore, the growing number of cases of drug overdose in hospitals, the need for personal disease monitors, and pollutants and toxins in the environment, all defined the requirement for economical and portable innovations (Li et al. 2012a, b).

Being able to satisfy all demands and more, screen-printed electrodes (SPEs) quickly became a well-established part of electrochemistry and biosensing. Its innate design of easily modifiable ink, and reference, working and counter electrodes gave SPEs an unrivaled power of adaptability and excellent accuracy. Fabrication of SPEs often follows a similar concept in which the functional electrodes are made using hard and heat-resistive substrates such as certain plastics and ceramics and the electrode components are created by printing different inks in successive layers to create the finished product. A final curing process is often necessary to anneal the layers of ink together with the underlying substrate for better stability of the electrode (Washe et al. 2013).

Commonly used materials for the previously mentioned inks are silver, gold and carbon for the conductive components of the electrode and other insulating polymers used for electrical shielding, protection and to increase shelf life. Although carbon inks have remained a top choice in terms of cost, other metals such as gold have their own advantages. For example, the affinity between thiol moieties and gold allows SPEs with gold working electrodes to be easily modified with the formation of self-assembled monolayers (SAM) (Mizutani et al. 1999; Retna and Ohsaka 2002). This significantly increases the applicability of gold SPEs in electrochemistry and biosensor research.

Pretreatment of SPEs can also be done to remove contaminants from the ink. This pretreatment process is performed in order to enhance the SPE's abilities. Pretreatments can be separated into electrochemical treatments (a two-step chemical), a potential cycling method, and an electrochemical pre-anodization method (Metters et al. 2011). To further demonstrate the SPE's adaptability, printing pastes were used in order to reduce electron transfer resistance. Increase in electron transfer resistance was a problem stemmed from insulating polymer found within the inks. This issue was easily nullified by printing pastes, with the added composition of noble gases, which acted as excellent electrocatalysts. However, the increased cost requirement for noble gases was a disadvantage and alternatives such as MnO_x and bismuth oxide were chosen instead (Dominguez Renedo et al. 2007).

The unrivaled versatility of SPEs remained one of their greatest assets. Modifications involving the deposition of substances such as metal films, polymers and enzymes on the surface of the electrodes were often utilized due to the increased functionality they gave. For example, metal-based electrodes, such as those printed with the previously mentioned gold and silver inks, were used by Laschi et al. (2006) to detect Pb and other hazardous chemicals. Initially, Wang and Tian (1992) pioneered the use of Hg film-modified electrodes in their work to determine Pb at ppb levels. This resulted in Hg film-modified electrodes being a much-used modification for SPEs. However, Hg remains highly toxic to the environment, and thus posed a serious problem for the SPE's disposability.

Alternatively, in the field of biosensors, modification of SPEs with enzymes remains a popular choice. Despite the cost of purification, isolation and extraction, enzyme electrodes are still popular in the study of clinically relevant biological processes and the determination of environmental chemical contaminants, pesticides and herbicides (Dock and Ruzgas 2003; Albareda-Servent et al. 2000). Due to their fast response and specificity towards biorecognition of their substrates, enzyme modification continues to play a role in biosensor design. A method utilizing differential pulse voltammetry on gold disposable SPEs was developed by Dounin et al. (2010) for the detection of paraoxon and carbonfuran by monitoring their inhibition of acetylcholine esterase activity. It was observed that their enzyme-modified electrode was able to produce results that rivaled those obtained from far more expensive technologies. Detection limits for paraoxon and carbonfuran reached as low as 10 ppt and 8 ppb, respectively. It was concluded that with small revisions of the method, it could prove to be an inexpensive and efficient way of detecting pesticides in real samples. For medical applications, SPEs have been widely used for glucose analysis to monitor diabetes (Luong et al. 2008; Hu 2009; Scognamiglio 2013). Deposition of the enzyme glucose oxidase (GOx) leads to a redox reaction which relays electric currents proportional to the level of glucose in the sample. Later developments have produced glucose dehydrogenase (GDH) disposable sensors as an alternative.

Immunoassay based on SPEs was another breakthrough for SPE technology and still today plays a major role in medicine, genetics, medical biology etc. The incorporation of SPEs revolutionized immunoassays to give rise to compact, hand-held devices which were inexpensive, yet highly sensitive and reliable. Such devices were mostly used for hormone analysis; however, recent research has discussed their usage for the detection of pathogenic bacteria such as salmonella. Current developments have progressed to the incorporation of new printed materials and carbon nanomaterials. Carbon nanomaterials, with the common example of carbon nanotubes (CNTs), have attracted interest in biosensor design. Two common types of CNTs are single-walled carbon nanotubes (SWCNTs) prepared from one sheet of graphite rolled into a tube, and multiwalled carbon nanotubes (MWCNTs) that contain multiple sheets of graphene (with sp^2 hybridized carbon) rolled into concentric tubules similar to that found in tree trunks (Baughman et al. 2002) . A wide range of modifications can be applied to carbon nanotubes, which make them ideal candidates for use in biochemical sensing systems (Chikae et al. 2008a, b; Katz and Willner 2004a, b; Kagan et al. 2011; Rivas et al. 2007).

In terms of clinical diagnostics, electrochemical devices have been used most frequently as they provide the clinician with a simple yet sensitive tool for disease analysis. Ease of miniaturization is another advantage offered by electrochemical biosensors, which can allow for development of at-home and bedside sensor types. In this chapter, the analytical applications of carbon nanomaterials will be discussed with respect to their involvement in electrochemical biosensing technology.

13.2 Application of Carbon Nanotubes in Biosensors and Integration into Screen-Printed Electrodes

The addition of carbon nanotubes (CNTs) to biosensors has been shown to improve analyte detection via an increased sensitivity to surface perturbations and modifications (Yao and Shiu 2007) and their fast electron transfer rates (Hu et al. 1999). The two most common types of CNTs are multi-walled nanotubes (MWCNT) and single-walled nanotubes (SWCNT). Promotion of electron transfer is possible due in part to their structurally dependent metallic character, addition of functional groups or the presence of metal ions at their surface.

Researchers have experimented and pushed carbon nanomaterials to realize their full capability since they were discovered by Iijima in 1991 (Iijima 1991; Iijima and Ichihashi 1993). This on-going research is exemplified by the large numbers of publications in this field (Sassolas et al. 2012; Valentini et al. 2013; Bhaskar et al. 2013; Heister et al. 2013). Furthermore, CNT-modified surfaces have been shown to be able to accumulate nucleic acids as well as alleviate surface fouling effects without compromising their remarkable sensitivity as nanoscale sensors (Wang and Kawde 2002). Further functionalization can be done on acid-treated CNTs with carboxylic moities on the surface. Such modifications include the attachment of enzymes or the use of CNTs as molecular wires to mediate electron transfer from the active site of an enzyme to the electrode, bypassing the use of mediator molecules. Although advances have been made, application of CNTs is still difficult due to multiple steps in the purification process and a lack of control of the CNTs' physical properties (Lahiff et al. 2010). Furthermore, it is difficult to limit the formation of double-walled or multi-walled CNTs during the production of SWCNT, thus adding another variable awaiting optimization during the production of CNTs.

13.2.1 CNT-Paste Electrodes

Compared to other methods of integrating CNTs into the fabrication of biosensors, CNT-paste electrodes (CNTPEs) are the simplest of them all. This type of electrode was first prepared by Norman Adams in 1958 while investigating the properties of dropping carbon electrodes as an alternative to dropping mercury electrodes for the analysis of anodic oxidation of organic molecules (Adams 1958). This paste electrode was prepared by mixing carbon and bromoform (serving as the binder) into a thick consistency, and then an electrode surface was formed by forcing the paste into a chosen capillary-like reservoir. The surface of the electrode was smoothed and inserted into the sample solution in an inverted manner, ensuring that the only electrical contact with the solution was through the carbon paste surface. Excellent performance was displayed by the CPEs prepared by Adams in his report and was comparable to previously developed carbon rod electrodes. The ease of preparation and renewal of the electrode surface makes this type of electrode superior to its predecessors (Adams 1958).

Although many types of new carbon materials have been developed, the popularity of CPEs has allowed them to remain useful for investigations of new sensors and detectors. In addition, the basic components of CPE—carbon (in the form of graphite powder) combined with a non-electroactive binder (popular choices include mineral/paraffin oil and nujol)—have not changed significantly over the years and the design, as shown in Fig. 13.2, still remains in the CPEs used today. Carbon is a common choice for electroanalysis as it is chemically inert, allows for a large potential window for analysis (as the oxidation of carbon is slow relative to the analyte under consideration), is relatively low-cost, has low background profile, has an easily regenerable surface and is a compatible material for sensing various types of samples. By adding CNTs into the CPE mixture, the new CNTPE still retains the electrochemical properties of its former CPE nature, with the added advantage of faster electron transfer of reactions occurring on the electrode surface through the aid of CNTs.

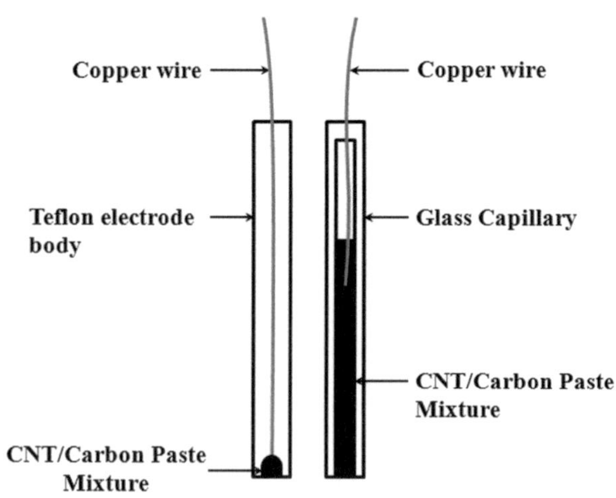

Fig. 13.2 Cross-sectional illustration of two commonly used carbon paste electrodes (CPE). On the *left*, the CNT/carbon paste mixture is forced into a small opening at the tip of a Teflon electrode body. On the *right*, the CNT/carbon paste mixture is packed into a glass capillary and a copper wire is inserted to allow for electrical conductance

Many reports have described the use of CNTs as an electrode material which allowed for the immobilization of enzymes to create enzyme biosensors. The first successful CNT-modified electrode was described by Britto and coworkers (1996) for the detection of dopamine. In their experiment, MWCNTs were thoroughly mixed with bromoform and packed into a glass tube with either a platinum or copper wire attachment. They were able to achieve a smooth and reproducible response to the oxidation of dopamine to dopaminequinone and found that CNT-modified electrodes out-performed other forms of carbon electrodes. The enhancement in detection is due to the added aspect ratio with the addition of CNTs (i.e. the inherent dimensions of the tubes and pores created by packing the CNTs into the paste electrode), which behaved as a steric regulator for more efficient oxidation reactions.

Similar to the electrodes described by Britto and coworkers (1996), CNTs have been commonly applied in the fabrication of enzyme biosensors in the form of a carbon paste electrode (CPE). Other early reports of protein incorporation into CNT electrodes were given by Davis, Coles and Hill in 1997, where horse heart cytochrome *c* and a copper-containing protein, azurin, were immobilized onto or within CNTs by first agitating a concentrated solution of protein with a suspension of CNTs in a phosphate buffer. The mixture was later packed into a glass capillary and it was shown that after such CNT modification, the proteins still retained their activity, thus leading the way for novel applications of CNTs for biosensing devices (Davis et al. 1997). In addition, work done by Davis and coworkers (Davis et al. 1997, 1998) showed that not only can CNTs be adapted as a useful electrode material, but also the surface and inside of CNTs can be modified with cytochrome *c* and azurin without denaturation.

A few years later in 2003, Rubianes and Rivas reported that enzymes can retain their catalytic activity as well and demonstrated the use of a CNT-CPE for the detection of hydrogen peroxide (Rubianes and Rivas 2003, 2004). In the development of oxidase-based biosensors, the accurate measurement of hydrogen peroxide is very important. In glucose biosensors, the immobilized glucose oxidase catalyzes the oxidation of glucose to gluconolactone, in which hydrogen peroxide is subsequently released. In previous generations of glucose biosensors, interference of easily oxidizable species, such as ascorbic and uric acid, was evident, but with the use of CNTs, this problem was alleviated and negligible interference was observed even though the catalytic activity of CNTs for these compounds was not diminished. Furthermore, a 50-fold enhancement in sensitivity was achieved with the prepared CNT-CPE as compared to CPE. Furthermore, the addition of CNTs produced a large electrocatalytic effect for the reduction of hydrogen peroxide, comparable to that of metallized carbon electrodes (Rubianes and Rivas 2003; Miscoria et al. 2005). A later report by the same authors showed that CNT-CPEs are also compatible with other enzymes such as lactate oxidase, polyphenol oxidase and alcohol dehydrogenase. The increase reactivity of the added CNTs permitted a higher detection sensitivity of lactate, phenols, catechols and alcohols (Rivas et al. 2007; Rubianes and Rivas 2004).

The enhancing ability of MWCNTs was tested in a study by Hung and Kerman (2011) for the detection of homocysteine (HCys). MWCNT carbon paste electrodes (CNT-CPEs) were optimized with the addition of gold nanostructures at the electrode surface. The combination of CNT-CPEs and gold nanostructures greatly increased signal intensity (more than 500-fold). As expected, the addition of gold greatly increased electron transduction from redox processes. Detection of protein aggregation in neurodegenerative diseases is another research area in which CNT-modified electrodes have been useful. Aggregation of amyloid-β (Aβ) in the presence of metals in vitro was directly monitored in a study by Hung et al. (2012). Electrochemical results were compared with the Thioflavin-T fluorescence assay and shown to be a capable alternative. Furthermore, Hung et al. (2012) emphasized that this assay platform was fast, sensitive and applicable to observing the aggregation kinetics of other clinically relevant peptides. Chikae et al. in 2008 designed a three-step electrochemical sensor for the detection of saccharide–protein interactions between sialic acid derivatives and Aβ. Specific interaction of sialic acid was shown to capture Aβ and, with differential pulse voltammetry, they were observed the oxidation signal of Tyr found intrinsically in Aβ peptides, in which the signal intensity can be indirectly used to infer the state of Aβ aggregation Chikae et al. (2008).

Nucleic acids are another category of biomolecules that can be accurately measured with a CNT-CPE. This type of electrode, combined with a suitable electrochemical pretreatment procedure, allowed for excellent adsorptive stripping measurements of trace levels of samples. It is thought that the pretreatment created a hydrophobic layer at the electrode surface which enhanced adsorption of guanine bases to the electrode surface. In addition, the adsorbed nucleic acid layer displayed substantial stability even after being transferred to a blank solution (Pedano and Rivas 2005). These properties allow CNT-CPEs to have excellent applications in trace detection of nucleic acids by providing a large signal enhancement of the oxidation signal in guanine bases in oligonucleotides and polynucleotides. Given their well-established advantages in biosensors using CPEs, the same procedure can easily be adapted to increase the performance of SPEs.

Carbon composite electrodes as described above offer a simple methodology for creating bulk modifications in biosensor development. Such experimental designs offer several advantages including the ease of surface renewal combined with a continual source of chosen bio-catalytic material, and a reduction in the required supporting material.

13.2.2 CNT-Coated Electrodes (Adsorption Method)

Another simple method of integrating CNTs into the working electrode of an electrochemical biosensor is by physical adsorption of the nanotubes onto the electrode surface. A logical process for this methodology would be to create a solution or suspension of CNTs which is used to coat the surface of an electrode (later the solvent would be allowed to evaporate), thus leaving an evenly coated layer of CNTs on the electrode surface. Due to the long lengths of unprocessed CNTs, the cumulative van der Waals forces become very strong, which causes them to have a propensity to orient themselves in a parallel manner and eventually form crystalline ropes (Girifalco et al. 2006).

Successful dissolution of CNTs in common organic solvents has been previously achieved after covalent functionalization of the nanotubes. Chen et al. (2003) showed that post-functionalized SWCNTs are soluble in tetrahydrofuran and 1,2-dichlorobenzene, and partially soluble in dichloromethane and chlorobenzene. The stability of a SWCNT coating produced from dimethylformamide dissolution was evaluated by cyclic voltammetry of the film-modified glassy carbon electrode after a water and ethanol rinse, followed by a 24-h air exposure. The voltammogram of a pH 6.9 Britton–Robinson buffer solution before and after rinsing and air exposure showed minute variations in the peak potentials and peak currents, proving that SWCNT films were fairly stable. Prior to the use of organic solvents, previous experimentation that used CNT-modified electrodes required the CNTs to be casted as sulphuric acid mixtures (Zen et al. 2003; Ho and Mikami 2011; Wang et al. 2003a, b).

Although this technique was able to accomplish the task of dispersing CNTs onto an electrode surface, the harsh conditions were not appropriate for further modifications with biomolecules as many organic solvents lead to their denaturation. To overcome this problem, aqueous dispersion of CNTs with the aid of surfactants was investigated. Incorporation of CNTs using Triton X-100 or sodium dodecyl sulfate (SDS), two commonly used surfactants (Fig. 13.3b, c, respectively), was able to achieve stable colloidal nanotube suspensions in water (Liu et al. 1998). This technique is now widely used for electrochemical studies due to its simplicity and success. Sensitive detection of the neurotransmitters dopamine, epinephrine and norepinephrine was achieved with a carbon-fiber nanoelectrode modified by a CNT-SDS solution, and good reproducibility was achieved at these electrodes after 0, 50, 500 and 1,000 cyclic voltammograms and after a 15-day hiatus (Chen et al. 2003). The choice of surfactant also affects the solubility of CNTs within water, as discussed by Islam et al. (2003); it is suspected that SDS may have a weaker interaction with the nanotube surface as compared to other surfactants such as sodium dodecylbenzene sulfonate (NaDDBS, Fig. 13.3a) and Triton X-100 due to the lack of a benzene functional group (an important functional group for participation in π-like stacking).

Stable polymers such as Nafion and Teflon and the biopolymer chitosan (Fig. 13.4) are also able to form well-dispersed suspensions of CNTs. As reported by Wang and coworkers, redox activity of hydrogen peroxide at a Nafion-CNT film-modified glassy carbon electrode was dramatically increased and was suitable for the preparation of an oxidase-modified amperiometric biosensor (Wang et al. 2003a, b). Previous reports on the use of SWCNTs reported significant improvement in signal

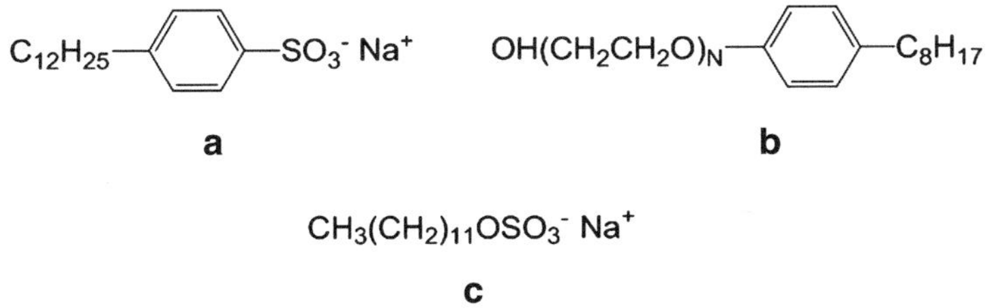

Fig. 13.3 Common surfactants used to increase the solubility of CNTs in aqueous environments. (**a**) Sodium dodecylbenzene sulfonate (NaDDBS), (**b**) Triton X-100, (**c**) sodium dodecyl sulfate (SDS)

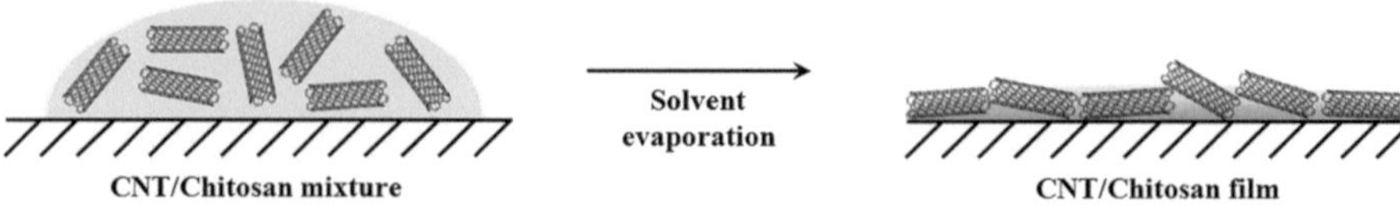

Fig. 13.4 Preparation of CNT/chitosan-coated electrodes involves the drop-casting of a CNT/chitosan mixture on an electrode surface, followed by solvent evaporation to allow for the formation of a CNT/chitosan film

detection over bare electrodes. This was exemplified by Kerman et al. (2005) when they monitored single-stranded binding protein (SSB) DNA interactions with the use of intrinsic nucleic acid and protein oxidation signals. The experiment aimed to take advantage of a unique binding between *Escherichia coli* SSB and single-stranded DNA (ssDNA) where the binding of SSB to single-stranded oligonucleotides (probe) decreased the voltammetric signal from guanine oxidation. These probes were attached to a SWCNT-modified SPE. With signal detection enhanced by numerous folds after the addition of SWCNTs, observations concluded that the CNTs provided a large active-electrode surface and also direct electron transfer. The redox signal of $K_3FE(CN)_6$ achieved near-linear improvement with respect to the amount of added SWCNTs.

13.2.3 Vertically Aligned CNT Sensors

Characteristics of CNTs such as good electrical conductivity and promotion of electron transfer in reactions can be used to describe bulk clusters of CNTs as well as individual moieties. For CNTs to be successfully used as a nanosensing array, they first must be aligned properly such that the spacing is substantially larger than the diameter (Fig. 13.5). This condition must be satisfied in order to avoid overlap of the diffusion layer between neighboring nanotubes (Morf and De Rooij 1997).

The advantages offered by vertically aligned CNTs can only be obtained by proper fabrication. Methods of synthesis include guided growth (with the assistance of a seeded surface, shown in Fig. 13.6), self-assembly, gas-flow-induced and high-temperature-induced assembly. Successful large-scale and high-density fabrication of aligned SWCNTs via guided chemical-vapor deposition (CVD) growth onto a single-crystal quartz substrate has been demonstrated by Kocabas and co-workers (2005). Traditional methods of CNT synthesis via CVD or arc discharge were able to create long CNTs but in a highly tangled fashion. Gao et al. was able to achieve various alignments of CNTs on an electrode surface by the application of electrical biases at different polarities during laser-assisted CVD CNT growth (Gao et al. 2009). Vertically-aligned CNTs produced on the cathode from this method were observed to have a neat and clean edge.

Chemical assembly of vertically aligned CNTs that involves the addition of functional groups to the tip end of CNTs is another popular method of synthesis.

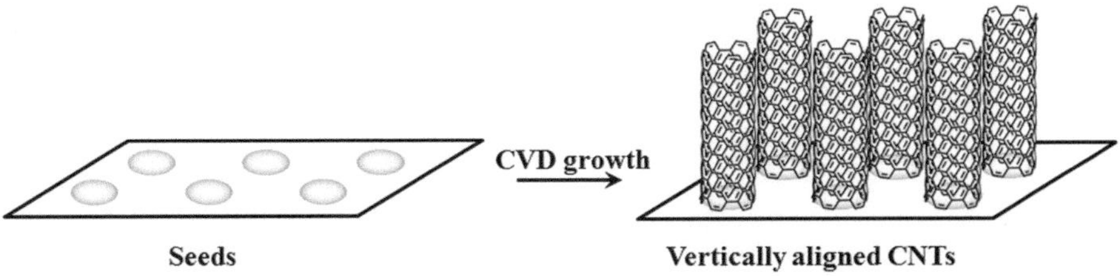

Fig. 13.5 Fabrication of vertically aligned CNTs via surface assembly of functionalized CNTs to a modified substrate surface. Chemical assembly methods can include surface condensation of amides and esters, gold-to-sulfur bond formation and electrostatic interaction (which does not require a linkage molecule)

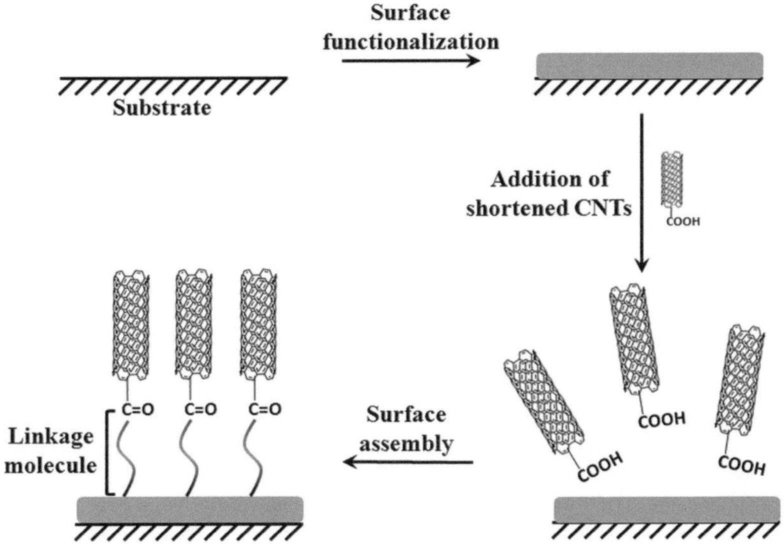

Fig. 13.6 Schematic of guided growth of CNTs on a "seeded" substrate surface through chemical-vapor deposition (CVD)

This scheme of synthesis (Fig. 13.5) allowed for the use of shortened CNTs, which gave researchers the opportunity to create a tailored sensing surface. Liu et al. (2000) demonstrated that the affinity between gold- and sulphur-containing functional groups can be used to create self-assembled monolayers (SAM) similar to those obtained by alkanethiols when exposed to a gold surface. Other types of chemical assembly techniques include surface condensation that relies on the formation of amide bonds or ester bonds (Diao et al. 2002). The overall surface condensation process remains largely the same for various substrates. For glass and silicone substrates, a silanization step with a compound that contains a terminal amino group was performed to create the required SAM. Ester bond formation can also achieve the same condensation process used to attach CNTs to a surface.

This coupling reaction occurs between a surface hydroxyl group and a carboxylic acid functional group at the tip end of the CNTs. Chemical assembly of vertically aligned CNT using the above described reactions gives well-aligned CNTs perpendicular to the plane of the substrate (Diao et al. 2002).

Similar orientation of CNTs can be achieved by electrostatic interaction. In this preparation, the carboxylic acid moieties at the tip end of CNTs were ionized to acquire a negative charge. As such, the substrate surface would require a positive charge for the interaction to take place. This attraction formed the basis of how the nanotubes were orientated on the surface. Furthermore, the surface of a substrate can be modified with a layer of metal oxide that is able to chelate with the ionized carboxylic acid group on the nanotubes and immobilize them onto the substrate. Chattopadhyay and coworkers reported the use of this technique for a metal-assisted monolayer formation of shortened CNTs and multilayer forest assemblies (Chattopadhyay et al. 2001). In their report, the surface was first modified by dipping the substrate in an aqueous acidic $FeCl_3$ solution for 15 min followed by immersion in a basic solution containing SWCNT for 30 min. This basic environment reacted with the surface-immobilized Fe^{3+} layer to form a Fe^{3+} hydroxide layer, which then set the platform for metal-assisted chelation between the substrate and the CNTs and allowed the production of SWCNT monolayers or high-density SWCNT forests. An advantage of this mode of synthesis is that the production of $Fe(OH)_3$ is relatively simple and therefore can be employed on a variety of substrates such as silicon and graphite.

As mentioned previously, the orientation of vertically aligned CNTs (v-CNTs) is spatially important for them to come into close proximity with the redox centers of enzymes. Many redox proteins do not have ideal charge transfer properties with an electrode surface even if there is direct contact. Therefore, the use of v-CNTs offers the chance for the production of biosensors using redox proteins that normally do not have sufficient electron transfer rates. Electrochemical properties of v-CNTs were studied by (Diao and Liu in 2002) in an effort to better design electrochemical sensors (Diao et al. 2002). They demonstrated that SWCNTs immobilized via amide bond formation on an insulated-gold substrate displayed similar electrochemical properties to many closely packed microelectrodes. Later reports showed that immobilized SWCNTs retained acceptable conductivity and were able to function as nanowires. However, sidewall additions induced local damage to the innate p-conjugation system of the CNTs and lowered electron transfer rates. Redox species that are directly attached to the tip end allow for better electron flow that is advantageous for proteins and other biomolecules with slow electron transfer rates. This methodology was used in the modification of electrode surfaces with enzymes, redox active macromolecules and ferrocene, for the development of cutting-edge electrochemical biosensors (Katz and Willner 2004a, b; Willner et al. 1996; Zayats et al. 2002; Patolsky et al. 2004). Glucose oxidase is a popular enzyme and has been investigated frequently in various methods of electroanalysis and novel biosensors. Liu et al. in 2005 studied the electrochemistry of glucose oxidase (GOx) immobilized on an aligned nanotube array using the self-assembly method (Liu et al. 2005). They found that by simply attaching GOx to the nanotube array, very few molecules were actually attached and direct electron transfer was not efficient.

Improvements were made to the design (as shown in Fig. 13.7) by directly attaching the redox center of GOx, then later reconstituting the apo-GOx over the bound redox co-factor, FAD, similar to the approach first described by Willner and co-workers and used in many of their sensor designs (Willner et al. 1996; Baravik et al. 2009; Yan et al. 2007, 2008). Their experiments showed that direct electron transfer was improved by the use of the apo-enzyme, but it is unknown whether the folded enzyme was electrochemically assessable.

Using aligned SWCNT-modified SPEs, Okuno et al. (2007a, b) successfully detected total prostate-specific antigen (T-PSA, a prostate cancer marker) using a label-free technique (Fig. 13.8). Detection was carried out using differential pulse voltammetry by monitoring the intrinsic oxidation signals of tyrosine and tryptophan in T-PSA. The detection limit was observed to be 0.25 ng/mL. The sensitivity of this assay was also tested by exposure to non-specific antigen BSA. Very slight deviations were detected from the BSA interference, thus highlighting the resilience

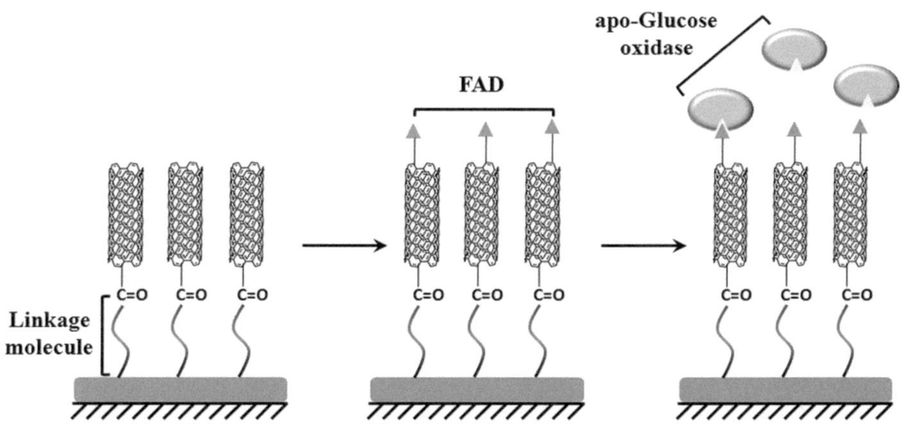

Fig. 13.7 Schematic example of a vertical arrangement of CNTs modified with redox-active components. Shown here, the co-factor flavin adenine dinucleotide (FAD) is coupled to the free end of the CNTs. Reconstitution of apo-glucose oxidase completes the assembly of an enzyme-modified CNT biosensor

Fig. 13.8 Setup for single-walled carbon nanotube (SWCNT)-modified platinum microelectrode functioning as the working electrode, a platinum wire serving as the counter electrode and a Ag/AgCl reference electrode (the latter two electrodes are above in solution and not displayed in the illustration). Adapted from Okuno et al. (2007a, b)

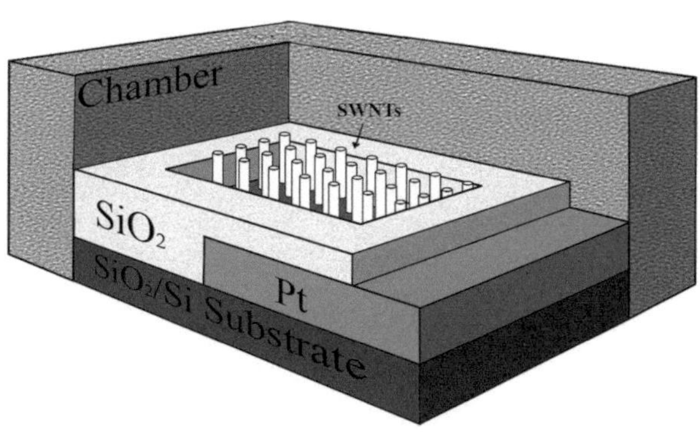

of the method to non-specific absorption. The results of the experiment showed promise for aligned-SWCNT microelectrodes. Additionally, Okuno et al. (2007a, b) created a SWCNT-array microelectrode chip with a Pt surface. Detection of L-cysteine, L-tryptophan and L-tyrosine was performed using differential pulse voltammetry and indicated a 100-fold increase in comparison to bare Pt-array microelectrodes.

Applications of v-CNTs, as discussed above, are only a few examples from the literature. Current fabrication techniques for v-CNTs are not easily combinable with screen-printing machinery as it is difficult to align them in the correct orientation required. However, this is not to say that such technology will not be available in the future, and will fulfill such requirements. Innovative SPEs offer a low-cost mass-production possibility, and combined with the advantages of v-CNTs, they open possibilities for the creation of novel biosensors. As technology for the creation of v-CNTs onto SPEs becomes more easily accessible, its utilization in biosensing is undoubtedly expected to increase.

13.3 Conclusions

As discussed in this chapter, various reports have demonstrated the possibility of merging CNT-composite materials with screen-printing fabrication techniques to create electrodes having improved sensitivity and stability (Wang and Musameh 2004). Furthermore, screen-printing allows for reproducible electrode surfaces to be made en-masse, which would normally require prior modification of the electrode surface—as in the case of drop-casting techniques. Customization of the working electrode is another advantage offered by screen-printing. As the inks can be varied to fit the need of the user, the percent composition of CNTs in the inks is just one of the parameters that can easily be manipulated. As shown by Li et al. (2002), different mass proportions of MWCNTs gave rise to higher responses to the target analyte. Biological moieties can also be incorporated into CNT mixtures for added functionality of the SPEs. As demonstrated by Sanchez et al. (2007), horseradish peroxidase enzyme was added to the working electrodes via a phase inversion technique.

Aside from carbon nanomaterials, metal nanoparticles also present a novel material that can be successfully incorporated into the sensing element of biosensors (Rusling 2012; Chikkaveeraiah et al. 2009, 2012; Malhotra et al. 2012; Venkatanarayanan et al. 2012; Sardesai et al. 2011). In conjunction with CNTs, the addition of metal nanoparticles (Fig. 13.9) has been shown to increase detection sensitivity by 100-fold, as reported by Hung and Kerman (2011). Numerous transition metals have been used in conjunction with CNTs to increase the surface area of the sensor and to optimize electron transfer between biomolecules and the electrode surface. Similar to CNTs, other forms of nanomaterials have also been successfully incorporated into screen-printed electrodes. It is expected that with the discovery of new nanomaterials, researchers will quickly harvest their potential and incorporate them into novel biosensors.

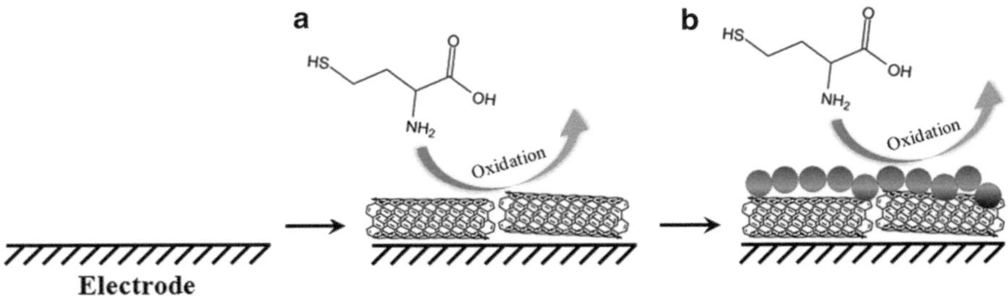

Fig. 13.9 Enhanced detection of homocysteine in solution was achieved by first modifying the electrode surface with (**a**) MWCNTs, then (**b**) electrodeposition of gold nanoparticles

References

Adams RN (1958) Carbon paste electrodes. Anal Chem 30(9):1576

Albareda-Servent M, Merkoci A, Alegret S (2000) Configurations used in the design of screen-printed enzymatic biosensors: a review. Sens Actuat B Chem 69(1):153–163

Baravik I, Tel-Vered R, Ovits O et al (2009) Electrical contacting of redox enzymes by means of oligoaniline-cross-linked enzyme/carbon nanotube composites. Langmuir 25(24):13978–13983

Baughman RH, Zakhidov AA, De Heer WA (2002) Carbon nanotubes - the route toward applications. Science 297(5582):787–792

Bhand S, Soundararajan S, Suguriu-Warnmark I et al (2010) Fructose-selective calorimetric biosensor in flow injection analysis. Anal Chim Acta 668(1):13–18

Bhaskar A, Deshmukh V, Prajapati L (2013) Carbon nanotube as a drug delivery system: a review. Int J Pharm Technol 5(2):2695–2711

Bohrn U, Stutz E, Fuchs K et al (2012) Monitoring of irritant gas using a whole-cell based sensor system. Sens Actuat B Chem 175:208–217

Britto PF, Santhanam KSV, Ajayan PM (1996) Carbon nanotube electrode for oxidation of dopamine. Bioelectrochem Bioenerg 41(1):121–125

Chattopadhyay D, Galeska I, Papadimitrakopoulos F (2001) Metal-assisted organization of shortened carbon nanotubes in monolayer and multilayer forest assemblies. J Am Chem Soc 123(38):9451–9452

Chen RS, Huang WH, Tong H et al (2003) Carbon fiber nanoelectrodes modified by single-walled carbon nanotubes. Anal Chem 75(22):6341–6345

Chikae M, Fukuda T, Kerman K et al (2008a) Amyloid-β detection with saccharide immobilized gold nanoparticle on carbon electrode. Bioelectrochem Bioenerg 74(1):118–123

Chikae M, Fukuda T, Kerman K et al (2008b) Amyloid-β detection with saccharide immobilized gold nanoparticle on carbon electrode. Bioelechem 74(1):118–123

Chikkaveeraiah BV, Bhirde A, Malhotra R et al (2009) Single-wall carbon nanotube forest arrays for immunoelectrochemical measurement of four protein biomarkers for prostate cancer. Anal Chem 81(21):9129–9134

Chikkaveeraiah BV, Soldà A, Choudhary D et al (2012) Ultrasensitive nanostructured immunosensor for stem and carcinoma cell pluripotency gatekeeper protein NANOG. Nanomedicine (Lond) 7(7):957–965

Clark LC (1956) Monitor and control of blood and tissue oxygen tensions. ASAIO J 2(1):41–48

Clark LC, Lyons C (1962) Electrode systems for continuous monitoring in cardiovascular surgery. Ann N Y Acad Sci 102:29–45

Davis JJ, Coles RJ, Allen H (1997) Protein electrochemistry at carbon nanotube electrodes. J Electroanal Chem 440(1–2):279–282
Davis JJ, Green M, Hill H et al (1998) The immobilisation of proteins in carbon nanotubes. Inorg Chim Acta 272(1–2):261–266
Diao P, Liu Z, Wu B et al (2002) Chemically assembled single-wall carbon nanotubes and their electrochemistry. Chemphyschem 3(10):898–991
Dock E, Ruzgas T (2003) Screen-printed electrodes modified with cellobiose dehydrogenase: amplification factor for catechol vs reversibility of ferricyanide. Electroanalysis 15(5):492–498
Dominguez Renedo O, Alonso-Lomillo MA, Arcos Martinez MJ (2007) Recent developments in the field of screen-printed electrodes and their related applications. Talanta 73:2002–2219
Dounin V, Veloso AJ, Schulze H et al (2010) Disposable electrochemical printed gold chips for the analysis of acetylcholinesterase inhibition. Anal Chim Acta 669(1–2):63–67
Endo T, Kerman K, Nagatani N et al (2005) Label-free detection of peptide nucleic acid-DNA hybridization using localized surface plasmon resonance based optical biosensor. Anal Chem 77(21):6976–6984
Gao Y, Zhou YS, Xiong W, Mahjouri-Samani M et al (2009) Controlled growth of carbon nanotubes on electrodes under different bias polarity. Appl Phys Lett 95(14):143117
Girifalco LA, Hodak M, Lee RS (2006) Carbon nanotubes, buckyballs, ropes, and a universal graphitic potential. Phys Rev B 62(19):13104–13110
Heister E, Brunner E, Dieckmann G et al (2013) Are carbon nanotubes a natural solution? Applications in biology and medicine. ACS Appl Mater Interfaces 5(6):1870–1891
Ho S, Mikami Y (2011) Porous carbon layers for counter electrodes in dye-sensitized solar cells: recent advances and a new screen-printing method. Pure Appl Chem 83(11):2089–2106
Hu J (2009) The evolution of commercialized glucose sensor in China. Biosens Bioelectron 24(5):1083–1089
Hu T, Odom TW, Lieber CM (1999) Chemistry and physics in one dimension: synthesis and properties of nanowires and nanotubes. Acc Chem Res 32:435–445
Hung VWS, Kerman K (2011) Gold electrodeposition on carbon nanotubes for the enhanced electrochemical detection of homocysteine. Electrochem Commun 13(4):328–330
Hung VWS, Masoom H, Kerman K (2012) Label-free electrochemical detection of Amyloid beta aggregation in the presence of iron, copper and zinc. J Electroanal Chem 681(1):89–95
Iijima S (1991) Helical microtubules of graphitic carbon. Nature 354:56–58
Iijima S, Ichihashi T (1993) Single-shell carbon nanotubes of 1-nm diameter. Nature 363:603–605
Islam MF, Rojas E, Bergey DM et al (2003) High weight fraction surfactant solubilization of single-wall carbon nanotubes in water. Nano Lett 3(2):269–273
Kagan D, Campuzano S, Balasubramanian S et al (2011) Functionalized micromachines for selective and rapid isolation of nucleic acid targets from complex samples. Nano Lett 11(5):2083–2087
Katz E, Willner I (2004a) Integrated nanoparticle-biomolecule hybrid systems: synthesis, properties, and applications. Angew Chem Int Ed Engl 43(45):6042–6108
Katz E, Willner I (2004b) Biomolecule-functionalized carbon nanotubes: applications in nanobioelectronics. Chemphyschem 5(8):1084–1104
Kerman K, Kraatz HB (2009) Electrochemical detection of protein tyrosine kinase-catalysed phosphorylation using gold nanoparticles. Biosens Bioelectron 24(5):1484–1489
Kerman K, Morita Y, Takamura Y et al (2003) Label-free electrochemical detection of DNA hybridization on gold electrode. Electrochem Commun 5(10):887–891
Kerman K, Kobayashi M, Tamiya E (2004a) Recent trends in electrochemical DNA biosensor technology. Meas Sci Technol 15:R1–R11
Kerman K, Matsubara Y, Morita Y et al (2004b) Peptide nucleic acid modified magnetic beads for intercalator based electrochemical detection of DNA hybridization. Sci Technol Adv Mater 5:351–357
Kerman K, Saito M, Morita Y et al (2004c) Electrochemical coding of single-nucleotide polymorphisms by monobase-modified gold nanoparticles. Anal Chem 76(7):1877–1884

Kerman K, Morita Y, Takamura Y et al (2005) Escherichia coli single-strand binding protein-DNA interactions on carbon nanotube-modified electrodes from a label-free electrochemical hybridization sensor. Anal Bioanal Chem 381(6):1114–1121

Kerman K, Vestergaard M, Tamiya E (2007) Label-free electrical sensing of small-molecule inhibition on tyrosine phosphorylation. Anal Chem 79(17):6881–6885

Kerman K, Song H, Duncan JS et al (2008) Peptide biosensors for the electrochemical measurement of protein kinase activity. Anal Chem 80(24):9395–9401

Khan M, Park S (2014) Liquid crystal-based proton sensitive glucose biosensor. Anal Chem 86(3):1493–1501

Kim C, Choi B, Seo J et al (2013) Mussel adhesive protein based whole cell array biosensors for detection of organophosphorus compounds. Biosens Bioelectron 41(1):199–204

Kocabas C, Hur SH, Gaur A et al (2005) Guided growth of large-scale, horizontally aligned arrays of single-walled carbon nanotubes and their use in thin-film transistors. Small 1(11):1110–1116

Lahiff E, Lynam C, Gilmartin N et al (2010) The increasing importance of carbon nanotubes and nanostructured conducting polymers in biosensors. Anal Bioanal Chem 398(4):1575–1589

Laschi S, Palchetti I, Mascini M (2006) Gold-based screen-printed sensor for detection of trace lead. Sens Actuat A 114(1):460–465

Li M, Li YT, Li DW et al (2012a) Recent developments and applications of screen-printed electrodes in environmental assays—a review. Anal Chim Acta 734:31–44

Li Y, Zhang L, Li M et al (2012b) A disposable biosensor based on immobilization of laccase with silica spheres on the MWCNTs-doped screen-printed electrode. Chem Cent J 6(1):103

Liu J, Rinzler AG, Dai H et al (1998) Fullerene pipes. Science 280(5367):1253–1256

Liu Z, Shen Z, Zhu T et al (2000) Organizing single-walled carbon nanotubes on gold using a wet chemical self-assembling technique. Langmuir 16(8):3569–3573

Liu G, Lin Y, Tu Y et al (2005) Ultrasensitive voltammetric detection of trace heavy metal ions using carbon nanotube nanoelectrode array. Analyst 130(7):1098–1101

Luong J, Male K, Glennon J (2008) Biosensor technology: technology push versus market pull. Biotech Adv 26(5):493–500

Malhotra R, Patel V, Chikkaveeraiah BV et al (2012) Ultrasensitive detection of cancer biomarkers in the clinic by use of a nanostructured microfluidic array. Anal Chem 84(14):6249–6255

Mikkelsen SR (1993) Sequence-selective biosensor for DNA based on electroactive hybridization indicators. Anal Chem 65(17):2317–2323

Millan KM, Saraullo S, Mikkelsen SR (1994) Voltammetric DNA biosensor for cystic fibrosis based on a modified carbon paste electrode. Anal Chem 66(18):2943–2948

Minunni M, Tombelli S, Scielzi R et al (2003) Detection of β-thalassemia by a DNA piezoelectric biosensor coupled with polymerase chain reaction. Anal Chim Acta 481(1):55–64

Mirkin CA, Letsinger RL, Mucic RC et al (1996) A DNA-based method for rationally assembling nanoparticles into macroscopic materials. Nature 382:607–609

Miscoria SA, Barrera GD, Rivas GA (2005) Enzymatic biosensor based on carbon paste electrodes modified with gold nanoparticles and polyphenol oxidase. Electroanalysis 17(17):1578–1582

Mizutani F, Sato Y, Yabuki S et al (1999) Enzyme electrodes based on self-assembled monolayers of thiol compounds on gold. Electrochim Acta 44(21):3833–3838

Morf W, De Rooij NF (1997) Performance of amperometric sensors based on multiple microelectrode arrays. Sens Actuat B 44(1–3):538–541

Okuno J, Maehashi K, Kerman K et al (2007a) Label-free immunosensor for prostate-specific antigen based on single-walled carbon nanotube array-modified microelectrodes. Biosens Bioelectron 22(9–10):2377–2381

Okuno J, Maehashi K, Matsumoto K et al (2007b) Single-walled carbon nanotube-arrayed microelectrode chip for electrochemical analysis. Electrochem Commun 9(1):13–18

Park SJ, Taton TA, Mirkin CA (2002) Array-based electrical detection of DNA with nanoparticle probes. Science 295(5559):1503–1506

Park S, Boo H, Chung TD (2006) Electrochemical non-enzymatic glucose sensors. Anal Chim Acta 556(1):46–57

Patolsky F, Weizmann Y, Willner I (2004) Long-range electrical contacting of redox enzymes by SWCNT connectors. Angew Chem Int Ed 43(16):2113–2117

Pedano ML, Rivas GA (2005) Immobilization of DNA at glassy carbon electrodes: a critical study of adsorbed layer. Sensors 5(6):424–447

Retna R, Ohsaka T (2002) Analytical applications of functionalized self-assembled monolayers on gold electrode: voltammetric sensing of DOPAC at the physiological level. Electroanalysis 14(10):679–684

Rivas GA, Rubianes MD, Rodríguez MC et al (2007) Carbon nanotubes for electrochemical biosensing. Talanta 74(3):291–307

Roberts G (2006) History's influence on screen printing's future. Screen Print 96(2):22–25

Rubianes MD, Rivas GA (2003) Carbon nanotubes paste electrode. Electrochem Commun 5(8):689–694

Rubianes MD, Rivas GA (2004) Enzymatic biosensors based on carbon nanotubes paste electrodes. Electroanalysis 17(1):73–78

Rusling JF (2012) Nanomaterials-based electrochemical immunosensors for proteins. Chem Record 12(1):164–176

Sanchez S, Pumera M, Cabruja E et al (2007) Carbon nanotubes/polysulfone composite screen-printed electrochemical enzyme sensor. Analyst 132:142–147

Sardesai NP, Barron JC, Rusling JF (2011) Carbon nanotube microwell array for sensitive electrochemiluminescent detection of cancer biomarker proteins. Anal Chem 83(17):6698–6703

Sassolas A, Blum L, Leca-Bouvier B (2012) Immobilization strategies to develop enzymatic biosensors. Biotech Adv 30(3):489–511

Scognamiglio V (2013) Nanotechnology in glucose monitoring: advances and challenges in the last 10 years. Biosens Bioelectron 47:12–25

Tombelli S, Mascini M, Scherm B et al (2009) DNA biosensors for the detection of a aflatoxin producing aspergillus flavus and *A. Parasiticus*. Montash Chem 140(8):901–907

Updike SJ, Hicks GP (1967) The enzyme electrode. Nature 214(5092):986–988

Valentini F, Carbone M, Palleshi G (2013) Carbon nanostructured materials for applications in nano-medicine, cultural heritage, and electrochemical biosensors. Anal Bioanal Chem 405(2):451–465

Venkatanarayanan A, Crowley K, Lestini E et al (2012) High sensitivity carbon nanotube based electrochemiluminescence sensor array. Biosens Bioelectron 31(1):233–239

Vestergaard M, Kerman K, Tamiya E (2007) An overview of label-free electrochemical protein sensors. Sensors 7(12):3442–3458

Wandemur G, Rodrigues D, Allil R et al (2014) Plastic optical fiber-based biosensor platform for rapid cell detection. Biosens Bioelectron 54:661–666

Wang J, Kawde AN (2002) Amplified label-free electrical detection of DNA hybridization. Analyst 127(3):383–386

Wang J, Musameh M (2004) Carbon nanotube screen-printed electrochemical sensors. Analyst 129:1–2

Wang J, Tian B (1992) Screen-printed stripping electrodes for decentralized testing of trace lead. Anal Chem 64:1706–1709

Wang J, Kawde AN, Musameh M (2003a) Carbon-nanotube-modified glassy carbon electrodes for amplified label-free electrochemical detection of DNA hybridization. Analyst 128(7):912–916

Wang J, Musameh M, Lin J (2003b) Solubilization of carbon nanotubes by Nafion: towards the preparation of amperometric biosensors. J Am Chem Soc 125(9):2408–2409

Washe A, Lozano-Sanchez P, Bejarano-Nosas D (2013) Facile and versatile approaches to enhancing electrochemical performance of screen printed electrodes. Electrochim Acta 91:166–172

Willner I, Heleg-Shabtai V, Blonder R et al (1996) Electrical wiring of glucose oxidase by reconstitution of FAD-modified monolayers assembled onto Au-electrodes. J Am Chem Soc 118(42):10321–10322

Yan YM, Yehezkeli O, Willner I (2007) Integrated, electrically contacted $NAD(P)^+$-dependent enzyme-carbon nanotube electrodes for biosensors and biofuel cell applications. Chemistry 13(36):10168–10175

Yan YM, Baravik I, Yehezkeli O et al (2008) Integrated electrically contacted glucose oxidase/carbon nanotube electrodes for the bioelectrocatalyzed detection of glucose. J Phys Chem C 112(46):17883–17888

Yao Y, Shiu KK (2007) Electron-transfer properties of different carbon nanotube materials, and their use in glucose biosensors. Anal Bioanal Chem 387(1):303–309

Zayats M, Katz E, Willner I (2002) Electrical contacting of flavoenzymes and $NAD(P)^+$-dependent enzymes by reconstitution and affinity interactions on phenylboronic acid monolayers associated with Au-electrodes. J Am Chem Soc 124(49):14724–14735

Zen J, Kumar A, Tsai D (2003) Recent updates of chemically modified electrodes in analytical chemistry. Electroanalysis 15(13):1073–1087

Zhang Y, Tadigadapa S (2004) Calorimetric biosensors with integrated microfluidic channels. Biosens Bioelectron 19(12):1733–1743

Chapter 14
Specialized Nanoneedles for Intracellular Analysis

Ryuzo Kawamura, Yaron R. Silberberg, and Chikashi Nakamura

Abstract Here, we introduce a novel approach to the detection of intracellular molecules by measuring direct interactions with an ultrathin probe, i.e. nanoneedle, which is mounted on an atomic force microscope (AFM). Standard AFM probes were sharpened, using a focused ion beam (FIB), to form high-aspect-ratio nanoneedles, which were then specifically functionalized and inserted into live cells. The insertion could be precisely detected using the resulted force–distance AFM curves, and no effect on cell viability was observed, even after repeated insertions. In addition, thanks to the high sensitivity of the AFM, distinct intermolecular unbinding events could be analyzed, which provided real-time information on the cytoskeleton state of the live cell. Following specific coatings and functionalization of the nanoneedles, various intracellular molecules could be detected and even inserted into live cells. The results presented here demonstrate the delivery of DNA vectors and the detection of mRNA and cytoskeletal proteins in live cells. Further advances in this technology, such as new developments in molecular functionalization options and improvements in the scale and accuracy of force measurements, will open possible new fields and applications for this diverse and powerful tool.

Keywords Atomic force microscopy • Force spectroscopy • Gene delivery • Intracellular measurements • Intracellular protein • Live cell analysis • Nanoneedle

14.1 Introduction

In the near future, stem cells with pluripotent differentiation capabilities are believed to become a major contributor to the advancement of regenerative medicine. Applications will include, among others, auto-transplantation therapies that overcome the problem of immune rejection and the use of model tissues for drug testing. Since the discovery of induced pluripotent stem (iPS) cells in 2006 (Takahashi and Yamanaka 2006), new technologies have been developed to allow cultivation,

R. Kawamura • Y.R. Silberberg • C. Nakamura (✉)
Biomedical Research Institute, National Institute of Advanced Industrial Science and Technology (AIST), Central 4, 1-1-1 Higashi, Tsukuba, Ibaraki 305-8562, Japan
e-mail: chikashi-nakamura@aist.go.jp

M.C. Vestergaard et al. (eds.), *Nanobiosensors and Nanobioanalyses*,
DOI 10.1007/978-4-431-55190-4_14

differentiation induction, imaging, and separation, with the aim of establishing practical applications for pluripotent stem cells. Currently, one of the major challenges in the field is the identification and separation of potent tumorigenic undifferentiated cells from differentiated cultures, and much effort is directed towards establishing efficient and reliable cell separation techniques.

Atomic force microscope (AFM) is a versatile tool for analysis of living cells. AFM allows for the investigation of the mechanical properties of a cell surface with high sensitivity and spatial resolution under physiological cell culture conditions (Costa 2006). By indenting the cell surface with an AFM probe, and fitting the acquired force–distance curves with the Hertz model (Hertz 1881), cell elasticity, or Young's modulus, can be obtained. Such measurements have been performed since the 1990s (Radmacher et al. 1996) using a variety of cell lines. Studies showed that some cancer cells show relatively lower elasticity (Haghparast et al. 2013), and so do undifferentiated mouse ES cells in comparison to differentiated cells (Pillarisetti et al. 2011). Such differences in elasticity may be used as a way of distinguishing differentiated pluripotent stem cells from undifferentiated ones. Due to its high sensitivity, AFM can also be used for the detection of single molecules. A variety of molecules have been previously investigated in live cells, but these were mainly limited to extracellular membrane proteins, due to the nature of AFM measurements. However, most of the important molecules that can be used as markers for distinguishing between cell states are intracellular. Therefore, for identifying the differentiation state of a cell, the cell interior needs to be investigated. The diameter of a suspended cell is approximately 10–15 μm, and it is separated from the extracellular space by a lipid bilayer membrane. The intracellular space is filled with a large number of complex organelles and biomolecules. Thus, molecular handling techniques with high spatial and temporal precision are required in order to allow manipulation of the intracellular components of a cell. In this chapter, we demonstrate techniques for the detection of intracellular markers using an AFM and a nanoneedle that can be inserted into the live cell and used as an intracellular probe.

14.2 Insertion of the Nanoneedle into Live Cells

Recent research has led to the development of nanoneedle technologies for use in the examination of living cells. Ultra-thin needle-shaped materials, including fabricated carbon nanotubes, have been developed for direct delivery of substances into mammalian cells with extremely minimal cellular damage associated with needle insertion (Chen et al. 2007; Yum et al. 2009). Single or multi-functional needle-shaped nanoprobes have also been developed for investigating the intracellular content of the cell (Sun et al. 2008; Singhal et al. 2011; Yan et al. 2012).

When using such needle-shaped probes, the most important factor for living cell analysis is efficiency of insertion. Needless to say, no information can be obtained when a probe does not reach the cytosol beyond the plasma membrane. The nanoneedle developed by our group is made of a silicon AFM tip etched to a

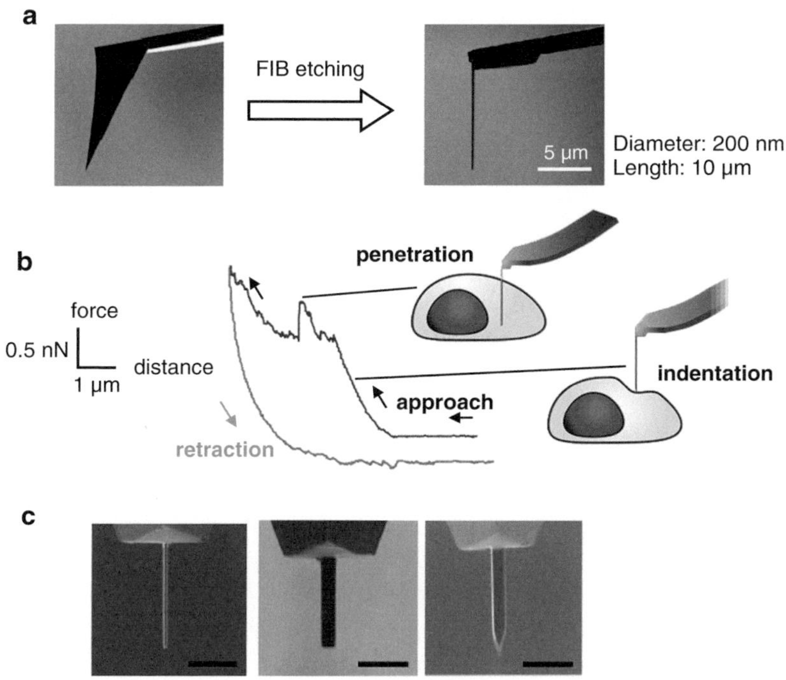

Fig. 14.1 (**a**) SEM images of AFM cantilever and nanoneedle. (**b**) Force–distance curve of nanoneedle insertion into a living cell. (**c**) *Left*: nanoneedle of diameter 200 nm; *Centre*: nanoneedle of diameter 800 nm; *Right*: nanoneedle of diameter 800 nm with conical tip end

high-aspect-ratio (over 50) cylindrical shape with a diameter of only 200 nm (Fig. 14.1a), and has a high mechanical stiffness, which enables repeated insertions through the cell membrane with minimal cell damage (Han et al. 2005a, b; Obataya et al. 2005a, b; Nakamura et al. 2008; Mieda et al. 2012; Ryu et al. 2013). Evidently, over 50 repeated insertions of the nanoneedle into a live cell did not affect the cell doubling time (Mieda et al. 2012). In addition, by employing a perpendicular approach with our AFM nanoneedle, extremely high insertion efficiencies were achieved. Figure 14.1b shows a typical force–distance curve obtained during insertion and retraction of the nanoneedle. During approach, the repulsive force gradually increases as the nanoneedle indents up to about 1 μm into the cell, before penetrating through the plasma membrane, leading to a sharp peak in the force curve. One of the advantages of using an AFM is that success or failure in nanoneedle insertion can be estimated from the force curve in each trial. Nanoneedle insertion efficiency varies depending on the diameter and end shape of the nanoneedle, as shown in Fig. 14.1c (Obataya et al. 2005a, b). The insertion efficiency of 200-nm diameter nanoneedles (left) into human epidermal melanocytes was the highest (92 %). 800-nm nanoneedles with an acute pyramidal tip end (right) showed a much lower efficiency (21 %) than nanoneedles with a cylindrical end and the same diameter (56 %, center). These results suggested that a sharp-edged nanoneedle is not advantageous for penetration through the cell membrane. The shear stress on the cell membrane using a cone-shaped or pyramidal tip might be smaller than that using a cylindrical tip at the same indentation depth.

Insertion efficiencies of nanoneedles vary with cell species (Kagiwada et al. 2010). In addition, observations showed that the actin cytoskeleton, including both stress fibers and the meshwork structure of the plasmalemmal actin undercoat, is very important in facilitating insertion of the nanoneedle into cells, although the mechanical properties of cells do not correlate obviously with the success rate of nanoneedle insertion. Some cell lines have showed low insertion efficiency; for example, insertion efficiency to mouse embryonic fibroblast Balb3T3 was about 40 %. Neural cells also showed relatively low efficiency. Improvement in the insertion efficiency of a nanoneedle into such cells is a significant challenge for nanoneedle-based cell manipulation and analysis. We have recently found that the formation of nanofilms improved insertion efficiency of the nanoneedle into fibroblast and neural cells and it was suggested that nanofilms, together with the mesh structure, directly contribute to the improved insertion efficiency (Amemiya et al. 2012).

Nanoneedle insertion can be applied to high-efficiency molecular delivery. A plasmid containing the GFP gene was adsorbed on a poly-lysine-modified nanoneedle surface, which was then inserted into primary cultured single human mesenchymal stem cell (MSCs) (Han et al. 2005a, b, 2008). A highly efficient gene delivery of over 70 % was achieved, which compared favorably with other major nonviral gene delivery methods (lipofection ~50 %, microinjection ~10 %).

High-efficiency DNA delivery can be applied for single cell diagnostics. We analyzed hormonal response in single breast cancer cells by reporter gene assay using the nanoneedle technology (Han et al. 2009). The proliferation of breast cancer cells is promoted by estrogen stimulation and antagonists, which bind to the estrogen receptor (ER) competitively and are utilized to reduce the growth of breast cancer cells. Some patients experience no therapeutic effect from such hormonal drugs and even suffer from serious side effects. This is often attributed to incorrect diagnosis, in which only the expression of ER is analyzed from a patient's cells. The effects of hormonal drugs on each patient should be tested before medication is given; however, the number of isolated cells is in short supply for further tests. Single-cell manipulation techniques using a nanoneedle are suitable for this type of test. An estrogen–ER complex interacts directly with an estrogen response element (ERE) resulting in enhancement of gene expression. We have attempted to evaluate estrogenic response activity in a human breast cancer cell line (MCF-7) from the expression level of estrogen-responsive GFP reporter vector delivered using a nanoneedle. Constructed estrogen-responsive GFP reporter vectors were first transferred into single MCF-7 cells. The fluorescent intensity of GFP was decreased to 46 % within 24 h by antagonist treatment, demonstrating the antagonistic activity in the operated cell. We were able to successfully determine the hormonal drug effect in a single MCF-7 cell within 48 h after gene delivery using this technique.

14.3 Use of Nanoneedles for mRNA Detection

The state of a cell, such as cell cycle, cell differentiation, or cell transformation, is generally controlled via gene regulation. Monitoring of intrinsic mRNAs in living cells is crucial for understanding these processes. However, investigative tools for monitoring of intrinsic mRNAs are usually limited to visualization in living cells with the help of fluorescent probes (e.g., molecular beacons, fluorescent oligo-DNA probes and fluorescent proteins) (Abe and Kool 2006; Bratu et al. 2003; Mhlanga et al. 2005; Nitin et al. 2004; Ozawa et al. 2007; Tsuji et al. 2000). Although these probes are very powerful tools for investigating the localization of mRNAs in living cells, they need to be expressed in or introduced into the cells by chemical or physical methods. However, these methods can only be used with certain cell types and growth conditions, and may affect cell viability due to invasive damage to cells. Moreover, the introduced or expressed probes cannot be removed from the cells. Using probes immobilized on a nanoneedle, we can ensure removal of the probes upon evacuation of the nanoneedle from the cell.

We have developed a molecular-beacon-modified nanoneedle for the detection of intrinsic mRNA in a living cell (Fig. 14.2a). A molecular beacon is an oligonucleotide with complementary sequences on either end of the probe; this enables the molecular beacon to fold into a hairpin configuration, in which a fluorophore and a quencher can be held in close proximity (Bratu et al. 2003; Tyagi and Kramer 1996). The formation of a probe–target hybrid abolishes the hairpin configuration, shifts the fluorophore away from the vicinity of the quencher, and restores the probe's fluorescence. The nanoprobe can be inserted into living cells and safely evacuated from the cells, leaving no probe residues in the cell. This method thus enables real-time investigation of target mRNAs in the living cell with minimal cell damage.

The GAPDH gene is a housekeeping gene that is highly expressed in various cell types. We have developed a method of detecting GAPDH mRNA in a single living cell as a proof-of-concept for quantifying the amounts of various mRNA species in a single living cell. A molecular beacon targeting the human GAPDH mRNA was designed as follows:

5′-(Alexa488)-ACGACGGAGTCCTTCCACGATACCACGTCG-bT-(BHQ-1)-3′, where underlined sequences indicate bases in the self-hybridizing stem, and bT indicates the biotinylated thymine base, GAPDH complementary sequence was designed by referring to a previous report (Tsourkas et al. 2003). A nanoneedle was chemically modified with 3-aminopropyl triethoxysilane and EZ-Link Sulfo-NHS-LC-LC-Biotin. The avidin-biotin bridging was utilized for immobilization of the molecular beacon. A molecular-beacon-functionalized nanoneedle was inserted into a HeLa cell using AFM and was kept inside for 2 min. A fluorescence image of the nanoneedle was then obtained using confocal laser scanning microscope. The nanoneedle was then evacuated from the cell and a second fluorescence image was obtained after 1 min.

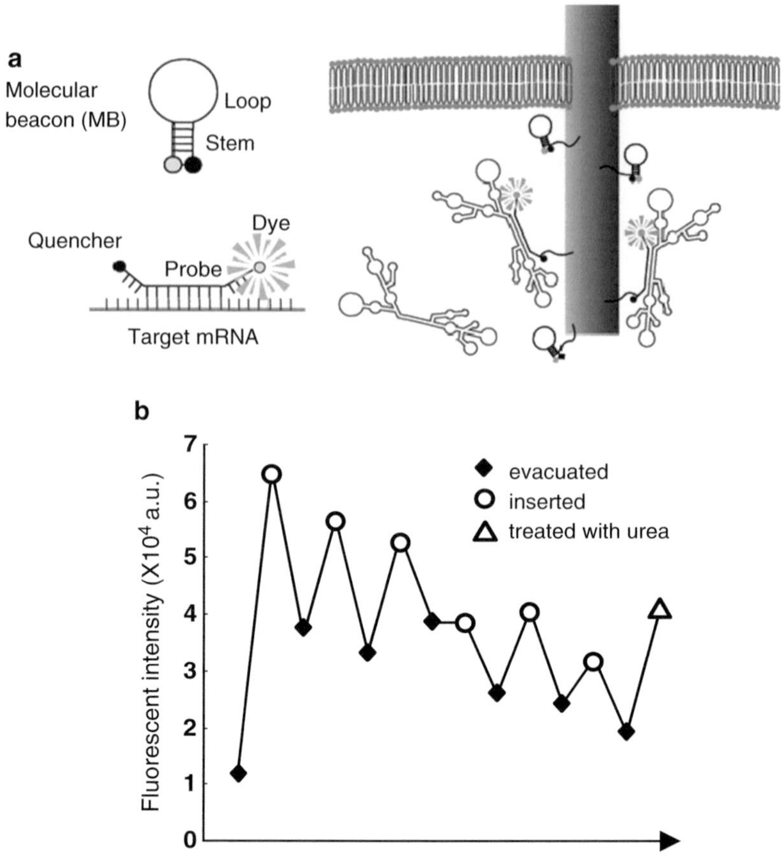

Fig. 14.2 (**a**) Schematic drawing of mRNA detection using a molecular-beacon-functionalized nanoneedle. (**b**) Change of fluorescence intensity of molecular-beacon-functionalized nanoneedle inside or outside of the cell

We confirmed insertion of a molecular-beacon-modified nanoneedle by real-time monitoring of the force–distance curves, as previously described. In the extracellular environment, the fluorescence intensity of the nanoneedle slightly varied in each experiment; however, it increased by a factor of 1.88 ± 0.80 following insertion into the HeLa cell. The variation in the fluorescence intensity of the nanoneedle for each measurement was assumed to be due to a variation in the surface modification with the molecular beacons. The results demonstrate that human GAPDH mRNA in a single HeLa cell could be successfully detected using the nanoneedle.

Successive insertions were carried out with multiple cells. The increment of fluorescent intensity on the nanoneedle was observed in five of six tested cells, as shown in Fig. 14.2b. The reproducible response showed that the nanoneedle method allowed the detection of mRNA in multiple cells using the same nanoneedle. Thus, we established that the immobilized molecular beacon could reversibly react with mRNAs.

Although GAPDH mRNA is about 1,300 bases long, the hybridized sequence is only 19 bases long. When extracting the nanoneedle from the cell, the mechanical friction between the mRNA and plasma membrane prevents extraction of mRNA from the cell. Therefore, most mRNAs hybridized with the molecular beacon remained in the cell. One more point of interest is the specificity of this detection method: there is a difference of three bases between mouse and human in the targeted sequences GAPDH sequences. A molecular-beacon-modified nanoneedle was also inserted into a mouse mesenchymal stem cell, C3H10T1/2, resulting in no significant response. Thus, the molecular beacon immobilized on the nanoneedle maintained high specificity, allowing discrimination between cell species.

14.4 Mechanical Detection of Intracellular Proteins

Beside optically-based techniques, alternative methods may be useful for the investigation of changes in the mechanical properties of cells, since many cell functions related to cytoskeletal dynamics have been demonstrated. One example is the use of optical tweezers (Guck et al. 2001), which has been employed for the diagnosis of cancer cells (Remmerbach et al. 2009). The main disadvantage of this technique is the inability to directly detect intracellular proteins. Concerning this issue, the nanoneedle probe combined with AFM has the advantage of allowing direct access to intracellular molecules. Detection of intracellular molecules is not limited to visualization using optical microscopy, but, in addition, force-based detection is possible using this layout. AFM has been extensively employed for the investigation of molecular interactions, using tips that are functionalized with various molecules, with the main advantage being the ability to measure forces at pico-newton precision. Using standard AFM cantilevers, force detection is limited to extracellular membrane-bound proteins. However, using the nanoneedle-shaped probe, after appropriate surface modifications and antibody immobilization, allows for the force-based detection of intracellular proteins. Unbinding forces between the antibody-immobilized nanoneedle and intracellular cytoskeletal proteins can be measured during nanoneedle retraction from the cell (Fig. 14.3a) (Obataya et al. 2005a, b;

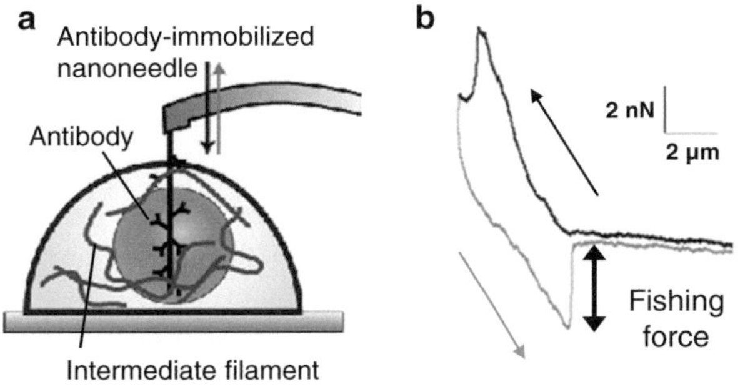

Fig. 14.3 (**a**) Schematic illustration of force-based detection of intracellular proteins using antibody-immobilized nanoneedle. (**b**) Fishing force obtained by detection of intracellular proteins

Mieda et al. 2012). An example of a force curve is shown in Fig. 14.3b. The peaks appearing during retraction of the nanoneedle from the cell represent unbinding events between the antigens and antibodies. This method allows direct analysis of intracellular proteins in the live cell without the need to introduce potentially harmful marker molecules and with only a transient penetration which does not affect cell viability (Mieda et al. 2012). Since AFM allows accurate manipulation of the probe, local measurement of cells adhering to the substrate is possible. Hence, the use of antibody-immobilized nanoneedles has an obvious advantage for diagnosis and research applications. Appropriate targets for this type of measurement are cytoskeletal proteins, which play an important role in fundamental functions of the cell such as motility, deformation, differentiation and intracellular transportation. The main components of the cytoskeleton are actin, intermediate filament and microtubules. Actin and microtubules are commonly expressed in all type of eukaryotic cells, while intermediate filaments are cell-type-specific. In the following section, mechanical detection of each of the three cytoskeletal components will be introduced.

Based on the sequence homology, intermediate filaments can be categorized into five different types. In humans, they are coded by over 70 different genes and their expression is often regulated by alternative splicing (Zehner 1991; Hol et al. 2003; Herrmann et al. 2009). Various types of intermediate filaments are differently expressed in a tissue-specific manner or under developmental regulations, although they have a common domain of coiled-coil at the centre of the sequences. Thus, intermediate filaments are often employed as cell-type-specific markers. Nestin, a type IV intermediate filament protein, was reported to show a transient increase in its expression in neural stem cells during the developmental process (Michalczyk and Ziman 2005). Discrimination of neural stem cells is an important step in their clinical use, such as the treatment of neurogenic diseases and healing of spinal cord and brain injuries (Iwanami et al. 2005; Ogawa et al. 2002; Okada et al. 2008). In a previous report, we successfully performed mechanical detection of nestin in nestin-positive P19 mouse embryonic carcinoma cells (Mieda et al. 2012). Antibodies were covalently immobilized onto a nanoneedle using 2-methacryloyloxyethyl-phosphorylcholine (MPC) polymer, which helps in reducing nonspecific adsorption of cellular substances. The peak pulling force during nanoneedle retraction from the cell following its insertion, termed the "fishing force", was measured for adherent P19 cells. Fishing forces of over 4 nN were measured for the nestin-positive P19 cells. Meanwhile, the fishing forces for mouse NIH3T3 fibroblast cells, which are nestin-negative, were limited to below 1 nN (Fig. 14.4). A threshold of fishing force was set as the average value plus four standard deviations (Av + 4SD) of the nestin-negative results and applied to NIH3T3 for discriminating nestin-positive cells. All of the measured P19 cells were judged as nestin-positive and NIH3T3 as nestin-negative. Furthermore, the antibody-immobilized nanoneedle underwent antibody-blocking testing in order to confirm that the fishing force was indeed caused by specific antibody–antigen binding. Blocking of the needle with recombinant Nes256 protein showed successful suppression of fishing forces to the level of nestin-negative cells. The performance of nestin detection by the antibody-immobilized nanoneedle can be also confirmed using P19 cells that have been differentiated to neurons by

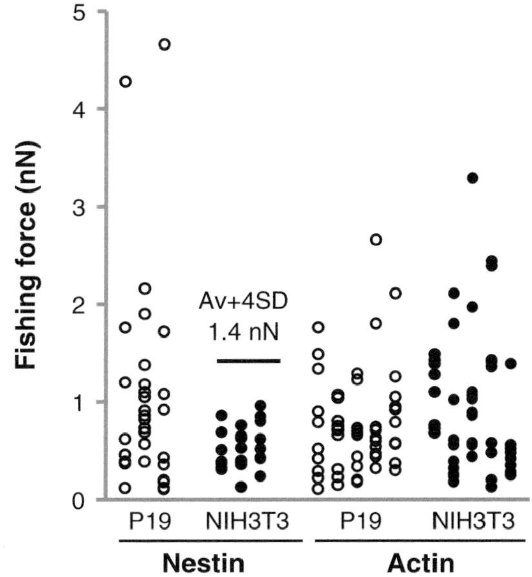

Fig. 14.4 Detection of nestin and actin from P19 and NIH3T3 cells. Fishing forces measured in single cells are plotted vertically

retinoic acid treatment. The differentiated cells are known to have decreased nestin expression. For those cells, the fishing forces were significantly decreased down to the level of nestin-negative cells. In addition, similar strong fishing forces for nestin-immobilized nanoneedles were also obtained on a series of mouse breast cancer cells. Since it is reported that nestin is often over-expressed in cancer cells such as breast, prostate and pancreatic cancers, which is thought to be related to the migration of cancer cells, such a quantitative analysis of nestin may be beneficial for unraveling the mechanisms of cancer metastasis and tumor progression.

Actin is known to form dynamic network structures, which depend on cell state and cell cycle. Traditionally, the state of the intracellular cytoskeleton is investigated using various imaging techniques, including the commonly used method of flow cytometry, which measures the intensity of pre-labeled cells for the analysis of various conditions such as hemolytic anemia (King et al. 2000). Mechanical analyses of cell deformability, such as the use of optical stretchers, are also used in the investigation of various diseases, including the identification of cancer, by live cell mechanical phenotyping (Guck et al. 2001; Remmerbach et al. 2009). These techniques, however, do not allow for the direct detection of intracellular proteins in a living cell. A method of analytically diagnosing the condition of the actin cytoskeleton in the live cell would therefore be desirable. The antibody-immobilized nanoneedle has great advantage for this purpose, since its high aspect ratio and mechanical stiffness allow repeated penetration through the membrane of the live cell with minimal cell damage (Obataya et al. 2005a, b; Han et al. 2005a, b). With the use of the anti-actin antibody-immobilized nanoneedle, which was prepared in the same way as in the case of nestin detection, actin filaments were successfully detected in the NIH3T3 cell line (Silberberg et al. 2013). The fishing force detected during nanoneedle evacuation reached a maximum of over 3 nN, while remaining usually around 0.5 nN when a bare, un-immobilized nanoneedle was used. As an additional control experiment, the antibody-immobilized nanoneedle was blocked with free globular actin (G-actin) in vitro prior to its insertion into the cell, showing decreased fishing forces down to the same level as the bare nanoneedle, pointing to the specificity of actin detection using this method. As the assembly and disassembly of filamentous actin can be controlled by drug treatments, the high sensitivity of this method was also shown through the measurement of the fishing force during actin disruption treatment. NIH3T3 cells were treated with 0.25 μM cytochalasin D, a drug that specifically disrupts actin filaments, and the time-dependent change in the fishing forces was successfully detected. With this relatively low dose of cytochalasin D, most of the actin stress fibers were retained yet some decrease in fishing forces could be detected, pointing to the high sensitivity of this method. A similar effect of another drug, Y-27632, a rho-kinase (ROCK) inhibitor known to disrupt actin stress fibers, could be also detected at low doses using this method (Narumiya et al. 2000). The mechanical detection of actin with the antibody-immobilized nanoneedle was found to be sensitive enough to distinguish minute changes in the cytoskeleton condition, which were otherwise difficult to detect by conventional fluorescence microscopy techniques. The availability of a high-sensitivity, real-time measurement in living cells, without the need for labeling or transfection procedures, has clear benefits for the research and diagnosis of cytoskeleton-related diseases.

Microtubules are the thickest of the cytoskeleton components, with a diameter of approximately 25 nm. They are hollow, cylindrical tubes which are composed of a unit of α- and β-tubulin heterodimers. Microtubules in cells are radially arranged to the periphery from the centrosome that is near the nucleus, spanning the entire cell. They play important roles in organelle transport and organization, in cell division and chromosome distribution, and in mechanical stabilization of the cell (Ingber 1993; Wang et al. 1993). Microtubules have been shown to be implicated in several neurodegenerative diseases such as Alzheimer's disease, the most common cause of dementia among elderly people, in which microtubule disintegration plays a major role (Salinas et al. 2007; Matsuyama and Jarvik 1989; Iqbal et al. 1986; Cash et al. 2003). Thus, the precise analysis of microtubule dynamics and their mechanical properties in the live cell has clear benefits in health and disease. By immobilizing anti-tubulin antibodies onto nanoneedles, microtubules could be successfully detected in a similar way as was done for actin (Silberberg et al. 2014). When nocodazole, a microtubule-disrupting agent, was added to the NIH3T3 culture, a clear time-dependent decrease in the measured microtubule fishing forces was observed, in a similar fashion to that observed for cytochalasin-treated cells in the case of actin detection. The high specificity and high sensitivity of microtubule targeting analysis again showed the ability of this technique to detect real-time changes in the cytoskeleton of live cells.

Antibody-immobilized nanoneedles were successfully inserted into living cells without causing disturbance to the membrane nor influencing cell viability (Mieda et al. 2012). Specific fishing force were successfully detected for three different types of cytoskeletal components, nestin, actin and microtubules. To enhance the versatility of this technique, improvements in the efficiency of immobilized antibody will be beneficial for facilitating large numbers of antigen–antibody interactions. Indeed, fishing forces of over 10 nano-newtons against intermediate filaments were detected in our recent research (data not shown) by optimizing the conditions for immobilizing antibodies to the nanoneedle. With such improvements, a wider variety of proteins including various types of intermediate filaments, actin-/microtubule-associated proteins and proteins related to cell adhesion, could be targeted in the near future.

14.5 Summary

The nanoneedle technology presented here is a novel and versatile tool, and can be applied for the detection of a wide variety of intracellular molecules. One of the main challenges left to tackle is the low throughput of the technique, which is currently limited to the analysis of one cell at a time. Some success in multiple cell analysis was reported previously using nanopillar or nanowire arrays (Berthing et al. 2012; Xie et al. 2012). However, these do not allow for actual force measurement, a clear benefit of the nanoneedle–AFM combination described here. The future development of nanoneedle arrays with high aspect ratio may provide an attractive tool with various applications, such as the selective high-throughput sorting of stem cells for medical transplantation.

References

Abe H, Kool ET (2006) Flow cytometric detection of specific RNAs in native human cells with quenched autoligating FRET probes. Proc Natl Acad Sci U S A 103:263–268

Amemiya Y et al (2012) Formation of nanofilms on cell surfaces to improve the insertion efficiency of a nanoneedle into cells. Biochem Biophys Res Commun 420:662–665

Berthing T et al (2012) Cell membrane conformation at vertical nanowire array interface revealed by fluorescence imaging. Nanotechnology 23:415102

Bratu DP, Cha BJ, Mhlanga MM, Kramer FR, Tyagi S (2003) Visualizing the distribution and transport of mRNAs in living cells. Proc Natl Acad Sci U S A 100:13308–13313

Cash AD et al (2003) Microtubule reduction in Alzheimer's disease and aging is independent of tau filament formation. Am J Pathol 162:1623–1627

Chen X, Kis A, Zettl A, Bertozzi CR (2007) A cell nanoinjector based on carbon nanotubes. Proc Natl Acad Sci U S A 104:8218–8222

Costa KD (2006) Imaging and probing cell mechanical properties with the atomic force microscope. Methods Mol Biol 319:331–361

Guck J et al (2001) The optical stretcher: a novel laser tool to micromanipulate cells. Biophys J 81:767–784

Haghparast SM, Kihara T, Shimizu Y, Yuba S, Miyake J (2013) Actin-based biomechanical features of suspended normal and cancer cells. J Biosci Bioeng 116:380–385

Han S, Nakamura C, Obataya I, Nakamura N, Miyake J (2005a) Gene expression using an ultrathin needle enabling accurate displacement and low invasiveness. Biochem Biophys Res Commun 332:633–639

Han SW, Nakamura C, Obataya I, Nakamura N, Miyake J (2005b) A molecular delivery system by using AFM and nanoneedle. Biosens Bioelectron 20:2120–2125

Han S et al (2008) High-efficiency DNA injection into a single human mesenchymal stem cell using a nanoneedle and atomic force microscopy. Nanomedicine 4:215–225

Han SW, Nakamura C, Imai Y, Nakamura N, Miyake J (2009) Monitoring of hormonal drug effect in a single breast cancer cell using an estrogen responsive GFP reporter vector delivered by a nanoneedle. Biosens Bioelectron 24:1219–1222

Herrmann H, Strelkov SV, Burkhard P, Aebi U (2009) Intermediate filaments: primary determinants of cell architecture and plasticity. J Clin Invest 119:1772–1783

Hertz H (1881) Über dieBerührung faster elastischer Körper. J für die reine und angewandte Mathematik 92:156–171

Hol EM et al (2003) Neuronal expression of GFAP in patients with Alzheimer pathology and identification of novel GFAP splice forms. Mol Psychiatry 8:786–796

Ingber DE (1993) Cellular tensegrity: defining new rules of biological design that govern the cytoskeleton. J Cell Sci 104(Pt 3):613–627

Iqbal K et al (1986) Defective brain microtubule assembly in Alzheimer's disease. Lancet 2:421–426

Iwanami A et al (2005) Transplantation of human neural stem cells for spinal cord injury in primates. J Neurosci Res 80:182–190

Kagiwada H et al (2010) The mechanical properties of a cell, as determined by its actin cytoskeleton, are important for nanoneedle insertion into a living cell. Cytoskeleton 67:496–503

King MJ et al (2000) Rapid flow cytometric test for the diagnosis of membrane cytoskeleton-associated haemolytic anaemia. Br J Haematol 111:924–933

Matsuyama SS, Jarvik LF (1989) Hypothesis: microtubules, a key to Alzheimer disease. Proc Natl Acad Sci U S A 86:8152–8156

Mhlanga MM, Vargas DY, Fung CW, Kramer FR, Tyagi S (2005) tRNA-linked molecular beacons for imaging mRNAs in the cytoplasm of living cells. Nucleic Acids Res 33:1902–1912

Michalczyk K, Ziman M (2005) Nestin structure and predicted function in cellular cytoskeletal organisation. Histol Histopathol 20:665–671

Mieda S et al (2012) Mechanical force-based probing of intracellular proteins from living cells using antibody-immobilized nanoneedles. Biosens Bioelectron 31:323–329

Nakamura C, Kamiishi H, Nakamura N, Miyake J (2008) A nanoneedle can be inserted into a living cell without any mechanical stress inducing calcium ion influx. Electrochem 76:586–589

Narumiya S, Ishizaki T, Uehata M (2000) Use and properties of ROCK-specific inhibitor Y-27632. Methods Enzymol 325:273–284

Nitin N, Santangelo PJ, Kim G, Nie S, Bao G (2004) Peptide-linked molecular beacons for efficient delivery and rapid mRNA detection in living cells. Nucleic Acids Res 32:e58

Obataya I, Nakamura C, Han S, Nakamura N, Miyake J (2005a) Nanoscale operation of a living cell using an atomic force microscope with a nanoneedle. Nano Lett 5:27–30

Obataya I, Nakamura C, Han S, Nakamura N, Miyake J (2005b) Mechanical sensing of the penetration of various nanoneedles into a living cell using atomic force microscopy. Biosens Bioelectron 20:1652–1655

Ogawa Y et al (2002) Transplantation of in vitro-expanded fetal neural progenitor cells results in neurogenesis and functional recovery after spinal cord contusion injury in adult rats. J Neurosci Res 69:925–933

Okada Y et al (2008) Spatiotemporal recapitulation of central nervous system development by murine embryonic stem cell-derived neural stem/progenitor cells. Stem Cells 26:3086–3098

Ozawa T, Natori Y, Sato M, Umezawa Y (2007) Imaging dynamics of endogenous mitochondrial RNA in single living cells. Nat Methods 4:413–419

Pillarisetti A et al (2011) Mechanical phenotyping of mouse embryonic stem cells: increase in stiffness with differentiation. Cell Reprogram 13:371–380

Radmacher M, Fritz M, Kacher CM, Cleveland JP, Hansma PK (1996) Measuring the viscoelastic properties of human platelets with the atomic force microscope. Biophys J 70:556–567

Remmerbach TW et al (2009) Oral cancer diagnosis by mechanical phenotyping. Cancer Res 69:1728–1732

Ryu S et al (2013) Nanoneedle insertion into the cell nucleus does not induce double-strand breaks in chromosomal DNA. J Biosci Bioeng 116:391–396

Salinas S, Carazo-Salas RE, Proukakis C, Schiavo G, Warner TT (2007) Spastin and microtubules: functions in health and disease. J Neurosci Res 85:2778–2782

Silberberg YR et al (2013) Evaluation of the actin cytoskeleton state using an antibody-functionalized nanoneedle and an AFM. Biosens Bioelectron 40:3–9

Silberberg YR et al (2014) Detection of microtubules in vivo using antibody-immobilized nanoneedles. J Biosci Bioeng 117:107–112

Singhal R et al (2011) Multifunctional carbon-nanotube cellular endoscopes. Nat Nanotechnol 6:57–64

Sun P et al (2008) Nanoelectrochemistry of mammalian cells. Proc Natl Acad Sci U S A 105:443–448

Takahashi K, Yamanaka S (2006) Induction of pluripotent stem cells from mouse embryonic and adult fibroblast cultures by defined factors. Cell 126:663–676

Tsourkas A, Behlke MA, Rose SD, Bao G (2003) Hybridization kinetics and thermodynamics of molecular beacons. Nucleic Acids Res 31:1319–1330

Tsuji A et al (2000) Direct observation of specific messenger RNA in a single living cell under a fluorescence microscope. Biophys J 78:3260–3274

Tyagi S, Kramer FR (1996) Molecular beacons: probes that fluoresce upon hybridization. Nat Biotechnol 14:303–308

Wang N, Butler JP, Ingber DE (1993) Mechanotransduction across the cell-surface and through the cytoskeleton. Science 260:1124–1127

Xie C, Lin ZL, Hanson L, Cui Y, Cui BX (2012) Intracellular recording of action potentials by nanopillar electroporation. Nat Nanotechnol 7:185–190

Yan R et al (2012) Nanowire-based single-cell endoscopy. Nat Nanotechnol 7:191–196

Yum K, Na S, Xiang Y, Wang N, Yu MF (2009) Mechanochemical delivery and dynamic tracking of fluorescent quantum dots in the cytoplasm and nucleus of living cells. Nano Lett 9:2193–2198

Zehner ZE (1991) Regulation of intermediate filament gene expression. Curr Opin Cell Biol 3:67–74

Chapter 15
Plasmonic Sensors for Analysis of Proteins and an Oncologic Drug in Human Serum

Jean-Francois Masson and Sandy Shuo Zhao

Abstract Nanobiosensors exploiting the optical phenomenon of surface plasmon resonance (SPR) are ideal candidates for the design of optical sensors for clinical diagnostics. The label-free nature and sensitivity of SPR biosensors to binding events makes them generally applicable to the detection of a broad class of biomolecules. Recent advances in instrument design, surface chemistry, nanomaterials and biosensing strategies have enabled numerous applications of SPR biosensors. This chapter will explore in greater detail the challenges and solutions developed recently for the analysis of proteins and drugs in crude biofluids. Specifically, surface chemistry has been investigated thoroughly to minimize the interference of nonspecific adsorption from biofluids, while nanomaterials have been exploited to increase the sensitivity of SPR biosensors, with biosensing strategies involving nanoparticles allowing for the analysis of small molecules. Additionally, miniaturization and optimization of instrumental design have paved the way towards point-of-care diagnostics. The successful detection of biologically relevant molecules directly in biofluids relies on all of these recent advances. In this chapter, they will be contextualized for the analysis of proteins and an oncologic drug, methotrexate, in crude serum samples using SPR sensing.

Keywords Biosensing • Instrumentation • Nanomaterials • Surface chemistry • Surface plasmon resonance

15.1 Introduction

Medicine is pushing strongly towards personalization of treatment and decentralization of patient care. This paradigm shift brings new sets of challenges for clinical biochemistry laboratories, as high throughput and large laboratory equipment may not be the ideal solution for these situations. Rapid and on-site analytical measurements may provide an accurate and sufficiently fast response to allow patients at

J.-F. Masson (✉) • S.S. Zhao
Department of Chemistry, University of Montreal, Montreal, QC, Canada
e-mail: jf.masson@umontreal.ca

M.C. Vestergaard et al. (eds.), *Nanobiosensors and Nanobioanalyses*,
DOI 10.1007/978-4-431-55190-4_15

home or at proximity healthcare facilities to monitor their susceptibility to develop common diseases or to monitor the progression or treatment of a disease. Novel analytical techniques are thus needed to combine portability, sensitivity, selectivity and quantification of a single or small set of markers or drugs administered to patients.

Biosensing relies on simple principles of signal transduction of molecular recognition events induced from the capture, hybridization or reaction of an analyte (such as proteins, DNA, enzymes and metabolites) by a host receptor. Transducers can rely on electrochemistry, optics, mass detection, magnetism or radioactivity to detect molecular events occurring on the surface of the sensor. Several biosensing architectures were discussed in Part II of this book and have advantages in terms of sensitivity, selectivity, simplicity, cost, accuracy and many other factors. Nanomaterials and nanotechnologies are increasingly exploited in biosensing architectures to improve the sensitivity, selectivity and simplicity of an analysis. Part III of this book reviews some strategies related to nanomaterials and their functionalization, and each sensing strategy, nanomaterial and surface modification have niche applications for which they are advantageous.

While biosensors can be applied to a variety of analytes, this chapter will concentrate on two important classes of molecules: proteins and oncologic drugs. Proteins are often biomarkers indicative of the state and progression of a disease. Monitoring the presence or concentration of specific proteins is important in establishing clinical diagnoses (Polanski and Anderson 2006). ELISA, a workhorse of clinical diagnostic laboratories, currently performs most protein analyses. While ELISA has high throughput, sensitivity and reliability, it is time- and labor-consuming and must be performed in centralized laboratories. In the case of critical emergencies, faster response times can be life-saving. Analysis at the sample point-of-collection (bedside, general practitioner's office, ambulance or patient's home) by a rapid technology can address these issues with current practices. Biosensors address these challenges by providing a simple and rapid methodology for measuring biomarkers.

Chemotherapy treatments use a cocktail of drugs administered at high enough concentrations to ensure efficacy of treatment, while minimizing the side effects. The metabolism of individual patients varies according to a set of factors (such as genetic variations, gender, age, among others (Dasgupta 2012)), and it has been demonstrated that regular monitoring of the serum level of oncologic drugs is beneficial to the patients, to avoid under- or overdosing, and for maximum treatment efficacy (Dasgupta 2012). While beneficial, long-term monitoring of a patient is difficult due to the lack of point-of-care analytical techniques to monitor this important class of drugs. Current analytical techniques for monitoring drug levels in patients require centralized clinical facilities. Liquid chromatography coupled to mass spectrometry (LC-MS) or dedicated analyzers (such as fluorescence polarization immunoassays) are the current state-of-the-art methodologies used to analyze oncologic drugs. These techniques are costly, require trained staff to operate and thus are only found in large, central clinical laboratories of major hospitals. The need is thus well established for the development of an analytical technique for the successful analysis of oncologic drugs. This is a great challenge for biosensing technologies.

15.2 Challenges of Biosensing in Human Biofluids

Biofluids often constitute the analytical samples for clinical diagnostics. In the case of sensing in crude biofluids, the transducer must be insensitive to the optical, physical and chemical properties of serum, urine, cell lysate or saliva. As the most common and often the most representative biofluid for clinical purposes, serum consists of a complex mixture of a high concentration of protein, metabolites and salts, in addition to being a scattering and absorptive optical media. The transducer must be insensitive to (or allow compensation for) these potential interferences.

Electrochemical sensors are very selective and perform relatively well in serum samples, such that the glucometer represents the success story of the field to date. However, they also represent a challenging solution for protein or drug detection due to the lack of electrochemical activity of these classes of molecules. Classical optical techniques are of limited scope in serum due to the scattering and absorption properties of this biofluid, but evanescent wave techniques are minimally perturbed in these conditions. While fluorescence and infrared (IR) spectroscopies are adapted for evanescent wave interrogation, fluorescence requires extrinsic labeling of the analyte and IR is poorly selective for proteins and suffers from a large background response from water.

Surface plasmon resonance (SPR) is an evanescent wave technique relying on the optical properties of thin metallic films deposited on a prism. Surface plasmons (SP) are generated by the interaction of light undergoing total internal reflection with free electrons in noble metals deposited as a thin film (generally Au) on the optical waveguide (Zayats et al. 2005). The frequency and wavevector of light interacting with SP depends on the optical density (or refractive index) in the vicinity of the thin Au film. This sensitive nature of SP to the immediate physical environment of the thin Au film makes it an ideal technique for the label-free and sensitive detection of analytes. SPR biosensors are based on the specific interaction or capture of an analyte with a selective molecular receptor immobilized to the Au film (Fig. 15.1).

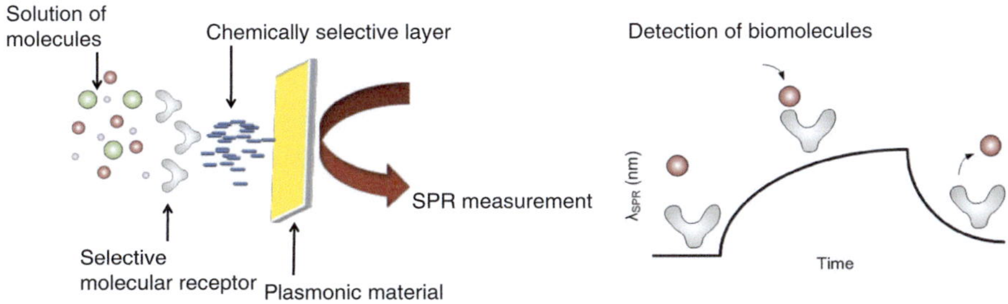

Fig. 15.1 Composition of SPR biosensors. The plasmonic material is modified with a chemically selective layer, capable of minimizing nonspecific adsorption and immobilizing highly active molecular receptors. The molecular receptor can be chemically bound to the layer, chelated or physisorbed. It then captures the analyte from a solution, which is converted into a change in the SPR signal that is observed in the sensorgram. Reproduced from Couture et al. (2013b) with permission from the PCCP Owner Societies

Binding events are accompanied by a refractive index change, which is transduced in a change of the SPR coupling conditions. Thus, the resonance angle or wavelength shifts and serves to quantify analytes. Several literature articles review the principles and applications of SPR sensing (Homola et al. 2005; Hoa et al. 2007; Sharma et al. 2007; Homola 2008; Abbas et al. 2011; Bolduc and Masson 2011; Couture et al. 2013b).

15.3 Minimizing Nonspecific Adsorption with Surface Chemistry

The molecular receptors immobilized on the SPR sensors impart most of the selectivity to the device. However, SPR surfaces are not fully covered by molecular receptors and the Au film is partly exposed to biofluids. These regions are prone to nonspecific interactions with the matrix of the biofluid, which depend on the nature of the specific biofluid being analysed. Generally, nonspecific adsorption is triggered by protein adsorption driven by hydrophobic interactions. Biofluids are often protein-rich environments and so nonspecific adsorption is particularly problematic in this media. SPR sensors are quasi-universal refractive index detectors, such that any proteins or biomolecules binding to the surface lead to a measurable change in the SPR signal. This nonspecific response is of the same order of magnitude as the specific response of the SPR sensors, limiting their use in biofluids. Nonspecific adsorption has thus been one of the major barriers for the use of SPR sensors in clinical analyses (Blaszykowski et al. 2012).

Understanding the fundamental processes driving nonspecific adsorption has been instrumental in mitigating this limiting factor of biosensors. Surface chemistry has been utilized in the past decade to explore the physical processes driving nonspecific adsorption and to attempt to minimize this problem (Mrksich and Whitesides 1996). The chemical group exposed at the surface of a sensor plays an important role in nonspecific adsorption. The Whitesides group has performed numerous studies on the nonspecific adsorption of two proteins, fibrinogen and lyzosyme (Ostuni et al. 2001a, b). These studies concluded that surfaces with more hydrophilic groups, such as ethylene glycols, are very efficient at preventing nonspecific adsorption, but other chemical groups can also minimize the effect. These include amine, amides and amino acids (Table 15.1). More hydrophobic groups performed significantly

Table 15.1 Average nonspecific adsorption from various chemical groups of monolayers[a]

	Hydro-phobic	Ether	EG[b]	Amine	Amide	Amide/amino acid	Crown ether	Sugar	Nitrile
% ML[c]	100	31	0.8	13	16	8.8	24	30	48

[a]Data collected from Ostuni et al. (2001a)
[b]EG: ethylene glycol
[c]% ML is the percentage of a monolayer created in 3 min by 1 mg/mL fibrinogen

worse in minimizing the nonspecific adsorption of these proteins. They also discovered that adsorption does not differ significantly between the two proteins tested on the various surfaces. The investigation of the underlying principles of nonspecific adsorption revealed that hydrogen bond acceptors, neutral and polar molecules impart protein-resistant properties to surfaces. It was also noted that zwitterionic surfaces fulfill these criteria (Ostuni et al. 2001b). The surfaces promoting a hydration layer and exclusion of proteins in this hydration layer were found to be protein-resistant (Kane et al. 2003).

Serum is a complex mixture of proteins composed essentially of hemoglobin, albumin, complements and factors, immunoglobulins, transferrin and fibrinogen (Polanski and Anderson 2006). While the groundbreaking work of Whitesides elucidated the underlying principles of protein adsorption, the complexity of serum may hide the interplay between proteins adsorbing to surfaces. Thus, studies were undertaken to comprehend the sequence of reactions ongoing in plasma (Green et al. 1999). It was revealed that albumin was the first protein to bind to the surface, but was then displaced by immunoglobulin G (IgG) and then by fibrinogen in a matter of minutes (Green et al. 1999). These experiments clearly demonstrated the cascade of protein adsorption that blood-based biofluids undergo on surfaces, and the influence of the presence of several molecules in the reaction mixture. Thus, there has been a driving force to study surfaces in actual biofluids to evaluate the nonspecific adsorption resistance of self-assembled monolayers.

Dextran is one of the most common surface chemistries in SPR studies. It provides high-density immobilization of molecular receptors, resulting in large SPR responses in the detection of analytes. Depending on the molecular weight and surface modifications, dextran can be made quite stable in biofluids (Frazier et al. 2000). Dextrans of lower molecular weight are less prone to nonspecific adsorption of serum on SPR sensors and immobilization of antibodies does not change the nonspecific adsorption properties (Masson et al. 2004a). However, it is important to measure the capability of surface chemistries in creating an effective biosensor and in protecting against nonspecific adsorption for assessing the performance of different surfaces for SPR biosensors. For example, dextrans of low molecular weight do not provide sufficient binding sites to detect low concentrations of cardiac Troponin I (Masson et al. 2005). It was observed that dextran of 500 kDa offered moderate protection in serum and provided relatively large signals in a biosensing experiment. However, the performance in crude serum was insufficient for the use of these sensors in biofluids.

Polyethylene glycol (PEG) surfaces provide better protection against nonspecific adsorption than dextran. However, PEG surfaces suffer from lower sensitivity due to fewer binding sites on the surface of the SPR sensor and more limited dynamic range in a myoglobin sensor (Masson et al. 2005). We recently revisited the performance of PEG monolayers in biosensing assays and confirmed that PEG was underperforming for the detection of IgG in comparison to mercaptoalkanoic acid and ionic liquid monolayers (Ratel et al. 2013). PEG monolayers immobilized three to five times fewer antibodies using an identical surface chemistry, and thus detection of IgG was reduced by an equivalent factor. Self-assembled monolayers based on mercaptoalkanoic acids have previously been used to improve the performance of sensors for myoglobin, with reasonable stability in serum (Masson et al. 2006a). These monolayers offered a good compromise between stability to nonspecific adsorption and sensing performance. Thus, they were employed in the detection of proteins in crude serum (Masson et al. 2007) and cell culture medium (Battaglia et al. 2005). These sensors, however, must be pre-incubated in the media prior to sample analysis, which may be limiting in the analysis of actual clinical samples, and therefore it was important to investigate alternative surface chemistries.

In the past decade new surface chemistries have been developed that further improve the performance of SPR sensors in crude biofluids (Blaszykowski et al. 2012). One of the very promising approaches to simultaneously minimizing nonspecific adsorption and improving biosensing exploits zwitterionic betaine layers. The betaine motif is composed of a quaternary alkylammonium with an acid group on one of the chains. The acid group can serve to anchor antibodies in biosensing assays. A series of papers published by the Jiang group demonstrated the ultralow fouling properties of sulfo- and carboxy-betaine monolayers in plasma and serum (Ladd et al. 2008; Vaisocherova et al. 2008, 2009b; Zhang et al. 2008; Cheng et al. 2009). They were able to achieve biosensing in crude serum for activated-leukocyte cell adhesion molecule (ALCAM) with less than 3 ng/cm^2 residual nonspecific adsorption on a classical SPR instrument (Vaisocherova et al. 2008) and with SPR imaging (Brault et al. 2013). These surfaces were also employed in long-term glucose sensing in whole blood with an excellent limit of detection and stability (Yang et al. 2012). These surfaces were recently used to protect implants imbedded in mice for extended periods of time of at least 3 months (Zhang et al. 2013). These excellent properties of betaine hold great promise in sensing applications.

In parallel with the development of betaine surface chemistry, peptide monolayers were also proposed to both minimize nonspecific adsorption and immobilize molecular receptors in biosensing schemes. Peptide and peptidomimetic monolayers are based on a modified peptide with a thiol on either the C- or N-terminus of the peptide. The group of Messersmith was first to propose a peptidomimetic monolayer with excellent properties to protect against nonspecific adsorption (Statz et al. 2005). They revealed that the side chain has an influence on nonspecific adsorption of serum, fibrinogen and lysozyme (Statz et al. 2008). The peptidomimetic compound with an ethyleneglycol side chain was more efficient at reducing nonspecific adsorption with 15 ng/cm^2 of residual nonspecifically bound proteins from serum (Statz et al. 2008). The potential of this class of molecule for reducing nonspecific adsorption was thus clearly demonstrated.

The early work by Whitesides revealed that amide compounds with amino acids protect against nonspecific adsorption (Table 15.1). Peptide coupling chemistry is simple to perform between an amino acid or peptide monolayer and a protein, usually with minimal impact on the activity of proteins. The potential of peptide monolayers to reduce nonspecific adsorption was demonstrated by the groups of Masson and Jiang. A survey of the nonspecific adsorption properties of amino acid monolayers demonstrated that the small, polar and charged amino acids were most efficient at reducing nonspecific adsorption (Fig. 15.2) (Bolduc and Masson 2008).

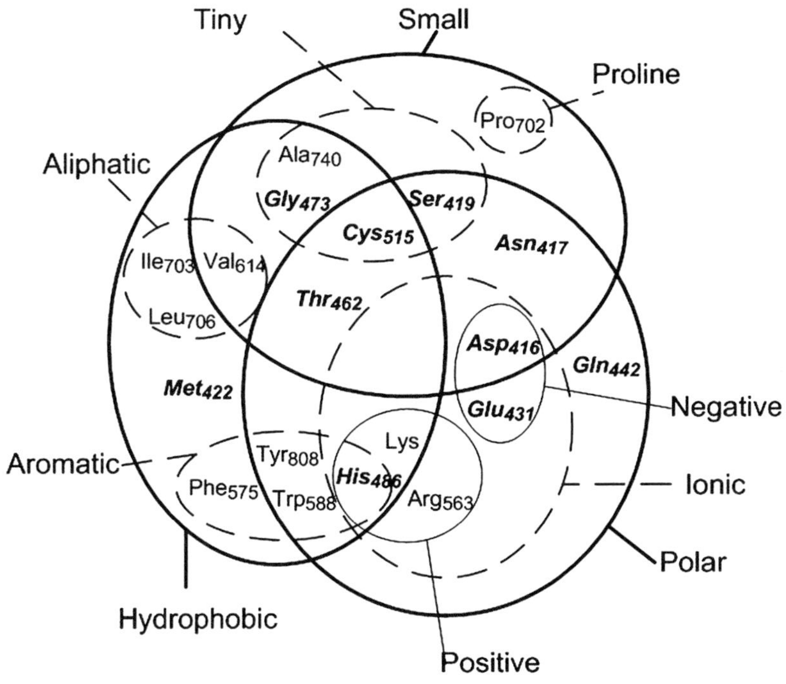

Fig. 15.2 Clustering of the nonspecific adsorption on 3-MPA-amino acids monolayer around a few physico-chemical properties represented by the Venn diagram. The surface coverage induced by bovine serum proteins is represented by the *subscript* next to each amino acid (ng/cm^2). Amino acids in *bold* have a surface coverage below the average of 551 ng/cm^2, while the others have a surface coverage above the average. Reprinted with permission from Bolduc and Masson (2008). Copyright (2008) American Chemical Society

While amino acids were moderately effective at reducing nonspecific adsorption of serum (about 400–800 ng/cm^2 of nonspecific adsorption), constructing homopeptides of increasing chain length improved resistance to nonspecific adsorption by up to 70 % (Bolduc et al. 2009a). Increasing the complexity of the peptide structure led to further improvement of nonspecific adsorption to 23 ng/cm^2 for binary patterned peptide monolayers (Bolduc et al. 2010) and to 12 ng/cm^2 for 3-MPA-LHDLHD-OH (Bolduc et al. 2011). These values report nonspecific adsorption of crude serum on the SPR sensors. Jiang et al. proposed a peptide with alternating charges that was able to reduce nonspecific adsorption of lysozyme and fibrinogen to less than 0.3 ng/cm^2 (Chen et al. 2009). The applicability of peptide monolayers was demonstrated in biosensing schemes by immobilizing enzymes, histidine-tagged biomolecules and antibodies to the peptide monolayer (Bolduc et al. 2011).

While serum and blood-based biofluids have been the main focus of nonspecific adsorption studies, other important biofluids for analysis include urine, cell lysate or saliva. We recently measured the nonspecific adsorption of urine from a single donor (the consent of the donor was obtained). Centrifugation was performed at

Table 15.2 Properties of common surface chemistries and low-fouling surfaces

Surface	Sensitivity	Nonspecific adsorption	Detection in serum	Commercially available
Dextran	High	Average to high	Diluted	Yes
Alkanethiols	Average	Average to high	Possible	Yes
PEG	Low	Low	Possible	Yes
Betaines	High	Ultralow	Yes	No
Peptides	High	Low	Yes	No

1,500 rpm for 5 min to remove cells. Nonspecific adsorption was measured at 26 ng/cm^2 with a monolayer of 3-MPA-LHDLHD-OH (unpublished data), demonstrating that measurements in undiluted urine are possible with SPR.

The rupture of the cell membrane by a mechanical or chemical force creates cell lysate, a medium rich in lipids and proteins. This biofluid is especially interesting in the study of the intracellular matrix for monitoring cell culture processes or in the analysis of biomolecules in biopsies. Peptide monolayers (Bolduc et al. 2009a; Aubé et al. 2013) and PEG (Kyo et al. 2005) were evaluated in crude cell lysate for biosensing. It was revealed that the negative charge on the monolayer favored high nonspecific adsorption. Hydrophobic and positively charged peptide monolayers were most effective at reducing nonspecific adsorption. However, the performance of these monolayers remains average in cell lysate, with approximately 150 ng/cm^2 of material nonspecifically bound to the surface (Aubé et al. 2013). The high concentration of lipids contained in cell lysate explains the difference in physicochemical properties required to minimize nonspecific adsorption. Mass spectrometric studies confirmed the presence of a high concentration of lipids on the SPR sensors when exposed to cell lysate (Aubé et al. 2013).

In conclusion, many of the common surface chemistries fail in crude biofluids due to either high protein/lipid adsorption or low sensitivity in bioassays (Table 15.2). Employing peptide monolayers or betaine surface chemistries can lower nonspecific adsorption to less than 15 ng/cm^2, a level sufficiently low to analyze biomolecules in crude biofluids. Sensitivity can be improved with detection schemes involving secondary amplification of the signal or by using enhanced plasmonic materials in the fabrication of the SPR sensors (see Sect. 15.4). While challenges remain for the analysis of biofluids other than serum, the recent progress in minimizing serum nonspecific adsorption provides encouragement that novel surface chemistries will be developed for crude cell lysate and other biofluids.

15.4 Enhancing Sensitivity with Nanomaterials

The performance of SPR biosensors is often assessed from the sensitivity to bulk refractive index. Bulk sensitivity of novel plasmonic materials is often compared for predicting analytical performance. While this parameter is important, surface

Table 15.3 Signal amplification strategies in SPR sensing

Method	Material	Amplification principles	Applications
Secondary detection/ competition assays	Antibodies, analogs	Mass augmentation	Proteins, metabolites, small molecules
Enzyme amplification	Enzymes	Multiple detection of single binding events, mass augmentation	DNA, protein, bacteria
Nanoparticles	Silica, polymer	Mass augmentation	General
Plasmon coupling	Au or Ag	Mass augmentation and changes in optical properties	General
Plasmonic materials	Gratings, hole arrays	Changes in optical properties, high sensitivity	General

sensitivity (i.e. binding of a molecule) and plasmon enhancing effect (i.e. plasmon coupling) are also critical in improving sensitivity. Thus, the ideal plasmonic material must combine high bulk sensitivity and short penetration depth (high surface sensitivity) and promote plasmon enhancing effects.

SPR sensing is generally performed on a thin Au film. Au has the advantages of being inert to oxidation, forming chemical bonds with thiols and forming self-assembled monolayers with thiolated compounds. It exhibits a high sensitivity to bulk refractive index and it is simple to manufacture. However, detection limits with Au films are generally in the high picomolar to low nanomolar range, sufficient for selected applications but not applicable to many biomarkers or DNA detection where low picomolar to high femtomolar detection limits are desirable.

Several sensing strategies have been developed to improve the detection limits of SPR sensors (Table 15.3). SPR sensors are sensitive to changes of refractive index, which is greater for molecules or nanomaterials of greater mass. SPR is thus sometime referred to (rightfully or not) as a mass sensor. A widely popular strategy involves mass augmentation of the detection step with large molecules or nanostructures. A larger signal-to-noise ratio is obtained by increasing the refractive index or mass of the detection step. Secondary detection employing antibodies improves detection limits. Antibodies are large (150 kDa) biomolecules for which SPR is very sensitive. Improvement of the limit of detection of about one order of magnitude is typically observed with antibody secondary detection steps. This strategy relies on established methods developed for ELISA assays, making it simple and amenable to SPR analysis.

Small molecules such as metabolites are difficult to detect with SPR. Their low masses are synonymous with low refractive index change and poor SPR sensitivity. To circumvent this problem, a competition assay between the metabolite or small molecule and an analog with a mass tag is created to enhance the response in SPR. The response of the analog with the mass tag is large in SPR, which decreases in proportion to the concentration of the metabolite or small molecules. In the case of DNA detection, enzyme amplification creates a cascade of reactions to digest the capture probe in the presence of the target (Goodrich et al. 2004; Lee et al. 2005) and the digestion of the probe decreases the mass bound to the sensor (Fig. 15.3).

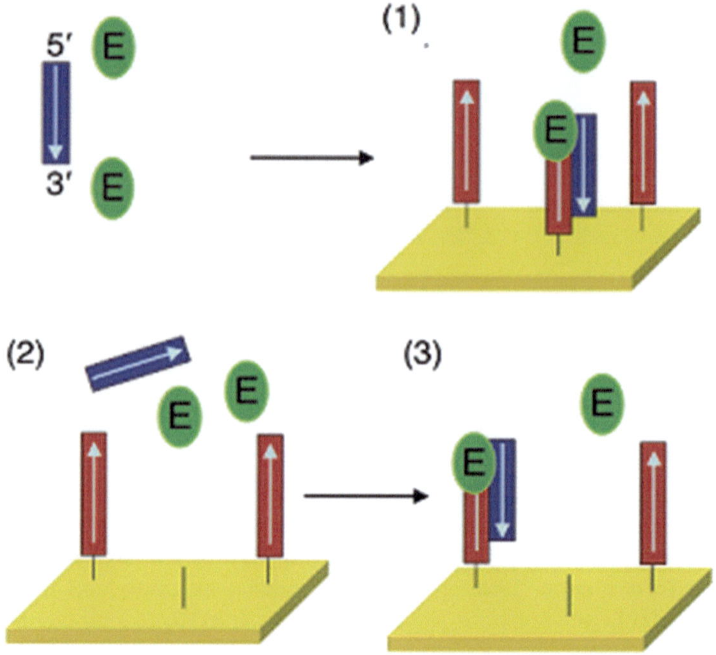

Fig. 15.3 Schematic of the enzyme-amplified detection of DNA. The enzyme digests the capture probe only when the target DNA analyte has hybridized, after which the analyte is released. The target is then recaptured on the surface by another probe, and the cycle is repeated. The digestion of the capture probes decreases the mass bound to the sensor, leading to SPR detection. Reprinted with permission from Lee et al. (2005). Copyright (2005) American Chemical Society

This sensing scheme has a theoretical detection limit of a single molecule, since a single molecule can react with all probes immobilized on the sensor if given enough time.

Enzymes are also exploited to increase the detection of proteins. The enzyme serves to precipitate a compound to the surface (generally 3,3′,5,5′-tetramethylbenzidine; TMB) of the SPR sensor (Li et al. 2007), which causes a significant mass increase on the sensor. In the first step, the analyte is captured by an aptamer or an antibody immobilized on the SPR sensor (Fig. 15.4—step a). The enzyme is bound to a primary antibody, captured by the analyte bound to the SPR sensor (Fig. 15.4—step b). The enzyme then catalyzes the reaction of TMB to form a dark blue precipitate on the SPR sensor. The large augmentation in mass due to the precipitation process results in a larger SPR response and lower detection limits. This process was also applied to the detection of bacteria in food samples (Linman et al. 2010). Using a process of in situ polymerization, Cheng succeeded in augmenting the mass detected for cholera toxin on the SPR sensor and lowering the detection limit by nearly two orders of magnitude (Liu et al. 2010).

Nanoparticles are large entities in comparison to molecules. Their relatively large weight and refractive index change at the surface of the SPR sensors makes them

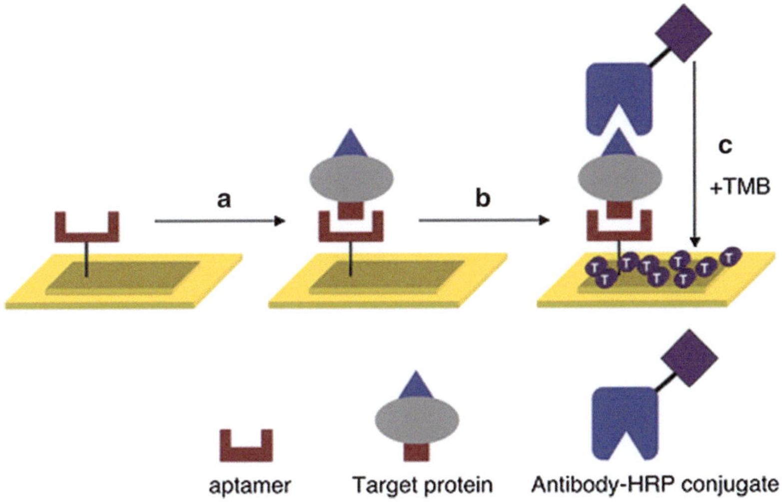

Fig. 15.4 Schematic representation of the enzyme-amplified precipitation of TMB on the SPR sensor. Reprinted with permission from Li et al. (2007). Copyright (2007) American Chemical Society

ideal for amplifying the response in SPR. Schasfoort et al. reported that the limit of detection for direct prostate-specific antigen (PSA) detection was improved from 300 ng/mL to single digit nanograms per milliliter with the use of secondary detection with nanoparticles (Besselink et al. 2004). Latex (Besselink et al. 2004), and silica (Zhou et al. 2011a) nanoparticles were employed in biosensing schemes where primary antibodies or DNA strands are bound to the nanoparticle. Following capture of the analyte onto the surface of the biosensor, the antibody–nanoparticle conjugate is reacted with the captured analyte and leads to large mass changes on the surface of the SPR sensor. The use of metallic nanoparticles can further improve detection limits in biosensing schemes in SPR (Besselink et al. 2004). Metallic nanoparticles undergo plasmonic coupling with metallic surfaces, further amplifying the changes in refractive index and lowering the detection limit. Porter et al. elegantly demonstrated that Au nanoparticles, when located in close proximity to the SPR sensor, have a large influence on the plasmonic properties (Driskell et al. 2006). The high sensitivity of Au nanoparticles as a secondary detection in SPR sensing schemes explains the great popularity of this methodology. A review by Bedford et al. details the recent advances in the use of Au nanoparticles in SPR sensing technology (Bedford et al. 2012).

SPR sensors are heralded as simple and label-free techniques for the detection of biomolecules. Many researchers are actively working in making SPR instruments smaller, cheaper and simpler to operate. However, secondary detection with nanoparticles is a labeling technique that limits the scope of these advantages of SPR sensing. In recent years, researchers have been interested in developing novel SPR substrates for improving the sensitivity of SPR sensors adapted to portable SPR sensing (Breault-Turcot and Masson 2012). Sensitivity to bulk refractive index and sensitivity to surface processes are important in SPR sensing. The bulk sensitivity

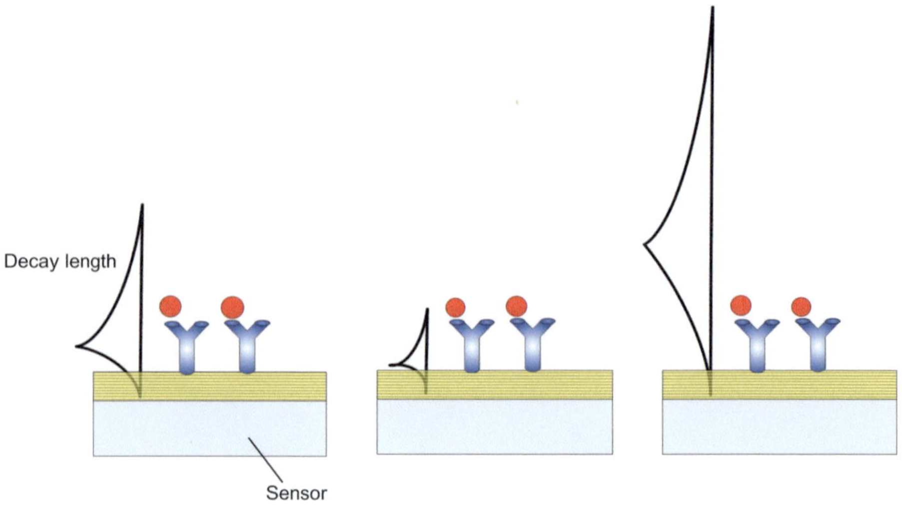

Fig. 15.5 Illustration of the effect of the decay length of the plasmon field on sensing. The optimal scenario (*left panel*) consists of having a decay length slightly larger than the total thickness of material deposited on the sensor during the bioassay. Longer decay lengths (*right panel*) will lead to small changes in the refractive index of the sensing volume, while smaller decay lengths (*center panel*) will result in binding of the analyte outside the most sensitive region of the SPR sensor

and the penetration depth of the plasmon field control the surface sensitivity. Small penetration depth is advantageous to concentrate the SPR field in the region where the molecule binds (Fig. 15.5). This leads to larger SPR shifts, a phenomenon exploited in localized SPR sensing.

Nanostructures are becoming ubiquitous in SPR sensing (Bolduc and Masson 2011; Roh et al. 2011; Breault-Turcot and Masson 2012; Couture et al. 2013b). The use of grating and grating-like SPR substrates currently represents one of the best options for enhancing the sensitivity of SPR sensors. The light is coupled to the plasmon using the grating modes of these nanostructures, changing the sensitivity (Homola et al. 1999) and penetration depth of the sensor (Gao et al. 2010). Sensitivity is generally smaller for grating structures than for SPR sensors based on a thin Au film (Homola et al. 1999; Couture et al. 2013b). Grating structures have been used in SPR sensing to simplify the detection scheme (direct excitation is possible, thus eliminating the need for prism coupling) (Piliarik et al. 2009). In this configuration, light usually travels through the solution for excitation of the plasmon, causing optical interference with absorption and scattering of the solution and increasing the noise in the system (Hoa et al. 2007). Similarly, nanosensors can be excited using the instrumental configuration of transmission spectroscopy, such as for Au nanoparticles or nanohole arrays. They represent an interesting alternative to classical SPR, as nanosensors exploiting the localized or the propagating surface plasmon phenomenon have similar analytical performance (Haes and Van Duyne 2004). Nanohole arrays are gaining significant interest for the design of chemical and biological sensors (Masson et al. 2010). It has been recently reported that nanohole arrays exhibit superior analytical performance in the prism-coupling configuration in comparison to transmission measurements (Couture et al. 2012). We recently found that tuning the grating coupling conditions has a severe impact on the plasmon properties of nanohole arrays and that the sensitivity can be significantly improved by using angled excitation (Couture et al. 2013a). With non-zero excitation angles, the penetration depth and the resonance wavelength of the surface plasmon decreases, resulting in a higher sensitivity towards binding events. Nonetheless, these sensors are also subject to optical interference of biofluids. To circumvent this problem, 1D or 2D grating structures have been recently investigated in the prism-coupling configuration.

Theoretical analysis of implementing a grating structure on a thin metal film was performed using rigorous coupled wave analysis (RCWA) modeling (Kyung Min et al. 2008). These simulations predicted that the coupling of the localized and bulk surface plasmon modes should lead to enhanced sensitivity. In addition, it is possible to locate the molecular receptors in the locations of highest sensitivity on the grating structure to further enhance the SPR response (Hoa et al. 2009). The light entering resonance with nanowire gratings undergoes a phase shift upon binding of the analytes to the grating (Zhou et al. 2011b). This phase shift is very sensitive to binding events and leads to lower detection limits (for example, a 20-fold reduction in thrombin detection was reported by Corn et al. (Zhou et al. 2011b)). The coupling of localized and propagating surface plasmon modes was thoroughly investigated by Bartlett et al. (Kelf et al. 2006; Sugawara et al. 2006). His team revealed that the co-excitation of the localized and propagating modes leads to enhanced sensitivity of the electric field. These fundamental studies prove the potential of grating-based structures.

Based on these fundamental advances, Masson has investigated the prism-coupling configuration of SPR and the plasmonic properties of microhole arrays. A periodicity of a few microns, with hole diameters in the same order of magnitude and a metal thickness of 50–200 nm, characterizes microhole arrays (Fig. 15.6). The localized and propagating properties of hole arrays were studied in detail with wavelength interrogation in the Kretschmann SPR configuration (Live and Masson 2009; Live et al. 2009, 2010). It was observed that localized and propagating SPR modes are accessible with prism coupling.

The optical properties of microhole arrays change significantly according to the ratio of hole diameter to periodicity. For large ratios, the substrate corresponds to triangle arrays and exhibits localized surface plasmon resonance properties. The optical properties are similar to propagating surface plasmon resonance for small hole diameter/periodicity ratios, corresponding to a hole array substrate. The transition region between triangle and hole arrays is particularly interesting. In the transition region, the structure is a combination of triangle and hole arrays. With these optimal specifications of 3.2 μm periodicity, 1.8 μm hole diameter and 65 nm thickness/depth, the plasmonic substrate exhibits both localized and propagating surface plasmon resonances (Live et al. 2012). This is characterized by the presence of two plasmon bands in the dispersion curve (Fig. 15.7), a precursor condition for enhanced sensitivity in SPR sensing.

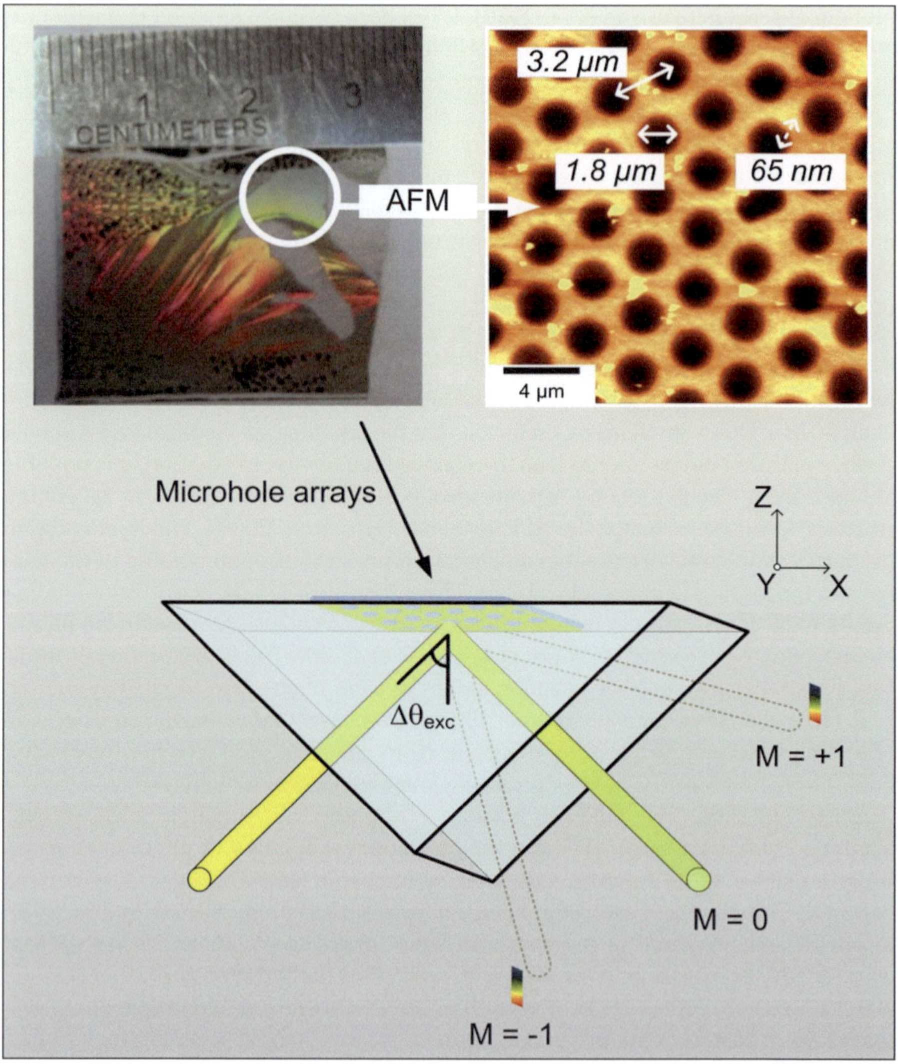

Fig. 15.6 Microhole arrays are excited in the Kretschmann configuration of SPR. The periodicity is characterized by the center-to-center distance between adjacent holes (3.2 μm), the hole diameter is in the order of 1.8 μm and the metal thickness is 65 nm. While the spectral analysis of microhole arrays focuses on the $m=0$ mode, the $m=\pm1$ modes are visible in the reflection spectrum. Reprinted with permission from Live et al. (2012). Copyright (2012) Springer

Similar to nanovoid arrays (Kelf et al. 2006), the co-excitation of the localized and propagating plasmon bands at 72–73° leads to enhanced bulk refractive index sensitivity of the SPR sensor (Live et al. 2012). This enhanced sensitivity results in greater sensitivity for the detection of proteins, as demonstrated for IgG (Fig. 15.8). An overall improvement by a factor of 3 in the SPR response was observed with the

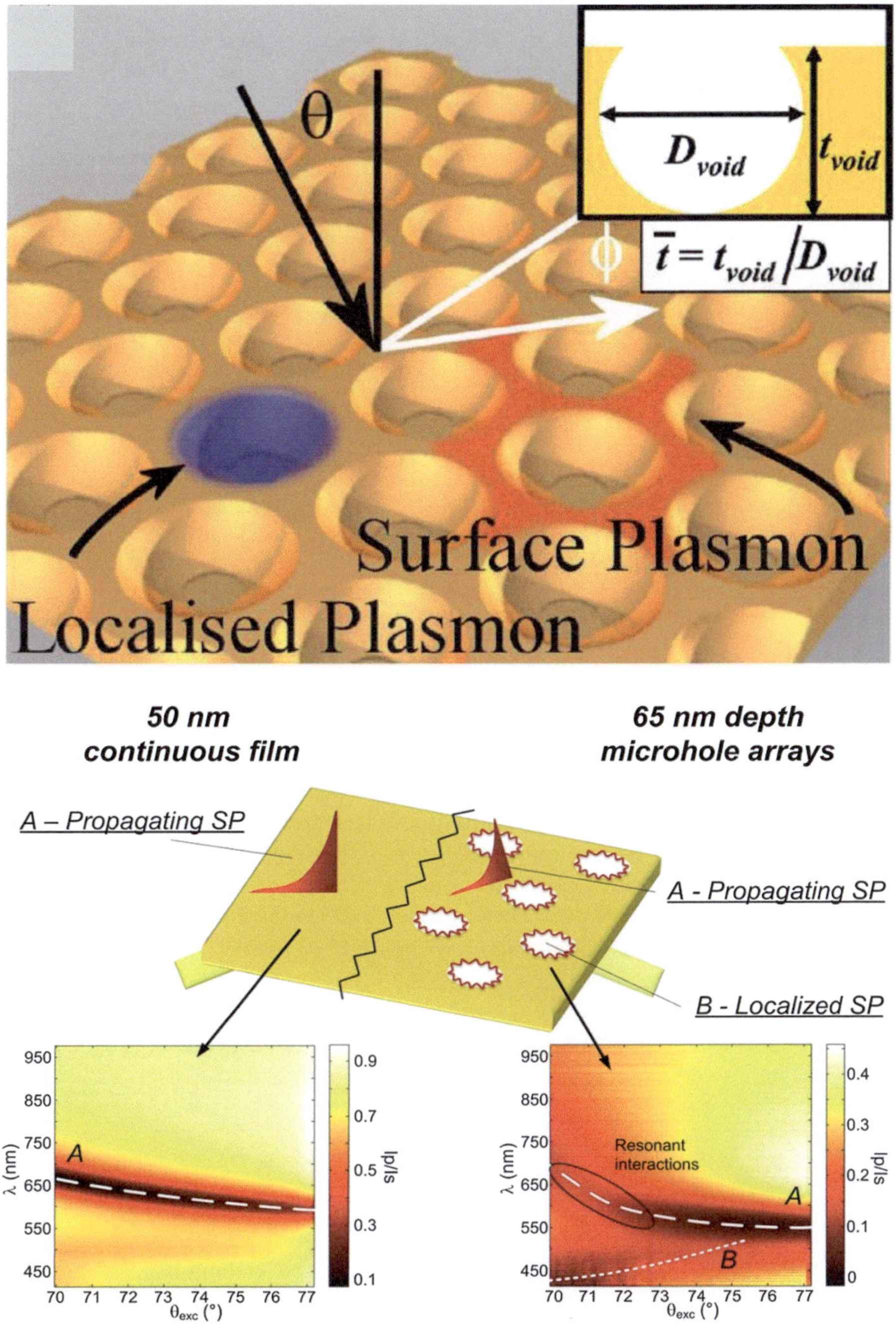

Fig. 15.7 *Top panel*: Representation of localized and propagating surface plasmons on nanovoid arrays. Reprinted with permission from Kelf et al. (2006). Copyright (2006) by the American Physical Society. *Lower panel*: Plasmon dispersion curves for thin Au film used in classical SPR sensing and for microhole arrays. The dispersion curve of thin Au film exhibits a single band that changes resonance wavelength according to excitation angle, while the microhole arrays have two surface plasmon bands corresponding to localized (lower wavelength) and to the propagating surface plasmon mode at higher wavelength. The coexcitation occurs at approximately 72–73° excitation angle. Reprinted with permission from Live et al. (2012). Copyright (2012) Springer

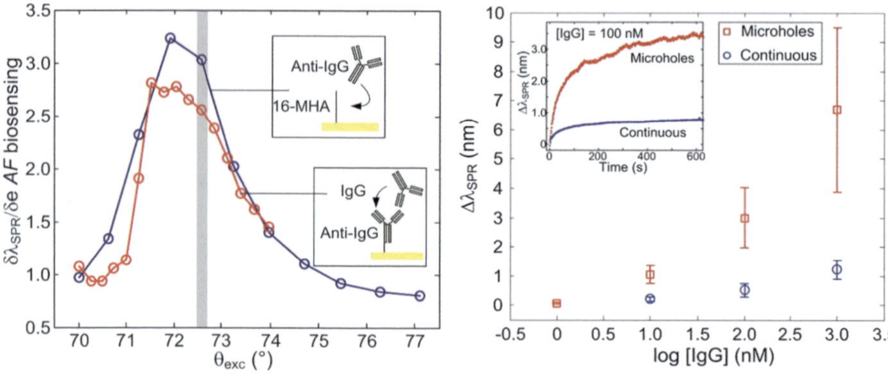

Fig. 15.8 *Left*: Amplification factor of the SPR response for the detection of antibody immobilization and of IgG detection. The amplification factor is defined by the ratio of the response to the biodetection event for the microhole array and the response obtained with a thin Au film under the same experimental conditions. *Right*: Calibration curve for IgG using a microhole array and thin Au film in the Kretschmann configuration of SPR. The detection limit was improved with the microhole arrays. Reprinted with permission from Live et al. (2012). Copyright (2012) Springer

microhole arrays analyzed in experimental conditions favoring the co-excitation of localized and propagating surface plasmons. Microhole arrays are thus providing superior analytical performance in biosensing schemes for SPR sensing.

Several strategies have been designed for enhancing the response of detection events in SPR sensing. Enhanced sensitivity can be obtained by improving the plasmonic material or by augmenting the mass of the detection event with either nanoparticles or precipitation events mediated with enzymes. Each strategy is suited to different challenges. For applications where rapidity and simplicity of analysis is critical, sensitivity is enhanced using novel nanomaterials such as gratings or hole arrays. If ultralow detection limits are required and analysis time is not an issue, amplification with the secondary detection techniques are more appropriate. SPR has become an established bioanalytical technique for the analysis of a wide range of biomolecules and these applications will be explored in the next section.

15.5 Applications

SPR sensors are becoming versatile and widely applicable technologies in chemical and biological sensing (Homola 2008; Couture et al. 2013b), for affinity measurements (Scarano et al. 2010), toxin detection (Hodnik and Anderluh 2009), food science (Situ et al. 2010) and for small molecules (Shankaran et al. 2007; Mitchell 2010). The broad applicability of SPR sensing is due to the quasi-universal refractive index transduction. Protein sensing is one of the strongholds of SPR sensing, as proteins are large biomolecules to which SPR is sensitive and low detection limits can be achieved using a direct assay. Direct assays in SPR sensing involve a single

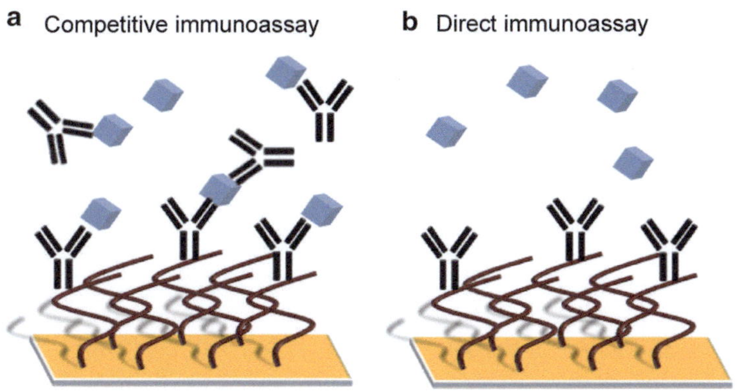

Fig. 15.9 Schematic representation of competitive immunoassays (**a**) and direct detection (**b**) with SPR sensors. The competitive immunoassay involves the binding of the free analyte to a molecular receptor. The competition is established with an analog of the analyte, which is tagged with a molecule or nanoparticle that will significantly enhance the mass detection. Capturing the analyte with a molecular receptor performs direct detection in SPR sensing. Reproduced from Breault-Turcot and Masson (2012) with permission from Springer

capture and detection step performed in real time (Fig. 15.9b), enabling rapid and sensitive measurements. The current workhorse in biochemistry laboratories is the ELISA assay, which requires multiple steps following the capture of the analyte to detect and quantify the concentration of proteins. SPR sensors can significantly decrease analysis time and should in the near future provide a portable instrumentation for point-of-measurement applications (see Sect. 15.6). Protein detection and gas sensing were among the first applications developed for SPR biosensing (Liedberg et al. 1983). Protein detection and quantification with SPR sensing has been applied to cancer biomarker screening, general health markers and disease diagnostics. In addition, SPR has elucidated binding kinetics and thermodynamic data for binding events and is able to screen for specific binding partners. Most methodologies developed for protein detection using ELISA assays were translated for SPR sensing, such that it is currently one of the most advanced applications of SPR sensing. Section 15.5.1 will oversee the challenges and potential applications for protein sensing in crude biofluids.

Small molecule detection is difficult with direct immunoassays, as the small molecular weight of this class of molecules does not induce significant changes in the refractive index during detection events. Most manufacturers of SPR instruments report the detection of small molecules to be in the 200–1,000-Da range. However, direct detection pushes the limits of the system and does not provide high signal-to-noise ratio. To circumvent this problem, small molecules have been detected with competitive immunoassays (Fig. 15.9a) or with inhibition immunoassays to improve signal-to-noise and provide quantitative results (Shankaran et al. 2007). The development of SPR sensors for analytes of low molecular weight is specifically interesting for drug discovery, therapeutic drug monitoring, and food and environmental analysis. The speed, sensitivity, selectivity and reliability of an

SPR analysis with a portable and miniaturized SPR instrument would benefit this series of important applications (Shankaran et al. 2007). In Sect. 15.5.2 we will overview the challenges of developing a sensor for therapeutic drug monitoring.

15.5.1 Protein Sensing

Serum is one of the most accessible biofluids for clinical chemistry. It is available in relatively large quantities (mL) and is obtained through standard blood testing procedures. In addition, the clinical relevance of a serum sample is high. Several diseases lead to elevated or decreased levels of proteins in serum which can be indicative of a specific disease (Polanski and Anderson 2006) and it is thus important to measure biomarkers in serum. To circumvent nonspecific adsorption, markers in serum were previously detected in diluted serum (often using a 1:10 dilution ratio) (Situ et al. 2008; Schlichtiger et al. 2013). While this can work for markers present at sufficiently high concentrations in serum, most markers are found at very low concentrations, and losing one order of magnitude in sensitivity due to dilution could be problematic.

The detection of biomarkers can be made possible in serum by removing nonspecifically bound proteins after the detection step. A membrane cloaking strategy was designed to remove nonspecifically bound proteins on an SPR sensor, leaving only the analyte bound to the molecular receptor at the surface. This strategy was employed for the detection of undiluted serum spiked with IgG (Phillips et al. 2007). Nonspecific adsorption can also be minimized using a wash buffer that removes nonspecifically adsorbed proteins. A series of papers by the Lechuga group has exploited this method to detect pituitary hormone (Trevino et al. 2009a) and human growth hormone (Trevino et al. 2009b), and to distinguish between the isoforms of human growth hormone (22 and 20 K) (de Juan-Franco et al. 2013) in serum. Secondary detection using either Au nanoparticles (Uludag and Tothill 2012) or antibodies modified with enzymes (Garay et al. 2010) enabled the detection of CK-MB (Garay et al. 2010) and PSA (Uludag and Tothill 2012) in serum samples. The use of specifically designed polymers in conjunction with BSA blocking and secondary detection using antibodies allowed for the detection of PSA at 5 ng/mL in 20 % serum samples (Kyprianou et al. 2013). While these methods work relatively well in laboratory settings, the extra steps required for the detection of the biomarkers in serum are labor-intensive.

Direct detection of biomarkers was demonstrated in cell culture medium (Battaglia et al. 2005) and in serum (Masson et al. 2007) by pre-incubating the sensor in an "analyte-free" biofluid. Detection of TNF-α, IL-1, IL-6, cTNI and myoglobin was achieved at the ng/mL level in undiluted biofluids. Recent developments in surface chemistries further reduced nonspecific adsorption and allowed quantification of analytes in undiluted serum. Peptide monolayers were recently used to detect MMP-3 in crude serum using SPR sensing (Bolduc et al. 2010). The Jiang group has developed an anti-fouling layer based on polycarboxybetaine capable of significantly suppressing nonspecific adsorption (Yang et al. 2009). They successfully reduced nonspecific binding from serum and plasma to barely detectable levels, allowing for detection of markers directly in crude biofluids. They applied this methodology to detect carcinoembryonic antigen autoantibodies (Ladd et al. 2009), CD166/ALCAM (Vaisocherova et al. 2009a), streptavidin (Vaisocherova et al. 2009b) and ALCAM (Vaisocherova et al. 2008). Figure 15.10 demonstrates the low fouling and quantitative nature of biosensors working in crude biofluids, utilising adequate surface chemistry. PEG failed to detect low concentrations of ALCAM, while the carboxybetaine surfaces were successful under the same conditions. These examples highlight the potential of SPR sensing in crude biofluids, with no or minimal sample preparation.

15.5.2 Drug Sensing

Patients undergoing treatment for diseases can be administered either a single drug or a cocktail of drugs. A patient's response to the treatment depends on a series of factors such as age, sex and genetic variations (Dasgupta 2012). Therapeutic drug monitoring (TDM) closely monitors the effective dose in the patient's biofluids, with the benefit of safer and more effective therapy. Thus, there is an increasing need to develop novel techniques for TDM. Surface plasmon resonance, in conjunction with Au nanoparticles, has the potential to significantly improve the detection of small molecules (Bedford et al. 2012) and could be applicable in TDM.

Competitive immunoassays function by establishing a competition between the analyte and an analog (or the analyte) tagged with a nanoparticle. Due to the high mass of the analog tagged with the nanoparticle, competition assays offer increased sensitivity for the detection of small molecules. The case of methotrexate (MTX) sensing will be examined in greater detail in this section.

Methotrexate is an anticancer agent commonly used in chemotherapy. It is very effective in treating cancer by halting the cascade of reactions occurring in DNA synthesis. MTX is a slow and tight binding inhibitor of the human dihydrofolate reductase enzyme (hDHFR), at the center of the purine and thymidine synthesis process essential to DNA replication. While blocking this enzyme, methotrexate also acts on healthy cells and causes side-effects. The safe and effective concentration window for methotrexate is relatively narrow, and as such TDM would be extremely beneficial.

Methotrexate provides an excellent model system for validating SPR biosensors in TDM. Direct detection leads to poor sensitivity (in the high micromolar range) and a competition assay had to be developed whereby physisorbing folic acid modified Au nanoparticles were able to compete with hDHFR (Fig. 15.11). It was realized that the stoichiometry of the reaction, driven by the concentration of Au NPs and the concentration of hDHFR, played a significant role in the performance and dynamic range of the assay (Zhao et al. 2012). While the K_D of methotrexate for hDHFR is in the low picomolar range, the assay could be tuned from the low

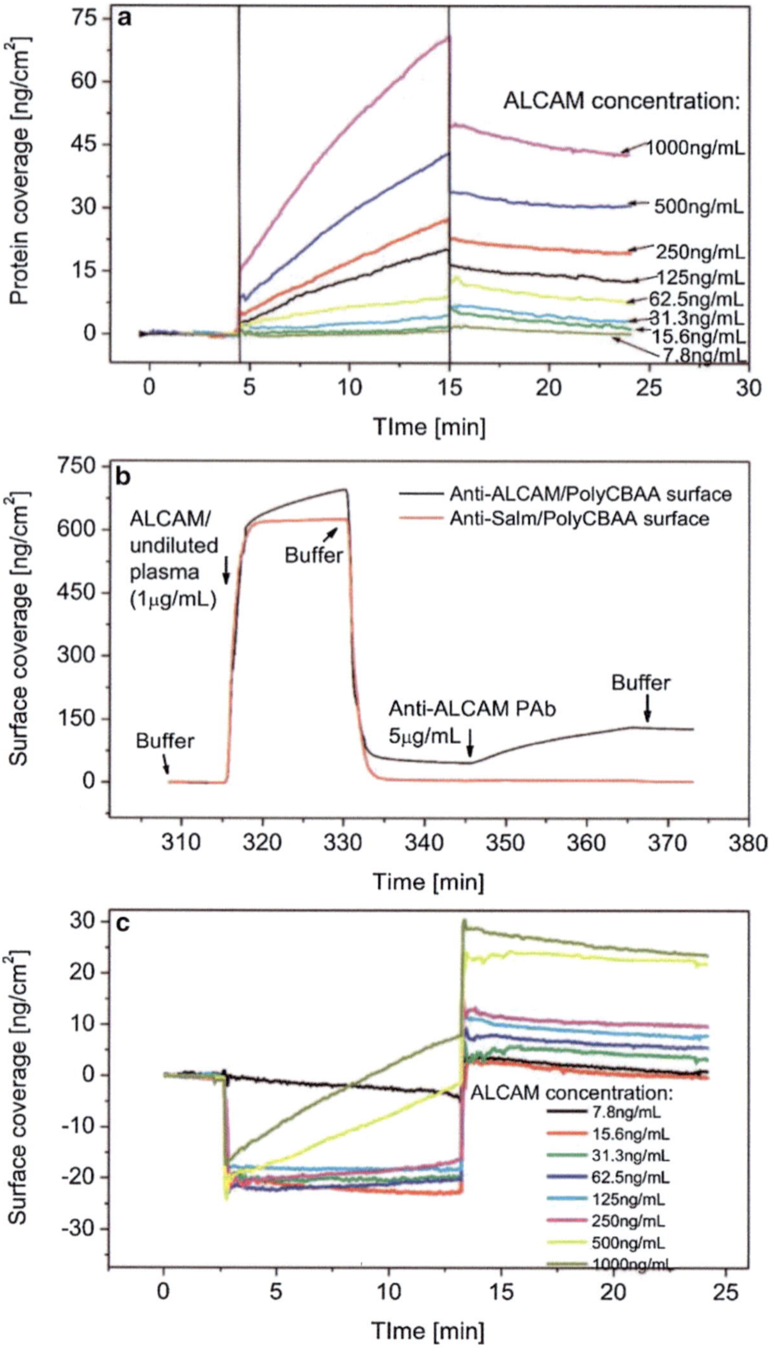

Fig. 15.10 Demonstration of the detection of ALCAM in undiluted plasma. *Panel A* shows the sensor response with reference compensation to remove the effect of bulk refractive index and temperature drift. *Panel B* shows the raw data of the SPR sensor constructed with the carboxybetaine surfaces. *Panel C* demonstrates that PEG surfaces lead to undetectable responses for low concentrations, under the same conditions as for the carboxybetaine surfaces in panel A. Reprinted with permission from Vaisocherova et al. (2008). Copyright (2008) American Chemical Society

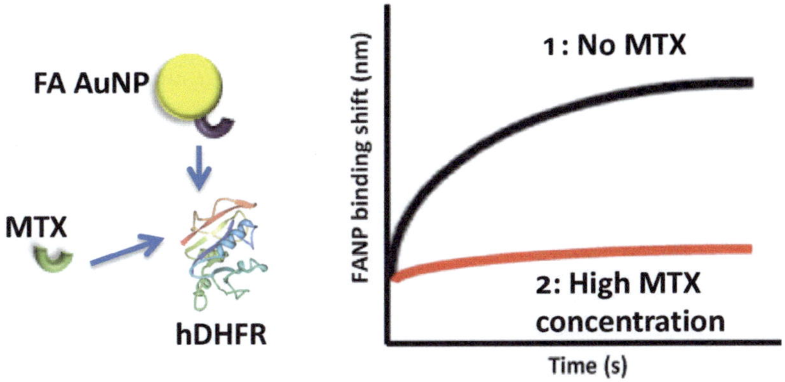

Fig. 15.11 Principles of nanoparticle-based SPR sensors for methotrexate detection. Folic acid Au nanoparticles (FA AuNP) competes with methotrexate (MTX) for the molecular receptor (hDHFR). A high methotrexate concentration blocks the enzyme from reacting with the AuNP and leads to no signal. A low methotrexate concentration favors the binding of hDHFR with the AuNP and leads to a high signal in SPR

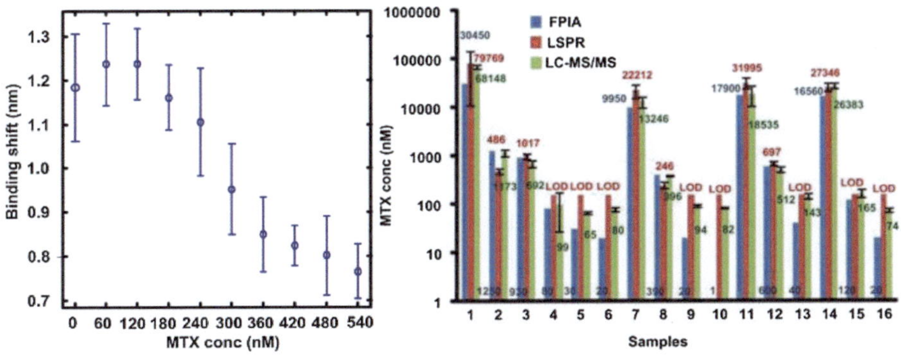

Fig. 15.12 Calibration curve for methotrexate in serum (*left*) which is within the clinical range. Comparison of the quantification of actual clinical samples from patients undergoing chemotherapy with the SPR assay (*right*). LC-MS/MS and fluorescence polarization immunoassay were selected due to their wide availability in clinical chemistry laboratories

nanomolar to the high nanomolar by simply changing the concentration of hDHFR. Calibration curves were established for the sensor in solution (as a localized surface plasmon resonance sensor). However, as previously mentioned, the optical interference of serum can be prohibitive for direct analysis of serum samples from patients.

To circumvent the issue of serum scattering and adsorption, a sample pretreatment step was added to the experimental setup for methotrexate quantification. A solid-phase extraction cartridge was used to remove proteins and isolate methotrexate from the complex matrix. Elution of methotrexate created an interference-free matrix for analysis. Calibration in serum was thus undertaken (Fig. 15.12) and the analysis of actual clinical samples was compared to established techniques. LC-MS/MS, fluorescence polarization immunoassays and SPR were in good agreement. This indicated the suitability of the competitive assay for the quantification of methotrexate.

While the competition assay works well in solution as a localized SPR test, sample preparation remains an issue with this methodology. It is impractical and time-consuming to perform solid-phase extraction of all samples. The use of SPR with the Kretschmann configuration is advisable for clinical analysis, as demonstrated in the section on protein analysis. Thus, the methodology for methotrexate detection was adapted on the SPR sensor (Bukar et al. 2014; Zhao et al. 2015). The assay was optimized to allow immobilization of active hDHFR on the surface of the SPR sensor (Bolduc et al. 2011; Ratel et al. 2013). hDHFR was histidine-tagged to allow chelation with the SPR sensor with a modified self-assembled monolayer (either 16-mercaptohexadecanoic acid or peptide monolayer) capable of chelating His-tags. hHDFR is prone to denaturation when immobilized using standard EDC-NHS chemistry (Bolduc et al. 2011), and thus this modified peptide monolayer is critical in the success of the assay. By using the binding shift (at equilibrium), detection of methotrexate was performed in the nanomolar range in less than 20 min. The detection time can be significantly improved by using the binding rate of the first 60 s, with equivalent results in calibration curves. This example shows the potential application of SPR sensing within clinical chemistry.

15.6 Towards Small and Portable Instrumentation for Biosensing

Miniature SPR sensors are being developed to combine sensitivity, simplicity and cost-effectiveness. Early attempts at SPR miniaturization relied on simple optics for angle interrogation SPR, which resulted in lower resolution. In addition, these systems were proposed without fluid handling components, requiring a great deal of user input in order to create a fully integrated system. For example, the Spreeta SPR instrument comprised optical and electronic parts only (Chinowsky et al. 2003b), with refractive index resolution about one order of magnitude worse than laboratory-based instruments. Nonetheless, they were successfully deployed in airborne biosensing (Chinowsky et al. 2003a), for the detection of *Staphylococcus aureus* enterotoxin B (Naimushin et al. 2002), for the measurement of serum albumin in urine (Navratilova and Skladal 2003) and for the detection of *E. coli* (Spangler et al. 2001). These results paved the way for the development of portable sensors.

Fiberoptic-based SPR sensors were evaluated for sensing biomolecules. The SPR sensor is located on the side of a fiberoptic in which the core has been exposed (Gentleman et al. 2004; Masson et al. 2004b) or at the distal end of a fiberoptic (Masson et al. 2006b). Detection of proteins at nanomolar concentrations was achieved in serum samples (Masson et al. 2007) and in cell culture medium (Battaglia et al. 2005). These SPR sensors rely on wavelength interrogation, which requires a single excitation wavelength for optimal resolution. While simple to use and to manufacture in a small-sized instrument, the multiple modes propagating in the fiberoptic lead to broadening of the plasmon resonance and poorer resolution. SPR sensors based on fiberoptics are also subject to mode mixing due to environmentally induced vibrations. This increases the noise level such that fiberoptic SPR sensing is difficult to implement in portable sensors. Recent developments in tilted fiber Bragg gratings has progressed towards solving these challenges (Shevchenko and Albert 2007; Shao et al. 2010; Caucheteur et al. 2011; Shevchenko et al. 2011).

Recently, portable SPR instruments have been proposed by several commercial sources: SPR micro (KMAC) (http://www.kmac.to/eng/product/php?cid=sprlab, accessed 16 Dec 2012), SPIRIT (Seattle sensing systems) (Chinowsky et al. 2007), Smart SPR SS-1001 (Mebius advance technology) (http://www.ntt-at.com/products_e/handy_spr/, accessed 16 Dec 2012) and Biosuplar 6 (Analytical μ-systems/Mivitec) (http://www.biosuplar.de/, accessed 16 Dec 2012). Fluidics and often injection systems have been included in the more recent versions of these SPR systems. The resolution of portable SPR systems remains in the 10^{-6} RIU range in comparison to 10^{-7} RIU for laboratory-based instruments (Chinowsky et al. 2007). A small and compact SPR instrument was designed based on the dove prism configuration (Bolduc et al. 2009b), and was recently augmented to a four-channel instrument (Zhao et al. 2015). The bioassay for methotrexate detection was adapted to the four-channel SPR instrument and tested in the clinical chemistry laboratory of a Montreal hospital (Fig. 15.13).

Spiking the pooled sera of six healthy patients with methotrexate was used to calibrate the SPR sensors with clinical samples. Triplicate analysis was performed in the calibration process. Using the multichannel system decreased the coefficient of variation from 31 % with the single channel instrument, to as low as 5 % with a multichannel instrument. Actual clinical samples were analyzed with the system. Sample preparation consisted of a ten-fold dilution in the folic acid Au nanoparticle solution, necessary to perform the competition assay. A series of clinical samples

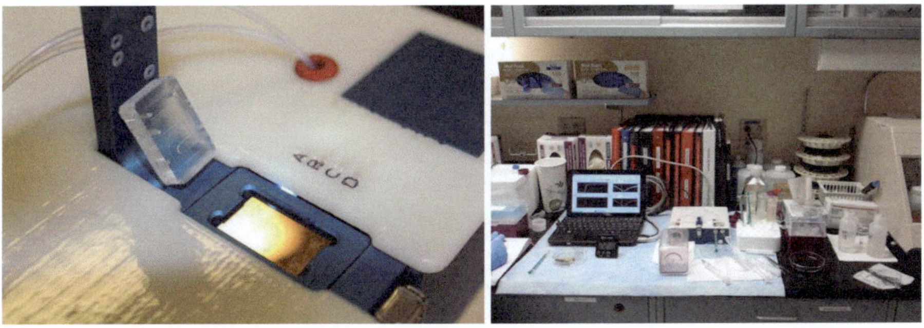

Fig. 15.13 *Left*: The four-channel SPR instrument was designed based on a 20×12 mm dove prism. A fluidic cell delivers the sample to three channels for a triplicate measurement and a fourth channel serves as the reference. *Right*: The system takes little space in a laboratory. The total space needed to run experiments is less than 1 m^2. No external power is required to run the instrument, making it portable. Reproduced from Zhao et al. (2015) with permission from Elsevier

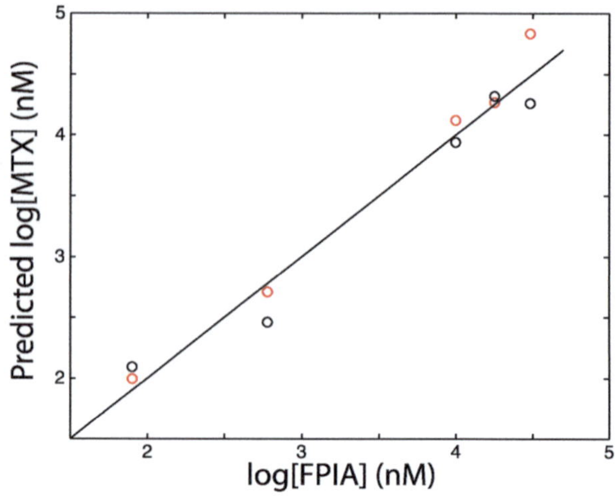

Fig. 15.14 Correlation study of methotrexate quantification in clinical samples. FPIA and LC-MS/MS results were compared with an SPR sensor. The *black line* represents a perfect correlation between FPIA and LC-MS or SPR. Data points in *black* were obtained with SPR and those in *red* with LC-MS

were provided to the researchers in a blind assay. Results from the methotrexate detection were compared with FPIA and LC-MS/MS. The results were in good agreement with the reference techniques (Fig. 15.14), showing the possibility of using the SPR instrument in quantification of a drug in TDM. Several users performed measurements with the SPR instrument with coefficients of variation between 6 and 23 %. Feedback from clinical chemists was positive, while they noted that automated injection might be more appropriate in a clinical chemistry environment. These results show that SPR sensors are suited for use in clinical chemistry laboratories (Zhao et al. 2015).

15.7 Conclusions

SPR sensing has matured as a viable technique for the analysis of a plethora of biomolecules. Examples illustrating the use of SPR in relevant biofluids are increasing and the technique is advancing towards use within clinical settings. Smaller SPR instruments are more integrated and perform well in comparison to previous generations of portable SPR systems. To achieve sensing in crude biofluids, new surface chemistries and novel nanomaterials have been developed to minimize nonspecific adsorption and to maximize sensitivity. Future developments in SPR will build on these achievements to refine surface chemistry in order to further decrease nonspecific adsorption, to further improve the sensitivity of the system and to increase the number of applications in clinical chemistry. It is also foreseeable that surface plasmon coupled techniques such as Raman and fluorescence spectroscopies will gain significant attention due to greater sensitivity.

References

Abbas A, Linman MJ, Cheng Q (2011) New trends in instrumental design for surface plasmon resonance-based biosensors. Biosens Bioelectron 26:1815–1824

Aubé A, Breault-Turcot J, Chaurand P, Pelletier JN, Masson J-F (2013) Non-specific adsorption of crude cell lysate on surface plasmon resonance sensors. Langmuir 29:10141–10148

Battaglia TM, Masson JF, Sierks MR, Beaudoin SP, Rogers J, Foster KN et al (2005) Quantification of cytokines involved in wound healing using surface plasmon resonance. Anal Chem 77:7016–7023

Bedford EE, Spadavecchia J, Pradier C-M, Gu FX (2012) Surface plasmon resonance biosensors incorporating gold nanoparticles. Macromol Biosci 12:724–739

Besselink GAJ, Kooyman RPH, van Os PJHJ, Engbers GHM, Schasfoort RBM (2004) Signal amplification on planar and gel-type sensor surfaces in surface plasmon resonance-based detection of prostate-specific antigen. Anal Biochem 333:165–173

Blaszykowski C, Sheikh S, Thompson M (2012) Surface chemistry to minimize fouling from blood-based fluids. Chem Soc Rev 41:5599–5612

Bolduc OR, Masson JF (2008) Monolayers of 3-mercaptopropyl-amino acid to reduce the nonspecific adsorption of serum proteins on the surface of biosensors. Langmuir 24:12085–12091

Bolduc OR, Masson J-F (2011) Advances in surface plasmon resonance sensing with nanoparticles and thin films: nanomaterials, surface chemistry, and hybrid plasmonic techniques. Anal Chem 83:8057–8062

Bolduc OR, Clouthier CM, Pelletier JN, Masson J-FO (2009a) Peptide self-assembled monolayers for label-free and unamplified surface plasmon resonance biosensing in crude cell lysate. Anal Chem 81:6779–6788

Bolduc OR, Live LS, Masson JF (2009b) High-resolution surface plasmon resonance sensors based on a dove prism. Talanta 77:1680–1687

Bolduc OR, Pelletier JN, Masson J-F (2010) SPR biosensing in crude serum using ultralow fouling binary patterned peptide SAM. Anal Chem 82:3699–3706

Bolduc OR, Lambert-Lanteigne P, Colin DY, Zhao SS, Proulx C, Boeglin D et al (2011) Modified peptide monolayer binding His-tagged biomolecules for small ligand screening with SPR biosensors. Analyst 136:3142–3148

Brault ND, White AD, Taylor AD, Yu Q, Jiang S (2013) Directly functionalizable surface platform for protein arrays in undiluted human blood plasma. Anal Chem 85:1447–1453

Breault-Turcot J, Masson J-F (2012) Nanostructured substrates for portable and miniature SPR biosensors. Anal Bioanal Chem 403:1477–1484

Bukar N, Zhao SS, Charbonneau D, Pelletier JN, Masson JF (2014) Influence of the Debye length on the interaction of a small molecule-modified Au nanoparticle with a surface-bound bioreceptor. Chem Comm 50:4947–4950

Caucheteur C, Shevchenko Y, Shao L-Y, Wuilpart M, Albert J (2011) High resolution interrogation of tilted fiber grating SPR sensors from polarization properties measurement. Opt Express 19:1656–1664

Chen S, Cao Z, Jiang S (2009) Ultra-low fouling peptide surfaces derived from natural amino acids. Biomaterials 30:5892–5896

Cheng G, Li G, Xue H, Chen S, Bryers JD, Jiang S (2009) Zwitterionic carboxybetaine polymer surfaces and their resistance to long-term biofilm formation. Biomaterials 30:5234–5240

Chinowsky T, Naimushin A, Soelberg S, Spinelli C, Kauffman P, Yee S et al (2003a) Airborne surface plasmon resonance biosensing. In: VoDinh T, Gauglitz G, Lieberman RA, Schafer KP, Killinger DK (eds) Environmental monitoring and remediation III, vol 5270, pp 182–188

Chinowsky TM, Quinn JG, Bartholomew DU, Kaiser R, Elkind JL (2003b) Performance of the Spreeta 2000 integrated surface plasmon resonance affinity sensor. Sens Actuat B Chem 91:266–274

Chinowsky TM, Soelberg SD, Baker P, Swanson NR, Kauffman P, Mactutis A et al (2007) Portable 24-analyte surface plasmon resonance instruments for rapid, versatile biodetection. Biosens Bioelectron 22:2268–2275

Couture M, Live LS, Dhawan A, Masson J-F (2012) EOT or Kretschmann configuration? Comparative study of the plasmonic modes in gold nanohole arrays. Analyst 137:4162–4170

Couture M, Liang Y, Poirier Richard H-P, Faid R, Peng W, Masson J-F (2013a) Tuning the 3D plasmon field of nanohole arrays. Nanoscale 5(24):12399–12408

Couture M, Zhao SS, Masson J-F (2013b) Modern surface plasmon resonance for bioanalytics and biophysics. Phys Chem Chem Phys 15:11190–11216

Dasgupta A (2012) Therapeutic drug monitoring - newer drugs and biomarkers. Elsevier Academic, Amsterdam

de Juan-Franco E, Rodriguez-Frade JM, Mellado M, Lechuga LM (2013) Implementation of a SPR immunosensor for the simultaneous detection of the 22 K and 20 K hGH isoforms in human serum samples. Talanta 114:268–275

Driskell JD, Lipert RJ, Porter MD (2006) Labeled gold nanoparticles immobilized at smooth metallic substrates: systematic investigation of surface plasmon resonance and surface-enhanced Raman scattering. J Phys Chem B 110:17444–17451

Frazier RA, Matthijs G, Davies MC, Roberts CJ, Schacht E, Tendler SJB (2000) Characterization of protein-resistant dextran monolayers. Biomaterials 21:957–966

Gao HW, Yang JC, Lin JY, Stuparu AD, Lee MH, Mrksich M et al (2010) Using the angle-dependent resonances of molded plasmonic crystals to improve the sensitivities of biosensors. Nano Lett 10:2549–2554

Garay F, Kisiel G, Fang A, Lindner E (2010) Surface plasmon resonance aided electrochemical immunosensor for CK-MB determination in undiluted serum samples. Anal Bioanal Chem 397:1873–1881

Gentleman DJ, Obando LA, Masson JF, Holloway JR, Booksh KS (2004) Calibration of fiber optic based surface plasmon resonance sensors in aqueous systems. Anal Chim Acta 515:291–302

Goodrich TT, Lee HJ, Corn RM (2004) Direct detection of genomic DNA by enzymatically amplified SPR imaging measurements of RNA microarrays. J Am Chem Soc 126:4086–4087

Green RJ, Davies MC, Roberts CJ, Tendler SJB (1999) Competitive protein adsorption as observed by surface plasmon resonance. Biomaterials 20:385–391

Haes AJ, Van Duyne RP (2004) A unified view of propagating and localized surface plasmon resonance biosensors. Anal Bioanal Chem 379:920–930

Hoa XD, Kirk AG, Tabrizian M (2007) Towards integrated and sensitive surface plasmon resonance biosensors: a review of recent progress. Biosens Bioelectron 23:151–160

Hoa XD, Kirk AG, Tabrizian M (2009) Enhanced SPR response from patterned immobilization of surface bioreceptors on nano-gratings. Biosens Bioelectron 24:3043–3048

Hodnik V, Anderluh G (2009) Toxin detection by surface plasmon resonance. Sensors 9:1339–1354

Homola J (2008) Surface plasmon resonance sensors for detection of chemical and biological species. Chem Rev 108:462–493

Homola J, Koudela I, Yee SS (1999) Surface plasmon resonance sensors based on diffraction gratings and prism couplers: sensitivity comparison. Sens Actuat B Chem 54:16–24

Homola J, Vaisocherova H, Dostalek J, Piliarik M (2005) Multi-analyte surface plasmon resonance biosensing. Methods 37:26–36

Kane RS, Deschatelets P, Whitesides GM (2003) Kosmotropes form the basis of protein-resistant surfaces. Langmuir 19:2388–2391

Kelf TA, Sugawara Y, Cole RM, Baumberg JJ, Abdelsalam ME, Cintra S et al (2006) Localized and delocalized plasmons in metallic nanovoids. Phys Rev B 74:245415

Kyo M, Usui-Aoki K, Koga H (2005) Label-free detection of proteins in crude cell lysate with antibody arrays by a surface plasmon resonance imaging technique. Anal Chem 77:7115–7121

Kyprianou D, Chianella I, Guerreiro A, Piletska EV, Piletsky SA (2013) Development of optical immunosensors for detection of proteins in serum. Talanta 103:260–266

Kyung Min B, Shuler ML, Sung June K, Soon Joon Y, Donghyun K (2008) Sensitivity enhancement of surface plasmon resonance imaging using periodic metallic nanowires. J Lightwave Technol 26:1472–1478
Ladd J, Zhang Z, Chen S, Hower JC, Jiang S (2008) Zwitterionic polymers exhibiting high resistance to nonspecific protein adsorption from human serum and plasma. Biomacromolecules 9:1357–1361
Ladd J, Lu H, Taylor AD, Goodell V, Disis ML, Jiang S (2009) Direct detection of carcinoembryonic antigen autoantibodies in clinical human serum samples using a surface plasmon resonance sensor. Colloids Surf B Biointerfaces 70:1–6
Lee HJ, Li Y, Wark AW, Corn RM (2005) Enzymatically amplified surface plasmon resonance imaging detection of DNA by exonuclease III digestion of DNA microarrays. Anal Chem 77:5096–5100
Li Y, Lee HJ, Corn RM (2007) Detection of protein biomarkers using RNA aptamer microarrays and enzymatically amplified surface plasmon resonance imaging. Anal Chem 79:1082–1088
Liedberg B, Nylander C, Lundstrom I (1983) Surface-plasmon resonance for gas-detection and biosensing. Sens Actuat 4:299–304
Linman MJ, Sugerman K, Cheng Q (2010) Detection of low levels of *Escherichia coli* in fresh spinach by surface plasmon resonance spectroscopy with a TMB-based enzymatic signal enhancement method. Sens Actuat B Chem 145:613–619
Liu Y, Dong Y, Jauw J, Linman MJ, Cheng Q (2010) Highly sensitive detection of protein toxins by surface plasmon resonance with biotinylation-based inline atom transfer radical polymerization amplification. Anal Chem 82:3679–3685
Live LS, Masson JF (2009) High sensitivity of plasmonic microstructures near the transition from short-range to propagating surface plasmon. J Phys Chem C 113:10052–10060
Live LS, Murray-Methot MP, Masson JF (2009) Localized and propagating surface plasmons in gold particles of near-micron size. J Phys Chem C 113:40–44
Live LS, Bolduc OR, Masson JF (2010) Propagating surface plasmon resonance on microhole arrays. Anal Chem 82:3780–3787
Live L, Dhawan A, Gibson K, Poirier-Richard H-P, Graham D, Canva M et al (2012) Angle-dependent resonance of localized and propagating surface plasmons in microhole arrays for enhanced biosensing. Anal Bioanal Chem 404:2859–2868
Masson JF, Battaglia TM, Kim YC, Prakash A, Beaudoin S, Booksh KS (2004a) Preparation of analyte-sensitive polymeric supports for biochemical sensors. Talanta 64:716–725
Masson JF, Obando L, Beaudoin S, Booksh K (2004b) Sensitive and real-time fiber-optic-based surface plasmon resonance sensors for myoglobin and cardiac troponin I. Talanta 62:865–870
Masson JF, Battaglia TM, Davidson MJ, Kim YC, Prakash AMC, Beaudoin S et al (2005) Biocompatible polymers for antibody support on gold surfaces. Talanta 67:918–925
Masson JF, Battaglia TM, Cramer J, Beaudoin S, Sierks M, Booksh KS (2006a) Reduction of nonspecific protein binding on surface plasmon resonance biosensors. Anal Bioanal Chem 386:1951–1959
Masson JF, Kim YC, Obando LA, Peng W, Booksh KS (2006b) Fiber-optic surface plasmon resonance sensors in the near-infrared spectral region. Appl Spectrosc 60:1241–1246
Masson JF, Battaglia TM, Khairallah P, Beaudoin S, Booksh KS (2007) Quantitative measurement of cardiac markers in undiluted serum. Anal Chem 79:612–619
Masson J-F, Murray-Methot M-P, Live LS (2010) Nanohole arrays in chemical analysis: manufacturing methods and applications. Analyst 135:1483–1489
Mitchell J (2010) Small molecule immunosensing using surface plasmon resonance. Sensors 10:7323–7346
Mrksich M, Whitesides GM (1996) Using self-assembled monolayers to understand the interactions of man-made surfaces with proteins and cells. Annu Rev Biophys Biomol Struct 25:55–78
Naimushin AN, Soelberg SD, Nguyen DK, Dunlap L, Bartholomew D, Elkind J et al (2002) Detection of *Staphylococcus aureus* enterotoxin B at femtomolar levels with a miniature integrated two-channel surface plasmon resonance (SPR) sensor. Biosens Bioelectron 17:573–584

Navratilova I, Skladal P (2003) Immunosensor for the measurement of human serum albumin in urine based on the Spreeta surface plasmon resonance sensor. Supramol Chem 15:109–115
Ostuni E, Chapman RG, Holmlin RE, Takayama S, Whitesides GM (2001a) A survey of structure-property relationships of surfaces that resist the adsorption of protein. Langmuir 17:5605–5620
Ostuni E, Chapman RG, Liang MN, Meluleni G, Pier G, Ingber DE et al (2001b) Self-assembled monolayers that resist the adsorption of proteins and the adhesion of bacterial and mammalian cells. Langmuir 17:6336–6343
Phillips KS, Han JH, Cheng Q (2007) Development of a "membrane cloaking" method for amperometric enzyme immunoassay and surface plasmon resonance analysis of proteins in serum samples. Anal Chem 79:899–907
Piliarik M, Vala M, Tichy I, Homola J (2009) Compact and low-cost biosensor based on novel approach to spectroscopy of surface plasmons. Biosens Bioelectron 24:3430–3435
Polanski M, Anderson NL (2006) A lis of candidate cancer biomarkers for targeted proteomics. Biomark Insights 2:1–48
Ratel M, Provencher-Girard A, Zhao SS, Breault-Turcot J, Labrecque-Carbonneau J, Branca M et al (2013) Imidazolium-based ionic liquid surfaces for biosensing. Anal Chem 85:5770–5777
Roh S, Chung T, Lee B (2011) Overview of the characteristics of micro- and nano-structured surface plasmon resonance sensors. Sensors 11:1565–1588
Scarano S, Mascini M, Turner APF, Minunni M (2010) Surface plasmon resonance imaging for affinity-based biosensors. Biosens Bioelectron 25:957–966
Schlichtiger A, Baier C, Yin M-X, Holmes AB, Maruyama M, Strasser R et al (2013) Covalent attachment of functionalized cardiolipin on a biosensor gold surface allows repetitive measurements of anticardiolipin antibodies in serum. Anal Bioanal Chem 405:275–285
Shankaran DR, Gobi KVA, Miura N (2007) Recent advancements in surface plasmon resonance immunosensors for detection of small molecules of biomedical, food and environmental interest. Sens Actuat B Chem 121:158–177
Shao L-Y, Shevchenko Y, Albert J (2010) Intrinsic temperature sensitivity of tilted fiber Bragg grating based surface plasmon resonance sensors. Opt Express 18:11464–11471
Sharma AK, Jha R, Gupta BD (2007) Fiber-optic sensors based on surface plasmon resonance: a comprehensive review. IEEE Sens J 7:1118–1129
Shevchenko YY, Albert J (2007) Plasmon resonances in gold-coated tilted fiber Bragg gratings. Opt Lett 32:211–213
Shevchenko Y, Francis TJ, Blair DAD, Walsh R, DeRosa MC, Albert J (2011) In situ biosensing with a surface plasmon resonance fiber grating aptasensor. Anal Chem 83:7027–7034
Situ C, Wylie ARG, Douglas A, Elliott CT (2008) Reduction of severe bovine serum associated matrix effects on carboxymethylated dextran coated biosensor surfaces. Talanta 76:832–836
Situ C, Buijs J, Mooney MH, Elliott CT (2010) Advances in surface plasmon resonance biosensor technology towards high-throughput, food-safety analysis. Trends Anal Chem 29:1305–1315
Spangler BD, Wilkinson EA, Murphy JT, Tyler BJ (2001) Comparison of the Spreeta (R) surface plasmon resonance sensor and a quartz crystal microbalance for detection of Escherichia coli heat-labile enterotoxin. Anal Chim Acta 444:149–161
Statz AR, Meagher RJ, Barron AE, Messersmith PB (2005) New peptidomimetic polymers for antifouling surfaces. J Am Chem Soc 127:7972–7973
Statz AR, Barron AE, Messersmith PB (2008) Protein, cell and bacterial fouling resistance of polypeptoid-modified surfaces: effect of side-chain chemistry. Soft Matter 4:131–139
Sugawara Y, Kelf TA, Baumberg JJ, Abdelsalam ME, Bartlett PN (2006) Strong coupling between localized plasmons and organic excitons in metal nanovoids. Phys Rev Lett 97:266808
Trevino J, Calle A, Miguel Rodriguez-Frade J, Mellado M, Lechuga LM (2009a) Surface plasmon resonance immunoassay analysis of pituitary hormones in urine and serum samples. Clin Chim Acta 403:56–62

Trevino J, Calle A, Rodriguez-Frade JM, Mellado M, Lechuga LM (2009b) Determination of human growth hormone in human serum samples by surface plasmon resonance immunoassay. Talanta 78:1011–1016

Uludag Y, Tothill IE (2012) Cancer biomarker detection in serum samples using surface plasmon resonance and quartz crystal microbalance sensors with nanoparticle signal amplification. Anal Chem 84:5898–5904

Vaisocherova H, Yang W, Zhang Z, Cao Z, Cheng G, Piliarik M et al (2008) Ultralow fouling and functionalizable surface chemistry based on a zwitterionic polymer enabling sensitive and specific protein detection in undiluted blood plasma. Anal Chem 80:7894–7901

Vaisocherova H, Faca VM, Taylor AD, Hanash S, Jiang SY (2009a) Comparative study of SPR and ELISA methods based on analysis of CD166/ALCAM levels in cancer and control human sera. Biosens Bioelectron 24:2143–2148

Vaisocherova H, Zhang Z, Yang W, Cao Z, Cheng G, Taylor AD et al (2009b) Functionalizable surface platform with reduced nonspecific protein adsorption from full blood plasm-aterial selection and protein immobilization optimization. Biosens Bioelectron 24:1924–1930

Yang W, Xue H, Li W, Zhang J, Jiang S (2009) Pursuing "zero" protein adsorption of poly(carboxybetaine) from undiluted blood serum and plasma. Langmuir 25:11911–11916

Yang W, Bai T, Carr LR, Keefe AJ, Xu J, Xue H et al (2012) The effect of lightly crosslinked poly(carboxybetaine) hydrogel coating on the performance of sensors in whole blood. Biomaterials 33:7945–7951

Zayats AV, Smolyaninov II, Maradudin AA (2005) Nano-optics of surface plasmon polaritons. Phys Rep 408:131–314

Zhang Z, Zhang M, Chen S, Horbetta TA, Ratner BD, Jiang S (2008) Blood compatibility of surfaces with superlow protein adsorption. Biomaterials 29:4285–4291

Zhang L, Cao Z, Bai T, Carr L, Ella-Menye J-R, Irvin C et al (2013) Zwitterionic hydrogels implanted in mice resist the foreign-body reaction. Nat Biotech 31:553–556

Zhao SS, Bichelberger MA, Colin DY, Robitaille R, Pelletier JN, Masson J-F (2012) Monitoring methotrexate in clinical samples from cancer patients during chemotherapy with a LSPR-based competitive sensor. Analyst 137:4742–4750

Zhao SS, Bukar N, Toulouse J, Pelechacz D, Robitaille R, Pelletier JN, Masson JF (2015) Miniature multi-channel SPR instrument for methotrexate monitoring in clinical samples. Biosens Bioelectron 64:664–670

Zhou W-J, Chen Y, Corn RM (2011a) Ultrasensitive microarray detection of short RNA sequences with enzymatically modified nanoparticles and surface plasmon resonance imaging measurements. Anal Chem 83:3897–3902

Zhou W-J, Halpern AR, Seefeld TH, Corn RM (2011b) Near infrared surface plasmon resonance phase imaging and nanoparticle-enhanced surface plasmon resonance phase imaging for ultrasensitive protein and DNA biosensing with oligonucleotide and aptamer microarrays. Anal Chem 84:440–445

Chapter 16
Scanning Electrochemical Microscopy for Imaging Single Cells and Biomolecules

Yasufumi Takahashi, Hitoshi Shiku, and Tomokazu Matsue

Abstract This article presents an overview of the recent progress in scanning electrochemical microscopy (SECM) for imaging single cells and biomolecules. SECM is a technique for characterizing the local electrochemical nature of various materials by scanning a probe microelectrode. The probe current reflects the electrochemical processes occurring in the small space surrounded by the probe and the substrate. The spatial resolution of SECM is inferior to conventional scanning probe microscopes such as scanning tunneling microscopy (STM) and atomic force microscopy (AFM), since the fabrication of the probe microelectrodes with nanometer sizes is quite difficult. However, recent progress in the fabrication of nanometer-scale electrodes and the development of electrode–sample distance control systems has greatly enhanced the capacity of SECM systems to solve problems in cell biology. The topics reviewed include the following: enzyme activity evaluation, electrochemical enzyme-linked immunosorbent assays, membrane permeability evaluation, respiratory activity measurements, reporter gene assays, membrane protein imaging, and neurotransmitter detection.

16.1 Introduction

Electrochemical analysis has been used to characterize various electroactive species, including in situ, in vivo, and in vitro measurements of biologically important species. Small electrodes are sometimes called ultramicroelectrodes (UMEs) to distinguish them from the traditionally termed microelectrodes with millimeter-sized domains (Matsue 2012; Matsue 2013). UMEs have unique characteristics such

Y. Takahashi • H. Shiku • T. Matsue (✉)
Advanced Institute for Materials Research (WPI-AIMR), Tohoku University, Aramaki 6-6-11, Aoba, Sendai 980-8579, Japan

Graduate School of Environmental Studies, Tohoku University, Katahira 2-1-1, Aoba, Sendai 980-8577, Japan
e-mail: matsue@bioinfo.che.tohoku.ac.jp

M.C. Vestergaard et al. (eds.), *Nanobiosensors and Nanobioanalyses*, DOI 10.1007/978-4-431-55190-4_16 as the capability to perform localized measurements, low double-layer charging currents, low ohmic drops, and fast mass transport. These characteristics enable the following:

- Simplified electrochemical measurements under steady-state conditions.
- Detection of chemical species in localized volumes such as living cells.
- Electrochemical measurements in highly resistive media.
- Monitoring of electrochemical processes over short time domains.

A steady-state, hemispheric diffusion region is formed on the UME in short time domains (less than 500 ms in the case of 10 μm radius), and the voltammogram of the UME shows a diffusion-controlled current expressed as follows (Amphlett and Denuault 1998; Shao and Mirkin 1998):

$$i = 4nFDCa \tag{16.1}$$

where F is Faraday's constant; D and C are the diffusion coefficient and the bulk concentration of the redox species, respectively; and a is the radius of the UME. This equation is particularly useful in the determination of biomolecules for characterizing localized events in biological samples. In this chapter, we summarize our research on the characterization of biomaterials and the sensing of biologically important species by scanning electrochemical microscopy (SECM).

16.2 Principles of Scanning Electrochemical Microscopy

SECM, a type of scanning probe microscopy (SPM), has been used to characterize and image the local electrochemical nature of various materials by scanning sample surfaces with a UME (Fig. 16.1) (Edwards et al. 2006; Sun et al. 2007; Mirkin et al. 2011). The spatial resolution of SECM is primarily dependent on the tip size of the UME and the distance between the UME and sample. SECM has unique characteristics that cannot be achieved with conventional SPM methods; for example, SECM can image and induce localized chemical reactions in a controlled manner. SECM has been widely applied for analyzing numerous systems involving electrochemistry, such as electrode surfaces, polymers, biomaterials, and liquid/liquid interfaces (Mirkin et al. 2011). There are two basic measurement modes for SECM: generation/collection (G/C) and feedback (FB).

16.2.1 Instrumentation

Although commercial instruments for SECM are available, many instruments are constructed by individual research groups and tailored to specific applications. The basic SECM system consists of four major components: an electrochemical cell (including UME, counter electrode, and reference electrode), a current detector (current amplifier or potentiostat), a micro/nano-positioner, and a data acquisition

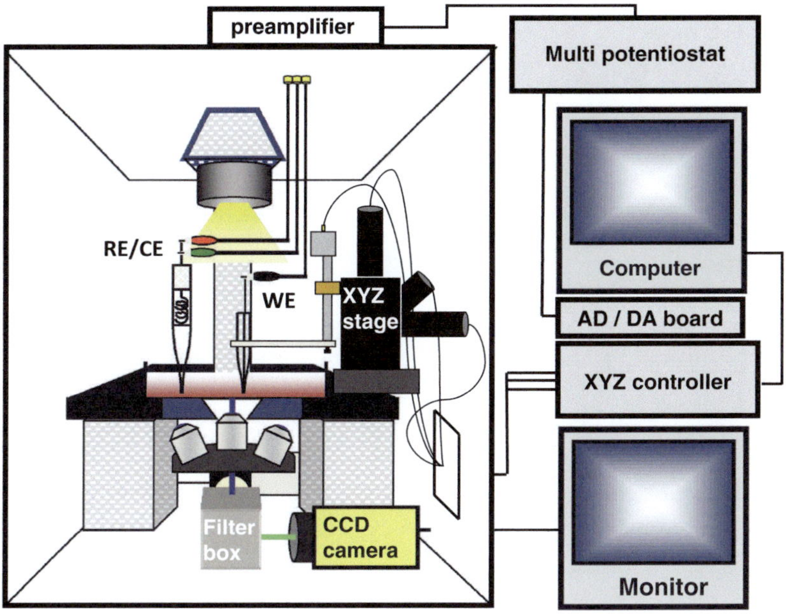

Fig. 16.1 Block diagram of the SECM measurement

system (Fig. 16.1). A vibration-free table is necessary to provide adequate vibration isolation for high-resolution measurements. In a typical SECM measurement, a two-electrode setup is employed because the current flow is extremely small (pA–nA), and therefore, the IR drop is negligible. If the sample needs external bias, a bipotentiostat is required. The UME is held on a micro/nano-positioner, which allows movement and positioning relative to the interface under investigation. Various positioners have been employed in SECM instruments, where the best choice depends on the type of measurement and spatial resolution required (Guell et al. 2012; Takahashi et al. 2012). To acquire the data, conventional AD/DA boards are available. However, in case of constant-distance-mode SECM, high-speed data transfer without imposing a load on the CPU is desired. We have used a field-programmable gate array board, which contains a matrix of reconfigurable gate array logic circuitry and can perform complex and discrete signal processing algorithms with clock rates at a maximum of 40 MHz, to enhance feedback distance regulation and decrease imaging time (Takahashi et al. 2010a).

16.2.2 Generation/Collection Mode

The substrate-generation/tip-collection (G/C) mode, as depicted in Fig. 16.2a, is used for imaging redox species concentration profiles and evaluating the redox species flux generated from the sample. When the UME is scanned through the thick

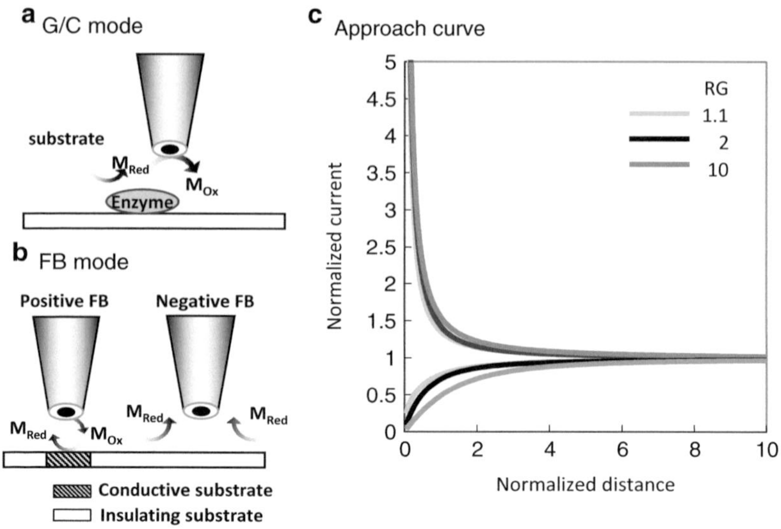

Fig. 16.2 Schematic illustration of the (**a**) generation collection mode, (**b**) feedback mode, (**c**) approach curves toward conducting (*top curves*) and insulating (*bottom curves*) substrates. RG is defined as the ratio between the insulator thickness and the radius of the UME

diffusion layer produced by the sample, the current changes reflect local variations in the concentration of the redox species. In contrast to the FB mode, the G/C mode offers high sensitivity measurements because the background current is lower (discussed later). This is because the solution contains no redox species except those generated by the sample. The G/C mode measurements have also been applied to detect short-lived active species, chemical release, and chemical consumption. Bard and coworkers used SECM to evaluate a half-life time of 0.2 ms for tri-*n*-propylamine radical ($TPrA^{\bullet+}$), which is generated during electrogenerated chemiluminescence process (Miao et al. 2002).

16.2.3 Feedback Mode

The FB mode, as depicted in Fig. 16.2b, is used for imaging and evaluating the redox activity of the sample surface. A mediator is oxidized or reduced at the UME and diffuses to the sample surface where it is converted back to its original form by a surface redox reaction if the sample surface promotes regeneration. The mediator at the sample surface diffuses again to the UME, thereby establishing a feedback cycle and enhancing the current through the UME. This situation is called positive feedback. If the sample surface is an inert electrical insulator, the species generated at the UME cannot react at the surface. In this case, the closer the UME is to the sample surface, the smaller the measured response. This situation is called negative feedback.

In the FB mode, either positive or negative, the probe current largely depends on the probe–substrate distance (Fig. 16.2c). Therefore, the FB mode has a high spatial resolution compared to the GC mode. The FB mode is particularly useful for studying redox enzymes such as glucose oxidase (GOD) and diaphorase (Dp) immobilized at surfaces. The FB mode is also used for observing living cells, which contain intracellular redox enzymes.

16.2.4 Fabrication of UMEs

Various methods for miniaturizing UMEs have been advocated to improve the resolution of SECM imaging. The reliable fabrication of nanoelectrodes with a small ratio of electrode insulation to active electrode is of particular importance for improving the spatial resolution of SECM. In particular, various approaches have been adopted to create small electrodes with a thin insulating coating, such as photolithography, chemical vapor deposition, electrodeposited paint methods, laser pulling techniques, and pyrolytic carbon deposition. Mezour and coworkers have recently developed a reproducible procedure for the fabrication of Pt disk-shaped microelectrodes with characteristic dimensions ranging from 50 nm to 1 μm in diameter by using a laser pulling technique (Mezour et al. 2011). We have developed a fabrication technique, pyrolytic carbon deposition inside glass pipettes, which enables fabrication of nanoelectrodes with insulation of excellent integrity (Takahashi et al. 2011b; Takahashi et al. 2012). For the fabrication of the pyrolytic carbon electrode, a quartz glass capillary (O.D. 1.2 mm, I.D. 0.9 mm; Sutter Instrument, USA) was pulled using a CO_2 laser puller (model P-2000, Sutter Instrument, USA). Butane was passed through the quartz capillary using a Tygon tube (O.D. 2.4 mm, I.D. 0.8 mm). The pipette was pressurized internally with butane gas to deposit the carbon. The taper of the pipette was inserted into another quartz capillary (O.D. 1.0 mm, I.D. 0.7 mm; Sutter Instrument, USA), which was filled with argon gas to prevent oxidation of the carbon layer and bending of the capillary at high temperatures. This approach also protected the pipette aperture from closing due to the softening of the quartz pipette walls. To form a pyrolytic carbon layer inside the capillary, the pipette taper was then heated with a Bunsen burner for times ranging from 0.5 s for a 100-nm radius electrode to 3 s for a 1-μm radius electrode.

16.2.5 Electrode–Sample Distance Control

Control of the distance between the electrode and sample is critical to achieve electrochemical measurements that are free from topographical artifacts. Various feedback distance-control systems have been developed, including those based on atomic force microscopy (Macpherson and Unwin 2000; Kranz et al. 2001), shear force (Hengstenberg et al. 2000; Takahashi et al. 2006; Takahashi et al. 2009b), impedance (Alpuche-Aviles and Wipf 2001; Kurulugama et al. 2005; Zhao et al. 2010), faradaic current (Fan and Bard 1999; Kurulugama et al. 2005), ion current (Comstock et al. 2010; Takahashi et al. 2010b; Takahashi et al. 2011b), and electrochemical signaling (Williams et al. 2009; Lai et al. 2011; Guell et al. 2012). Impedance feedback and constant-current distance control are effective methods for obtaining high-resolution topography images because they do not require additional modification of the electrode for distance control. However, the impedance-feedback mode has resolution limitations due to the measurement principle (Kurulugama et al. 2005). The constant-current mode has been shown to be capable of high-resolution imaging, but only for topography to date (Fan and Bard 1999).

Shear force-based tip–sample distance control is a successful feedback mechanism for high-resolution imaging. In particular, the feedback mechanism using a tuning fork has become the standard system for controlling the probe position with a relatively simple and low-cost design (Takahashi et al. 2009b). We successfully imaged the topography of a single cell and the respiration activity using a conical optical fiber electrode (Takahashi et al. 2006). For the successful imaging of living cell topographies using scanning force microscopy, the magnitude of the force interaction between the probe and cell is important (Takahashi et al. 2009b). We optimized the operating conditions of the vibrating probe to detect shear forces against living cells when the probe approached different oscillation amplitudes.

Scanning ion conductance microscopy (SICM) is a promising technique for non-contact topographical analyses under physiological conditions (Korchev et al. 1997; Novak et al. 2009). SICM uses a nanopipette as a scanning probe to detect ionic current between an electrode inside the pipette and one located in a bath. The pipette–sample distance is regulated by the ionic currents used as feedback signals. The magnitude of the ionic current depends on the pipette–sample distance; when the pipette approaches the sample surface, the resistance between the two electrodes increases, and the ionic current decreases. Therefore, the ionic current can be used as a feedback signal to maintain a constant distance. We have developed a combined SECM–SICM system to study living cells in a physiological environment (Takahashi et al. 2010b; Takahashi et al. 2011b).

The voltage-switching mode (VSM) SECM was developed to achieve constant-distance mode measurements with the possibility of simultaneous (electro) chemical flux measurements at the same location (Takahashi et al. 2012). The electrode was moved toward the sample surface while detecting the distance-dependent current for the hindered diffusion of a redox-active solute to the tip. The probe potential was biased to activate the cathodic process. Compared to the bulk solution, when the current diminished by a set amount, the probe was at a desired distance from the surface or structure of interest (typically one electrode radius away). The probe was then stopped and the z-position of the piezoelectric actuator was registered. By conducting this process at numerous points over the surface, the topography was mapped. When the electrode reached the desired position at each point, the applied voltage could be switched to a positive value for electrochemical (flux) imaging of the sample surface.

The probe-scanning program is also important in the imaging of living cell topographies. The hopping mode, in which the probe repetitively approaches the sample surface and retracts at each point, is a promising method for imaging the convoluted surface topography of living cells (Novak et al. 2009; Takahashi et al. 2010b; Takahashi et al. 2011b; Takahashi et al. 2012). Such image acquisition is very time-consuming compared with conventional lateral scanning methods. However, this limitation was resolved by changing the step size during measurement and pre-scanning the area of interest (Novak et al. 2009).

16.3 Applications of Scanning Electrochemical Microscopy

16.3.1 Enzyme Activity Evaluation

The FB mode has been used for the characterization of the kinetic constant of localized patterned enzymes. Pierce and coworkers immobilized GOD on a nylon and hydrogel membrane and studied the kinetics of GOD and mediators on the basis of zero- and first-order electron-transfer kinetics (Pierce et al. 1992). They proposed the following empirical equation to estimate the detection limit of the FB-mode SECM for a zero-order heterogeneous enzymatic reaction:

$$J_{\mathrm{m}} > 10^{-3} DC / a,$$

where J_{m} is the flux of the mediator at the surface; D and C are the diffusion coefficient and the bulk concentration of the mediator, respectively; and a is the radius of the probe microelectrode. This relationship suggests that high spatial resolution can only be attained for enzymes with high activities. The activity of a monolayer of GOD was found insufficiently high to yield a detectable response using a microelectrode with a radius of several microns; however, the characterization of a Dp monolayer on a glass surface in the FB mode was possible (Shiku et al. 1995).

The G/C mode has also been used for the investigation of localized enzyme reactions (Wittstock and Schuhmann 1997). Besides GOD, enzymes typically characterized with this mode are alkaline phosphatase (AP) and β-galactosidase (β-GAL). AP catalyzes the hydrolysis of *p*-aminophenylphosphate (PAPP) to *p*-aminophenol (PAP), which can be detected with a microelectrode located close to AP. Wijayawardhana and coworkers used SECM for sensing an AP-labeled antigen (Wijayawardhana et al. 2000). We have also used AP and β-GAL as reporter proteins to detect gene expression with SECM, as described later.

16.3.2 Electrochemical Enzyme-Linked Immunosorbent Assay

The combination of SECM with an enzyme-linked immunosorbent assay (ELISA) offers a novel assay system. We have characterized carcinoembryonic antigen (CEA) microspotted on a glass substrate (Shiku et al. 1996). Sample preparation was similar to the conventional sandwich method utilizing horseradish peroxidase

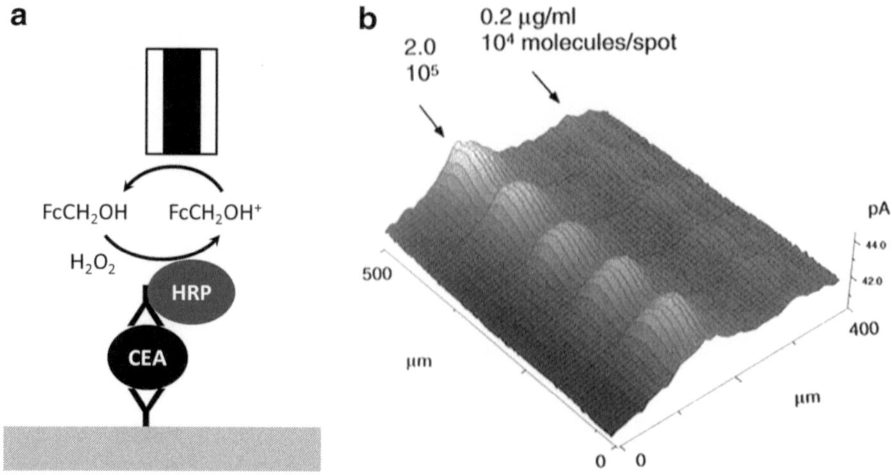

Fig. 16.3 (**a**) Schematic illustration of SECM-ELISA. (**b**) SECM image of a series of antigen–antibody immobilized spots. CEA concentration of the solution for making spots, 2.0 (*left*) and 0.2 μg/mL (*right*) (Reprinted with permission from Shiku et al. 1996, Copyright 1996 ACS)

(HRP)-labeled antibodies; however, the analyte solution containing CEA was microspotted in a 20-μm radius region with a tapered glass capillary. In the presence of H_2O_2 and ferrocenemethanol ($FcCH_2OH$), the probe detected the reduction current of $FcCH_2OH^+$ produced by the HRP-catalyzed reaction on the substrate. Using SECM, as few as 10^4 CEA molecules per spot were detectable (Fig. 16.3). The SECM–ELISA method was applied to assay leukocidin, a toxic protein produced by methicillin-resistant *Staphylococcus aureus* (MRSA), in the pg/mL concentration range (Kasai et al. 2000). With the aim of improving the sensitivity of SECM–ELISA, the procedure to form the sandwich structure was refined by optimizing the preparation conditions such as pH. In addition, protein A and bovine serum albumin were used as the base layer of the sandwich structure and as the blocking layer for nonspecific adsorptions, respectively. The size of the microelectrode tip was also optimized. Through these modifications, the detection limit of 5.25 pg/mL leukocidin was achieved. The sensitivity attained here would facilitate early diagnosis of MRSA infections. The SECM–ELISA technique was also applied as a sensing system for antibody arrays to detect several proteins in a small sample volume simultaneously (Shiku et al. 1997).

16.3.3 Membrane Permeability Evaluation

SECM affords a microscopic view of the heterogeneous flux distribution across porous membranes. The probe senses a particular redox species by choosing a suitable electrode potential; moreover, the local flux can be quantitatively determined by analyzing SECM images. Bath and coworkers imaged the transportation of redox species through porous membranes (Bath et al. 1998). The local fluxes of a particular redox species through the membrane pores were measured using a probe located close to the membrane surface. The G/C-mode SECM imaging allows the quantitative estimation of the flux through an individual pore and the pore size. Macpherson and coworkers measured osmotically driven redox fluxes through porous laryngeal cartilage (Macpherson et al. 1997). The topographic image was subtracted to yield the two-dimensional flux distributions. The bilayer lipid membrane (BLM) permeation process for hydrophobic ferrocene derivatives has been investigated by microvoltammetry (Yamada et al. 1991). Linear sweep voltammetry was conducted using a microelectrode positioned close to the BLM. The results clearly reflect the differences in BLM permeability to different redox species. Digital simulations analyzed the permeation phenomena by considering several processes such as transmembrane mass transfer and diffusion in the vicinity of the BLM at each side of the membrane. A similar experimental procedure was applied to estimate the permeation of several redox species through a cell membrane of an algal protoplast (Yasukawa et al. 1998). The ion transport across voltage-gated ion channels was investigated using the BLM method (Yamada et al. 1991; Matsue et al. 1994). Alamethicin, a polypeptide antibiotic, forms barrel-type ion channels penetrating through the BLM when membrane potentials are applied. Figure 16.4 shows a schematic of this experimental system. A redox species was added to the *cis*-side of the membrane. A microelectrode soaked in the *trans*-side was positioned close to the BLM incorporating alamethicin channels. Then, a positive membrane potential pulse was applied at the working electrode on the *cis*-side, where the total ionic current was measured. Simultaneously, the probe microelectrode detected the redox current of the electroactive species, which permeated from the *cis*- to the *trans*-side through the alamethicin ion channels. The chronoamperometric responses were recorded using the microelectrode and analyzed by digital simulation to obtain the permselectivity of the channels for several redox species. The transportation of $Fe(CN)_6^{4-}$ through the channels was found to be restricted compared with that of cationic $Ru(NH_3)_6^{3+}$, because of its negative charge.

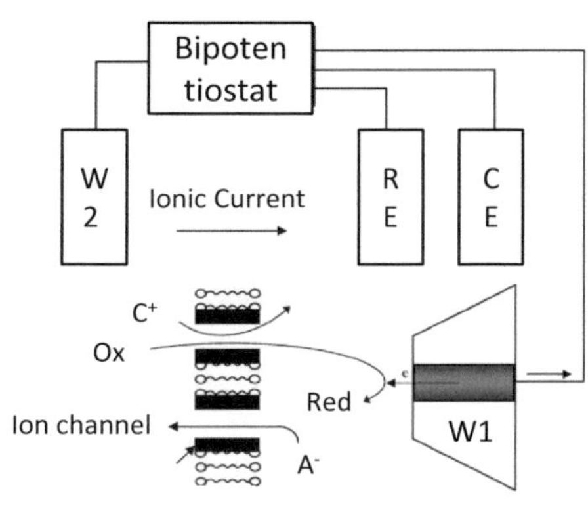

Fig. 16.4 Ionic and redox current simultaneous measurements of transport across a BLM (Reprinted with permission from Matsue et al. 1994, Copyright 1994 ACS)

16.3.4 *Respiratory Activity Measurements*

An SECM image of the oxygen concentration around a cell directly reflects its respiration activity and indicates the status of the cell. The quantitative and noninvasive characterization of the respiration activity of living cells has received considerable attention.

Photosynthetic oxygen generation from protoplasts was imaged using SECM (Yasukawa et al. 1999). In this experiment, we used dual-Pt microdisks to obtain the topography and concentration profile of oxygen simultaneously. Figure 16.5 shows dual-SECM images of the topography and photosynthetic activity of a single living protoplast (radius, 25 μm). The dark area with low oxidation currents for $Fe(CN)_6^{4-}$ (Fig. 16.5a) coincided with the location of the protoplast in an optical microscopic image. The oxidation current decreased because of the blocking of diffusion when

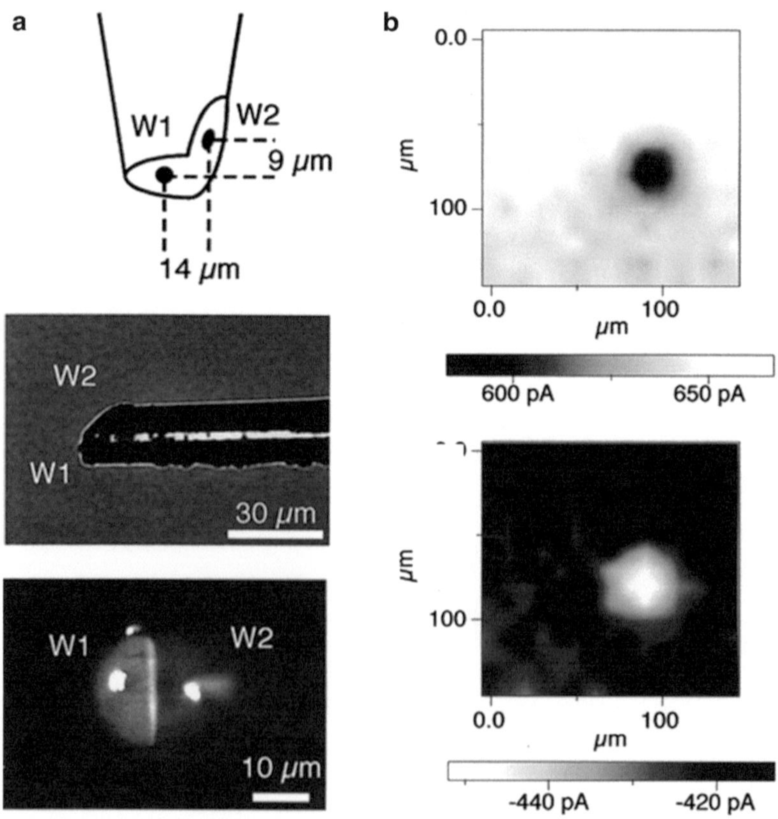

Fig. 16.5 (**a**) Horizontal and vertical microscopic images of dual-microdisk electrodes. (**b**) Dual-SECM images of a single, living protoplast based on (*top*) the oxidation current for $Fe(CN)_6^{4-}$ and (*bottom*) the reduction current for oxygen (Reprinted with permission from Yasukawa et al. 1999, Copyright 1999 ACS)

the tip approached the protoplast surface. Therefore, the image based on the oxidation current for $Fe(CN)_6^{4-}$ shows topographic information at the single-cell level. The SECM image based on the reduction current for oxygen (Fig. 16.5b) shows the opposite image. The image of the cell appears as an area with a large reduction current because of the photosynthetic generation of oxygen from the single protoplast. Thus, this image provides information about the photosynthetic activity of individual cells.

The evaluation of the respiratory activity of patterned mammalian cells (HeLa cells) has been performed by SECM (Nishizawa et al. 2002). The patterning of living cells has been extensively investigated by several research groups since the spatial control of mammalian cell adhesion and growth is a critical issue in many areas of biotechnology. The SECM results proved that the micropatterned HeLa cells were viable because of the oxygen uptake. The control of the degree of cell spreading by patterning was found to affect cellular respiratory activity.

The evaluation of single embryo activity has also been characterized by SECM (Shiku et al. 2001). The oxygen concentration profiles near the embryos were in good agreement with the theoretical spherical diffusion. When an embryo reached the stage of a morula with a 74-μm radius on day six after in vitro fertilization, the oxygen concentration difference between the bulk solution and the morula surface was 7 μM. The oxygen consumption rate of a single morula was estimated to be 1.4×10^{-14} mol/s. After the SECM analysis, the embryo was continuously cultured for another 2 days and grew to the stage of a blastocyst with a 10-μm radius. For the blastocyst, the oxygen consumption was found to be in the range $(2.50\pm0.46)\times10^{-14}$ mol/s. The oxygen consumption of the morulae on day six after in vitro fertilization was strongly related to the morphological quality of the embryo. The morulae showing a larger oxygen consumption rate developed into blastocysts of a larger size.

Drug sensitivity assays have also been performed by SECM (Torisawa et al. 2003). Cell-containing collagen droplets were immobilized in an array of microholes machined in a silicon wafer. Two types of cancer cells, the human erythroleukemia cell line (K562) and its adriamycin-resistant sub-line (K562/ADM), were spatially addressed within the silicon microstructure for the comparative characterization of their sensitivity to the anticancer drug ADM. The results obtained by the SECM respiratory activity assay showed results comparable to the conventional colorimetric succinic dehydrogenase inhibition assay. This indicates that the change in the number of living cells (cell proliferation rate) is predominantly responsible for the observed change in the degree of oxygen consumption.

16.3.5 Reporter Gene Assays

An SECM-based assay has been applied to modern genetic engineering. Among the various technologies currently used in biotechnology facilities, the reporter gene system has been recognized as a rapid and convenient method for mRNA detection to analyze gene expression. Reporter gene systems are frequently used in gene expression studies by the incorporation of a vector plasmid or fusion of a promoter gene.

The gene *lacZ*, which encodes β-GAL, is a widely used reporter. Since β-GAL catalyzes the hydrolysis of *p*-aminophenyl β-D-galactopyranoside to yield electroactive PAP, *lacZ* has been used as a reporter gene for electrochemical detection (Kaya et al. 2004). PAP was oxidized at an SECM scanning probe set at 0.30 V vs. Ag/AgCl. The gene expression activity of recombinant cells can be easily imaged and monitored with conventional SECM systems. Our group also developed an electrochemical microbial chip for mutagen screening (Matsui et al. 2006). Recombinant *E. coli* carrying *lacZ* were exposed to mutagen solutions and embedded in a microcavity (5 nL) on a glass substrate using collagen gel (Fig. 16.6). The β-GAL expression on the microbial chip was electrochemically monitored by SECM. This system has several advantages compared with the conventional *umu* test: drastic reduction of the required sample volume, much faster β-GAL detection, and a lower detection limit for the three mutagens.

AP has also been used as a reporter protein that dephosphorylates PAPP to electroactive PAP. However, the expressed proteins, in general, remain inside the cells and only react with the substrate that permeates through the cell membrane. Since the permeability of PAPP through cell membranes is usually low, the electrochemical signals of PAP are also small. Therefore, we used secreted placental alkaline phosphatase (SEAP) as a reporter protein. SEAP is a useful reporter because it is secreted into the culture medium. The cellular secretion of SEAP is directly proportional to the changes in the intracellular SEAP mRNA. This characteristic allows the continuous electrochemical quantification of gene expression in living cells. Our group investigated in situ cellular signal transduction of three-dimensional cultured cells (Torisawa et al. 2006). These cells were genetically engineered to produce SEAP and immobilized in an array of microholes on a chip. Cellular SEAP expression was triggered by exposure to tumor necrosis factor α and was continuously monitored by employing an SECM-based assay.

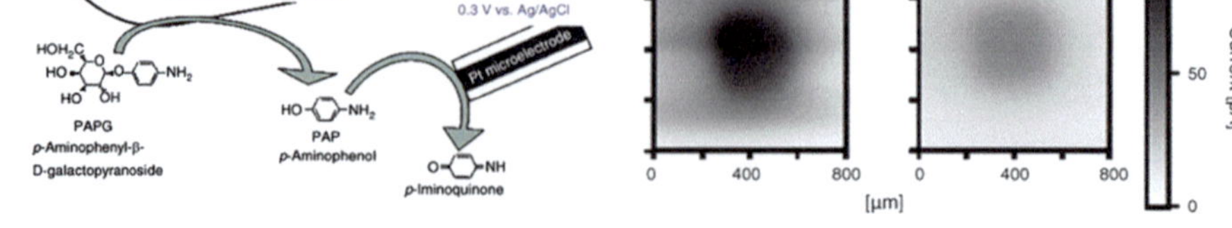

Fig. 16.6 (**a**) Schematic illustration of electrochemical mutagen screening. (**b**) SECM images of the *E. coli*-collagen gel micro pattern (*left*) with and (*right*) without 5.0×10^{-1} µg/mL mutagen (AF-2) (Reprinted with permission from Matsui et al. 2006, Copyright 2006 Elsevier)

16.3.6 Membrane Protein Imaging

SECM has been used to visualize receptor proteins on living cell membranes. The measurement of membrane proteins by SECM has several advantages. First, a single adherent cell can be measured without removing it from the culture dish; second, the optimization of the labeling antibody concentration is possible because the faradaic current is suitable for quantitative estimation; and third, a faradaic current image corresponding to the expression state of the measured membrane protein is available at the single-cell level. We characterized the expression level of epidermal growth factor receptor (EGFR) by SECM (Takahashi et al. 2009a). EGFR is a key membrane protein associated with cancer. It elicits various cell-type-specific responses, leading to cell proliferation, differentiation, apoptosis, and migration. To estimate EGFR expression levels by SECM, EGFR was labeled with AP via an antibody (Fig. 16.7a). The oxidation current of PAP produced by the AP-catalyzed reaction was monitored to estimate the density of EGFR on the cell surface. The decrease in the expression level of EGFR induced by the EGF-triggered endocytosis was estimated by comparing the faradaic current responses of the cells with and without EGF stimulation. Figure 16.7b shows the SECM images of a single EGFR-expressing CHO cell and a wild-type CHO cell. The genetically engineered CHO shows high electrochemical signals, indicating that EGFR is expressed at the cell surface. We optimized the concentration of the labeling antibody for EGFR at the cell surface and confirmed distinct differences in EGFR expression levels among different cell types. SECM measurements were found to be comparable with the results of flow cytometry experiments.

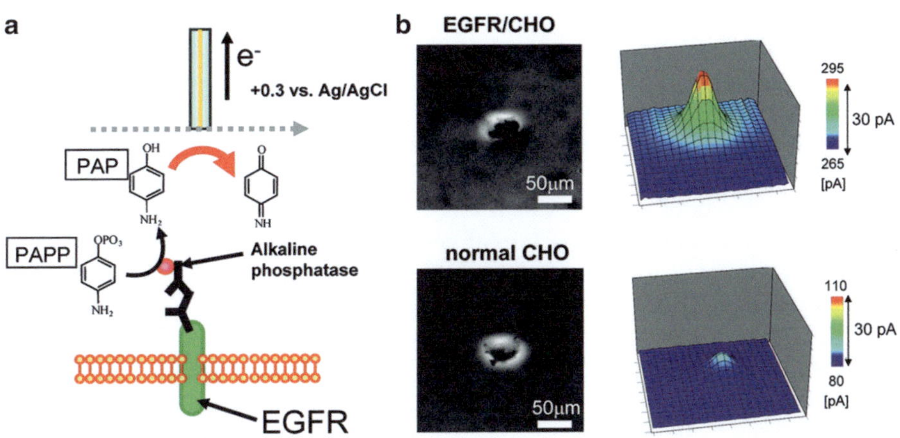

Fig. 16.7 (**a**) Schematic illustration of EGFR detection using SECM. (**b**) Single CHO cell SECM image of EGFR/CHO cell and normal CHO cell (Reprinted with permission from Takahashi et al. 2009b, Copyright 2009 ACS)

We have also reported an SECM-based receptor-mediated endocytosis detection method (Takahashi et al. 2011a). Receptor-mediated endocytosis is an important process that negatively regulates the receptor-mediated signals by reducing the surface concentration of the receptor itself (down-regulation), and controls the strength and duration of the signals downstream. Then, the receptor is recycled back into the plasma membrane. EGF binds to EGFR, and the activated EGFR initiates the signaling cascades, thereby promoting cell proliferation, differentiation, apoptosis, and migration. This signaling is controlled by EGF-triggered endocytosis, which reduces the number of EGFR molecules exposed to the outside medium. Since SECM detected the ALP activity on the outer membrane, the procedure helped discriminate EGFR on the outer membrane from the intracellular EGFR involved in endocytosis. SECM showed a marked decrease of 93 % in the current responses generated due to the AP activity with the addition of the EGF, clearly indicating that EGF triggered the endocytosis, which led to the withdrawal of most EGFRs from the outer membrane.

16.3.7 Neurotransmitter Detection

The high temporal and spatial resolution of the carbon-fiber microelectrode has facilitated the study of exocytosis by single neural cells. It enables the detection of the quantities of neurotransmitters from a single neural vesicle as a transient current spike in the electrode response (Chen et al. 1994; Troyer et al. 2002). SECM is a powerful tool for investigating the spatial distribution of neurotransmitter release and for quantitatively analyzing single vesicles. Hengstenberg and coworkers detected the neurotransmitters using shear-force distance control SECM (Hengstenberg et al. 2001). Liebetrau and coworkers imaged differentiated and undifferentiated PC12 cell topographies using several mediators that were selected for their biocompatibility from a larger pool of candidates (Liebetrau et al. 2003). They also performed topography imaging of differentiated and undifferentiated PC12 cells using constant-current mode or constant-impedance mode (Kurulugama et al. 2005). The exposure of undifferentiated PC12 cells to nerve growth factor induced the growth of neuritis, which was also imaged by constant-distance-mode SECM. Because of the thin structure of neurites compared to the cell height, they are difficult to image using constant-height mode. The constant-impedance mode was applied to image the differentiated PC12 cells. An advantage of the impedance-based constant-distance mode is that the images can be recorded in the growth media without an added mediator, facilitating long-term imaging during growth and development. By combining amperometry and constant-height impedance, SECM has even allowed the simultaneous mapping of topography and the detection of vesicular release events while moving the tip across a cell.

We developed SECM–SICM (Takahashi et al. 2011b) and VSM-SECM (Takahashi et al. 2012), both of which have visualized differentiated PC12 topography and detected neurotransmitter release events (Fig. 16.8). An advantage of SECM–SICM is that the probe filled with electrolyte can be used to apply different

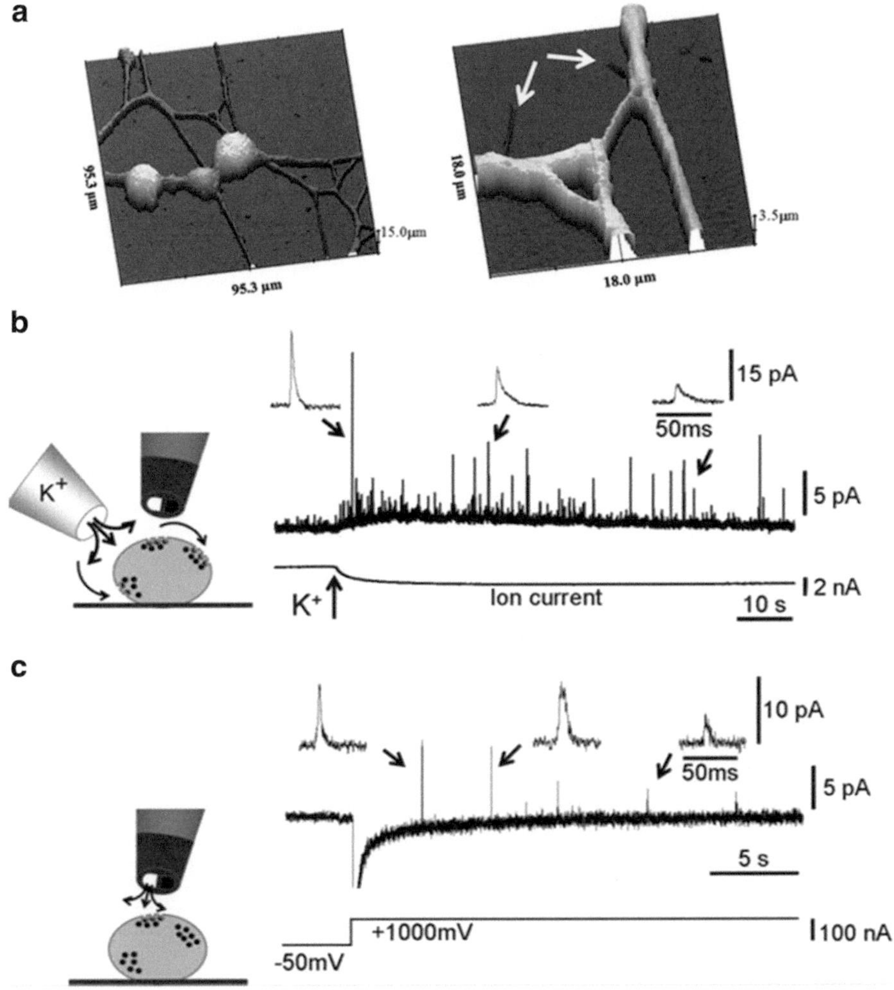

Fig. 16.8 (**a**) Nanoscale topography images of differentiated PC12 cells using SECM–SICM. The *arrows* showed the dendritic structures. A series of current spikes corresponding to neurotransmitter release detected after (**b**) whole cell stimulation with 105 mM K^+ using another micropipette and (**c**) voltage-driven delivery of K^+ ions using a SICM barrel (Reprinted with permission from Takahashi et al. 2011a, b, Copyright 2011 John Wiley and Sons)

reagents for the local stimulation of the cell. Therefore, voltage-driven application of K^+ ions was realized by SECM–SICM itself to achieve both the local depolarization of the cell membrane and simultaneous detection of the neurotransmitter. With local stimulation, we always detected a low frequency of current spikes compared to the entire cell stimulation. This finding suggests that SECM–SICM can induce and detect localized releases of neurotransmitters over the cell surface, thus opening up possibilities to perform mapping of neurotransmitter release sites.

16.4 Conclusions and Outlook

The fabrication of nanometer-scale electrodes and the development of electrode–sample distance control systems have greatly enhanced the capacity of SECM systems to solve problems in cell biology. We have introduced impressive studies using SECM to examine biological molecules, such as enzyme activity evaluation, electrochemical enzyme-linked immunosorbent assays, membrane permeability evaluations, respiratory activity measurements, reporter gene assays, membrane protein imaging, and neurotransmitter detection. In the future, SECM will provide us with information on the heterogeneous distribution of chemicals at the cell surface and intracellular chemical reaction profiles, which will improve our understanding of many biological phenomena.

References

Alpuche-Aviles MA, Wipf DO (2001) Impedance feedback control for scanning electrochemical microscopy. Anal Chem 73(20):4873–4881

Amphlett JL, Denuault G (1998) Scanning electrochemical microscopy (SECM): an investigation of the effects of tip geometry on amperometric tip response. J Phys Chem B 102(49): 9946–9951

Bath BD, Lee RD, White HS, Scott ER (1998) Imaging molecular transport in porous membranes. Observation and analysis of electroosmotic flow in individual pores using the scanning electrochemical microscope. Anal Chem 70(6):1047–1058

Chen TK, Luo GO, Ewing AG (1994) Amperometric monitoring of stimulated catecholamine release from rat pheochromocytoma (Pc12) cells at the zeptomole level. Anal Chem 66(19):3031–3035

Comstock DJ, Elam JW, Pellin MJ, Hersam MC (2010) Integrated ultramicroelectrode-nanopipet probe for concurrent scanning electrochemical microscopy and scanning ion conductance microscopy. Anal Chem 82(4):1270–1276

Edwards MA, Martin S, Whitworth AL, Macpherson JV, Unwin PR (2006) Scanning electrochemical microscopy: principles and applications to biophysical systems. Physiol Meas 27(12):R63–R108

Fan FRF, Bard AJ (1999) Imaging of biological macromolecules on mica in humid air by scanning electrochemical microscopy. Proc Natl Acad Sci U S A 96(25):14222–14227

Guell AG, Ebejer N, Snowden ME, McKelvey K, Macpherson JV, Unwin PR (2012) Quantitative nanoscale visualization of heterogeneous electron transfer rates in 2D carbon nanotube networks. Proc Natl Acad Sci U S A 109(29):11487–11492

Hengstenberg A, Kranz C, Schuhmann W (2000) Facilitated tip-positioning and applications of non-electrode tips in scanning electrochemical microscopy using a sheer force based constant-distance mode. Chem Eur J 6(9):1547–1554

Hengstenberg A, Blochl A, Dietzel ID, Schuhmann W (2001) Spatially resolved detection of neurotransmitter secretion from individual cells by means of scanning electrochemical microscopy. Angew Chem Int Ed 40(5):905–908

Kasai S, Yokota A, Zhou HF, Nishizawa M, Niwa K, Onouchi T, Matsue T (2000) Immunoassay of the MRSA-related toxic protein, leukocidin, with scanning electrochemical microscopy. Anal Chem 72(23):5761–5765

Kaya T, Nagamine K, Matsui N, Yasukawa T, Shiku H, Matsue T (2004) On-chip electrochemical measurement of beta-galactosidase expression using a microbial chip. Chem Commun (2):248–249

Korchev YE, Bashford CL, Milovanovic M, Vodyanoy I, Lab MJ (1997) Scanning ion conductance microscopy of living cells. Biophys J 73(2):653–658
Kranz C, Friedbacher G, Mizaikoff B (2001) Integrating an ultramicroelectrode in an AFM cantilever: combined technology for enhanced information. Anal Chem 73(11):2491
Kurulugama RT, Wipf DO, Takacs SA, Pongmayteegul S, Garris PA, Baur JE (2005) Scanning electrochemical microscopy of model neurons: constant distance imaging. Anal Chem 77(4):1111–1117
Lai SCS, Dudin PV, Macpherson JV, Unwin PR (2011) Visualizing zeptomole (electro)catalysis at single nanoparticles within an ensemble. J Am Chem Soc 133(28):10744–10747
Liebetrau JM, Miller HM, Baur JE (2003) Scanning electrochemical microscopy of model neurons: imaging and real-time detection of morphological changes. Anal Chem 75(3):563–571
Macpherson JV, Unwin PR (2000) Combined scanning electrochemical-atomic force microscopy. Anal Chem 72(2):276–285
Macpherson JV, OHare D, Unwin PR, Winlove CP (1997) Quantitative spatially resolved measurements of mass transfer through laryngeal cartilage. Biophys J 73(5):2771–2781
Matsue T (2012) Development of biosensing devices and systems using micro/nanoelectrodes. Bull Chem Soc Jpn 85(5):545–557
Matsue T (2013) Bioimaging with micro/nanoelectrode systems. Anal Sci 29(2):171–179
Matsue T, Shiku H, Yamada H, Uchida I (1994) Permselectivity of voltage-gated alamethicin ion-channel studied by microamperometry. J Phys Chem 98(43):11001–11003
Matsui N, Kaya T, Nagamine K, Yasukawa T, Shiku H, Matsue T (2006) Electrochemical mutagen screening using microbial chip. Biosens Bioelectron 21(7):1202–1209
Mezour MA, Morin M, Mauzeroll J (2011) Fabrication and characterization of laser pulled platinum microelectrodes with controlled geometry. Anal Chem 83(6):2378–2382
Miao WJ, Choi JP, Bard AJ (2002) Electrogenerated chemiluminescence 69: the tris(2,2′-bipyridine)ruthenium(II), ($Ru(bpy)_3^{2+}$)/tri-*n*-propylamine (TPrA) system revisited - A new route involving $TPrA^{•+}$ cation radicals. J Am Chem Soc 124(48):14478–14485
Mirkin MV, Nogala W, Velmurugan J, Wang YX (2011) Scanning electrochemical microscopy in the 21st century. Update 1: five years after. Phys Chem Chem Phys 13(48):21196–21212
Nishizawa M, Takoh K, Matsue T (2002) Micropatterning of HeLa cells on glass substrates and evaluation of respiratory activity using microelectrodes. Langmuir 18(9):3645–3649
Novak P, Li C, Shevchuk AI, Stepanyan R, Caldwell M, Hughes S, Smart TG, Gorelik J, Ostanin VP, Lab MJ, Moss GWJ, Frolenkov GI, Klenerman D, Korchev YE (2009) Nanoscale live-cell imaging using hopping probe ion conductance microscopy. Nat Methods 6(4):279–281
Pierce DT, Unwin PR, Bard AJ (1992) Scanning electrochemical microscopy.17. Studies of enzyme mediator kinetics for membrane-immobilized and surface-immobilized glucose-oxidase. Anal Chem 64(17):1795–1804
Shao YH, Mirkin MV (1998) Probing ion transfer at the liquid/liquid interface by scanning electrochemical microscopy (SECM). J Phys Chem B 102(49):9915–9921
Shiku H, Takeda T, Yamada H, Matsue T, Uchida I (1995) Microfabrication and characterization of diaphorase-patterned surfaces by scanning electrochemical microscopy. Anal Chem 67(2): 312–317
Shiku H, Matsue T, Uchida I (1996) Detection of microspotted carcinoembryonic antigen on a glass substrate by scanning electrochemical microscopy. Anal Chem 68(7):1276–1278
Shiku H, Hara Y, Matsue T, Uchida I, Yamauchi T (1997) Dual immunoassay of human chorionic gonadotropin and human placental lactogen at a miocrofabricated substrate by scanning electrochemical microscopy. J Electroanal Chem 438(1–2):187–190
Shiku H, Shiraishi T, Ohya H, Matsue T, Abe H, Hoshi H, Kobayashi M (2001) Oxygen consumption of single bovine embryos probed by scanning electrochemical microscopy. Anal Chem 73(15):3751–3758
Sun P, Laforge FO, Mirkin MV (2007) Scanning electrochemical microscopy in the 21st century. Phys Chem Chem Phys 9(7):802–823
Takahashi Y, Hirano Y, Yasukawa T, Shiku H, Yamada H, Matsue T (2006) Topographic, electrochemical, and optical images captured using standing approach mode scanning electrochemical/optical microscopy. Langmuir 22(25):10299–10306

Takahashi Y, Miyamoto T, Shiku H, Asano R, Yasukawa T, Kumagai I, Matsue T (2009a) Electrochemical detection of epidermal growth factor receptors on a single living cell surface by scanning electrochemical microscopy. Anal Chem 81(7):2785–2790

Takahashi Y, Shiku H, Murata T, Yasukawa T, Matsue T (2009b) Transfected single-cell imaging by scanning electrochemical optical microscopy with shear force feedback regulation. Anal Chem 81(23):9674–9681

Takahashi Y, Murakami Y, Nagamine K, Shiku H, Aoyagi S, Yasukawa T, Kanzaki M, Matsue T (2010a) Topographic imaging of convoluted surface of live cells by scanning ion conductance microscopy in a standing approach mode. Phys Chem Chem Phys 12(34):10012–10017

Takahashi Y, Shevchuk AI, Novak P, Murakami Y, Shiku H, Korchev YE, Matsue T (2010b) Simultaneous noncontact topography and electrochemical imaging by SECM/SICM featuring ion current feedback regulation. J Am Chem Soc 132(29):10118–10126

Takahashi Y, Miyamoto T, Shiku H, Ino K, Yasukawa T, Asano R, Kumagai I, Matsue T (2011a) Electrochemical detection of receptor-mediated endocytosis by scanning electrochemical microscopy. Phys Chem Chem Phys 13(37):16569–16573

Takahashi Y, Shevchuk AI, Novak P, Zhang YJ, Ebejer N, Macpherson JV, Unwin PR, Pollard AJ, Roy D, Clifford CA, Shiku H, Matsue T, Klenerman D, Korchev YE (2011b) Multifunctional nanoprobes for nanoscale chemical imaging and localized chemical delivery at surfaces and interfaces. Angew Chem Int Ed 50(41):9638–9642

Takahashi Y, Shevchuk AI, Novak P, Babakinejad B, Macpherson J, Unwin PR, Shiku H, Gorelik J, Klenerman D, Korchev YE, Matsue T (2012) Topographical and electrochemical nanoscale imaging of living cells using voltage-switching mode scanning electrochemical microscopy. Proc Natl Acad Sci U S A 109(29):11540–11545

Torisawa YS, Kaya T, Takii Y, Oyamatsu D, Nishizawa M, Matsue T (2003) Scanning electrochemical microscopy-based drug sensitivity test for a cell culture integrated in silicon microstructures. Anal Chem 75(9):2154–2158

Torisawa YS, Ohara N, Nagamine K, Kasai S, Yasukawa T, Shiku H, Matsue T (2006) Electrochemical monitoring of cellular signal transduction with a secreted alkaline phosphatase reporter system. Anal Chem 78(22):7625–7631

Troyer KP, Heien MLAV, Venton BJ, Wightman RM (2002) Neurochemistry and electroanalytical probes. Curr Opin Chem Biol 6(5):696–703

Wijayawardhana CA, Wittstock G, Halsall HB, Heineman WR (2000) Spatially addressed deposition and imaging of biochemically active bead microstructures by scanning electrochemical microscopy. Anal Chem 72(2):333–338

Williams CG, Edwards MA, Colley AL, Macpherson JV, Unwin PR (2009) Scanning micropipet contact method for high-resolution imaging of electrode surface redox activity. Anal Chem 81(7):2486–2495

Wittstock G, Schuhmann W (1997) Formation and imaging of microscopic enzymatically active spots on an alkanethiolate-covered gold electrode by scanning electrochemical microscopy. Anal Chem 69(24):5059–5066

Yamada H, Matsue T, Uchida I (1991) A microvoltammetric study of permeation of ferrocene derivatives through a planer bilayer lipid-membrane. Biochem Biophys Res Commun 180(3):1330–1334

Yasukawa T, Uchida I, Matsue T (1998) Permeation of redox species through a cell membrane of a single, living algal protoplast studied by microamperometry. Biochim Biophys Acta 1369(1):152–158

Yasukawa T, Kaya T, Matsue T (1999) Dual imaging of topography and photosynthetic activity of a single protoplast by scanning electrochemical microscopy. Anal Chem 71(20):4637–4641

Zhao XC, Diakowski PM, Ding ZF (2010) Deconvoluting topography and spatial physiological activity of live macrophage cells by scanning electrochemical microscopy in constant-distance mode. Anal Chem 82(20):8371–8373

Chapter 17
Field-Effect Transistors: Current Advances and Challenges in Bringing Them to Point-of-Care

Shrey Pathak and Pedro Estrela

Abstract Portable, facile and accurate detection of biomarkers is essential for the development of clinically relevant and commercially viable point-of-care diagnostic devices. Such diagnostic solutions have and are constantly transforming the healthcare, environmental monitoring and food safety industries. The ever-growing demand for devices with higher sensitivity along with cost-effective packaging has put tremendous pressure on the field of biosensors. This demand for ever more parallel detection with lower manufacturing costs has to be satisfied by employing semiconductor technologies. Field-effect transistors have played an instrumental role in the development of various biosensing techniques, both as sensors and as enablers for other electronic and electrochemical techniques. This chapter reviews the application of field-effect transistors as transducer elements for biosensing applications (BioFETs), recent advances in BioFETs using novel semiconductor technologies and nanomaterials, the role of standard FETs in addressing large arrays of electrochemical sensors, and the challenges in circuitry and integration faced by electronic biosensing arrays.

Keywords Device integration • Field-effect transistors • Impedimetric biosensors • Potentiometric biosensors

17.1 Introduction

There is a growing need for point-of-care (POC) technologies to be more miniaturized, portable, accurate and disposable. Such devices are critical in settings where laboratory facilities are minimal and time is a crucial factor—e.g. in emergency wards, for military applications and for use in developing countries. Moreover the

---

S. Pathak • P. Estrela (✉)
Department of Electronic and Electrical Engineering, University of Bath, Bath BA2 7AY, UK
e-mail: p.estrela@bath.ac.uk

M.C. Vestergaard et al. (eds.), *Nanobiosensors and Nanobioanalyses*,
DOI 10.1007/978-4-431-55190-4_17

origin of such POC devices has to be attributed to the driving forces in the areas of molecular analysis, bio-defence, molecular biology and electronics. But it is the area of micro- and nano-electronics which has provided the tools that have truly helped this area to go from bulky instrumentation to miniaturized hand-held devices. POC devices using label-free techniques have the added advantage of reducing costs and avoiding the need for sample pre-treatment.

Over the past few decades, efforts have been devoted to exploiting semiconductor field-effect transistors (FETs) in chemical and biological sensors due to their highly characterized behaviour and the ease of implementation in portable instrumentation. Additionally, electrical detection of bimolecular interactions is highly desirable due to its suitability for low-cost portable sensors that can be used in the field by non-specialized personnel. Most of the efforts have been devoted to the development of ion-sensitive field-effect transistors (ISFETs) thanks to the pioneer work carried out by Bergveld in the 1970s (Bergveld 1970). Research in this area has focussed mostly on the development of ISFETs for the detection of specific ions and analytes using appropriate ion-selective or enzymatic membranes. Selectivity of ISFETs can be induced by the appropriate incorporation of certain pH-sensitive insulators or ion-selective membranes. One of the main advantages of such FET microsensors is their ability to operate in equilibrium conditions.

Another attribute that makes electrical measurements so attractive is the ability to parallel-process information, thus enabling faster analysis. The current gold standards are fluorescence DNA and protein microarrays. In terms of scalability, most optical technologies suffer from the diffraction limit of light creating an upper limit for the probe and thus compromising the quality of information received. In that regard, electrical methods based on FETs offer an attractive alternative, which in contrast to most methods (e.g. optical, piezoelectric, etc.) is downsizable—in theory down to the molecular scale.

Besides these potentiometric FET-based methods, a series of other electrochemical techniques can be applied to the detection of biomolecular interactions. Depending on the desired dynamic detection range and the specific properties of the system under study, techniques such as electrochemical impedance spectroscopy, voltage step capacitance measurements, amperometry, differential pulse voltammetry, squarewave voltammetry, AC voltammetry and chrono-potentiometric stripping analysis can be used for label-free detection of DNA, proteins and other biomolecules. Often these techniques require the use of redox mediators. Electrochemical impedance spectroscopy (EIS), in particular, is a very promising technique for DNA biosensing.

Of special interest for FET-based chemical and biological sensors is the use of thin-film transistors (TFTs). For example, polycrystalline silicon (poly-Si) TFTs, which can provide the drive logic as well as the switching matrix, are a very interesting technology for the development of low-cost, disposable biosensors, with a large number of parallel channels. By employing poly-Si TFTs, a microarray of over 10^5 channels, with integrated logic drivers, would require only a few tens of electrical connections to the rest of the system. These could be provided by edge connectors, thereby enabling easy insertion and removal of the sensor array from the external reading system and, therefore, single use of a complex microarray. Furthermore, poly-Si TFTs with a special extended gate structure have been used as potentiometric sensors for DNA hybridization. The construction of TFT-addressed biosensor microarrays with integral scan and readout circuits constitutes, in our opinion, one of the great future challenges for TFT-integrated electronics.

17.2 Field-Effect Transistors

One of the major advantages of employing FETs in sensor applications is their mature manufacturing technology. Due to the development of the microelectronics industry, microfabrication processes have been well established, allowing FETs to be mass-produced with extremely high yield. Thin films of different materials (metals, dielectrics) can be deposited on large areas of substrates and device patterns can be created by lithography. The area it occupies on the silicon wafer during the fabrication run mainly determines the cost of each die, and thus complex sensor arrays can be fabricated at affordable cost. This is especially attractive for biosensor applications, as disposability is a highly emphasized feature to avoid contamination.

17.2.1 Field-Effect Device Technologies

In order to understand the operating principle of a field-effect transistor, we must first examine one of the most basic and commonly employed FETs, the metal–oxide–semiconductor FET (MOSFET). It consists of a p-type single crystal silicon semiconductor substrate with two heavily doped n-type regions, namely the source and drain, a dielectric layer, usually SiO_2, and a top metal layer, namely the gate deposited on the dielectric in contact with both the source and the drain (in essence depicting two back-to-back diodes), as shown in Fig. 17.1a. Another popular variation of the

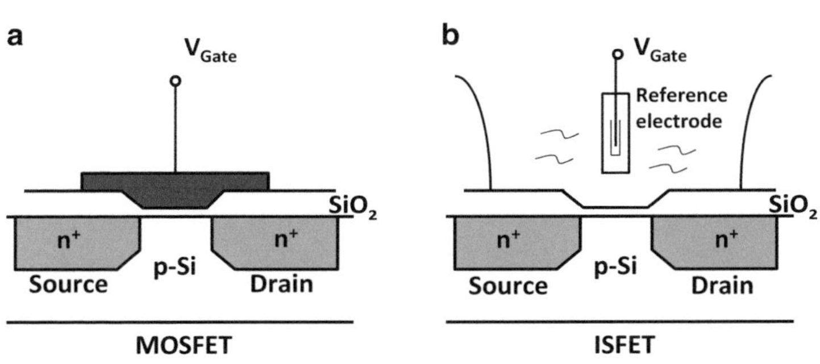

Fig. 17.1 Structure of (**a**) a metal–oxide–semiconductor field-effect transistor (MOSFET) and (**b**) an ion-sensitive field-effect transistor (ISFET)

MOSFET that has attracted lot of attention in the area of point-of-care diagnostics and potentiometric biosensing is the ion-sensitive FET, shown in Fig. 17.1b. In an ISFET, the gate metal is replaced by an electrolyte with an external reference electrode and sensing takes place via changes at the dielectric–electrolyte interface.

The flow of charge between the source and the drain of a FET is modulated by the magnitude of the voltage applied at the gate electrode. When the metal gate voltage with respect to the source, V_{GS}, is lower than the threshold voltage V_T, the p–n junction between the drain and the substrate is reverse-biased, resulting in no current flow between the source and drain. When V_{GS} is greater than V_T, the induced electric field across the dielectric is large enough to convert a small area of the lightly doped p-type silicon substrate into n-type (inversion—minority carriers of the p-type silicon). This results in the formation of a continuous channel at the gate allowing electrons (majority carriers of the n-doped region) to flow between source and drain. Due to the presence of an insulating dielectric layer, no current is allowed to flow from the gate into the semiconductor—although, depending on the nature and the thickness of the oxide, leakage currents of the order of 10^{-15} A should be expected. Typical characteristics of the drain current (I_D) with respect to the voltage between the drain and source (V_{DS}) are shown in Fig. 17.2 for different values of V_{GS}.

By its working principle, the MOSFET amplifies the input signal V_{GS} with an intrinsic gain given by the transconductance g_m. In the linear region where V_{GS} is small and in the saturation region where V_{GS} is sufficiently large, g_m is given by the following equations, respectively:

$$g_m = \left.\frac{\partial I_D}{\partial V_G}\right|_{V_D = \text{constant}} = \frac{W}{L}\mu C V_D = \frac{W}{L}\mu C\left(V_G - V_T\right)$$

where μ is the carrier mobility of the substrate material, C the gate capacitance per unit area and W and L the width and length of the conducting channel, respectively. Hence the amplification power of a MOSFET device is closely related to the mobility of the doped semiconductor material and can be tuned by the design of the transistor.

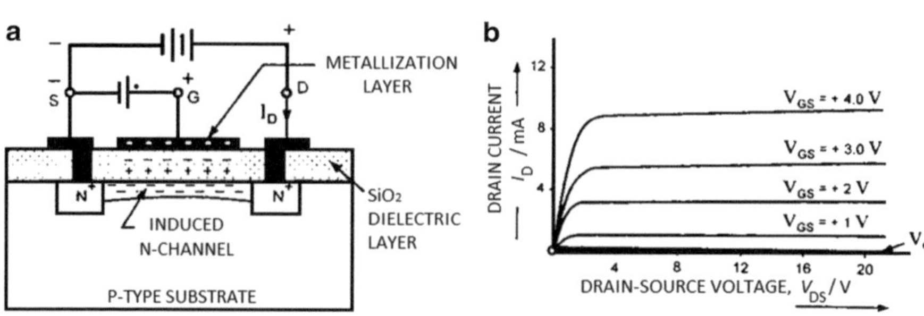

Operation of n-channel MOSFET *Output characteristics of a MOSFET*

Fig. 17.2 An n-channel MOSFEt along with the generated output characteristics

The sensitivity of the drain current to the charge on the gate electrode can hence be explored for sensor applications—i.e. the charge of a biolayer at the gate metal induces changes in the MOSFET. In the case of an ISFET, ions can adsorb on the surface generating an electric field, similar to applying a voltage at the metal gate (Bergveld 1970; Estrela et al. 2007). When an external gate voltage is applied through a reference electrode in the solution, the electrical field introduced by the adsorbed ions leads to a shift in the intrinsic characteristics of the device.

17.2.2 Single Crystalline Silicon and CMOS

Traditional FET transistors are fabricated on a single crystalline silicon wafer of a few hundred micrometres thickness. The silicon crystalline framework is homogeneous and continuous with very low levels of defects. Complementary metal–oxide–semiconductor (CMOS) is the standard single-crystal silicon-based semiconductor fabrication technology, which is distinguished from other types of fabrication technologies by providing both n-type and p-type MOSFETs on the same substrate. CMOS has been used predominantly in microprocessors, memories and other digital logic circuits due to its low power consumption and unmatched production yield. CMOS technology is also used for a wide variety of analog circuits such as image sensors, data converters and transceivers. Driven by the microelectronics industry, the CMOS fabrication process has been continuously refined to make smaller MOSFETs, which are both faster and more cost-efficient. Although the silicon MOSFET transistor does not have the best noise and speed performances compared with other semiconductor devices in the field of electronics, the well-established CMOS technologies certainly makes it an obvious choice for biosensor applications.

Despite the high performance of CMOS, its manufacturing process requires very high-cost equipment, clean-room facilities and expensive high-purity single-crystal silicon wafers. Those limitations have created a barrier to further reduction of the fabrication costs and hindered the use of CMOS technology in large areas of electronics such as displays.

17.2.3 Recent Advances

Besides using the CMOS process, which employs single crystalline silicon as a substrate, FETs can also be fabricated on thin films of semiconductors such as amorphous (α-Si) or polycrystalline silicon. A direct benefit of these technologies is the replacement of expensive single crystalline silicon wafers with cheaper insulators supporting a thin layer of deposited semiconductor as substrate, which substantially reduces the manufacturing costs. A thin-film transistor is a metal–insulator–semiconductor field-effect transistor (MISFET) fabricated on an insulating substrate by employing entirely thin-film constituents. The total thickness of the transistor is normally less than 1 μm (Li et al. 2012a). The following section highlights recent advances demonstrating how the different configurations are being employed in the area of biosensing.

17.2.3.1 Thin-Film Transistors

In the case of amorphous silicon TFT, the conducting channel is created in the amorphous silicon layer, in which the long-range order of lattice is absent and the atoms form a continuous random network. Due to this disordered nature of the material, α-Si has a high level of defects, which are normally passivated and reduced by hydrogen to prevent anomalous electrical behaviour. Consequently the electron mobility is reduced to 1–10 $cm^2\ V^{-1}\ s^{-1}$, compared with a few hundred for single crystalline silicon. This essentially ruled out amorphous silicon TFT for analog circuits and high-speed logic circuits, where high internal gain and large fan-out of transistors are required.

While most amorphous silicon TFTs suffer from low electronic performance, they are very flexible in application and manufacturing. One important advantage is that amorphous Si can be deposited at temperatures as low as 75 °C. This makes it possible for the device to be made not only on glass, but also on plastics. In addition, amorphous silicon can be deposited over very large areas by plasma-enhanced chemical vapour deposition (PECVD) with standard industrial equipment. Both these features make mass-scale production of amorphous silicon TFT-based devices relatively easy and economic. The main application for amorphous silicon TFTs is on liquid-crystal displays (LCDs), in which a TFT transistor individually drives each pixel. Despite its much reduced manufacturing cost and versatile form factor, the main drawback of TFTs compared with single crystalline silicon devices is their low electrical performance. This is a direct result of the low electron mobility of the semiconductor material employed for TFT fabrication.

Liquid-crystal displays normally employ a matrix of amorphous silicon TFTs to control the voltage applied to the individual pixels. Polycrystalline silicon TFTs have a much higher mobility (>100 $cm^2\ V^{-1}\ s^{-1}$) than α-Si TFTs and can therefore be used to provide the drive logic as well as the pixel transistors. It is this property that makes poly-Si TFTs a very interesting technology for the development of low-cost disposable biosensors, with a large number of parallel channels. A microarray of 100,000 channels with integrated logic drivers would require only a few tens of electrical connections to the rest of the system.

In recent years, organic or polymer semiconductor materials have been intensively researched to make TFTs. These organic TFTs can be made by processes which do not require clean-room facilities, making them suitable candidates for disposable biosensor applications. However, as the development of those devices is still in its infancy and the manufacturing processes have not been well established, the use of organic semiconductor TFTs for biosensing applications is still very rare and has not made a real impact as yet.

17.2.3.2 Carbon Nanotube Field-Effect Transistors

Since the discovery of carbon nanotubes (CNTs) in 1991 by Iijima (1991), a great deal of effort has been devoted to the fundamental understanding of their electrical, mechanical and chemical properties and of their use in a wide range of applications such as electronics and sensors. CNTs are divided into single-walled carbon nanotubes (SWNTs) and multi-walled carbon nanotubes (MWNTs). A SWNT can be formed by rolling a graphene sheet (hexagonal structure) into a cylinder and a MWNT is composed of concentric graphene cylinders with an interlayer spacing of 0.34 nm. Most CNTs are synthesized by arc discharge, laser ablation or chemical vapour deposition methods (Hu et al. 2010).

Semiconducting SWNTs play a central role in the operation of SWNT-based field-effect transistors (SWNT-FETs), first developed in the late 1990s (Tans et al. 1998; Martel et al. 1998). SWNT-FET devices are composed of individual SWNTs or random networks of SWNTs placed between a source (S) and a drain (D) electrode on a SiO_2/Si substrate. The Si layer can act as back gate, which is separated by the SiO_2 insulating layer. Since the work function of SWNTs is higher than that of most metals, the contact barrier between SWNTs and metals is usually a Schottky barrier (SB). The height of the SB in the SWNT-FET contact is determined by the work function of the electrode metal (Taillades et al. 1999). The conductance of SWNTs in devices can be modulated by applying a potential to the gate electrodes with a constant bias voltage V_{DS} (Hu et al. 2010).

Sensing gas or organic vapour molecules is central to environmental monitoring, control of chemical processes, health protection and agricultural applications—another application of POC devices. SWNT-FET sensors have been shown to be sensitive to gases such as NH_3, NO_2, H_2, CH_4, CO and H_2S, and some volatile organic compounds such as ethanol and methanol. Liu et al. (2006) first demonstrated a CNT-based gas sensor using a FET consisting of individual semiconducting SWNTs. The FET was made by growing SWNTs via CVD on SiO_2/Si substrates, then photo-lithographically patterning metal electrodes on a single SWNT. The conductance of the SWNT-FET varied dramatically under various gate voltages when the device was exposed to NO_2 or NH_3. Upon exposure to NO_2, the conductance of the FET increased by about three orders of magnitude, and the transfer characteristics showed a shift of +4 V in gate voltage. The response time, defined as the time duration for signal stability from the introduction of the sample, ranged from 2 to 10 s with 200 ppm of NO_2. As for sensing NH_3, the response time to 1 % NH_3 was about 1 or 2 min, and sensitivity from 10 to 100 ppm was achieved. More recently, Mattmann et al. (2009) demonstrated CNT-FETs with Al_2O_3 passivated contacts for NO_2 detection with very low limit concentration (80 ppb). Other work reported in this area has been on organic chemical agents such as toxins very harmful for human health and the environment. These chemicals usually exist in liquid form, but their vapours have the potential to be absorbed through the skin. Novak et al. (2003) demonstrated that a FET sensor based on a SWNT network can detect dimethyl methylphosphonate (DMMP), a stimulant for the nerve agent sarin. Roberts et al. (2009) fabricated thin-film transistor (TFT) sensors consisting of aligned, sorted nanotube networks. This device allowed them to achieve stable low-voltage operation under aqueous conditions. The authors developed a method of enriching semiconducting SWNTs and aligning them in a one-step solution deposition process by controlling substrate surface chemistry on silicon. These SWNT-TFTs were used to detect trace concentrations, down to 2 ppb, of dimethyl methylphosphonate (DMMP) and trinitrotoluene (TNT) in aqueous solutions. Along with reliable cycling underwater, the TFT sensors fabricated with aligned, sorted nanotube networks enriched with semiconductor SWNTs showed a higher sensitivity to analytes than those fabricated with random, unsorted networks with predominantly metallic charge transport. Peng et al. (2009b) systematically studied sensing mechanisms for CNT-based NH_3 sensors. The electrical conductance of a semiconducting SWNT was found to be sensitive to its environment and was shown to vary significantly with surface adsorption of various chemicals and biomolecules. This makes SWNT-FETs very promising candidates for label-free biosensing. The SWNT-FETs for biosensor applications are composed of SWNT networks or individual semiconducting SWNTs. SWNT-FET-based biosensors have been reported to detect various biological species such as DNA, proteins and cells (Hu et al. 2010). An exciting development was the detection of cancer in breath through the sensing of volatile organic compounds using random networks of CNTs (Peng et al. 2009a).

17.2.3.3 Graphene-Based Field-Effect Transistors

The discovery of graphene by Novoselov et al. (2004) led to a significant amount of work on trying to achieve the ultimate biosensing limit of detection: the single molecule (Hu et al. 2010). The graphene-based FET, has a non-functionalized single-layer graphene as the channel (between the source and the drain), exhibiting a linear increase in conductance dependent on the electrolyte pH—a potential use as a POC sensor element. Chen et al. (2005) reported a graphene gas sensor by using partially reduced graphene oxide. The graphene oxide transistor demonstrated little response to chemical gases such as NO_2, with the device showing a typical p-type transistor behaviour. However, the conductance of the sensor increased when annealed, thus exhibiting higher sensitivity to exposures to NO_2 from ~1.55 to 100 ppm. Ang et al. (2008) demonstrated solution-gated epitaxial graphene as a pH sensor. The sensor was fabricated from a few layers of graphene epitaxially grown on a SiC substrate. The electrochemical double layer at the graphene–electrolyte interface was observed to be very sensitive to pH and the conductance of the device responded accordingly.

Varghese et al. (2009) studied the noncovalent interaction of DNA nucleobases and nucleosides with graphene by isothermal titration calorimetry. The results demonstrated that in an aqueous solution the interaction energies of the nucleotide bases vary in the order guanine (G) > adenine (A) > cytosine (C) > thymine (T), while the interaction energies of A–T and G–C pairs are somewhere between those of the constituent bases. The conductance of graphene-based FETs has been reported to increase with adsorbed proteins (e.g. picomolar range of bovine serum albumin), which implies that graphene-based FETs could be used as highly sensitive electrical biosensors.

Mohanty and Berry (2008) developed a graphene-based single bacterium resolution biodevice and DNA FET, while Dong et al. (2010) demonstrated that CVD-grown large-sized graphene films consisting of monolayered and few-layered graphene domains could be used to fabricate liquid-gated transistors for DNA sensor with a detection concentration as low as 0.01 nM. The addition of gold nanoparticles on the surface of graphene devices could extend the upper limit of DNA detection due to the increase in loading of probe DNA molecules. Cohen-Karni et al. (2010) demonstrated recordings from electrogenic cells using single-layer graphene FETs as well as simultaneous recording using graphene and SiNW-FETs. Graphene-FET conductance signals were recorded from spontaneously beating embryonic chicken cardiomyocytes which yielded well-defined extracellular signals.

17.3 Microfluidics and Packaging

Microfluidics is an attractive technology for POC devices, in particular where low-volume samples need to be processed or when parallel sensing of different analytes is required (Parsa et al. 2008). The coupling of microfluidics with FET-type biosensor arrays can lead to better control of biorecognition layer immobilization processes, faster analysis times and lower cross-talk issues.

17.3.1 Traditional Packaging Techniques

Silicon was one of the first materials used to create microfluidic channels. Fabrication of silicon-based microfluidic devices was achieved by employing various etching methods (dry/wet) or some additive methods—e.g. chemical vapour deposition of metals (Iliescu et al. 2012). Although silicon surface chemistry is a highly developed area of research, one of the main reasons why such microfluidic devices were not employed was the fact that silicon has a relatively high elastic modulus (130–180 GPa) and thus it is hard to fabricate active fluidic components such as valves and pumps. For these reasons glass and polymers are preferred choices (Washburn et al. 2009; Anderson et al. 2011).

Initially glass emerged as the main alternative substrate. Microstructures could be created by etching into the glass through traditional means and channels could be created relatively easily (Iliescu et al. 2012). However, it quickly became apparent that glass in itself could only be employed as part of a hybrid device, thus requiring more sophisticated techniques such as plasma etching. Wu et al. (2010) demonstrated a hybrid device containing glass and polydimethylsiloxane (PDMS) microfluidics attached to electronic integrated circuits. Although glass has low background fluorescence and is biocompatible, the drawbacks include a relatively low nonspecific adsorption coefficient and impermeability of gases. Mellors et al. (2008, 2010) demonstrated all-glass devices that combined microcapillary electrophoresis (μCE) and electrospray ionization–mass spectrometry (ESI-MS) by directing the separation channel to a corner of the device to create a nanospray.

Another area which attracted significant interest in the early years of microfluidics was that of ceramics. Low-temperature co-fired ceramics (LTCC) were laminar in nature, thus helping in the construction of a three-dimensional device where a stacked or a layered structure could be build. LTCCs are aluminium oxide-based materials that come in laminate sheets and which can be patterned, assembled and then fired at elevated temperatures. Fakunle and Fritsch (2010) demonstrated that LTCC devices provide low levels of non-specifically adsorbed molecules in immunoassays such as the enzyme-linked immunosorbent assay (ELISA). Zhang and Eitel (2012) also investigated LTCC bio-stability by measuring material leaching in aqueous solutions; high rates of leaching occurred with acidic and basic solutions, while solutions near physiological pH resulted in low leaching rates. Electrodes can be deposited onto LTCC using expansion-matched metal pastes. In this manner, Almeida et al. (2011) used an LTCC device with potentiometric detection to analyze sulfamethoxazole and trimethoprim in fish farm water.

17.3.2 Non-conventional Polymer Techniques

Polymers such as organic-based long-chain materials are some of the most commonly employed materials in the area of microfluidics. They have gained significant interest in microfluidics in the past couple of decades. One of the main advantages of employing polymers for microfluidic applications is the fact that they are easy to fabricate, relatively inexpensive, amenable to mass production processes, and can be adapted to a varied degree for different applications, through formulation changes or chemical modifications.

PDMS is one of the most used polymers for microfluidic applications. Device molds can be created by employing conventional micro-machining or photolithography methods, and PDMS microstructures can be casted and cured on these molds. One of the main advantages of using PDMS is its relatively low elastic modulus (300–500 kPa) which enables the creation and integration of valves and pumps easily as well as complex microfluidic designs (Effenhauser et al. 1997). Another advantage that PDMS lends to the area of microfluidics is its ability, in the cured form, to allow gas to permeate through. This is highly advantageous as it proves to be an ideal material for cellular studies as it allows oxygen and carbon dioxide to permeate through. Thus there is no doubt that PDMS offers several advantages over silicon, glass and ceramics, but like all materials it also has its own disadvantages. Being a low-molecular-weight polymer, it has been found to leach into the solution, thus negatively impacting cellular studies (Araci and Quake 2012). PDMS is also a hydrophobic polymer, thus increasing the chances of nonspecific adsorption and permeation of hydrophobic molecules (Berthier et al. 2012). Although chemical modification by e.g. plasma exposure has the ability to negate the chances of nonspecific adsorption, it has been shown that the modification only lasts for a short time span (Roman et al. 2005; Bodas and Khan-Malek 2006).

17.3.3 Recent Advances

More recently, paper-based functionalized substrates have received much attention from the research community. They are proving to be valuable tools in the construction of microfluidic devices in rapid diagnostic test kits. This is because paper-based microfluidic devices offer four basic capabilities on a single analytical device: (1) distribution of a sample into segregated regions enabling the construction of high-throughput assays; (2) the ability to move samples by using capillary forces; (3) compatibility with small volumes of samples, an important characteristic for kits which use low volumes for diagnosis; and (4) easy elimination of hazardous waste as the devices can be easily disposed off (Martinez et al. 2008). Although they offer significant advantages over current technologies, paper-based microfluidics is still in a very early development stage (Li et al. 2012a, b; Ballerini et al. 2012; Liana et al. 2012).

17.4 Signal Extraction, Conditioning and Quantification

When an electrolyte is in contact with an electrode, an electrochemical double layer forms. In the Gouy–Chapman–Stern model of the electrochemical double layer (Bard and Faulkner 2001), it is assumed that the solvent provides a continuous dielectric medium with dielectric permittivity equal to its bulk value, that charges of discrete ions are smeared out into a continuous distribution of net charge density, and that ion–ion interactions can be neglected so that all ions in solution are free to contribute to the charge density. Due to their finite size, ions may not approach the electrode closer than the outer Helmholtz plane (OHP). Since there is no charge between the electrode and OHP, the electric field E is constant in this region and the electrostatic potential φ varies linearly. Outside the OHP, the potential may be determined by considering the solution to be divided into laminae parallel to the electrode. The laminae are in thermal equilibrium, but at differing energies due to the potential φ, so the concentration n_i of species i with valence z_i is related to its bulk concentration n_i^0 by the Boltzmann factor $n_i = n_i^0 \exp(-z_i e\varphi/kT)$.

The net charge density $\rho(x)$ is related to the potential by the Poisson equation

$$\rho(x) = \varepsilon\varepsilon_0 \frac{d^2\varphi}{dx^2}$$

where ε is the relative dielectric permittivity, ε_0 is the permittivity of free space and x is the distance from the electrode. Use of boundary conditions leads to the non-linear Poisson–Boltzmann equation. For $\varphi \ll kT/e$, the linearized Poisson–Boltzmann equation results. Alternatively, the non-linear Poisson–Boltzmann equation may be solved for a symmetrical electrolyte that contains only one cationic and one anionic species, both with charge magnitude z, giving the Grahame equation for the charge per unit area on the electrode σ_1:

$$\sigma_1 = -\varepsilon\varepsilon_0 \left.\frac{d\varphi}{dx}\right|_{OHP} = \sqrt{8kT\varepsilon\varepsilon_0 n^0}\ \sin h \frac{|z| e\varphi_{\text{OHP}}}{2kT}$$

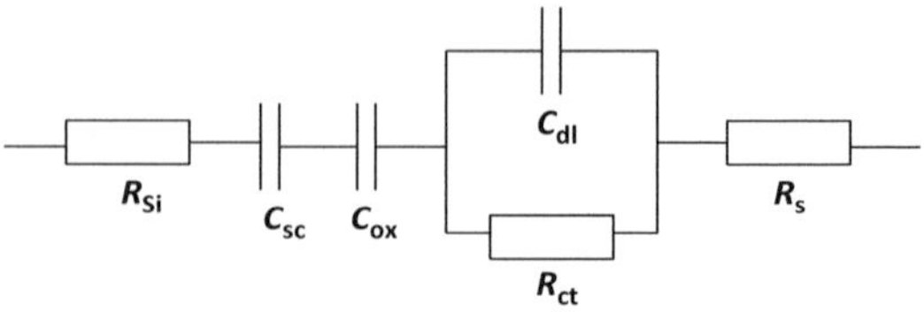

Fig. 17.3 Equivalent circuit for a field-effect device with gate immersed in solution

The impedance of a FET with the gate immersed in solution and potential applied to a reference electrode in solution may be represented by the equivalent circuit shown in Fig. 17.3. The circuit consists of the silicon resistance R_{Si}, space-charge capacitance C_{SC}, oxide capacitance C_{ox} of the FET and the Randles equivalent circuit for the double layer, where Z_W has been omitted since there are no redox molecules in solution. In the absence of redox molecules, R_{ct} is large and Z_{imag} can be considered to result from the series combination of the three capacitances.

When the biomolecular interaction happens at the solid–solution interface, it changes the value of C_{dl}. At fixed applied potential, this would introduce charge redistribution between C_{dl}.and C_{ox}, where the change of potential across C_{ox} depends on the ratio of the two capacitors, C_{dl} and C_{ox}. The value of this ratio is fixed when the biomolecular probe is immobilized directly on the gate dielectric or on the gate electrode directly on top of the dielectric. In an extended gate structure, a sensing pad is electrically connected to the gate electrode. The area of the sensing pad can be much larger than that of the transistor. In this configuration, the ratio C_{dl}/C_{ox} can be greatly improved by increasing the double-layer area, offering a larger voltage shift for the measurement (Estrela et al. 2007).

17.5 Promises

17.5.1 FETs Employed in DNA Sensing

DNA-based biologically-sensitive FETs (BioFETs) have been fabricated with very different approaches to immobilization strategies, hybridization, rinsing and measurement conditions. A change in the charge density of a biolayer immobilized on an electrode induces a change in the electrode surface charge density, σ_0, which in turn alters the surface potential, φ_0, that is, the open circuit potential (OCP). A change in the surface potential may be generated by a catalytic reaction product, surface polarization effects, or the change in dipole moments occurring with bio-affinity reactions. It can also be due to potential changes arising from biochemical processes in living systems, such as the action potential of nerve cells. The FET acts as a potentiometric transducer. In the case of DNA, the increase in negative charge in a layer of immobilized DNA probes upon hybridization with target oligonucleotides causes a significant change in φ_0. If immobilization is on the gate of an FET, hybridization causes a shift in the flat-band potential, V_{fb}, of the semiconductor. This causes a shift in the current–voltage (I_D versus V_{GS}) characteristic of the FET (Estrela et al. 2007).

These DNA FET sensors have had varying levels of success, achieving different immobilization densities, hybridization efficiencies, amount of non-specific binding, and stability. For a high sensitivity, a large voltage shift upon hybridization is needed. This requires a large increase in surface charge density upon hybridization, requiring a large surface density of probes that still allows high hybridization efficiency. To achieve a stable, high-density probe layer resulting in high-efficiency hybridization, end-tethered covalent attachment is necessary. Many designs are based upon functionalization of the gate dielectric of an ISFET. However, since the pH selectivity of the gate oxide is not required, functionalization of a gate metal is an option that allows immobilization using thiol chemistry. This enables easy and reproducible fabrication of high-density and highly stable mixed self-assembled monolayers of thiolated oligonucleotides, using only a single surface chemistry step.

Several groups have independently demonstrated the feasibility of such applications (Astier et al. 2006; Estrela and Migliorato 2007; Benner et al. 2007; Hornblower et al. 2007; Cockro et al. 2008) with some groups demonstrating, to a certain extent, the ability to read an entire genome. Matsumoto et al. (2012) demonstrated that primer extension reactions could be detected using an ISFET device. Their system demonstrated that under optimized conditions the primer extension reactions could be monitored with a resolution of a single base, thus leading to the development of the ability to read the entire sequence of the target DNA (Maehashi et al. 2007; Matsumoto and Miyahara 2013). Estrela and Migliorato (2007) have shown DNA detection on poly-Si TFTs. A mixed self-assembled monolayer of thiolated DNA probes and mercaptohexanol was immobilized onto the gold gate of an extended-gate poly-Si TFT. A shift of the I_D–V_{GS} characteristics of the order of 300 mV was obtained upon hybridization of the immobilized probe with a fully complementary strand. The shift was independent of the electrode area thus demonstrating that a microarray could be constructed where a known DNA probe is immobilized on each FET. The inherent miniaturization and compatibility with micro-fabrication technologies make the technique highly promising for the development of low-cost portable devices.

More recently, Rothberg et al. (2011) developed an ISFET array with 1.2 million wells that facilitated in independent sequencing reactions using three bacterial genomes, demonstrating a massively parallel array of semiconductor-sensing devices or an “ion chip”. Although they have demonstrated the sequencing of the bacterial genome, such a technology could well be employed for studying the human genome at the device level and could open other avenues in the area of medical diagnostics.

SWNT-FET sensors which measure DNA hybridization also hold great potential for large-scale genetic testing, clinical diagnosis and fast detection of biological warfare agents. Much attention has been given to the problem of interaction of CNTs and DNA, with the purpose of application in drug delivery and sensing. The promising application of functionalized SWNTs in monitoring DNA hybridization was demonstrated by Williams et al. (2002). A coupling peptide nucleic acid (PNA)—an uncharged DNA analogue—was covalently linked to the carboxyl-functionalized tip of SWNTs. Star et al. (2006) reported SWNT network FETs that function as selective detectors of DNA immobilization and hybridization.

SWNT network FETs with immobilized synthetic oligonucleotides have been shown to specifically recognize target DNA sequences. DNA hybridization with complementary target DNA sequences resulted in reduction of the SWNT-FET conductance. The sensing mechanism relies on the fact that single-stranded DNA adsorbed on sidewalls of SWNTs can result in electron doping to the SWNT semiconductor channels. This SWNT-FET DNA sensor can detect samples with picomolar to micromolar DNA concentrations. Dastagir et al. (2007) reported the unambiguous detection of a sequence of hepatitis C virus (HCV) at concentrations down to the fractional picomolar range using SWNT-FET devices functionalized with peptide nucleic acid sequences.

17.5.2 FETs Employed in Protein Sensing

The number of analytes that have been reportedly detected using enzyme-modified ISFETs (EnFETs) include a series of saccharides, urea, penicillin, ethanol, formaldehyde, lactose, lactate, organophosphorus pesticides, ascorbic acid, acetylcholine, creatinine and so on (Schöning and Poghossian 2002; Estrela and Migliorato 2007; Matsumoto and Miyahara 2013). Very recently, Bhalla et al. (2014) have used field-effect devices to measure protein phosphorylation. Another class of FETs that fall into the protein-coupled BioFET category is the immuno-reaction coupled FET (ImmunoFET). The analytes that have been reportedly detected using the different commercialized immunoassays relate to viruses (e.g. anti-HIV antibodies and p24 antigen for HIV), bacteria (e.g. anti-treponemal antibodies for syphilis and early secretory antigenic target 6 for tuberculosis) and parasites (e.g. histidine-rich protein 2 for malaria) (Chin et al. 2007). Such assay techniques have also been commonly used to detect diseases that are non-communicable, e.g. prostate cancer via measuring the levels of prostate-specific antigen, or heart diseases via the measurement of levels of B-type natriuretic peptide. The reference gold standard of immunoassays is ELISA, which employs multi-well substrates (von Lode 2005; Madhivanan et al. 2009).

Although success has been achieved in this area, there are still major issues in relation to detection: for example, issues relating to the double-layer capacitance or the Debye length. BioFET detection is inherently permitted only within a short distance, which corresponds to the Debye length, i.e. a few nanometers. Beyond this length counter-ion screening effects predominate and the charge detection is severely hampered. Some approaches have been presented that, to some extent, are successful although an increasing interest has developed in exploiting the ISFET as a means to examine proteins themselves. For example, it has been utilized as a platform for monitoring structural changes in proteins. Maltose-binding protein (MBP) undergoes a structural change upon stimulation with maltose. During this event, a significant decrease in the current on the basis of the charge-dependent capacitance measurement was observed in a manner sensitive to maltose but not to the control glucose. Such specific conformational changes can also be monitored using ISFETs. Other works have investigated the quantitative and kinetic aspect of the nonspecific protein adsorption (onto an extended gold electrode) based on the BioFET format. Estrela et al. (2010) have shown sub-picomolar protein detection using CMOS gold gate FETs. Self-assembled monolayers (SAMs) of alkanethiols with various terminal groups were prepared and tested for interactions with different types of proteins including bovine serum albumin (BSA) (Hu et al. 2010).

17.6 Challenges

17.6.1 Surface Interaction

Advancement in the area of FET-based biosensors is ultimately dependent on the search for optimal transducer materials, i.e. the gate electrode or dielectric material. The current standard CMOS technology employs aluminium as the top layer, which is not biocompatible. Another problem associated with Al is its high corrosion rate when in contact with any liquid. Some research groups have reportedly modified the Al layer by depositing an additional layer using clean-room facilities, but limited success has been reported using this technique. Graham et al. (2011) demonstrated a simple technique which could be carried out in a very minimalistic and cost-effective manner. They first anodized the aluminium electrode to create porous structures of Al_2O_3 which were then electroplated with gold chloride/cyanide solution. They then carried out successful characterization of nerve fibres on the modified surface. From the point-of-care perspective this seems to be a very attractive option in order to make the surface biocompatible, although limited studies have been reported using this method.

17.6.2 System Integration

With the currently available technology, a CMOS die measures only a few square millimetres to centimetres in area. With the additional demand for high-throughput assay, the integration of such miniature devices with microfluidics has been an issue. Although there are specialist clean-room facilities which can be used to integrate such systems, one of the major drawbacks of most POC diagnostic systems is their bulky nature dictated by the current micromachining technology, which adds further to the production cost thus pushing prices up. A key challenge in this area is to integrate chips into packages which are flexible and yet cheap to mass-produce. This demand has opened a new field of research which deals with flexible integrated circuit/microfluidics integration and packaging technology. Recently Zhang et al. (2013) reported the seamless integration of CMOS sensor chips with PDMS microfluidics into a flexible unit, thus achieving a true "lab-on-a-chip" system that has both CMOS functionalities and microfluidic sample manipulation on the same flexible substrate. They also employed "liquid metals" to create connections between the sensor chip and the external measurement setup. Although this area of research is still in its adolescence, it is believed that the potentials rewards outweigh the risks if such systems can be commercialized.

17.6.3 Signal Quantification

The signal levels generated due to biomolecular interactions can be extracted and quantified using various techniques. An understanding of the electrical and electrochemical properties of the biolayer is fundamental for the optimization of potentiometric BioFET measurements. Electrochemical impedance spectroscopy can, within certain constraints, provide an accurate description of the system. However, the description of the system in terms of an equivalent electrical circuit requires the use of constant phase elements (CPEs) instead of a double-layer capacitance element. This arises mainly from inhomogeneous protein or DNA distribution on the electrode surface. As the physical units of CPE depend on the parameter which defines the level of inhomogeneity of the system, direct comparison of the CPE values is not physically relevant (even though many reports seem to attempt to do so). There are different models which can be used for the calculation of the system's capacitance from the CPE values. The first model, proposed by Brug et al. (1984), is appropriate for surface distribution, whereas the model by Hsu and Mansfeld (2001) is for a normal distribution. Although the formulae are for very different conditions, they are generally found to be applied incorrectly, leading to flawed values of effective resistance and capacitance thus leading to macroscopic errors in many hypotheses. Hirschorn et al. (2010) demonstrated the subtle different between the models and illustrated the importance of using the correct formulae for a given system.

17.7 Conclusions

We have reviewed the field of FETs for point-of-care diagnostics. It is only fair to assume that over the last decade micro- and nano-devices employed for POC diagnostics have furthered the progress of research of both academic and industrial laboratories. Having said this, we see several challenges in this area before these technologies can be translated into ideal POC tools. Through the extensive research being carried out in this area, some groups have demonstrated the successful miniaturization of most peripheral instrumentation required for the readout and conditioning of miniaturized BioFET arrays (Lafleur et al. 2013). This could prove to be an important trend for the miniaturization and, most importantly, the cost reduction of the whole system, opening new avenues towards the development of biosensors for POC diagnostics. Such a low-cost requirement is an essential feature for sensors for the developing world, for example (Yager et al. 2006; Chin et al. 2007).

With regards to detection, the employment of various nanomaterials and nanostructures as elements for the signal-transduction of FETs is increasingly gaining research interest and this direction of research is firmly justified in terms of targeting nanoscale species. Unfortunately, efforts and the ability to maintain control at these levels of matter while maintaining reproducible signal and economically viable devices are still premature. To add more colour to this conundrum, the underlying mechanisms for the observed signal-transduction pathways are also not always evident, thus providing plenty of avenues for further research and development in this area.

Acknowledgments S.P. was funded from the European Union's Seventh Framework Programme for research, technological development and demonstration under grant agreement no. 278832 (project 'hiPAD').

References

Almeida SAA, Arasa E, Puyol M et al (2011) Novel LTCC-potentiometric microfluidic device for biparametric analysis of organic compounds carrying plastic antibodies as ionophores: application to sulfamethoxazole and trimethoprim. Biosens Bioelectron 30:197–203

Anderson RR, Hu W, Noh JW et al (2011) Transient deflection response in microcantilever array integrated with polydimethylsiloxane (PDMS) microfluidics. Lab Chip 11:2088–2096

Ang PK, Chen W, Wee ATS et al (2008) Solution-gated epitaxial graphene as pH sensor. J Am Chem Soc 130:14392–14393

Araci IE, Quake SR (2012) Microfluidic very large scale integration (mVLSI) with integrated micromechanical valves. Lab Chip 12:2803–2806

Astier Y, Braha O, Bayley H (2006) Toward single molecule DNA sequencing: direct identification of ribonucleoside and deoxyribonucleoside 5′-monophosphates by using an engineered protein nanopore equipped with a molecular adapter. J Am Chem Soc 128:1705–1710

Ballerini DR, Li X, Shen W (2012) Patterned paper and alternative materials as substrates for low-cost microfluidic diagnostics. Microfluid Nanofluid 13:769–787

Bard AJ, Faulkner LR (2001) Electrochemical measurements, fundamentals and applications, 2nd edn. Wiley, New York

Benner S, Chen RJA, Wilson NA et al (2007) Sequence-specific detection of individual DNA polymerase complexes in real time using a nanopore. Nat Nanotechnol 2:718–724

Bergveld P (1970) Development of an ion-sensitive solid-state device for neurophysiological measurements. IEEE Trans Biomed Eng 19:70–71

Berthier E, Young EWK, Beebe D (2012) Engineers are from PDMS-land, biologists are from polystyrenia. Lab Chip 12:1224–1237

Bhalla N, Di Lorenzo M, Pula G et al (2014) Protein phosphorylation analysis based on proton release detection: potential tools for drug discovery. Biosens Bioelectron 54:109–114

Bodas D, Khan-Malek C (2006) Formation of more stable hydrophilic surfaces of PDMS by plasma and chemical treatments. Microelectron Eng 83:1277–1279

Brug GJ, van den Eeden ALG, Sluyters-Rehbach M et al (1984) The analysis of electrode impedances complicated by the presence of a constant phase element. J Electroanal Chem 176: 275–295

Chen Z, Appenzeller J, Knoch J et al (2005) The role of metal-nanotube contact in the performance of carbon nanotube field-effect transistors. Nano Lett 5:1497–1502

Chin CD, Linder V, Sia SK (2007) Lab-on-a-chip devices for global health: past studies and future opportunities. Lab Chip 7:41–57

Cockro SL, Chu J, Amorin M et al (2008) A single-molecule nanopore device detects DNA polymerase activity with single-nucleotide resolution. J Am Chem Soc 130:818–820

Cohen-Karni T, Qing Q, Li Q et al (2010) Graphene and nanowire transistors for cellular interfaces and electrical recording. Nano Lett 10:1098–1102

Dastagir T, Forzani ES, Zhang R et al (2007) Electrical detection of hepatitis C virus RNA on single wall carbon nanotube-field effect transistors. Analyst 132:738–740
Dong X, Shi Y, Huang W et al (2010) Electrical detection of DNA hybridization with single-base specificity using transistors based on CVD-grown graphene sheets. Adv Mater 22:1–5
Effenhauser CS, Bruin GJM, Paulus A et al (1997) Integrated capillary electrophoresis on flexible silicone microdevices: analysis of DNA restriction fragments and detection of single DNA molecules on microchips. Anal Chem 69:3451–3457
Estrela P, Migliorato P (2007) Chemical and biological sensors using polycrystalline silicon TFTs. J Mater Chem 17:219–224
Estrela P, Keighley SD, Migliorato P (2007) Field-effect potentiometric biosensors. In: Ozoemena KI (ed) Recent advances in analytical electrochemistry. Transworld Research Network, Kerala, pp 199–230
Estrela P, Paul D, Song Q et al (2010) Label-free sub-picomolar protein detection with field-effect transistors. Anal Chem 82:3531–3536
Fakunle ES, Fritsch I (2010) Low-temperature co-fired ceramic microchannels with individually addressable screen-printed gold electrodes on four walls for self-contained electrochemical immunoassays. Anal Bioanal Chem 398:2605–2615
Graham AHD, Robbins J, Bowen CR et al (2011) Commercialisation of CMOS integrated circuit technology in multi-electrode arrays for neuroscience and cell-based biosensors. Sensors 11:4943–4971
Hirschorn B, Orazem ME, Tribollet B et al (2010) Determination of effective capacitance and film thickness from constant-phase-element parameters. Electrochim Acta 55:6218–6227
Hornblower B, Coombs A, Whitaker RD et al (2007) Single-molecule analysis of DNA-protein complexes using nanopores. Nat Methods 4:315–317
Hsu CH, Mansfeld F (2001) Concerning the conversion of the constant phase element parameter Y_0 into a capacitance. Corrosion 57:747–748
Hu PA, Zhang J, Li L et al (2010) Carbon nanostructure-based field-effect transistors for label-free chemical/biological sensors. Sensors 10:5139–5159
Iijima S (1991) Helical microtubules of graphitic carbon. Nature 354:56–58
Iliescu C, Taylor H, Avram M et al (2012) A practical guide for the fabrication of microfluidic devices using glass and silicon. Biomicrofluid 6:016505
Lafleur JP, Kwapiszewski R, Jensen TG (2013) Rapid photochemical surface patterning of proteins in thiolene based microfluidic devices. Analyst 138:845–849
Li P, Migliorato P, Estrela P (2012a) Application of field-effect transistors to label-free electrical DNA biosensor arrays. In: Özsöz M (ed) Electrochemical DNA biosensors. Pan Stanford Publishing, Singapore, pp 163–204
Li X, Ballerini DR, Shen W (2012b) A perspective on paper-based microfluidics: current status and future trends. Biomicrofluid 6:11301–11303
Liana DD, Raguse B, Gooding JJ et al (2012) Recent advances in paper-based sensors. Sensors 12:11505–11526
Liu Y, Yuan R, Chai YQ et al (2006) Direct electrochemistry of horseradish peroxidase immobilized on gold colloid/cysteine/nafion-modified platinum disk electrode. Sens Actuat B 115:109–115
Madhivanan P, Krupp K, Hardin J et al (2009) Simple and inexpensive point-of-care tests improve diagnosis of vaginal infections in resource constrained settings. Trop Med Int Health 14:703–708
Maehashi K, Katsura T, Kerman K et al (2007) Label-free protein biosensor based on aptamer-modified carbon nanotube field-effect transistors. Anal Chem 79:782–787
Martel R, Schmidt T, Shea HR et al (1998) Single- and multi-wall carbon nanotube field-effect transistors. Appl Phys Lett 73:2447–2449
Martinez AW, Phillips ST, Carrilho E et al (2008) Simple telemedicine for developing regions: camera phones and paper-based microfluidic devices for real-time, off-site diagnosis. Anal Chem 80:3699–3707
Matsumoto A, Miyahara Y (2013) Current and emerging challenges of field effect transistor based bio-sensing. Nanoscale 5:10702–10718
Matsumoto A, Miyahara Y, Kataoka K (2012) Intelligent surfaces for field-effect transistor-based nanobiosensing. In: Grandin H, Textor M (eds) Intelligent surfaces in biotechnology: scientific

and engineering concepts, enabling technologies, and translation to bio-oriented applications. Wiley, New York, pp 123–140
Mattmann M, Helbling T, Durrer L et al (2009) Sub-ppm NO_2 detection by Al_2O_3 contact passivated carbon nanotube field effect transistors. Appl Phys Lett 94:183502
Mellors JS, Gorbounov V, Ramsey RS et al (2008) Fully integrated glass microfluidic device for performing high-efficiency capillary electrophoresis and electrospray ionization mass spectrometry. Anal Chem 80:6881–6887
Mellors JS, Jorabchi K, Smith LM et al (2010) Integrated microfluidic device for automated single cell analysis using electrophoretic separation and electrospray ionization mass spectrometry. Anal Chem 82:967–973
Mohanty N, Berry V (2008) Graphene-based single-bacterium resolution biodevice and DNA transistor: interfacing graphene derivatives with nanoscale and microscale biocomponents. Nano Lett 8:4469–4476
Novak JP, Snow ES, Houser EJ et al (2003) Nerve agent detection using networks of single walled carbon nanotubes. Appl Phys Lett 83:4026–4028
Novoselov KS, Geim AK, Morozov SV et al (2004) Electric field effect in atomically thin carbon films. Science 306:666–669
Parsa H, Chin CD, Mongkolwisetwara P et al (2008) Effect of volume- and time-based constraints on capture of analytes in microfluidic heterogeneous immunoassays. Lab Chip 8:2062–2070
Peng G, Tisch U, Haick H (2009a) Detection of nonpolar molecules by means of carrier scattering in random networks of carbon nanotubes: toward diagnosis of diseases via breath samples. Nano Lett 9:1362–1368
Peng N, Zhang Q, Chow CL et al (2009b) Sensing mechanisms for carbon nanotube based NH_3 gas detection. Nano Lett 9:1626–1630
Roberts ME, LeMieux MC, Bao Z (2009) Sorted and aligned single-walled carbon nanotube networks for transistor-based aqueous chemical sensors. ACS Nano 10:3287–3293
Roman GT, Hlaus T, Bass KJ et al (2005) Sol–gel modified poly(dimethylsiloxane) microfluidic devices with high electroosmotic mobilities and hydrophilic channel wall characteristics. Anal Chem 77:1414–1422
Rothberg JM, Hinz W, Rearick TM et al (2011) An integrated semiconductor device enabling non-optical genome sequencing. Nature 475:348–352
Schöning MJ, Poghossian A (2002) Recent advances in biologically sensitive field-effect transistors (BioFETs). Analyst 127:1137–1151
Star A, Tu E, Niemann J et al (2006) Label-free detection of DNA hybridization using carbon nanotube network field-effect transistors. Proc Natl Acad Sci U S A 103:921–926
Taillades G, Valls O, Bratov A et al (1999) ISE and ISFET microsensors based on a sensitive chalcogenide glass for copper ion detection in solution. Sens Actuat B 59:123–127
Tans SJ, Verschueren ARM, Dekker C (1998) Room-temperature transistor based on a single carbon nanotube. Nature 393:49–52
Varghese N, Mogera U, Govindaraj A et al (2009) Binding of DNA nucleobases and nucleosides with graphene. Chemphyschem 10:206–210
von Lode P (2005) Point-of-care immunotesting: approaching the analytical performance of central laboratory methods. Clin Biochem 38:591–606
Washburn AL, Gunn LC, Bailey RC (2009) Label-free quantitation of a cancer biomarker in complex media using silicon photonic microring resonators. Anal Chem 81:9499–9506
Williams KA, Veenhuizen PTM, Torre BG et al (2002) Carbon nanotubes with DNA recognition. Nature 420:761–766
Wu A, Wang L, Jensen E et al (2010) Modular integration of electronics and microfluidic systems using flexible printed circuit boards. Lab Chip 10:519–521
Yager P, Edwards T, Fu E et al (2006) Microfluidic diagnostic technologies for global public health. Nature 442:412–418
Zhang W, Eitel RE (2012) Biostability of low-temperature co-fired ceramic materials for microfluidic and biomedical devices. Int J Appl Ceram Technol 9:60–66
Zhang BD, Quan K, Can EL (2013) Flexible packaging of solid-state integrated circuit chips with elastomeric microfluidics. Nat Sci Rep 3:1–8

Concluding Remarks

Through the preceding chapters, this book has constructed a panoramic picture of the fast-growing field of nanobiosensors and nanobioanalysis. From the inspiring examples and insightful exposition provided by the authors, who are preeminent researchers at the leading edge of this field, it has become palpable that nanotechnology has been, and will continue to be, playing an essential role in the revolutions occurring in the field of biosensing and bioanalysis. The emergence of novel nanomaterials, and the ability to manipulate structures at the nanoscale, have contributed tremendously to the development of ground-breaking biosensing and bioanalytical technologies, in addition to significantly improving the performance of traditional methods. It can be expected that breakthroughs in the field of nanobiosensors and nanobioanalysis will likely bring profound benefits and advancement to the improvements of people's health and quality of life. Nevertheless, along with the exciting opportunities, there exist huge challenges that call for continuous inputs and supplies of talents, efforts, and resources.

Following an insightful review in the first part, by Profs. Vestergaard and Tamiya, of the basic principles, main players, and detection strategies in the field of nanobiosensing and nanobioanalysis (Chap. 1), the four chapters in the second part of the book represent the most updated and frontier breakthroughs in this field, wherein nanotechnology and nanomaterials serve as an indispensable cornerstone. The analytical strategies developed by Razeeb et al. and described in Chap. 2 demonstrate that vertically aligned nanowires can be an excellent sensor platform, especially for the detection of hydrogen peroxide, glutamate, and glucose. In Chap. 3, Chen et al. show readers how SiNW-FETs and graphene-FETs are successfully used in various biosensing applications to provide important performance advantages. A novel Au-capped nanopillar LSPR chip developed by Saito and Yamanaka for the detection of anti-antigen reactions is demonstrated in Chap. 4, while Ünlü et al. present a multiplexed digital molecular detection strategy using metallic nanoparticle labels and taking advantage of the unique properties of the Interferometric Reflectance Imaging Sensor (IRIS) in Chap. 5. It is difficult to imagine that without the help of

M.C. Vestergaard et al. (eds.), *Nanobiosensors and Nanobioanalyses*,
DOI 10.1007/978-4-431-55190-4

these nanomaterials, such exceptional performances in terms of sensitivity, selectivity, and operational simplicity could be achieved.

Based on the recognition that nanomaterials play an essential role in the advancement of biosensing and bioanalysis, the third part of this book is devoted to a discussion of the functionalization and applications of nanomaterials commonly used in the field (gold nanoparticles in Chap. 6, nanocarbon film in Chap. 7, hybrid metallic nanoparticles in Chap. 8, silver nanoparticles in Chap. 9, quantum dots in Chaps. 10 and 11, and inorganic and organic nanoparticles in Chap. 12). From the plentitude and diversity of novel nanomaterials, and the availability of tools and methods for their functionalization, it is not difficult for one to realize that more opportunities for breakthroughs in this field are yet to be made, and that the marriage between nanotechnology and biosensing and bioanalytical technologies is bound to give birth to next generations of more exciting and influential technologies.

Focusing on the applications of nanobiosensor technologies, the fourth part of the book provides a series of vivid examples to give the reader a better understanding of the feasibility of these novel technologies. In this part, the authors demonstrate the applications of CNT-composite materials and metal nanoparticles (Chap. 13), nanoneedles (Chap. 14), SPR sensing (Chap. 15), nanometer-scale electrodes (Chap. 16), and FETs (Chap. 17) in areas such as point-of-care testing and diagnostics, detection of intracellular processes, SECM systems, and so on. We believe that after reading these chapters, the values of these new technologies and the capabilities enabled thereby will become apparent to the readers.

In conclusion, it is our belief that the readers of this book will be enabled to appreciate the tremendous and exciting opportunities brought by the integration of nanotechnologies into biosensing and bioanalytical platforms. On the other hand, the challenges that researchers encounter in this field are huge and intimidating, which is no doubt a result of the preciseness required for the manipulation of matter on the nanoscale, in combination with the inherent complexity of biological systems. Nothing less than consistent and focused efforts are required to overcome these challenges. Both the opportunities and the challenges lead one to conclude that a larger number of talents and more sizable resources need to be invested in the field of nanobiosensors and nanobioanalysis, a field that is bound to bring the next epoch-making technology to the world.

Index

M.C. Vestergaard et al. (eds.), *Nanobiosensors and Nanobioanalyses*,
DOI 10.1007/978-4-431-55190-4

MIX
Papier aus verantwortungsvollen Quellen
Paper from responsible sources
FSC® C105338

If you have any concerns about our products,
you can contact us on
ProductSafety@springernature.com

In case Publisher is established outside the EU,
the EU authorized representative is:
Springer Nature Customer Service Center GmbH
Europaplatz 3, 69115 Heidelberg, Germany

Printed by Libri Plureos GmbH
in Hamburg, Germany